Recovery and Restoration of Damaged Ecosystems

Recovery and Restoration of Damaged Ecosystems

Edited by J. Cairns, Jr.,

K. L. Dickson,

and E. E. Herricks

Proceedings of the International Symposium
on the Recovery of Damaged Ecosystems
held at Virginia Polytechnic Institute
and State University, Blacksburg, Virginia,
on March 23-25, 1975

University Press of Virginia

Charlottesville

THE UNIVERSITY PRESS OF VIRGINIA

First published 1977

Library of Congress Cataloging in Publication
Data

International Symposium on the Recovery of
Damaged Ecosystems, Virginia Polytechnic
Institute and State University, 1975.
Recovery and restoration of damaged eco-
systems.

1. Ecology–Congresses. 2. Environmental
protection–Congresses. I. Cairns, John, 1923–
II. Dickson, Kenneth L. III. Herricks, Edwin E.
IV. Title.
QH541.I575 1975 639′.9 76-49453
ISBN 0-8139-0676-8

Printed in the United States of America

The publication of this volume is
sponsored by the VPI Educational
Foundation, Inc.

Preface

A major concern in this country as well as the rest of the world is the environmental degradation caused by man's activities. Despite protective measures, environmental risks are steadily increasing. While countless discussions and innumerable papers have dealt with the causes and consequences of our damaging activities, this volume explores a different aspect of the topic: the prospects of recovery if damage does occur. To our knowledge no other book with this concern has yet been published.

An experienced group of international scientists has contributed case histories, presented theories, and raised important questions related to the restoration and recovery of damaged ecosystems. The book focuses on three major topics: the nature of recovery processes for various ecosystems; identification of the elements common to the recovery process for all ecosystems, as well as the unique attributes in different kinds of ecosystems; and the prospects for accelerated recovery and restoration by human intervention and management. Individual chapters discuss the recovery of streams and lakes by natural and artificial methods; the reintroduction of plants and wildlife to damaged ecosystems; the role of fire in the natural growth of forests; environmental factors in surface mine recovery; the effects of oil spills and recovery from them; the recovery of tropical forest systems, tundra, and taiga surfaces; the recovery of cities; and the political problems inherent in environmental protection legislation.

The book, which is an outgrowth of a symposium on the Recovery of Damaged Ecosystems held in March 1975 at Virginia Polytechnic Institute and State University, presents compelling evidence of the need for immediate action to develop the scientific data, technology, and international and national implementation programs to restore damaged ecosystems. We feel that the book will be useful to ecologists, sociologists, regional planners, and historians, as well as legislators, regulatory agencies, and environmental management agencies. Not only does the book present valuable information about how ecosystems recover after stress, but it presents a challenge to mankind to become a participant, not just a spectator.

This book would not have been possible without the support of Union Carbide Corporation and the National Science Foundation, Division of Environmental Systems and Resources. We are indebted to our editorial assistant, Darla Davis Donald, for her assistance in the editing and preparation of this publication.

John Cairns, Jr.
Kenneth L. Dickson
Biology Department and Center for Environmental Studies
Virginia Polytechnic Institute and State University
Edwin E. Herricks
Department of Civil Engineering
University of Illinois at Urbana-Champaign

Contents

Contributors

Sven Björk, Institute of Limnology, University of Lund, Sweden.

Daniel B. Botkin, Ecosystems Center, Marine Biological Laboratory, Woods Hole, Massachusetts.

John Cairns, Jr., and Kenneth L. Dickson, Biology Department and Center for Environmental Studies, Virginia Polytechnic Institute and State University.

Robert R. Curry, Sierra Club Research, San Francisco; current affiliation–Department of Geology, University of Montana.

Brian Dicks, Oil Pollution Research Unit, Pembrokeshire, Great Britain.

W. T. Edmondson, Department of Zoology, University of Washington.

Michael S. Foster, Department of Biological Science, California State University, AND Robert W. Holmes, Marine Science Institute and Department of Biological Sciences, University of California.

Arlo W. Fast, Aquatic Environmental Sciences, Union Carbide Corporation; current affiliation–Limnological Associates, Pacifica, California.

A. L. H. Gameson, Water Research Center, Stevenage Laboratory, England, AND Alwyne Wheeler, British Museum of Natural History, London.

David R. Goldfield, Division of Environmental and Urban Systems, Virginia Polytechnic Institute and State University.

P. A. Harcombe, Biology Department, Rice University.

Edwin E. Herricks, Aquatic Environmental Sciences, Union Carbide Corporation; current affiliation–University of Illinois, Urbana.

Ronald D. Hill and Elmore C. Grim, Mining Pollution Control Branch, Industrial Waste Treatment Research Laboratory, National Environmental Research Center, Environmental Protection Agency.

Anthony Nelson-Smith, Department of Zoology, University College of Swansea, South Wales.

Paul A. Opler, Office of Endangered Species, U.S. Department of the Interior, AND Herbert G. Baker, Department of Botany, University of California, AND Gordon W. Frankie, Department of Entomology, Texas A. & M. University.

Robert B. Platt, Department of Biology, Emory University.

Charles V. Riley, Department of Biological Sciences, Kent State University.

Martha Sager, Director of Environmental Systems Management, American University.

Keith Van Cleve, Forest Soils Laboratory, University of Alaska.

Richard J. Vogl, Biological Department, California State University, Los Angeles.

Recovery and Restoration of Damaged Ecosystems

Reinhabiting the Earth: Life Support and the Future Primitive

Robert R. Curry

Ecosystem Recovery

Organisms inhabiting a place virtually always change it. Only through this process of change has life evolved on earth. The planetary atmosphere of earth three to four billion years ago was nearly as forbidding as that of its sister planets today. But primitive anaerobic bacteria changed that greatly, and to them we owe our debt of existence. There is a current renewed public interest and growing awareness of the general hypothesis that organisms themselves may control the rate of biologic evolution on earth. Today's "Gaia hypothesis" of Lovelock and Epton (1975) is a natural outgrowth and restatement of the findings of Cloud and of Barghoorn in previous decades (Cloud, 1968; Cloud and Gibor, 1970). It is apparent that the organic uniqueness of the earth is the direct result of ecosystem modification by tenants of this planet. Darwinian selection cannot function unless niches and stresses are available and unless biochemical and metabolic systems within organisms coincide with the limitations and provisions of the sites that support those organisms. In other words, oxygen-based metabolism requires available oxygen. The rise of oxygen-releasing organisms provides the limiting and controlling factor for the rate of evolution and complexity of higher animals. Relative enrichment of the oxygen content of the earth's atmosphere has closely paralleled the evolutionary development of its life. The development of man with his highly oxygen-dependent central nervous system has now provided the noetic impetus to control ecosystem balances. We perceive the maintenance of ecosystem stability and diversity as necessary for the survival of the earth's present component of organisms, including ourselves. In a way, this is antievolutionary. Mankind perceives that it has evolved within a certain mosaic of ecosystems upon which it has slowly come to realize that it is dependent. But it also shows a biologically imperative pragmatism wherein we, albeit anthropocentrically, believe that the earth's present life-supporting capabilities provide the best opportunities for that component of organisms and that mosaic of ecosystems with which we most want to share our lives during our remarkably short period of tenureship on earth.

The topics covered in this volume reflect a growing awareness of the responsiveness of ecosystems to man-induced or man-triggered stresses. In general, we know so little about the natural, pre-impact states of most ecosystems that it is very difficult to do more than chronicle the rates and response patterns of those units during what we presume to be periods of recovery to stresses. Chapters in this volume include those dealing with recovery from individual stress conditions such as fire, marine oil spills, and surface mining and those dealing with general response patterns and mechanisms of certain ecosystems such as stream systems, lakes, tundra, and tropical forests. These chapters comprise an important first step in developing an understanding of the natural resiliency of ecosystems and thus in assessing the probable outcome of man's manipulation of those systems. In this introductory chapter I shall attempt to deal with some of the natural variations and stresses that have brought about the evolution of today's ecosystems through geological time and to try to relate these instabilities of ecosystems to man's dependence on them. All ecosystems are natural systems, with or without the presence of man. Because of his high degree of flexibility and adaptability, man can inhabit a very wide range of ecosystems, and even those few sites on this planet where he does not permanently reside and reproduce today, such as the high arctic or seafloors, will doubtless soon yield to his territorial expansiveness. It is therefore not my wish to emulate occasional current practice by using the term *human ecosystems* to describe those systems that have been altered by man to suit his immediate needs. This chapter will deal particularly with agronomic ecosystems and the general problem of life-support systems for mankind. The modified ecosystems are seen as stress states that are ephemeral when viewed over geologic time. This is small consolation to the resident of London or Los Angeles where none but the smallest relics of the original whole self-sufficient nonhuman populations can be found, but it remains true that the same ecosystem processes and response characteristics occur in greatly altered sites as are found in so-called natural sites. Mankind has been able to totally alter ecosystems and rates of response, but he has been essentially ineffectual so far in changing the biologic and geochemical processes by which recovery occurs.

Reinhabitation

The very title of this volume implies a quantum jump forward in maturity of philosophy of stewardship of the earth. We are attempting to begin to bring man back within the flow of organic wholeness, to reinhabit the earth's ecosystems within their ranges of tolerance to us—not by chance, but by understanding. We are finally just beginning to respond to North American poet Robinson Jeffers's admonition:

> . . . the greatest beauty is organic wholeness,
> the wholeness of life and things,
> the divine beauty of the universe.
> Love that, not man apart from that . . .

An understanding of the recovery of damaged ecosystems, whether damaged by man-caused perturbations or by natural events, can lead man to a philosophical reinhabitation of the earth—within the balances and dynamics of its ecosystems. Today we see this philosophical reinhabitation manifested in a natural philosophy of low-energy, nonconsumptive life-styles practiced by a few and espoused by many. This "future primitive" philosophy demands a very high degree of understanding of those ecosystems upon which one chooses, deliberately, to become reliant. A regional sense of identity places man within those ecosystems that he inhabits or otherwise relies upon directly and forces a balanced use of ecosystem resources. Such biogeographic identity demands a sound understanding of response of ecosystems to stress (Gorsline and House, 1974).

A prime example of both the theory and the practice of reinhabitation is revealed in the work of Wakkon (1974) on the Pacific Rim island of Suwanose, south of Japan at the edge of the East China Sea. Here the interdependence of a small community of organisms is chronicled from the human vantage point within the natural cycle of ecosystem changes (Fig. 1). Mankind is seen only within the larger community of organisms, climate, and ocean currents that serve to define and limit the responsive character of the ecosystems upon which that community is interdependent.

Reinhabitation and recovery of damaged ecosystems is also a timely subject because of current, somewhat fatalistic discussion of actual physical recolonization of severely damaged earth ecosystems from space (O'Neill, 1974; Chedd, 1974). The National Aeronautics and Space Administration (Steincamp, 1975) is now seriously investigating the establishment of extraterrestrial breeding human population support systems, that, among other justifications, will permit the establishment and maintenance of a viable earthbound population limited by the carrying capacity of existing ecosystems with excess population living in space and deriving all their energy and matériel needs from beyond earth. By extension, if nuclear holocaust, diversion of plutonium into the hydrosphere or atmosphere, or some other severe worldwide stress situation occurs to the point of loss of the present gene pool on earth, limited reinhabitation will be possible after recovery of damaged ecosystems.

Another related futurist research project of the same institution proposes to investigate the feasibility of constructing organisms, through genetic manipulation, that can survive and reproduce rapidly on other planets to effect an alteration of the "primitive" atmospheres of those places such that inoculants of life from earth could survive. Through repeated inoculation of genetically selected and constructed organisms, it is purported that one would be able to greatly speed up the process of organic evolution that took geologic millenia on earth because of biogeochemical and Darwinian evolutionary limitations, thus preparing the planets for man and his fellow earthbound organisms at greatly accelerated rates. The hazards of such gene engineering are unknown but probably enormous.

In studying recovery of damaged ecosystems, it becomes increasingly apparent that the system of natural preserves that exists today on earth is woefully inadequate upon which to base hopes for preservation of adequate gene pools for recovery of severely damaged ecosystems (Curry, 1970; Dasmann, 1972, 1975; Ehrlich, 1974). As man's ability to severely alter the life-support systems of sites on the planet, even for short periods of time, approaches the scale of size capable of wiping out entire breeding populations of organisms, it becomes clear that recovery of damaged ecosystems depends, in part, upon preservation of viable populations of all the organisms that are successionally needed to "reclaim" those damaged sites. Wilderness and nature preserve areas of the globe are today far too restricted to permit reinoculation of damaged ecosystems from undamaged sites. Despite the highest technological achievements of NASA in placing men on the moon, the complexity of the task of creating the substrate conditions, variability, and microclimatic conditions necessary to support any more than a simplified Noachian component of the earth's biota in space indefinitely is almost beyond comprehension. Since we cannot even reclaim our arid western strip-mined lands today, or even comprehend what was there before we started mining, it seems clearly premature to place reliance upon extraterrestrial biotic refugia. The recovery of damaged ecosystems with the same life forms upon which man relies demands that breeding populations representative of all mature and seral plant and animal communities be preserved. This means we need grassland, strandline, marine, subterranean, overgrazed, burnt, and thoroughly disturbed wilderness areas of refuges, as well as the pristine tropical or subalpine forest or high mountain ecosystems that we recognize as worth preserving today. Each time we lose a component of the biota at a site, we lose the combined efficiency of energy and material storage and transfer at that site and the resiliency of that site to adapt to climatic or other stress conditions. Although these two factors cannot be maximized simultaneously in any systems that obey the laws of thermodynamics, the full biotic complement is vital to maintenance of ecosystem balances. Although an unfilled niche will not remain so for long, it does not follow that the newly evolving or colonizing organism will bear the same synergistic relationships to the other components of the site biota as did the species that was lost. Darwinian fitness can be viewed as nothing more than the most efficient utilization of available energy and materials at a site of interspecies conflict through the range of stress states that may be externally imposed on that site. The paleontologic record reveals almost innumerable examples of evolutionary "dead ends" where species as well as communities interdependent upon species became extinct through Darwinian evolution toward maximization of energy efficiency (site utilization) to remain master of a niche without regard for adaptability to external stress states. The "Jack of all trades and master of none" such as the soil bacteria, the sponge, and the sea gull have the adaptability to survive, whereas man's increasing emphasis on energy dependence for maintenance of his living niches places him in an increasingly precarious and vulnerable position.

Since mankind is a part of the earth's biota and since there is no biologic reason to consider himself different from any of the other opportunistic species that botanists call weeds, the nature preserve concept can as well be applied to preservation of our own genetic and cultural diversity. Current generations of mankind might benefit much more from "rare and endangered cultures" legislation than they may from great expenditures of effort and faith directed toward the establishment of space colonies during these final decades of the second millenium A.D. By now we should have learned that whenever we are provided a promise of a prodigal son who will return to right the wrongs we have brought about, this savior cult affords us a justification for reprehensible housekeeping. With reproducing colonies of men in space, it will become increasingly easier to seriously consider activities with global-scale risk. Let us rather try to establish respect for maintenance of the world's ecosystems. This must include the very difficult development of nonpatronizing respect for so-called primitive human cultures as well as the nonhuman elements of the ecosystems upon which we all rely absolutely. Since men should not be forced to remain in any culture, it is not always prudent to place parks around people, but it should be possible to establish ecosystem refugia adjacent to human cultural centers wherein the Tassaday of the Philippines, the Xavante of Brazil, or the American Indians of southwestern United States can have the option of living the life-styles of their progenitors. This does not mean that the Havasupai should be allowed to build a hotel in the Grand Canyon any more than anyone else should build a hotel there or that colonists should log the Amazon forests relied upon by the Xavante. It merely means that the ecosystems still utilized by solar-based man should be set aside to provide reserves for recovery of damaged ecosystems elsewhere. It also means that certain human uses of these ecosystems can be accommodated so long as carrying capacity is not exceeded without a full system of checks and balances.

Natural Perturbations in Ecosystem Stresses

All ecosystems of the earth, including those of the deep ocean, are in part limited and defined by both today's climate and the history of climatic change at that site. This historical aspect is often overlooked by ecologists and even biogeographers in that it is often assumed that climatic conditions change more slowly than do biotic responses. What is seldom appreciated is that terrestial sites have a "climatic memory" in the biogeochemical status of their soils. In fact, our present understanding of climatic change in the recent geologic past suggests that climate may change very much faster than many animals and most plants can adapt.

Much of the understanding of the history of world climatic changes, upon which our concepts of rates of change are based, is derived from the pollen zonation that was first investigated in the Danish peat bogs. The picture of climatic change for the last 10,000 years of postglacial time that is derived from such work shows that, after a few final cycles of cold associated with readvance of mountain glaciers, temperate latitudes became progressively warmer through about 6000 years ago, and that since then there has been a cyclic and somewhat progressive cooling that has culminated in repeated little ice ages during the last 3000 or more years. In general, the severity and frequency of periods of ice buildup has been increasing through at least the nineteenth century.

A basic problem with this kind of analysis is that observed climatic change for the last 100 years of instrumental record is combined with observed changes in pollen-producing vegetation to synthesize a record for periods of time before manipulation of ecosystems by man. Since reconstructed climates at a site are based upon today's climates at sites where the species producing the fossil pollen exist today, and since radiocarbon estimations of the ages of the fossil pollen are not precise and tend to blur analysis of changes of but a few centuries' duration, it is not really possible to place time constants on the rates of climatic change that have accounted for the changes in biota revealed in the pollen. Thus, the picture that emerges from much pollen work is one of climatic change that may create an abrupt stratigraphic record but is assumed to have been gradual and progressive.

Several investigatory methods have given us a rather different picture in recent years. Careful climatic reconstructions of periods prior to instrumental records show that individual climatic events of considerable severity occurred that heralded changes within but a few years. Examples include historical accounts of the freezing of the Nile and the Dardanelles and ice blockage of Scandanavian, Russian, and Icelandic coasts. Such events were often associated with famines, crop failures over many years, and general sociocultural chaos that historians like to ascribe to the attributes of individual people or cultures (Lamb et al., 1966).

Work with deep-sea sediment samples shows that organisms living at the sea surface often change dramatically in species composition at an almost instantaneous point in the stratigraphic record. By looking at marine shoreline sediments that formed during the risings and fallings of sea level that were associated with sequential accumulation and melting of land ice during the Pleistocene epoch, we can tell that the great continental ice sheets of the high temperate latitudes formed rather slowly, over periods of perhaps several thousand years, but melted much more rapidly, yielding the bulk of stored water to the oceans in but a millenium or less. This has led to the idea that a more dramatic climatic overbalancing is associated with the termination of a glacial period than with its onset. From this, coupled with the pollen evidence and the general lobate, outflowing morphology of the glacial moraines that mark the outer limits of the past ice masses, a general picture of slow accumulation of ice in northern Sweden and eastern Hudson's Bay emerged. The textbooks implied that ice up to thousands of feet thick had to form at some subpolar latitude before it began to flow outward toward the temperate latitudes to scour across northern Europe and eastern North America. In fact, there is very little basis for such a theory. Although it can be demonstrated that the ice flowed generally from the region of greatest thickness toward its thin margins, it does not follow that the ice had to first build up near its flow center to flow out and cool its own path southward. If, as previously thought, there was to be a several-thousand-year lag between the onset of a glacial advance and the destruction of temperate ecosystems, then the nearly unanimous professional opinion that the next period of glacial advance was in sight was of little direct consequence to those who recognized mankind's dependence upon the current aeral extent and pattern of ecosystems. However, our present understanding of atmospheric dynamics, heat exchange, and paleoclimatology lends increasing support to a more disturbing alternate hypothesis. Although it may take thousands of years to build full ice volumes on land, it does not follow that this ice builds with a progressively expanding margin. More probably, winter snowfall simply does not melt over large areas of what we call the north temperate latitudes. Furthermore, long persistent snow and cooler temperatures greatly restrict the effective growing seasons over large areas adjacent to the areas of building snow and ice fields. There is no basic physical reason that this could not happen in a matter of decades or even less. Its effects upon ecosystems would be drastic (Kukla et al., 1972; Bryson, 1974; Imbrie, 1974).

We know that vegetational zonation and associated animal ranges were shifted equatorward during the twenty or more colder periods of the Pleistocene and preceding late Pliocene when glaciers were periodically waxing and waning. We also know that progressively increasing numbers of organisms became extinct through the Pleistocene, either because of the continued climatic stress and resulting reduced habitat, or because of the pressures of expanding human populations, or both. The most diverse ecosystems that we find on earth today are generally in sites where Pleistocene stresses were least and where geographic displacement was minor. The biota of the most disturbed subarctic and subalpine communities are characteristically made up of species that are highly adapted to soils that are very immature—that is, where variable conditions of geologic substrate limit and affect the biota it supports. Tropical ecosystems, on the other hand, may reflect more what plant ecologists have called climatic climax states, with less influence by soil factors.

In temperate latitudes, a severely stressed ecosystem may never recover its original character because that character was a function of biogeographic history and inherited soil chemistry. These often developed under climatic conditions that do not exist at the site today and will not exist again until after the completion of the next glacial cycle. Thus, after severe stress such as strip mining, intensive agriculture and silviculture, or urbanization, it may be meaningless to try to recover the original ecosystem. Instead, it may be necessary to accept another community that can develop in sequence with the slowly changing site geochemistry and induced microclimatic changes.

In equatorial latitudes, our present Pleistocene climatic model suggests that today's weathering conditions will form the same soils found there today if given very long periods of time. Disturbed tropical ecosystems have much greater potential for recovery to original conditions through geologic time if the disturbance is sufficient to strip away the biologically depleted weathering mantle that we recognize as mature tropical soils, or if the site happens to be just downwind from a volcanic source of natural mineral fertilizer (volcanic ash). However, the potential for efficient capture and utilization of solar energy in tropical ecosystems that are depleted without soil removal is much lower than in temperate latitudes because of climatic history and genotypic adaptability of the resident species.

There is thus an apparent paradox wherein one can envision more rapid and more successful recovery of severely damaged ecosystems by primary succession in high temperate and polar latitudes than in equatorial areas. Viewed in the perspective of geologic time, we can see that the mosaic of seral communities that might define one of the ecosystems of southeastern Alaska could be expected to become reestablished, although not in the same community pattern, within perhaps as little as 4000 years following its virtually complete destruction by regrowth of glaciers (Viereck, 1966). However, the relatively more minor forest removal from the Congo or Amazon basins could render the watersheds, their rivers, and the marine areas near their outlets drastically changed for tens of thousands of years, or periods of time that are long with respect to the phylogenetic life span of the inhabiting species.

In the subtropics, and probably in the tropics as well, there appears to be a history of climatic change that is contemporaneous with that of the more polar latitudes, although its meridional climatic characteristics seem to be directly out of phase with those in the higher latitudes. Thus, when ice was accumulating in central and northern Europe, tropical deserts expanded. Alternately, when north temperate precipitation was at its all-time low 5000 to 6000 years ago, the southern Sahara was apparently a grassland or steppe, the great Thar or Punjab Indian desert supported a highly developed Dravidian culture, and east African lakes reached the highest shoreline levels known.

We thus develop a picture of rapid and frequent shifts in precipitation patterns throughout most of the last three million years with alternate meridional compression and expansion of climatic zones and storm tracks. It is reasonable to assume that to survive, biotic elements had to have a very wide distributional range because, for example, only a relict disjunct community might survive to repopulate a grassland species of the North American Great Plains that was prevented from reproducing for 150 sequential years by simultaneous marked decrease of its growing season over its entire major range.

Soils in temperate and polar latitudes will generally develop chemically only when conditions of warmth and wetness coincide. Thus, some sites, particularly in temperate latitudes, are inhabited by communities that are adapted to a balanced nutrient capital inherited from weathering under previous vegetation and climate. This means that recovery of damaged ecosystems depends critically upon the state of balance between soil formation and the predisturbance biota. Where soils form nearly continuously, as in the tropics, this factor is not critical, but in marginal sites like the African Sahel or the North American western plains, it greatly limits the recovery potential of damaged ecosystems. In some areas of unusual geologic stability like central Australia, soils reflect climates of tens of millions of years ago and it is probable that most of the species inhabiting them have evolved since their formation. Only very depauperate communities of native organisms could be expected to recolonize such areas if they lost their soil, and more probably entirely new ecosystems would develop dominated by introduced species that were genetically prepared for the disturbed site conditions.

History of Man's Use of Ecosystems

Just as necessity is the mother of invention, so also did mankind develop alternative energy sources as his numbers began to exceed the solar-based carrying capacity of his progressively deforested chosen living areas. Indeed, the Neanderthals of Pleistocene Europe as well as earlier men relied upon short-term seasonal storage of solar energy by burning wood, and reliance upon that fuel energy supplement alone allowed development of precariously dependent sociopolitical systems such as the Greek and Roman empires that toppled, in part, when supplies of that commodity could not be maintained without development of dangerously overextended, far-ranging landholding and extractive systems. The net energy efficiency of wood carried from north of the Alps to Rome without benefit of logging trucks or charcoal trains was doubtless negative. But it was the development of fossil fuel sources that really opened the world population gates.

Some societies exceeded their carrying capacities sooner than others (some still have not), while still others adjusted through territorial expansion. Oceania today still has a population density of but $2/km^2$ compared to the world average of $28/km^2$. Carbon is fixed as fossil fuel at a rate of about 28×10^6 tons per year, but current consumption is close to 6×10^9 tons per annum, or about 200 times the rate of fixation (Portola Institute, 1974). Coal has a heat content of over 9 Kcal/g compared with only 4.2 for the average wood, and the annual yield from a mine is comparable to that from thousands of square kilometers of sustained-yield forest.

By using peat and dried dung, fuels could be had beyond the climatic limits of trees, and by using coal and lignite, population densities could exceed those that were limited by need for trapping solar energy in a food or fuel chain near the site of habitation. Thus, not at all by chance, did human populations explode forth beginning in the seventeenth century and continue to do so today within the limits of growth of energy sources. This is one of the underlying basic reasons why many of us see nuclear energy as a threat even if the waste hazard and other threats could be mitigated. Unlimited energy means essentially unlimited possibilities for ecosystem damage and eradication. The Etruscans and Greeks virtually deforested the lands bordering the Adriatic, and the Romans then taxed the land into agricultural misuse. As we can see from the sequentially higher flooding of the Arno River valley at Florence, the instability started by the Greeks and Etruscans is today progressing toward even greater soil loss rather than recovering as a forest ecosystem (Carter and Dale, 1974).

Man has always modified his living and hunting sites just as any organism does as it occupies a niche. Man's use of fire and irrigation are well-recognized examples of more effective ecosystem modification. Man is either a visitor or a tenant in an ecosystem, modifying it to serve his short-term needs. Almost by definition, this means that human alterations of ecosystems tend to make them more primitive, more growth-oriented but less stable, directed more toward production rather than protection. Each technologic advance from chopper to stone knife, from spear to bow to gun permitted solar-based man to occupy of "lay claim" to a greater portion of the earth as his domain, and thus to modify his resident and nearby ecosystems more thoroughly. While it remains debatable that man-the-hunter was solely responsible for the great number of Pleistocene mammal extinctions, it is certainly probable that for at least the last 300,000 years in both western Europe and probably in North America, major modification of the animal component of ecosystems has been caused by man. Climate alone would not selectively eliminate only the larger species used for man's food without simultaneously causing extinctions in some proportional number of smaller animals and plants (H. deLumley, 1972; M. deLumley, 1973; Steen-McIntyre et al., 1973; Laville, 1975).

Among the modifications of ecosystems that man has effected for his own needs, the development of agriculture and concomitant irrigation and fertilization provides an example of one of many extreme modifications that I feel should be raised in the context of recovery of damaged ecosystems. Agronomic ecosystems are no more or less "unnatural" than are those with fire suppression, urbanization, oil spills in tidewaters, or other severely disrupting efforts of man. We can see in the geologic record of Pleistocene deposits along the Santa Barbara Channel, for instance, that enormous discharges of heavy oil occurred on the order of 100,000 years ago, possibly associated with fault offset in some of the shallow submarine oil fields that were advertently tapped there by man in 1968. The death assemblages of tar-bound shellfish suggest that nearshore fauna and environments were essentially the same as those found there today. The characteristic of wide latitudinal variation of the narrow band of nearshore marine ecosystems provides a ready source of organisms to move into a disturbed area. So long as the damage is piecemeal and does not wipe out the whole length of the ecosystem within biogeographic range of feasible propagation, it is probable that geologic and particularly geochemical processes over long periods of time will render the site suitable for ultimate recolonization.

But what happens if we destroy a significant portion of the species that comprise the biota of an ecosystem? What if we cause the extinction of elements of the ecosystem? New ecosystems develop with new interrelationships between species. From a purely theoretical perspective, we may postulate that new ecosystems at sites that were highly stressed by Pleistocene climatic change and that possess immature soils may develop with energy efficiencies, ground cover, and biotic diversity that approach those of the original site inhabitants. However, in long-stable subtropical sites the adaptability of recolonizing species may limit diversity, stability, and efficiency of the new ecosystems compared to those existing before disturbance. In general, it must be said that, in fact, we do not know what will happen when major components of a biota are eliminated and cannot be reintroduced. Based upon study of Pleistocene extinctions, we can postulate that other organisms rapidly fill the niches released by others. There may be specific exceptions to this in the case of large mammals hunted to extinction in North America, but unfilled niches are rare. Blighted chestnuts in eastern United States were rapidly replaced by oaks. Fossil extinctions do not really tell the story because the vertebrate paleontology of the North American camel or saber-toothed tiger does little to reveal to us the dynamics of total ecosystem development at the habitats of these organisms after they were eliminated.

Agronomic Ecosystems

One of the major ecosystem modifications that man has effected, through energy subsidies, to increase the utility of his habitat is agriculture. Although we tend to visualize our voracious energy appetites as used to produce automobiles, air travel, and other luxuries of a fossil fuel economy, the world total consumption of energy is largely directed toward agricultural production and inhabitation of previously uninhabitable areas of the planet. Even the most primitive agriculture, with no energy subsidy except that of man and beast, constitutes a non-steady-state consumptive use of a resource. The resource is soil productivity, built by millenia of weathering under nonagronomic ecosystems to develop nutrient and energy storage. Soils were depleted long before tractors and fertilizers were developed (Carter and Dale, 1974).

Primary plant succession and ecosystem development, as presently conceived, are in reality just expressions of progressive changes in biogeochemical limitations within the soil base of biotic communities. Just as atmospheric geochemistry limited organic evolution on earth, so also does soil geochemistry limit ecosystem development at a site. It is true that plants alter and affect soil development just as they do atmospheric development, so one cannot distinguish causal agents. But, in general, the rates of geochemical development in a soil limit and define the rates of ecosystem support and development that can occur at that site. Soils develop through a general sequence of increasing stored, biologically available nutrient capital and increasingly complex looping of nutrient and material cycling and storage pathways. The more mature the ecosystem, the greater the degree of closure of nutrient and material cycling loops (Whittaker and Woodwell, 1967). Thus tropical ecosystems may have a large portion of their nutrient capital within the standing biomass and rapidly and efficiently recycle dead biomass through the substrate and back into plant and animal tissue. Since niches are filled, little material is lost during recycling. Immature high-temperature-latitude ecosystems are very different. Only small portions of the nutrients in the soil are found in the biotope it supports. Unfilled successional niches coupled with climatic seasonality combine to render losses of nutrients and stored energy great in comparison to the flux between organisms, soil, and sun. Nutrient cycling pathways are open so that freshly weathered nutrients may be used only once by an organism and then lost from that site through leaching or erosion. Fresh unused nutrient minerals are generally available in excess, and the biomass is generally limited by just a few substances such as nitrate or phosphorus. In the case of glacial loess and volcanic ash, these immature soil substances may provide some of the most fertile and tolerant substrates on earth since severe ecosystem damage will not greatly alter the geochemical availability and flux of soil nutrients.

But tropical soils, or those in temperate latitudes that developed under warm and wet conditions, such as the relict red soils of the San Joaquin Valley in California, some of the older interglacial gumbotil soils of the Great Plains, and the laterites of southeastern United States, behave very differently after ecosystem disturbance. These soils, having been depleted through previous geologic millenia, are low in biologically available principal nutrients. Unlike the immature soils of temperate ecosystems wherein nutrient storage sites are in part within clay minerals, many of the "overmature" soils have a large portion of the available nutrients in the biomass and in dead organic matter in the soil profile and litter layers. Upon disturbance, these soils generally release the stored portions of their nutrients, which then form a biologically available pool of dissolved ions. Such nutrients act as a very efficient fertilizer but can be carried off in surface runoff or by downward percolation of groundwaters.

Plowing fields and slash-burning are nothing more than deliberate ecosystem manipulations that cause release of nutrients in excess from their previously bound storage sites. When a grassland ecosystem is first disturbed by plowing, the delicately segregated nutrient cations that are stored in biologically exchangeable storage sites in clay minerals or as organic chelates in the rooting zone are immediately released through soil microbial and leaching action. Rather than being available to plants only upon demand when a root hair presents a hydrogen ion to the storage medium to displace a needed cation, the disturbed storage sites are subject to flushing by rainwaters or irrigation and flood forth in great excess over the amounts that could normally be used in fully developed ecosystems. Thus plowing or other surface disturbance temporarily greatly increases site productivity, but does so by consumptive disruption of biogeochemically segregated storage media (Curry, 1973, 1975).

A field that has been plowed for several decades, depending upon depth of plowing and local meteorologic conditions, may become depleted in some of the necessary critical nutrients like phosphorus and nitrogen. These can be supplemented through fertilization until other nutrients become limiting, but then those nutrients also must be supplemented. Additional energy inputs in the form of repeated tractor plowing, mulching, and intensive fertilization then are progressively necessary to maintain the site productivity. In sites of immature soils, such intensive management may continue to bear results for decades or centuries before soils are so depleted of needed nutrients that they are being used merely as hydroponic growth media. However, tropical soils or well-developed, highly productive temperate-latitude paleosoils cannot withstand such long ill use before they fail. With few available stored nutrients outside the standing biomass and severe limitations on some of those nutrients, old tropical soils offer little prospect for any more than short-term agronomic use without immense inputs of fertilizer and energy. If such soils can be thoroughly disturbed to bring up fresh mineral matter from beneath several meters of weathering profile, then warm-wet climates will render those nutrients available to a biotope in a short geologic period of time. But if the lands are flat and the soils are deep and without benefit of airborne volcanic ash infusions every century or so, or if the geologic substrate is deficient in critical nutrients even if exposed, then nonsubsidized agriculture will soon fail. In other words, contrary to general belief, secondary succession can actually be inhibited by a depleted mature soil in comparison to an immature soil because the surface layers in the former are low in nutrients.

Even if supplemental energy were to be considered as an unlimited resource, phosphatic fertilizer sources are severely limited. World use of fertilizers has risen enormously in the last three decades, from around 8 million metric tons in 1944-45 to over 60 million metric tons in 1974-75. About 30 kg of phosphate rock is used today for every person on the globe each year. This is, in essence, a post–World War II response to soil depletion. Consumptive use of fertilizers is rising much faster than food production per unit of fertilized land—a clear statement that a fertilizer junkie cannot maintain his habit indefinitely. The chemical drug dependence of modern agriculture portends the greatest challenge for recovery of damaged life-support ecosystems.

To transport and prevent deleterious effects of the great concentrations of excess nutrients that must be available to bathe ecosystems demanding hybrid seeds, nations using large amounts of fertilizer generally overuse water. The United States uses eight times the per capita, per day agricultural water that India uses. If the 13,000 liters per capita, per day agricultural water use rate of the United States were applied to the world's population, the world's total hydrologic budget would be exceeded just for intensive agriculture at U.S. standards. In reality, while much of the world's land is arable, the combination of soil and climatic factors rather severely limits sites that can sustain food production without enormous and increasing inputs of energy and materials. Technologically dependent agronomic systems touted by the world's affluent nations are, in reality, nothing more than glamorous roads toward ecosystem ruin, a well-meaning but myopic technical suggestion analagous to a proposal that we increase the output of a candle by cutting it into several smaller candles and lighting them all at the same time.

Damaged Agronomic Ecosystems

To illustrate some of the consequences of damage to agronomic ecosystems and prospects for recovery, the example of Palau is timely. The Palau group of the Caroline Islands is a rather typical assemblage of coral atolls lying at about 7° north latitude in the western Pacific. They are the westernmost of the Pacific Island Trust Territories of the United States, politically inherited at the close of the Second World War after Japanese occupation. The Palau group has a population of about 10,000 on a land area of about 460 km^2. A great portion of the partly solar-based economy is reliant upon the very large area of tropical seas that surround the islands.

During World War II, partly agricultural native populations were displaced and occupying forces took turns trying to destroy each other. One of the islands of about 12.4 km^2 area was the site of a Japanese airfield. This island, Peleliu, unlike some other Micronesian islands, is essentially a raised coral limestone platform without volcanic or alluvial sources of some essential nutrients. Agriculture is practiced on the lower boggy regions and sandy areas where soils have developed. Taro is the principal crop of the bog areas, and it can be sustained if fertilized in the traditional Pacific island method by periodic addition of leaves of certain plants to the muck soils. Sandy soils sustain coconut palms, cassava, sweet potato, maize, Polynesian arrowroot, and squash. The major life-supporting agricultural soil unit comprised but 1.7 km^2 of the island area. This was the flat smooth land that was selected for an airfield by the Japanese, and which was ultimately bombed literally into the pre-agricultural stone age by American aerial and naval forces beginning in September 1944.

Prewar Japanese efforts to fortify the island and house troops seriously disrupted the native and altered ecosystems of the island. The major airfield and full supporting facilities were constructed upon a foundation of quarried limestone. The prime agricultural areas were covered with limestone and then sprinkled with salt water to form semipermanent

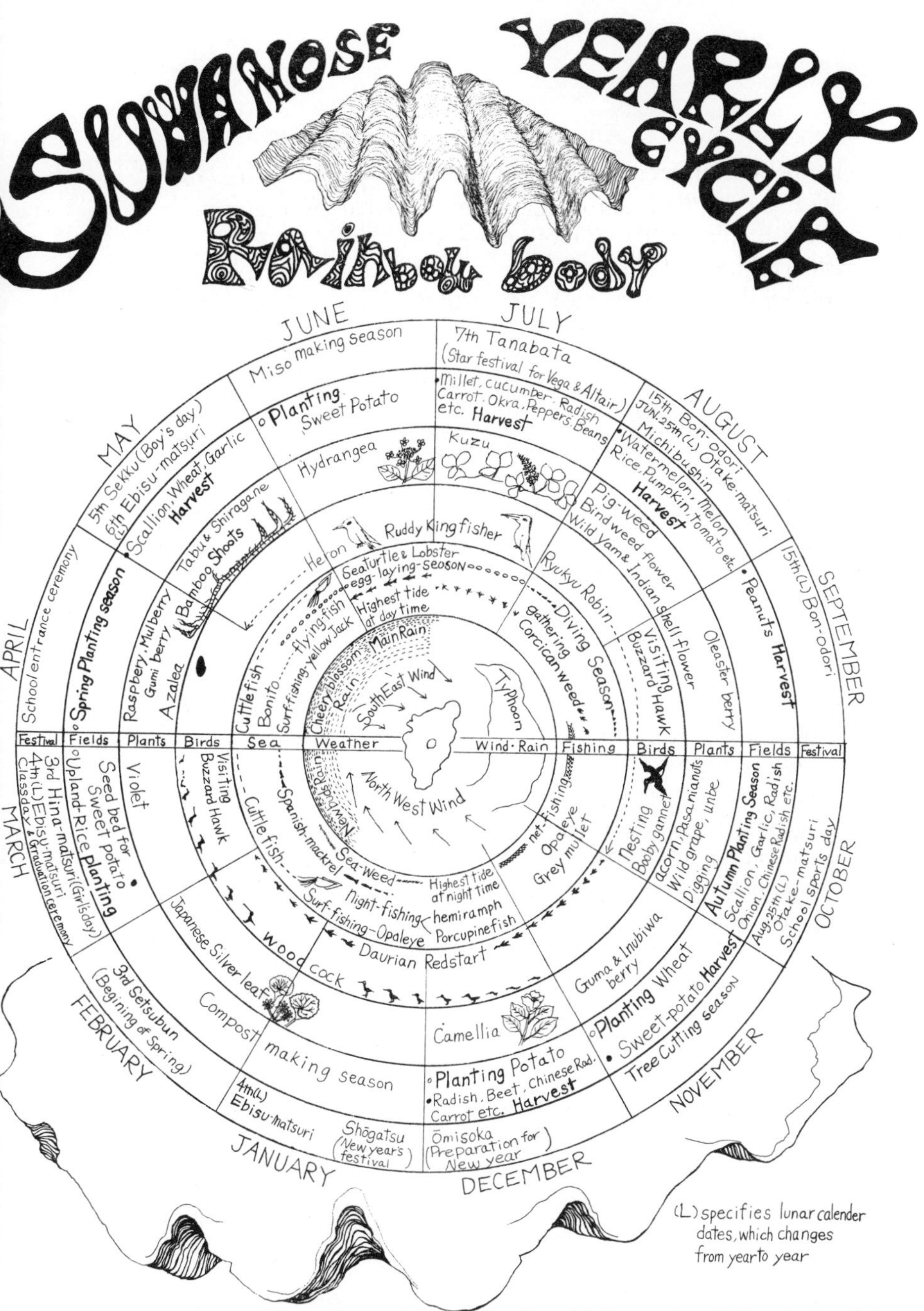

Fig. 1. Suwanose Island yearly cycle. A reflection of the regional identity and interdependence of resident man upon this western Pacific volcanic island ecosystem. From Wakkon, 1974, who writes (April 20, 1975) that the illustration is presently only fragmentary and incomplete and that another ten years will be necessary to "rough out" a fairly complete "web of life" for this small island. Wakkon admonishes readers of this volume to accept her sincere apologies for "much to be desired" in this figure and asks forgiveness for possible inaccuracies in English translation of Ryukyu plant names. Specific changes requested are: April-May, Birds—insert "Laying season for most wild birds"; September, Plants—change "Oleaster Berry to Gumi Berry"; October, Plants—omit "digging."

cementlike surfaces. After heavy bombing, the Americans moved in and refilled their bomb craters, enlarged the airfield, and occupied the bulk of the remaining agricultural lands with troop support facilities. These troops further quarried limestone and mixed it with salt water to pave their temporary quarters. About one half of the 1.7 km^2 of prime agricultural lands was severely disturbed and paved. The bog soils and much of the sandy soils were disturbed only by roads.

Today many Palauans consider that the Peleliu soils were somehow poisoned by American bombs, since agricultural productivity has not been able to be restored after 30 years of intense tropical weathering. When they returned to their island 2 years after the war, having been evacuated in 1942 by the Japanese, they found that the site had been essentially deforested, with the dead trees still standing. Despite repeated attempts to reestablish agricultural practices and patterns of prewar years, the damage has apparently been too great. In 1964 a severe typhoon destroyed all the homes of reintroduced people and washed salt water over some of the potential agricultural soils. Revegetation had been sparce and lacked vigor. Today's agricultural ecosystems on the island are much diminished from those of prewar years, and USDA efforts have not been able to alter the course of events significantly.

The Palau situation affords us a laboratory in which we may observe potentials for recovery of damaged agronomic ecosystems. The southeast Asian defoliated forests and cratered fields of Cambodia and Vietnam are presumably more thoroughly altered than the World War II efforts on Peleliu. Other Pacific islands receiving comparable treatment during that war have recovered a semblance of their prewar ecosystems, but parts of Peleliu have not. The other islands had additional sources of nutrients and their prewar biotas were possibly not as marginally supported as were those of Peleliu. Parts of the Mekong lowlands of southeast Asia similarly have alluvial sources of fresh mineral matter and nutrients, as long as flooding is allowed to continue without control by contemplated dams, but the bulk of the higher ground of Vietnam carries old, precarious, worn, deeply weathered tropical soils.

When a tropical forest ecosystem is deforested in a monsoon or typhoon climate, there exists a short period of time when the critical available nutrients are capable of being leached or carried out of the ecosystems. So long as the damage occurs to only a portion of the plant community, waterborne nutrients should be picked up by some component of adjacent communities, and thus those nutrients can be saved and eventually cycled back to their sites or origin. But if intense bombing with associated fire or more "modern" use of herbicides is considered, then a single destructive event can yield a great and long-persistent reduction in site productivitiy. This will greatly limit the return of natural or agricultural ecosystems. Until that flotsam of chemical weathering that we know of as tropical soils is lost through slow erosion, there is little opportunity for natural primary recovery of initial ecosystems. If fertilizers are used to supplement the lost nutrients, then recovery may be possible but for soils such as laterites, few nutrient ion storage sites are available for temporary fertilizer storage. Thus, one is forced to consider continuous fertilizer application until mature ecosystems with maximal standing biomass are developed. Only then can fertilizer application be gradually reduced so long as fertilizer-dependent species have not been introduced.

Fertilizer Damage

Agronomic systems can be damaged directly and indirectly by use of fertilizer. Much of the world's population gets its protein from grain and cereal crops. In the past decade, yields of these crops have been increased enormously through use of hybrid seed varieties that are dependent upon large quantities of nitrogen fertilizers, water, and pesticides. The ammonia-based nitrogen fertilizers are derived largely by combining atmospheric nitrogen with natural gas or petroleum. In 1972 the Philippines paid about $50/ton for fertilizers. By 1974 these prices had risen to about $225/ton largely as a result of the energy crisis.

Soil bacteria (*Nitrosomonas* spp.) oxidize the ammonia fertilizers to nitrite plus H+ ions. The nitrite is then further oxidized by other bacteria (*Nitrobacter* spp.) to nitrate, in which form it is readily available to plants and utterly water soluble, so that it cannot be retained in the soil unless taken up by plants immediately. The hydrogen ions liberated by the first oxidation then displace stored monovalent and divalent nutrient cations such as calcium, potassium, sodium, and magnesium from their insoluble storage sites in organic colloids or clay minerals to make those also available to plants. Thus fertilizers work in two fashions: one, by directly providing a nutrient to a plant, and another, by indirectly providing a substance that will cause release of stored nutrient capital. Through this fashion, soils can be "burnt-out" by repeated fertilizer use, even more efficiently than by repeated plowing or slash-burning. Thus a short-term chemical "fix" can lead to long-term damage to soil systems.

Old-field succession as a form of secondary recovery of a damaged ecosystem depends very much on the actual degree and nature of fertilizer use prior to abandonment of a farmland. The classic plant ecology old-field succession studies generally chronicle the recovery through geochemical resegregation and reconstitution of nutrient storage sites disturbed by plowing and "additive" fertilizer use such as manure. I am not aware of work that details the rates and processes of recovery of fields that have lost productivity through continued "subtractive" fertilizer use. One would expect that recovery would be a function of the amount of nutrient capital destroyed coupled with the rate of formation of new nutrients by primary rock weathering and nitrogen fixation. Long-continued, highly productive farming of lands with deep mature soils may render recovery upon abandonment more difficult and slower than would recovery of a younger soil with a shorter history of equally intensive management and damage.

Climatic Damage

Agronomic ecosystems are highly sensitive to climatic damage. Following the basic biologic axiom that high-productivity, low-diversity systems are less stable than higher diversity systems; and coupling this with the recent development of hybrid seed varieties that need great inputs of energy and materials to complete their highly time-dependent growth cycles, we are presented with the serious risk that human populations cannot be fed under natural conditions of climatic perturbation.

Most of the modern hybrid varietals were developed during the late 1950s and 1960s as part of our purported green revolution. This was a period of time of unusually mild, consistent global weather. The crop distribution patterns that were established based upon the reliable climatic parameters and special seeds frighten many Pleistocene scientists today (Kukla et al., 1972; Bryson, 1974; Bryson and Ross, 1974). The mid-latitude breadbaskets of the Soviet Union, Canada, the United States, and Europe are severely impacted by cooler periods of modern climates when westerly circulation patterns expand and site conditions become much more variable than during the drier, warmer times. Several sequential wet years may be followed by a local dry year that exceeds the drought of generally drier times. Years of long growing seasons are interrupted by years of wet springs and falls that hamper planting and harvesting. The extreme weather conditions of the period of 1972-74 are an example of such perfectly expectable "unexpected" weather.

Climatic change is the order, not exception. Ecosystems tend to develop so as to balance a maximization of site energy and material utilization against a diversity-maintained genotypic adaptiveness to climatic stresses.

Prolonged drought at sites of periodic retreat of monsoon rains like the sub-Sahara and India has tempered local ecosystems to a species composition that includes enough drought-adapted species to maintain some sort of integrity through the range of extremes that can be expected at that site. Such a compositional adaptation is the direct result of the biogeographic development of local ecosystems at a site through the range of climates that existed while the assemblage of organisms was developing. Thus all ecosystems must be dynamic communities of organisms that will continuously change relative species dominance and occasionally drop old species or add new species in response to stress. The existing composition of any ecosystem is a function of its developmental history. In this sense, the term *damaged* is inappropriate to describe ecosystems because it implies a scale of values that is not really available. All one can say from a historical perspective is that an ecosystem was *modified.*

Conclusions

Mankind is reliant upon the maintenance of a dynamic mosaic of life-supporting ecosystems. Technological advances of recent decades render it probable that total ecosystems can be severely damaged and component genetic material lost forever. To insure recovery of damaged ecosystems, one needs to conserve examples of a sufficient variety of ecosystems to be able to reconstitute or reinoculate any damaged site. The greatest single hazard to man's use of ecosystems is seen as his absurdly shortsighted and utterly suicidal modern agricultural practices. Unless we can reinhabit the earth with a future-primitive philosophy based on a sound understanding of regional ecosystem dynamics and limitations, we cannot expect to continue to live as a coinhabitant of the biotic communities that gave us our very existence. We must learn to appreciate the full implications of the realization that we cannot permanently alter our life-support systems without permanently altering our gene pool and hence ourselves.

Acknowledgment

Robinson Jeffers's poetry is taken from "The Answer" from *The Selected Poetry of Robinson Jeffers,* © 1959 by Random House, Inc., New York.

Literature Cited

Bryson, R. A. 1974. A perspective on climatic change. Science 184:753-760.

Bryson, R. A., and J. E. Ross. 1974. A pandemic of acute malnutrition (excerpt). *In* The Other Side, No. 2. The Environmental Fund, Washington, D.C. 4 pp.

Carter, V. G., and T. Dale. 1974. Topsoil and Civilization. 2nd ed. Univ. of Oklahoma Press, Norman. 292 pp.

Chedd, Graham. 1974. Man-made planets, seriously. CoEvolution Quarterly 1(4):70-72.

Cloud, Preston, Jr. 1968. Atmospheric and hydrospheric evolution on the primitive earth. Science 160:729-736.

Cloud, Preston, Jr., and Aharon Gibor. 1970. The oxygen cycle. Pages 59-68 *in* The Biosphere—A Scientific American Book. W. H. Freeman, San Francisco. 134 pp.

Curry, R. R. 1970. A proposal for ecological refugia. Intecol. Bull. 1:3-7.

_____. 1973. Geologic and hydrologic effects of even-aged management on productivity of forest soils, particularly in the Douglas-fir region. Pages 1-43 *in* R. K. Hermann, ed. Even-Aged Management. Oregon State Univ. Press, Corvallis.

_____. 1975. Biogeochemical limitations on western reclamation. Pages 18-47 *in* M. K. Wali, ed. Practices and Problems of Land Reclamation in Western North America. U.S. Bureau of Mines and University of North Dakota Press, Grand Forks.

Dasmann, R. F. 1972. Towards a system for classifying natural regions of the world and their representation by national parks and reserves. Biol. Conserv. 4:247-255.

_____. 1975. National parks, nature conservation and future primitive. Paper presented to South Pacific Conference on National Parks, Wellington, N.Z., Feb. 24-27, 1975.

DeLumley, H. 1972. La Grotte de l'Hortus. Les Chasseurs Néandertaliens et leur milieu de vie. Univ. de Provence, Lab. de Paléontologie Humaine et de Préhist., Marseilles. 668 pp.

DeLumley, Marie-Antoinette. 1973. Anténéandertaliens et Néandertaliens du Bassin Méditerranéen Occidental Européen. Univ. de Provence, Lab. de Paléontologie Humaine et de Préhist., Marseilles, Mem. 2.

Ehrlich, P. R. 1974. Human population and environmental problems. Environ. Conserv. 1:15-20.

Gorsline, J., and L. House. 1974. Future primitive. *In* P. Berg, ed. Planet-Drum, Bundle 3. San Francisco (Box 31251, San Francisco, CA 94131).

Imbrie, J. 1974. Lessons from the geologic and historic climatic records. CLIMAP, Brown Univ., Providence.

Kukla, G. J., R. K. Matthews, and J. M. Mitchell, Jr. 1972. The end of the present interglacial. Quat. Res. 2:261-269.

Lamb, H. H., R. P. W. Lewis, and A. Woodroffe. 1966. Atmospheric circulation and the main climatic variables between 8000 and 0 B.C.:

Meteorological evidence. Pages 174-217 *in* J. S. Sawyer, ed. World Climate from 8000 to 0 B.C. Royal Meteorological Soc., London. 299 pp.

Laville, H. 1975. Climatologie et chronologie du Paléolithique en Périgord: Étude sédimentologique de dépôts en grottes et sous abris. Univ. de Provence, Lab. de Paléontologie Humaine et de Préhist., Marseilles, Mem. 4.

Lovelock, J., and S. Epton. 1975. The quest for Gaia. New Sci. 65:304-306.

O'Neill, G. 1974. Colonization of space. Physics Today. Sept., pp. 32-40 (also published in New Sci., Oct. 24, 1974).

Portola Institute. 1974. Energy Primer. R. M. Merrill et al., eds. Portola Institute, Menlo Park, Calif. 200 pp.

Steen-McIntyre, Virginia, R. Fryxell, and H. Malde. 1973. Unexpectedly old age of deposits at Hueyatlaco archaeological site, Valsequillo, Mexico, implied by new stratigraphic and petrographic findings. Pages 820-821 *in* Abstract with Programs, Dallas Meeting, Geological Society of America, Boulder, Colorado.

Steincamp, J. 1975. Space Colonization by the Year 2000—An Assessment. Marshall Space Flight Center, Program Development Office, NASA, Jan. 15.

Viereck, L. A. 1966. Plant succession and soil development on gravel outwash of the Muldrow Glacier, Alaska. Ecol. Monogr. 36:181-199.

Wakkon. 40074. (1974). Suwanose yearly cycle. Pages 33-49 *in* Kurisu, ed. Om. English edition, Cosmic Child Community, Tokyo (Kokubunji Embassy, 2-664 Nishi-Kolgakubo, Kokubunji-shi, Tokyo, Japan).

Whittaker, R. H., and G. M. Woodwell. 1967. Evolution of natural committees. *In* J. A. Weins, ed. Ecosystem Structure and Function. Oregon State University Press, Corvallis. 176 pp.

Recovery of Streams from Spills of Hazardous Materials

John Cairns, Jr., and Kenneth L. Dickson

Abstract

Two case histories of streams (Clinch River and Roanoke River in Virginia) which were damaged by spills of hazardous materials (caustic, acid, and ethyl benzene-creosote) and subsequently recovered are discussed. From these and other studies of the biological recovery of damaged streams, four characteristics of ecosystems which relate to the recovery process have been identified: (1) ecosystem vulnerability, (2) elasticity, or ability to recover from damage, (3) inertia, of ability to resist displacement of structural and functional characteristics, and (4) resiliency, or the number of times a system can recover after stress. Each of these characteristics is described and methods of determining each are presented.

Introduction

To restore damaged ecosystems, one must understand both the processes which lead to disruption of ecosystem function and the processes which are involved in the restoration of structural and functional integrity. These processes must be understood before appropriate management practices can be developed for restoring and rejuvenating damaged aquatic ecosystems.

The purpose of this chapter is to discuss four characteristics of aquatic systems which to a large degree regulate the response of a system to stress and the rate and degree of recovery from stress. The chapter is directed toward preliminary approximations of (1) ecosystem vulnerability, (2) elasticity, (3) inertia, and (4) resiliency.

Aquatic ecosystems have the ability to assimilate a certain amount of waste material and maintain near normal function. With the constant use and reuse of water from our natural aquatic systems by industries, agriculture, and municipalities, the function may be seriously altered or disrupted if the assimilative capacity is exceeded. The ability of a stream or river to assimilate wastes is governed by the capacity of the system to transform them before they reach deleterious levels. If an overload occurs, the system is disrupted and the transforming capacity may be substantially reduced. Recovery may be rapid or slow depending upon a number of factors, including (1) severity and duration of the stress, (2) number and kinds of associated stress, (3) recolonization of the area by useful aquatic organisms, and (4) residual effects upon nonbiological units (e.g., substrate, etc.).

Industrial spills typically involve an abrupt release of a waste into a river. The waste usually passes downstream in a slug which lengthens as it proceeds due to mixing characteristics of the river and channel water. Exposure to peak concentration is usually short and will vary depending on velocity and other factors. This usually produces an acute stress which may eliminate sensitive organisms but has considerably less effect upon the populations of tolerant and moderately tolerant organisms. Once the stress is removed or reduced, the community will become reestablished through processes such as downstream drift and other methods of recolonization, although many of the species may be different from those originally present.

A brief description of the recovery from pollutional stress of the Clinch River in Virginia after spills of caustic and acid and of the Roanoke River in Virginia after a spill of a mixture of ethyl benzene-creosote follows. Since these have been published elsewhere in detail, only a condensation of the case histories is included with references for those who wish more information. These case histories, and those described by Herricks elsewhere in this book, were the stimuli for the discussion of ecosystem vulnerability, elasticity, inertia, and resiliency.

Case Histories

Clinch River

The aquatic environment of the Clinch River in southwestern Virginia has been acutely stressed by two spills of hazardous materials. In 1967 caustic wastewaters from a fly ash pond caused extensive damage to the aquatic biota. In 1971 a spill of sulfuric acid had a similar, though more restricted, effect. Both spills came from the Appalachian Power Company's power plant at Carbo, Virginia. In addition to the acute pollutional stress from the two spills, a limited portion of the river has been chronically stressed by the day-to-day operations of the power plant. Moreover, periodic flooding has had short-term deleterious effects on the biota.

The benthic macroinvertebrate community was sampled repeatedly in 1969, 1970, and 1971 at 21 ecologically similar stations to assess the effects of the two spills and to determine the effects of flooding. The methods used to describe the indigenous macrobenthic communities were numbers of organisms, density, diversity, and cluster analysis of presence-absence data.

Results from the 1969 surveys indicated that except for molluscs, the stream's macrobenthos had virtually recovered from the 1967 fly ash pond spill. Information from the 1970 surveys also tended to support this observation, although an unexpected acid spill seriously reduced biological diversity at four stations immediately below the power plant for approximately three months.

The 1971 surveys were similar to the surveys done in 1969 and 1970 except the fieldwork was completed two to three weeks after late spring flooding. Results from these collections indicated that flooding had an equalizing effect on the macrobenthos at the different stations; i.e., areas that were usually sparsely inhabited became richer in species. Stations previously very rich in species, on the other hand, tended to lose species so that after flooding all stations were faunally more similar than they were at other times of the year when flooding was less severe or not so close to the time of sampling.

References for detailed publications on the Clinch River are Cairns et al., 1971, 1973; Crossman et al., 1973; Kaesler et al., 1974.

Roanoke River

The effect of an abrupt release of acutely toxic material into a river is an immediate stress on the organisms in the river followed by dilution of the waste and eventual restoration of water quality. A spill of this nature occurred October 10, 1970, at Salem, Virginia, on the Roanoke River. Ethyl benzene mixed with creosote was spilled into the Roanoke River from a primary storage tank of the Koppers Company. Approximately 2000 gallons escaped and entered an open cooling water ditch which flowed into the Roanoke River. Approximately 400 to 600 gallons of the solvent entered the river in a period of one to two hours. River flow at the time of the spill was 19,000 gal/min, resulting in an estimated concentration of solvent at the point of discharge of 1000 ppm. The subsequent fishkill was reported by the Virginia Water Pollution Control Board; a total of 13,281 fish were killed, consisting of 7979 rough fish and 5302 sport fish. Initial estimates by the water control board indicated that biological damage extended for a distance of seven miles below the plant's outfall.

The exact toxicity of the ethylbenzene-creosote mixture is not known. However, using static bioassays, Cairns and Scheier (1959) found that the 96-hour TL_m (median tolerance limit) for *Lepomis macrochirus* (bluegill sunfish) was 10.0 ppm for creosol. Pickering and Henderson (1966) established the 96-hour TL_m for the bluegill using ethyl benzene, at 29 ppm under static conditions. The synergistic or antagonistic actions of these compounds, or their action with other compounds in the river, could have altered the acute toxicity of the spilled material in the Roanoke River.

Sampling stations were selected above and below the site of the spill and were sampled for both fish and aquatic macroinvertebrates. Sampling was done between 6 and 10 days after the spill for aquatic invertebrates and 9 to 11 days after the spill for fish. A follow-up bottom fauna and cursory fish survey was conducted April 1 and 2, 1971 (six months later), to determine the extent of recovery.

The results of the study of the recovery of the Roanoke River following a spill of ethyl benzene-creosote were:

1. The effects of an acute stress from an industrial spill of ethyl benzene-creosote were to decrease the diversity and density of both the fish and bottom fauna for approximately three miles below the site of the spill.
2. A differential response of the stress was apparent with all major groups of fish, except the minnows, being entirely eliminated in the stressed area. Perhaps the minnows avoided the stress by swimming into small isolated tributaries, or perhaps they rapidly reinvaded the area, or they may be more tolerant to this type of stress.
3. Mayflies, stone flies, caddis flies, and mussels did not survive the stress and were eliminated. However, mayflies and stone flies were present six months later, indicating improved water quality.
4. Riffle beetles, true flies, crayfish, some snails, and segmented worms apparently were able to survive the short-term exposure to the ethyl benzene-creosote stress.

A reference for a detailed discussion of the Roanoke River study is Cairns et al., 1971.

Studies of the biological recovery of the Clinch River and Roanoke River following spills of hazardous material stimulated us to try to develop a conceptual framework to understand better what characteristics of aquatic systems are altered by abrupt pollutional stress, the degree of damage caused by stress, and the rate and extent of recovery possible in an aquatic system. Circumstantial evidence indicated that some systems recover from stress more rapidly than others. The discussions which follow attempt to define four response characteristics of ecosystems essential to understanding the processes of restoration of structural and functional integrity.

Ecosystem Characteristics

Vulnerability to Irreversible Damage

All water ecosystems are not equally vulnerable to irreversible damage, nor are they subjected to the same intensity or frequency of perturbations of human origin. Irreversible damage is an exceedingly difficult characteristic to define because one might well hypothesize that given sufficient time, no damage is irreversible. However, it seems reasonable to be anthropocentric and define irreversible damage as requiring a recovery period longer than the human life span. Lakes and other water ecosystems regularly have irreversible changes. Some are clearly deleterious, but others are part of the normal evolutionary or successional process due to climatic cycles, etc.

Table 1 represents an attempt to categorize four water systems (rivers, lakes, estuaries, and oceans) in terms of their vulnerability to irreversible effects, the severity of the environmental stresses to which they are now exposed, and the frequency of the man-originated perturbations to which they are presently exposed. The rank ordering starts with 1, signifying the greatest vulnerability to irreversible changes, the highest severity of man-made stresses, or the greatest frequency of

Table 1. Relative vulnerability of four water ecosystems to irreversible effects; the severity of stress presently resulting from man-originated activities; frequency through both space and time of man-originated perturbations

	Irreversibility of effects	Stress severity	Frequency of man-originated perturbations	Total
Lakes	2 (20)	3 (80)	3 (80)	8 (180)
Rivers	3 (80)	2 (50)	1 (1)	6 (131)
Estuaries	4 (100)	1 (1)	2 (40)	7 (151)
Oceans	1 (1)	4 (100)	4 (100)	9 (201)

man-originated perturbations, with 4 representing the opposite extreme. In addition, Table 1 shows the same rank ordering on a scale of 1 to 100 with the same trend as the one before, that is, 1 represents the greatest irreversibility of effects and 100 the least. Use of this second scale allows for some weighting and grouping of systems. For example, on the irreversibility of effects, lakes and oceans are much closer to one another on the scale of 1 to 100 than to either rivers or estuaries. This permits more flexibility for the association of similar systems, and also introduces a higher degree of subjectivity. However, it is interesting to note that in using the second system, the final rank ordering remains the same and the relative distance of each system from the others remains approximately similar.

A brief discussion of the different operational and ecological characteristics of each system follows.

Lakes

Although lakes vary greatly both in origin and in ecological characteristics, it will simplify the conceptual approach if only temperate zone stratified lakes are considered. Most of the important lakes in the United States are in this category. It also should be noted that many of the larger reservoirs are stratified and share many ecological characteristics with temperate zone stratified lakes. One of the most important characteristics of lakes in this category is the periodicity of changes within the system. These lakes typically become thermally stratified during both summer and winter "stagnation." The three layers resulting from this stratification (the epilimnion, or surface layer; the thermocline, or intermediate layer; and the hypolimnion, or bottom layer) are typically quite distinct from each other in chemical, physical, and biological characteristics. During the spring and fall of each year most temperate zone lakes "turn over" (i.e., the water is usually circulated completely and strongly top to bottom). Since these turnover periods are accompanied by comparable biological changes and since the biota of the lakes is geared to a certain tempo of turnover, any interference with this pattern (such as might result from the discharge of heated wastewater from a power plant) might have consequences out of proportion to the numerical value of the temperature change. The second important characteristic of lakes is that they are nutrient traps in which phosphorous and other nutrient materials tend to accumulate and contribute to the aging or eutrophication process. The majority of lakes go through a typical aging process which may be partly reversible if a sufficient expenditure of energy is made, but one might safely say that there is a high degree of irreversibility of any part of the aging process. Although many of the environmental evangelists have categorized lakes in which the aging process has been accelerated (such as Lake Erie) as dead lakes, nothing could be further from the truth since the total biomass of organisms now present is demonstrably greater than in the historic past. On the other hand, there seems to be little question that the aging process has been considerably accelerated and the life expectancy of Lake Erie has been severely reduced. In addition, except for the turnover periods already mentioned, the environments of lakes tend to be considerably more stable than those of estuaries and rivers. As a consequence, the organisms which inhabit lakes are probably not as resistant to continual rapid short-term changes as those in rivers and estuaries. Because of the tendency for lakes to accumulate all sorts of materials and since the vulnerability to irreversible effects is quite high, lake protection should have a high priority.

Rivers

The major portion of all waste discharges into aquatic systems in the United States is initially placed into flowing waters. One of the strongest contributing factors to this practice is that the wastes are immediately carried away from the doorstep of the discharger. Also the ability of rivers to cleanse themselves of impurities is well-known. The benefits of this removal from one's doorstep are so advantageous to the discharger that the city of Chicago went to considerable trouble and expense to switch its discharges from Lake Michigan to the Mississippi drainage. Many of the rivers in the Lake Erie basin carry considerable nutrients to the lake, thus accentuating problems of eutrophication. The construction of hydroelectric, flood control, and other types of reservoirs which have been designed without consideration of the altered environmental conditions created by the discharge of wastes has permitted some serious problems to arise. For example, some reservoirs in Kansas have become eutrophic before reaching pool level as the result of extraordinary runoff from cattle feedlots which contributed enormous amounts of nutrients in a relatively short period of time. The factors contributing to the recovery of rivers and other water ecosystems will be discussed in a section of the chapter devoted specificially to that topic. Wastes discharged into rivers will sometimes be carried to lakes and/or reservoirs and, with few exceptions such as the Great Salt Lake drainage basin in Utah, ultimately reach the oceans. The fact that river water, particularly in an extensive drainage basin such as the Mississippi, may be used and reused many times before reaching the oceans is a compelling reason for lessening the waste discharges into these systems. Persistent wastes ultimately will affect both estuaries and oceans, as well as any lakes that are enroute to these. The remarkable elasticity of river ecosystems (or ability to recover) after stress may be unfortunate since the persistent materials in them ultimately are deposited in the oceans, which appear to have considerably less elasticity. Unfortunately most citizens are more aware of events in rivers.

Estuaries

Estuaries are probably the most productive ecological systems in the world. The gross primary productivity of 20,000 Kcal per M^2 per year of estuaries and reefs is matched only by the wet tropical and subtropical broad, leveled evergreen forests (Odum, 1971). By comparison, the coastal marine zones produce 2000 Kcal per M^2 per year and fuel-subsidized mechanized agriculture, 12,000. Deserts and tundras produce only 200 Kcal. Because of this extraordinarily high biological activity, estuaries represent an important means of degrading river-borne wastes before they reach the oceans. Because the retention time in most estuaries is substantial, it is frequently sufficient to accomplish the degradation of all but the most intractable compounds. In a sense an estuary represents the ultimate biofilter. Because estuaries are at the interface between river and ocean ecosystems, are subjected to strong tidal influences, and often have extensive tidal flats where temperature and other changes may be extreme, the organisms which are permanent residents in them are therefore usually markedly resistant to change. Such changes may include frequent and moderately prolonged exposure to air. Because of this, estuarine ecosystems probably have the greatest inertia of any water ecosystem in that they have an enormous ability to resist ecological disequilibrium. However, because the number of species is usually lower than in rivers and oceans, once an estuary is placed into disequilibrium, restoration of equilibrium may be quite slow.

Estuaries also represent important nursery grounds for a variety of species, some of which, such as shrimp, are commercially valuable. A variety of fish species utilize estuaries as nursery grounds for the young, and the loss of these nursery grounds would probably have severe effects upon food crops of the oceans as well as on the ocean ecosystems themselves.

The number of species in estuaries is usually appreciably less than in oceans or rivers, but the biomass is larger. This relationship is depicted in

Figure 1. As a result of this lowered diversity, estuaries probably have lower functional redundancy than the systems they connect.

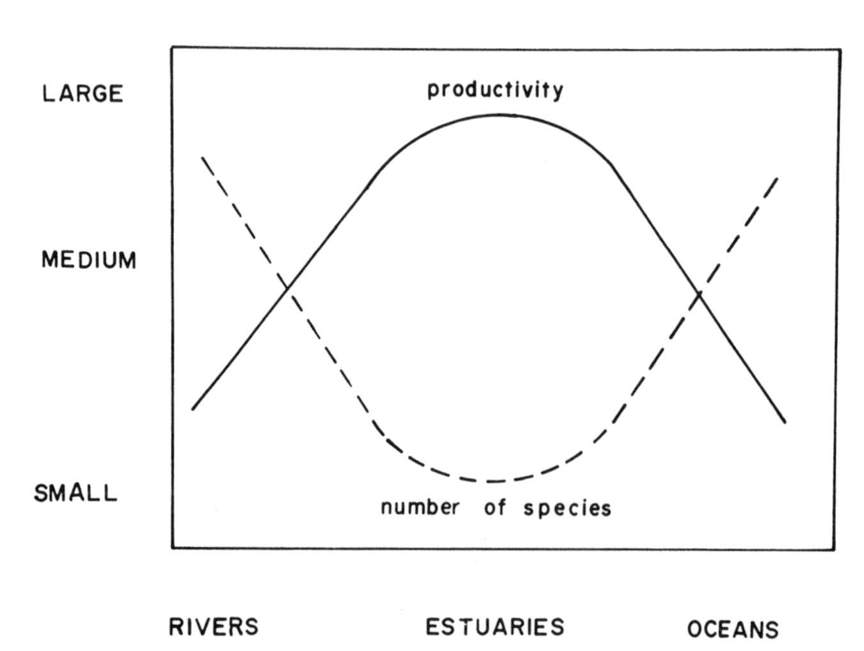

Fig. 1. Schematic illustrating the relationship between number of species and productivity in rivers, estuaries, and oceans

Oceans

The oceans represent the ultimate in climatic stability in water environments. Although some aquatic environments are more stable, such as constant temperature thermal springs and the like, certainly the oceans represent the most stable of the large-scale water ecosystems. Although many of the highly migratory species may be exposed to a variety of temperature conditions and other water quality changes, the sessile organisms are usually in a fairly constant and homogeneous environment and have been conditioned to this for countless generations. As a result they do not readily adapt to frequent or sudden changes and are not likely to tolerate them. Heilbrunn (1937) has noted that coral reef organisms may die at temperatures only 2°F above optimal. In addition Arthur L.

DeVries (pers. comms.) has shown that some antarctic fish may die at temperatures only a few degrees different from those which they regularly inhabit. He has demonstrated that there is a strong temperature relationship to the distribution of these fishes in the coastal region near McMurdo Sound and that the sensitivity to temperature change is so great that a vertical movement of only a few feet and a temperature change of only two or three degrees may cause death or severe physiological stress. Thus, a small temperature change routinely endured by most freshwater and estuarine fish would be lethal for these antarctic fish which are not highly migratory.

It appears highly probable that the vast ocean ecosystems are also highly fragile and protected primarily by their vastness, at least until now (with the exception of localized areas such as the "dead sea spot" off New York City), through the dilution of all introduced deleterious materials. The recovery of localized damaged areas of the oceans is somewhat speculative but probably will follow the same general principles to be discussed in the recovery section later. However, should an entire ocean be damaged, the time required for recovery staggers the imagination, especially since many of the organisms are highly vulnerable to change and therefore presumably would not be able to suffer the rigors of invading new areas far distant from their original habitat.

Ability to Recover, or Elasticity

Some means of estimating the elasticity of an ecosystem, or its ability to recover after displacement of structure and/or function to a steady state closely approximating the original, is badly needed (Cairns, in press). This information not only would be useful after accidental damage to an ecosystem to predict the rate of recovery, but would also be valuable in estimating the ecological vulnerability of various ecologically similar sites proposed for development. Some approximation of the relative vulnerability of the proposed sites could be made and incorporated into the series of factors considered in site selection. The factors which we consider important in the development of a recovery index follow, with a brief discussion of each.

a. Existence of nearby epicenters (e.g., for rivers these might be tributaries) for providing organisms to reinvade a damaged system. Rating system: 1–poor; 2–moderate; 3–good.

In the Clinch River study it was clear that in the initial stages of the recovery process some tributaries contributed more organisms than others. It was also evident that some of the organisms recolonizing damaged areas came from the headwaters of the main stream. These comparatively healthy tributaries and headwaters which furnished organisms to recolonize damaged areas were a key factor in the very rapid recovery which occurred. These tributary and headwater areas might be considered epicenters from which organisms departed to invade and subsequently colonize the damaged areas. A substantial reduction in the sources of potential recolonizing organisms would have undoubtedly altered the recovery pattern of the Clinch River. Therefore, the epicenters are a prime factor in the recovery pattern of a damaged ecosystem; without new potential colonizers, the process cannot occur.

b. Transportability or mobility of dissemules (the dissemules might be spores, eggs, larvae, flying adults which might lay eggs, or other stages in the life history of an organism which permits it either voluntarily or involuntarily to move to a new area). Rating system: 1—poor; 2—moderate; 3—good.

The Clinch River studies showed quite clearly that some groups of organisms have a greater potential for becoming reestablished in a damaged area than others. Fish, for example, moved into the damaged areas relatively soon after the fly ash pond spill occurred which temporarily destroyed the biota of over a hundred miles of the Clinch River. The same was true for the acid spill which occurred sometime later but which affected a more limited area. Aquatic insect larvae which "drift" downstream and are, therefore, good recolonizers also became reestablished rather soon. On the other hand, the molluscs are rather slow to reinvade damaged areas. At the time this chapter was being written, there were species of the molluscs which had not returned to the damaged areas although more than five years have passed since the last spill. If the damaged community consisted almost entirely of organisms with a high degree of transportability of dissemules, the prospects for rapid recovery would be high. If it consists primarily of those not easily transported and thus less likely to reinvade, the prospects of rapid recovery would be rather poor.

c. Condition of the habitat following pollutional stress. Rating system: 1—poor; 2—moderate; 3—good.

The fly ash pond spill and the acid spill on the Clinch River had marked effects on the indigenous biota, but they had only small short-term effects on the physical habitat of the river and on the chemical characteristics of the water. On the other hand, had extensive siltation blanketed the riffle areas, this would have resulted in marked alteration of the habitat which might persist for a substantial period of time.

d. Presence of residual toxicants following pollutional stress. Rating system: 1—large amounts; 2—moderate amounts; 3—none.

The two Clinch River spills changed the pH of the river—the first to a high pH, the second to a low. In both cases, the "slug" of water differing markedly from the ambient pH passed through the river system leaving no residual effects. On the other hand, the intrusion of biocides (e.g., dieldrin, aldrin, mercury, or lead) would almost certainly have left residuals which would probably have persisted in the system for a considerable length of time. In the Roanoke River study, the presence of ethyl benzene-creosote in the sediments caused prolonged toxicity to aquatic insects below the site of the spill. The presence of residual toxicants might well impair the recovery of a damaged ecosystem by maintaining toxic conditions unsuitable for potential colonizing organisms. Thus, the presence of such residual toxicants should diminish the recoverability potential of a system.

e. Chemical-physical environmental quality after pollutional stress. Rating system: 1—in severe disequilibrium; 2—partially restored; 3—normal.

Pollutional stress may put the chemical-physical environment of an ecosystem into severe disequilibrium either through alteration of the substratum (or other portions of the ecosystem) or through elimination of certain biota. For example, a reservoir or lake with a substantial algal growth normally might have a dissolved oxygen concentration at saturation during daylight hours and well above 2 or 3 ppm even during the longest periods of darkness. If, however, these plants were destroyed, the additional decaying organic load, together with the absence of the plants as a source of oxygen, might alter the system from aerobic to anaerobic. This change might be of considerable duration if recovery were left entirely to natural processes. However, artificial aeration would almost certainly accelerate return to an approximation of the original condition. In systems such as the Clinch River and Roanoke River where the flow-through rates are quite high, the restoration of the quality of the physical-chemical environment required only a few hours because the toxic materials were rapidly removed from the original spill site. On the other hand, a substantial portion of the river biota was damaged during the passage of the slug. The return to an approximation of the original chemical-physical conditions is an important prerequisite for the reestablishment of a community characteristic of that particular system.

f. Management or organizational capabilities for immediate and direct control of damaged area. Rating system: 1—none; 2—some; 3—strong enforcement possible.

In some cases river drainage authorities or other management groups exist which may be capable of aiding the recovery process. For example, if an oxygen disequilibrium exists in a reservoir or lake, an approximation of the normal oxygen regime might be achieved by artificial aeration (Fast, in this volume). This would presumably enhance the conditions for re-establishment of organisms characteristic to that system, and thereby, enable the natural balance to be restored more quickly than might otherwise occur. The cleanup of oil after spills (excluding ecologically damaging cleanup methods, e.g., Nelson-Smith, in this volume) by organizations charged with the management of a specific ecosystem is another example of such an activity. Reintroduction of certain types of species not likely to reinvade the area on their own is another rather simple example of management intervention in the recovery process. Probably the most valuable contribution a managerial organization might provide is the establishment of baseline or "normal" conditions so that the degree of disequilibrium can be documented when an accident occurs (Cairns and Dickson, 1974). When the displacement from normal is known, the necessary corrective steps are usually reasonably clear, and the resources available to aid the recovery process can be more efficiently directed to achieve the desired goals. As the organizations charged with ecosystem management become politically and operationally stronger, their role in the recovery process will become increasingly important.

The corrective actions used as illustrations are relatively simple and straightforward. However, knowing which corrective action is appropriate requires a fairly substantial knowledge of the system and a relatively large pool of background data regarding its "normal" condition. This is probably where most ecosystem management groups fail.

Using the characteristics listed above, and their respective rating systems, a recovery index can be developed. Using this index, one can arrive at a rather crude approximation of the probability of relatively rapid recovery. This would mean that somewhere between 40% and 60% of the species might become reestablished under optimal conditions in the first year after a severe stress, between 60% and 80% in the following year, and perhaps as much as 95% of the species by the third year. Natural processes, with essentially no assistance from a management or a river basin group, accomplished this after spills on the Clinch River and Roanoke River. These were studied by the Aquatic Ecology Group at Virginia Tech, and the usefulness of this estimate has also been checked with data provided by some acid mine drainage studies (Herricks and Cairns, 1972, 1974) and seems adequate in this regard as well. The equation for the recovery index follows:

$$\text{Recovery Index} = a \times b \times c \times d \times e \times f$$

400+ : chances of rapid recovery excellent
55–399 : chances of rapid recovery fair to good
less than 55: chances of rapid recovery poor

During the development of the simplistic equation considerably more complicated equations were considered and rejected because the refinements seemed meaningless in view of our present state of knowledge. However, there seems to be a very definite need to formalize the estimation of recovery, and one hopes that more precise equations, properly weighted, will evolve from this modest beginning.

Inertia, or Ability to Resist Displacement of Structure and Function

Inertia can be defined as the ability of an aquatic community or ecosystem to resist displacement or disequilibrium in regards to either structure or function. As was the case for the recovery index, the factors influencing inertia are listed in order of importance. This rank ordering is substantially less justifiable scientifically than for the recovery index. However, it seems quite clear that the factors are interacting and the justification for multiplying the ratings is considerably stronger than for the rank ordering. The time element is exceedingly difficult to address when evaluating inertia because there may be a substantial lag between the onset of stress and the subsequent symptoms of displacement recognizable to ecologists. Therefore, one should view this index with considerably more caution than the recovery index, which is only used following recognizable displacement.

Although time lag and displacements removed from the onset of stress are a serious problem, they are not as incurable as they might at first appear. One can make estimates of the probability of effects not surfacing initially by going through the protocol outlined in *Principles for Evaluating Chemicals in the Environment* (National Academy of Sciences, 1974). Such procedures would, with some modification of the methodologies, enable one to categorize other stresses and estimate the probability of a lag response. A brief discussion of the primary factors producing inertia follows.

a. Indigenous organisms accustomed to highly variable environmental conditions. Rating system: 1—poor; 2—moderate; 3—good.

One would intuitively expect organisms in an estuary, certain deserts, and other environments where temperature and other environmental conditions may shift rapidly to have developed various physiological, behavioral, and structural capabilities for resisting the deleterious effects of these stresses. Some, but not all, of these mechanisms, such as the ability of a clam to isolate itself from the environment for brief intervals, provide protection against both naturally occurring stresses and those induced by human activities. While it is highly unlikely that one would be able to categorize every organism within a system in regard to its tolerance of highly variable environmental conditions, a person knowledgeable about the system should be able to make a sound estimate of the degree of resistance to changing conditions for the majority of indigenous organisms.

b. System has high structural and functional redundancy. Rating system: 1—poor; 2—moderate; 3—good.

As Odum (1969) has shown, early stages of ecosystem evolution are relatively simple systems and as a consequence have relatively low functional redundancy. That is, the loss of a particular species might well mean the loss of a particular trophic level or function because no other species capable of fulfilling this role is present. On the other hand, a mature system with great complexity, and one in which the fractionation of activities and roles is substantial, is likely to have a high functional redundancy. This minimizes the impact of the loss of a single species (though the exact role or function may not be entirely replaced or taken over by other species present, a substantial portion may be). A system with a high functional redundancy therefore should be less vulnerable to a loss of a single component than one with low functional redundancy. Therefore, it should be better able to resist displacement of both structure and function.

c. Stream order, flow dependability, turbulent diffusivity, and flushing capacity. Rating system: 1—poor; 2—moderate; 3—good.

These characteristics essentially have to do with the volume of water available for dilution, the rapidity with which a waste or other form of stress would be dissipated, and the rate at which it would be removed from the system. A system in which these characteristics are very dependable would be less vulnerable to structural or functional displacement than one which periodically had substantial losses or reductions in one or more of these characteristics. In considering these characteristics in relation to the inertial stability of a system, one should not lose sight of the fact that wastes which are not degraded or transformed, but are merely mixed and carried away, will have an impact elsewhere. Those that are degraded and transformed also may have an impact, but most probably a lesser one.

d. Hard, well-buffered water antagonistic to toxic substances. Rating system: 1—poor; 2—moderate; 3—good.

The literature on water pollution is replete with illustrations of alterations of the impact of toxic chemicals on aquatic organisms due to differences in water quality such as hardness, pH, temperature, etc. Some hard, well-buffered waters may substantially reduce the impact of many toxic materials, whereas very soft, well-buffered waters may not. With some knowledge of the causative factors of pollutional stress, one should be able to estimate whether or not water quality will significantly affect the toxicity or dose-response curve. In addition to utilizing the literature for this purpose, it is always well to carry out the site-specific tests with indigenous organisms which also will enable one to define the relationship between the organisms, water quality, and dose-response curve to a particular pollutional stress.

e. System close to a major ecological transitional threshold (e.g., from a cold to a warm water fishery). Rating system: 1—close; 2—moderate margin of safety; 3—substantial margin of safety.

Considering thermal loading, one might say with reasonable confidence that the Columbia River is closer to a major transitional temperature threshold (which might cause the loss of salmonid fisheries) than the Savannah River (which would cause the loss of the channel catfish, bass, and other organisms characteristic of that system). Long-term continuing studies are necessary to estimate when a system is beginning to approach a major transitional threshold. For those systems where such studies do not exist, such an estimate will be extraordinarily difficult. Even when data do exist, the task will not be easy. It is clear that the inertial stability of a system approaching a major ecological transitional threshold will be less than one substantially away from this point.

f. Presence of a drainage basin management group with a water quality monitoring program. Rating system: 1—none; 2—weak organization; 3—strong organization.

The river basin management group can protect against displacement or disequilibrium in two primary ways:

(1) ongoing studies which will enable the investigators and decision makers to know when a major ecological transitional threshold is being approached, and

(2) the development and maintenance of an environmental quality control system with rapid information feedback to detect the onset of a pollutional stress which might cause displacement or equilibrium (Cairns et al., 1970; Cairns, 1972).

In order to be effective this management group would, of course, have to have the capability either of taking immediate remedial action before the waste enters the receiving system or, if it has already entered, of taking appropriate measures to reduce the impact upon the system. For thermal loadings, long-term studies to provide information about the system, which would enable the identification of ecological transitional thresholds, would appear to be the most important task. However, since power plants frequently use chlorine, slimicides, and other toxic materials, and since heavy metals may be found in the waste discharges, the biological-chemical-physical monitoring system which would enable the group to detect the appearance of toxic materials and take remedial action immediately is also important.

$$\text{Inertia Index} = a \times b \times c \times d \times e \times f$$

400+ : high degree of inertia
55–399 : moderate degree of inertia
less than 55: poor inertia

Resiliency

Resiliency was defined earlier as the number of times a system can snap back after displacement. A very informative illustration of resiliency has been provided by Brian Dicks (in this volume) in his investigation of the effect of oil spills on salt-marsh plants in Great Britain. Simulating a series of rather small oil spills restricted to a limited area, Dicks found that the salt-marsh organisms could recover relatively rapidly after two or three spills in reasonably close succession. Even four or five exposures in reasonably close sequence allowed subsequent recovery. However, when 16 or 17 insults were given with about the same frequency and at the same volumes to a comparable area, there was no recovery—at least within a time span considerably larger than that in which recovery had been previously demonstrated. This suggests that the salt-marsh ecosystem had a "reserve" which would carry the system through repeated stresses but that the number of stresses which could be tolerated was finite even though the area exposed and the volume of oil spilled were kept constant for each exposure. Thus, the resiliency of this system could be determined by a field bioassay with repeated exposure in which the exposure intervals and volumes of oil used in each were controlled but in which nothing else in the natural system was modified. Perhaps there are other systems for which resiliency could also be easily determined. It is quite evident that for the Great lakes, oceans, and other large and complex ecosystems, the determination of resiliency will be difficult. Recognition that this problem exists and even a crude estimate of the degree of resiliency may be enormously useful in developing management plans.

Conclusions

The ability of a stream or river to assimilate wastes is governed by the capacity of the system to transform them before they reach deleterious levels. If an overload occurs, the system is disrupted and the transforming capacity may be substantially reduced. Recovery may be rapid or slow depending upon a number of factors, including (1) severity and duration of the stress, (2) number and kinds of stresses, (3) residual effects on non-biological units (e.g., substrate), (4) presence of epicenters for recolonizing organisms, (5) innate vulnerability of the system, (6) inertia of the system, and (7) resiliency of the systems. Recognizing that aquatic systems differ in their ability to recover following stress, it is necessary to develop methods for predicting the recovery response of our aquatic systems so that we can establish priorities for their management and protection. In this chapter we have attempted to present a first effort at defining and measuring several characteristics of aquatic systems which are important in the recovery process. We hope this chapter will stimulate additional efforts to understand better the roles of ecosystem vulnerability, elasticity, inertia, and resiliency in the process of ecosystem recovery.

Literature Cited

Cairns, J., Jr. 1972. Rationalization of multiple use. Pages 421-430 *in* R. T. Oglesby, C. A. Carlson, and J. A. McCann, eds. River Ecology and Man. Academic Press, New York.

_____. In press. Heated wastewater effects on aquatic ecosystems. Proc. Second Thermal Ecology Symposium. United States Energy Research and Development Commission.

Cairns, J., Jr., J. S. Crossman, K. L. Dickson, and E. E. Herricks, 1971. The recovery of damaged ecosystems. Proc. Symp. on Recovery and Restoration of Damaged Ecosystems. Assoc. Southeastern Biol. Bull. 18:79-106.

_____. 1973. The effects of major industrial spills upon stream organisms. Proc. 26th Annual Purdue Industrial Waste Conf. Purdue Univ. Eng. Bull. 140 (1): 156-170.

Cairns, J., Jr., and K. L. Dickson. 1974. Guidelines for developing studies to determine the biological effects of thermal discharges. Aware. 43:9-12.

Cairns, J., Jr., K. L. Dickson, R. E. Sparks, and W. T. Waller. 1970. A preliminary report on rapid biological information systems for water pollution control. J. Water Pollut. Control Fed. 42:685-703.

Cairns, J., Jr., and A. Scheier. 1959. The relationship of bluegill sunfish body size to tolerance from some common chemicals. Proc. 26th Annual Purdue Industrial Waste Conf., Purdue Univ. Eng. Bull. 43(3):243-252.

Crossman, J. S., J. Cairns, Jr., and R. L. Kaesler, 1973. Aquatic invertebrate recovery in the Clinch River following hazardous spills and floods. Water Resour. Res. Center Bull. 63, Virginia Polytechnic Institute and State University, Blacksburg. 66 pp.

Dicks, B. In press. Changes in vegetation of an oiled Southampton water salt marsh. *In* this volume.

National Academy of Sciences. 1974. Principles for Evaluating Chemicals in the Environment. Washington, D.C. 454 pp.

Fast, A. W. In press. Artificial aeration of lakes as a restoration technique. *In* this volume.

Heilbrunn, L. V., 1937. An Outline of General Physiology. W. B. Saunders Co., Philadelphia. 748 pp.

———. 1943. General Physiology. W. B. Saunders Co., Philadelphia. 748 pp.

Herricks, E. E. and J. Cairns, Jr. 1972. The recovery of stream macrobenthic communities from the effects of acid mine drainage. Pages 370-398 *in* Fourth Symposium on Coal Mine Drainage Research, Ohio River Valley Water Sanitation Commission, Bituminous Coal Research, Inc.

———. 1974. Rehabilitation of Streams Receiving Acid Mine Drainage. Water Resour. Res. Center Bull. 66, Virginia Polytechnic Institute and State University, Blackburg. 284 pp.

Kaesler, R. L., J. Cairns, Jr., and J. S. Crossman. 1974. The use of cluster analysis in the assessment of spills of hazardous materials. Am. Midl. Nat. 92:94-114.

Nelson-Smith, A. In press. Recovery of some British rocky seashores from oil spills and cleanup operations. *In* this volume.

Odum, E. P. 1969. The strategy of ecosystem development. Science. 164:262-270.

———. 1971. Fundamentals of Ecology. W. B. Saunders Co. Philadelphia. 574 pp.

Pickering, Q. H., and C. Henderson. 1966. Acute toxicity of some important petrochemicals to fish. J. Water Pollut. Control Fed. 38(9):1419-1429.

Recovery of Streams from Chronic Pollutional Stress—Acid Mine Drainage

Edwin E. Herricks

Abstract

The recovery of stream communities from chronic pollutional stress is related to both distance from the point of stress introduction and a time-related decrease in stress intensity to levels where aquatic community structure and function are reestablished. This chapter identifies the damage caused by acid mine drainage (AMD), the effect of AMD on aquatic ecosystems, particularly macrobenthic communities, and the most significant mechanisms which relate to the restoration and recovery of stream ecosystems damaged by long-duration stress. In general, recovery from chronic pollutional stress can occur if (1) stress is reduced and damaged habitats restored, (2) souces of recolonizing organisms are available, and (3) seasonal variability in stream conditions do not preclude maintenance of stream communities. Stream restoration from AMD did occur, as shown by cluster and principal coordinate analysis of macrobenthic samples collected over a two-year period on Indian Creek, Fayette County, Pennsylvania. Recovery occurred within eight miles of the location of maximum stress, although aquatic communities in "high stress" portions of the stream were quite variable depending on seasonal factors such as discharge and water quality.

Introduction

Streams and rivers subjected to pollutional stress may respond in predictable ways based on both the interaction between physical, chemical, and biological components of the stream system and the type, intensity, and duration of the stress. Moreover, the recovery from stress can be predicted based on a knowledge of factors which can be assessed in terms of alteration of the structure and function of aquatic ecosystems. Similarly, recovery can be identified as restoration of the aquatic ecosystem to normal structure and function.

Long-duration, or chronic, discharges have the highest potential for causing damage, and recovery is related to reduction of stress to levels where normal (e.g., similar to upstream or other similar environments) structure and function can be reestablished. Generally, long-duration stress caused by a point source (or nonpoint sources where the location of the discharges is limited to one area) is reduced as a function of distance from the discharge source. The relationship of recovery through distance can be described by an expression which includes parameters relating to physical factors (e.g., discharge, watershed morphology, and geology), chemical factors (e.g., water quality or characteristics), and biological factors (e.g., presence and abundance of biota, toxicity, or sources of recolonizing organisms) of the receiving stream and characteristics of the discharge or event that increased stress and caused damage. This expression must also contain time-related parameters which may affect both the physical, chemical, and biological nature of the receiving stream (e.g., seasonal changes in flow, oxygen, or biological conditions) and the characteristics of the stress (e.g., degradable compounds whose effect is changed through time).

Extensive data are needed to formulate the expressions relating damage (i.e., excessive stress) and recovery, and any assumptions relating stress with damage are complicated by the fact that all elements of the biological system are under constant stress. The number of environmental variables precludes a condition where optimal levels of all variables can be maintained at all times; thus, any organism is under stress (graded from insignificant to extreme based on individual tolerance). A change in any environmental variable may be intolerable and cause dysfunction. This dysfunction can be observed at all levels of biological organization (e.g., subcellar to galactic). If the dysfunction results in elimination of a component of the system (e.g., death of an organism or alteration of higher levels of organization), then damage has occurred. It should be apparent that both natural and man-induced stress can cause damage, and the interpretation of damage is limited only by the precision of observation.

This chapter explores several of the parameters which make up part of the expression relating recovery of damaged benthic communities from long-duration acid mine discharges. Acid mine drainage (AMD) is a generic term for low pH discharges which contain high levels of dissolved salts caused by mining activities. These discharges are long-duration causes of stress and damage because once the pyritic minerals (FeS_2), and possibly other metal sulfides, are exposed to oxygen and water, acid production will continue until all of the pyritic or sulfide minerals have been exhausted. Because AMD does cause severe long-duration stress in aquatic ecosystems, it is ideally suited for study relating long-term damage caused by man's activities to the restoration and recovery of damaged aquatic ecosystems. It is the intent of this chapter to identify the causes of damage, determine the nature of its effect on benthic communities, and identify the most significant mechanisms which relate to restoration and recovery of the damaged stream ecosystem.

Recovery Models

The mechanisms which control restoration and recovery may best be described and understood through change in stress levels after discharge. Basic to understanding changes in stress levels in a stream is an understanding of the singular nature of the stream itself. A stream or river is a dynamic open system subject to instant change. While the stream provides the necessary habitat for the function of biotic systems associated with it, the stream continues to transport dissolved and solid materials through a gradient determined by continuing geologic processes (e.g., base level determination and gravity). This capacity for transport coupled with the instantaneous changes which can occur in the physical, chemical, or biological components of the stream provide a self-regenerative capacity as well as the capacity for damage.

The major causes of damage in a stream system are (1) destruction of habitats through alteration of the physical system, (2) reduction or elimination of any component of the physical, chemical, or biological system which is essential for continued biotic function, and (3) destruction or injury of the biota by addition of toxic materials. When a stress of any one of these types is added to the stream, the response of individuals, species, or communities may be variable. Although major changes may have taken place in structure or function, a moderation of the stress will lead to recovery through time or distance. Figure 1 relates changes in stress through distance. (Note: Since this discussion is directed at stream systems, distance and time relationships are considered together because even though stress may change through time, the flowing nature of the stream superimposes distance on this function.) As illustrated in the figure, intensity of stress is always greater than zero. Thus, when a change occurs in any component contributing to stress, changes will occur in the

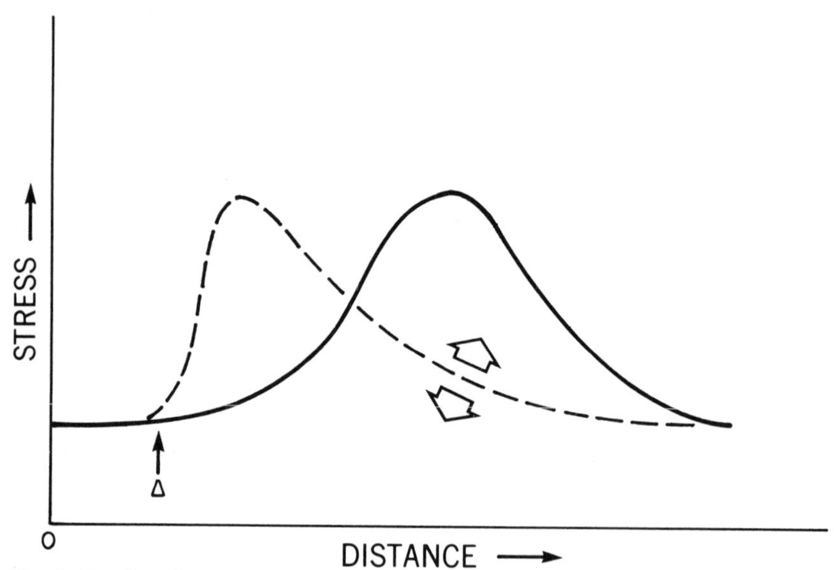

Fig. 1. Family of curves representing change in stress through time following a stress introduction

system. Since increased stress may be related to damage, this generalized model can relate maximum stress to damage, and the amelioration of stress can be related to recovery. Although only two curves are illustrated in Figure 1, the figure should be taken to represent a family of curves, each relating a change in stress through distance and each having its own effect on damage and recovery.

Using a generalized model of this type, Figure 2 has been drawn to illustrate the relationship between stress, caused by acid mine drainage, and distance along the mainstem of the stream system. AMD immediately produces all three types of stress in a stream. Physical changes occur in the stream due to deposition of metal hydroxide flocs which make habitats unsuitable for most aquatic organisms. If the mine drainage is poorly oxidized (e.g., from some underground sources), the oxygen demand created may eliminate those biota which require high oxygen levels. In addition, the metal salts in AMD are highly toxic. Thus, as illustrated in Figure 2, stress levels reach their maximum immediately after discharge. The reduction of this high level of stress is dependent on chemical reaction and dilution. In this generalized model, stress intensity will remain at a maximum until either upstream inputs are changed with no corresponding increase in the discharge (greater dilution of the source),

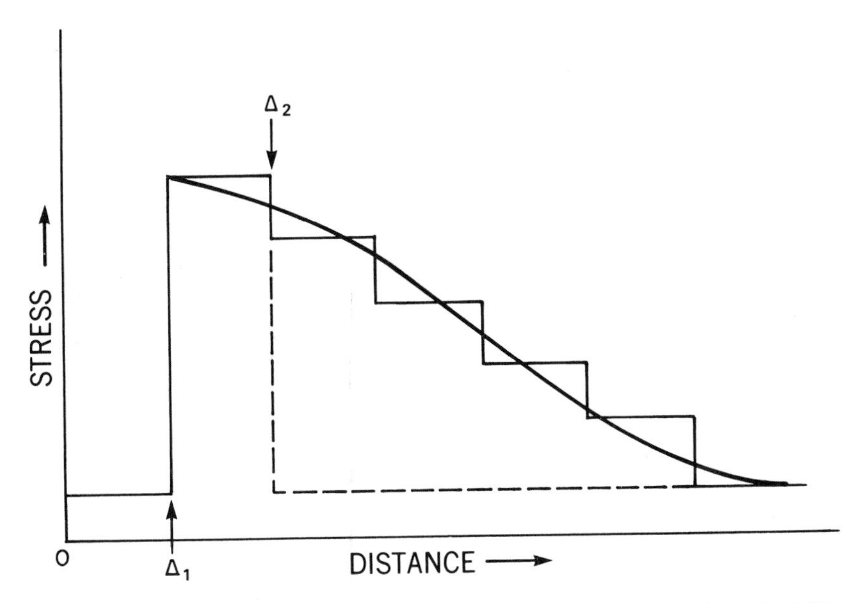

Fig. 2. Hypothetical changes in stress in a stream due to introduction of AMD

or downstream inputs occur (unpolluted tributaries which dilute or chemically react with stream waters). At each of these downstream inputs of unpolluted water, the stress intensity is decreased. Although all added stress may be reduced by one input (as illustrated), in most streams stress is reduced at each unpolluted tributary, represented as a stepwise reduction. Although this stepped reduction may be valid in theory, the relationship in nature may better be expressed by a continuum illustrating gradual reduction of stress through distance.

Although this general model may oversimplify the complex relationships between the stream system and the nature of the AMD, it is useful to initially simplify the complex interrelationships and provide a basis for further discussion. The following discussion will relate observations made on Indian Creek, Fayette County, Pennsylvania, which has been affected by AMD since the early 1900s.

Watershed Study Area

The complex interrelationships between the causes of stream damage and stream recovery demanded a watershed where detailed physical, chemical, and biological information could be obtained. The two most important criteria for selection were (1) significant damage to stream water quality and biota to AMD and (2) evidence of restored water quality and biology in downstream areas. The Indian Creek watershed, Fayette and Westmoreland counties, Pennsylvania, was selected as the primary study area.

The Indian Creek watershed lies in the southwestern Pennsylvania coalfields. Coal mining in Indian Creek valley began in the early 1900s near Indian Head and Melcroft (Fig. 3). Production reached a peak of

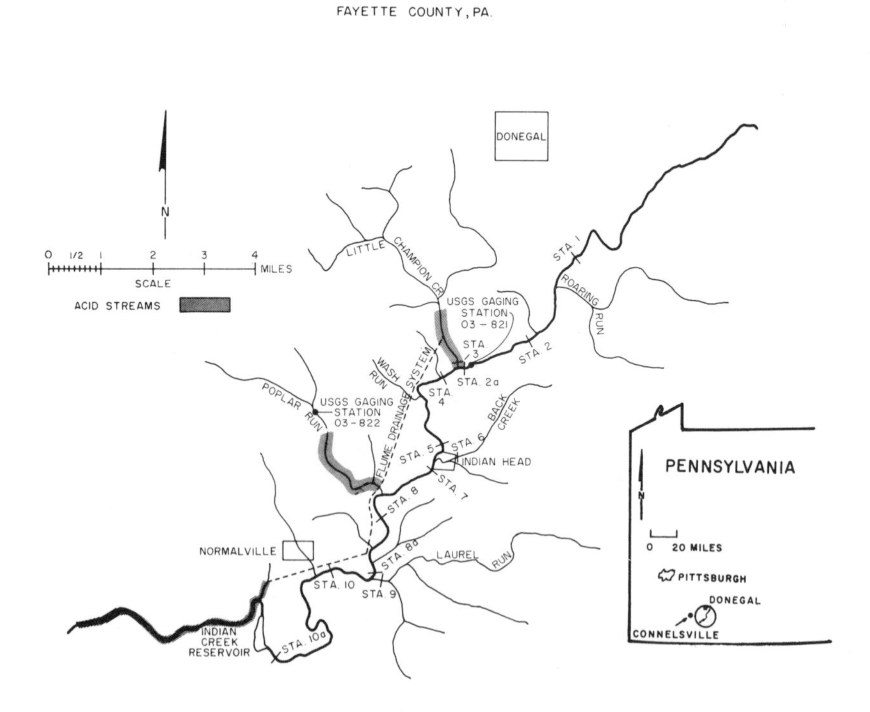

Fig. 3. Indian Creek drainage map

34 million tons in 1916 (Pennsyvania Dept. of Mines and Mineral Industries, 1969) and then steadily declined; only 1.2 million tons were mined in 1967. Although the last active deep mine, Melcroft 1 on Champion Run, was closed down in 1966, several surface mining operations were active in the headwaters of both Champion Run and Poplar Run during the time of this study.

AMD was noted in the watershed since mining began. Mine drainage in Indian Creek was the subject of a ruling made by the Pennsylvania Supreme Court in 1924 (Pennsylvania Dept. of Mines and Mineral Industries, 1969). It was claimed that mine drainage polluted the reservoir at Normalville, rendering the water unsuitable for use as boiler water for steam locomotives. As a result of this court action, a diversion system was constructed to carry mine drainage away from the upstream portion of Indian Creek and discharge it below the reservoir. This diversion system was constructed in 1924 and for the most part was still in operation during the study. Portions of the diversion system were old mine works and depended on regular pumping of mine drainage from collecting basins near Champion Run. Several of these pumping stations were no longer maintained after the Melcroft 1 mine closed down, and AMD was allowed to discharge into Champion Run, polluting Indian Creek. In addition to AMD from the abandoned underground works, a large quantity of AMD originated from gob piles from coal washing along the east bank of Champion Run. The severity of the AMD from both surface and subsurface sources, including active and abandoned mines, produced major acid mine drainage stress in Indian Creek, changing both water quality and the structure of aquatic communities.

Physical-Chemical Studies

Physical-chemical studies included both field sampling and computer simulation. Water quality parameters were monitored on a regular basis. These included analysis for pH, sulfate, heavy metals, total acidity, alkalinity, and hardness, dissolved oxygen, and conductance. A portion of these data, mean and standard deviations for total acidity and alkalinity, pH, and sulfate, are contained in Table 1. The remainder of the water quality data was published by Herricks and Cairns (1974).

Physical data are also of major importance in determining environmental impact. Key among the various physical parameters is stream flow or discharge. Values of discharge reflect the hydrologic response of the stream to changes occurring throughout the watershed, providing a single

Table 1. Mean and standard deviation values for four water quality parameters, Indian Creek, Fayette County, Pennsylvania

	Station											
	1	2	2a	3	4R	4L	5	6	7	8	9	10
pH												
mean	7.2	6.9	6.4	4.0	6.0	5.8	6.6	7.1	6.9	6.6	7.1	7.0
SD	.2	.4	2.2	.4	1.6	.6	.3	.2	.3	.9	.2	.4
Total acid												
mean	3.8	2.5	4.3	50.7	6.1	12.0	4.9	2.8	4.3	2.6	2.8	3.7
SD	3.6	2.3	3.4	24.3	5.8	8.9	3.6	3.0	4.0	2.7	2.8	2.7
Total alkalinity												
mean	23.6	21.3	22.3	0	16.5	6.5	10.4	21.0	12.7	16.3	16.8	11.5
SD	7.6	7.6	10.0	0	7.5	3.6	6.1	8.1	5.1	3.2	6.3	5.3
Sulfate												
mean	13.2	15.8	20.4	128.2	27.6	49.0	33.0	10.0	35.5	40.9	8.9	29.9
SD	2.0	4.7	6.2	82.0	11.3	16.8	11.4	3.1	9.4	15.3	1.4	7.1

parameter which relates changes in the physical-chemical conditions to changes in stream biota. Furthermore, information on discharge quality and quantity combined with AMD volumes and characteristics may provide an accurate estimate of the overall impact AMD may have on the environment. Unfortunately, streamflows were available for only one subbasin in the Indian Creek watershed. To supplement this data, hydrologic simulation techniques were used to generate flow data for all major subbasins which had no historical streamflow records and to provide basic data to allow development of water quality models to better define the impact of mine drainage on Indian Creek (Herricks et al., 1975).

Water Quality Results

Water quality was poor in some areas of the watershed, showing the impact of AMD on Indian Creek. The major source of AMD was the tributary Champion Run. The lower portion of Champion Run, approximately one mile from its mouth, was severely polluted. In addition to this major source of acid drainage, additional sources of mine drainage were found due to seepage upstream near Station 2 and ongoing mining in areas of the watershed, mainly from Poplar Run. Several gob piles, the waste from coal-washing operations which may produce AMD, were observed along Indian Creek, specifically near Stations 5, 8, and 10.

A summary of analyses of water quality performed for total acidity and alkalinity, pH, and sulfate during the two-year sampling effort have been summarized in Table 1. At Station 1 water quality was good. Alkalinity values were consistently high, while acidity values were consistently low. Sulfate values had a small range, the average value being below 15 ppm. (Note: Because sulfates are usually present in low concentrations in Appalachian waters [less than 20 ppm] and are found in high concentrations in AMD, the presence of elevated sulfate concentrations indicated AMD influence and accurately predicts AMD concentration.) Station 2 showed similar results, but Station 2a had higher mean values for acidity and sulfate, and a greater range for all four parameters. Champion Run, Station 3, had consistently high acidity and sulfate, no alkalinity, and low pH. The ranges for all parameters were large, indicating a variable but poor water quality. This may be attributed to variable discharge: AMD volume relationships in Champion Run. Station 4 was unmixed. The right-bank samples show lower mean acidity and sulfate values and higher mean alkalinity than the left bank samples. The range of values recorded for both right and left banks is large, indicating a variable water quality. This variability may also be attributed to changes in discharge relationships between the upstream portion of Indian Creek and Champion Run. Station 5 had higher mean acidity and sulfate values and lower mean alkalinity when compared with Stations 1 and 2. The water quality of the mainstream of Indian Creek improved considerably after the confluence with unpolluted tributaries downstream from Station 5. There were no major tributaries between Station 4 and Station 5. Station 6 and Station 9 were unpolluted tributaries. All four parameters listed in Table 1 were similar in value to mean and range values recorded for Station 1. The addition of these unpolluted tributary waters improved the mainstem water quality not only by dilution but also by chemical reaction. For example, at Station 7 mean values of alkalinity increased while mean values of acidity decreased when compared with Station 5, indicating neutralization of AMD. Water quality at Station 8 showed improvement. Poplar Run was another acid tributary; sulfate values were consistently high, as was mean acidity, while mean alkalinity was low. By Station 10 mean values of sulfate were lower. The range of values of acidity for Station 10 was less than Station 7, and the values of alkalinity, although lower than Station 8, had little variability.

In summary the water quality in the mainstem of Indian Creek illustrated the overall effect of a severe AMD influence. The upstream stations maintained good water quality throughout the study period. Although mine drainage from Champion Run was poorly mixed at Station 4, water quality was poor. With complete mixing of the AMD and no tributary influence, Station 5 had the poorest water quality. Although concentrations of sulfate increased throughout the stream, the addition of unpolluted tributary water improved overall water quality. The water quality at Station 7 was variable; this variability decreased by Station 10.

The effect of AMD from Champion Run in Indian Creek varied throughout the study due to seasonal and other influences. During periods of high discharge the effect of the AMD was decreased. This decrease can be explained by the following facts: (1) flow volume from the upstream portion of Indian Creek was at times large enough so that acid drainage in Champion Run was diluted; (2) higher flow volumes from mine drainage sources reduced the concentration of the AMD entering Indian Creek; and (3) higher discharges from tributaries including the upstream portion of Champion Run further diluted the AMD, reducing its influence on the downstream portions of Indian Creek. High flows such as those in March 1972 diluted the acid mine drainage but had little capacity for neutralization. Thus during periods of extended rainfall streamflow was predominantly direct runoff, and the alkalinity and pH of water samples from throughout the watershed was reduced.

During periods of low discharge the effect of mine drainage from Champion Run was intensified. The acid drainage from the Melcroft mines was only partially diluted by discharge volumes from the upstream portions of Champion Run and Indian Creek. The pH of Champion Run was generally lower, and concentrations of sulfate were higher when compared with mean station values. During these periods concentrations of sulfate on both the right and left banks at Station 4 were high, and there was little change in sulfate concentrations by Station 5. This indicated that the volume of mine drainage was sufficient to affect the total streambed across the full width of Station 4, and the acid drainage was only partially diluted and little neutralized by upstream flow. During these periods of low discharge the effect of the mine drainage extended farther downstream, affecting Station 7. Improvement of water quality occurred below each confluence with an unpolluted tributary. Even though the dilution ability based on streamflow was reduced, the neutralization capacity of the eastern tributaries was improved. Steamflow during periods of low discharge is predominantly from groundwater sources. Because groundwater in the headwaters of the eastern tributaries flows through calcareous rocks, the alkalinity of tributary waters was increased. Thus, the ability to neutralize the acid drainage was also increased.

The relationship between AMD concentration and streamflow is very complicated and is summarized in Figure 4. Because most mine drainage in Champion Run was from underground sources, these discharges (e.g., underground Q), on a seasonal basis, maintained a relatively constant AMD volume, although AMD concentration varied with season (Appalachian Regional Comm., 1969, pp. C32-33). The volume of mine drainage is dependent on infiltration of rainfall to underground areas. Although oxidation of pyrite is not appreciably changed by the amount of water present, the concentration of pyritic oxidation end products will vary with volume. Because the infiltration rate is greater during the winter, the volume of mine discharges is increased from January through May. Infiltration decreases during the summer months; thus mine discharge volumes also decrease (see Fig. 4).

The major source of acid in an underground mine are pyritic materials located above normal water levels. When the mine is flooded by high base flow (e.g., high infiltration rate), the pyritic oxidation is limited by oxygen transport (e.g., temperature-related saturation values) in the water, reducing overall AMD concentration. If flow through the mine has been low for some time, the oxygen-rich atmosphere of the mine allows rapid

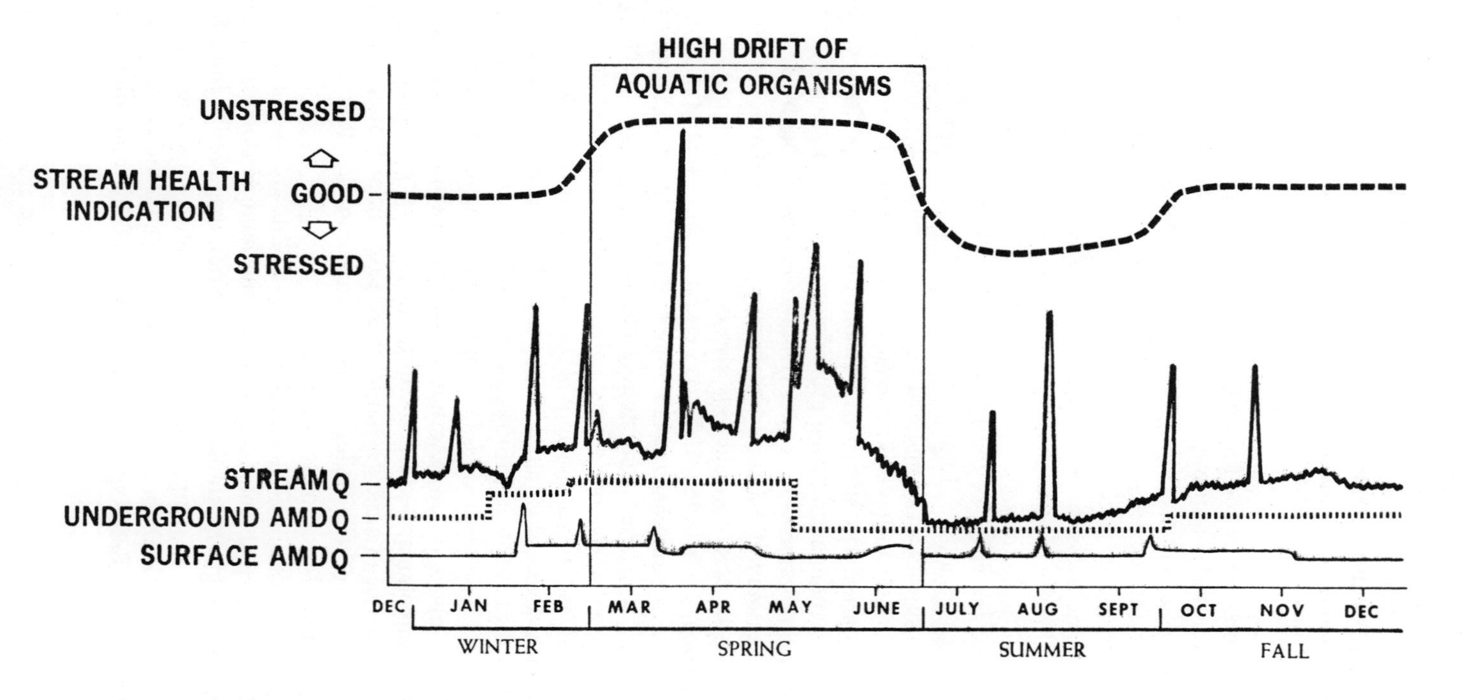

Fig. 4. Seasonally related changes in stream discharge, surface, and underground AMD discharges related to subjective evaluation of stream health

oxidation of pyrite, and large quantities of oxidation products may be present on unflooded surfaces. As flow through the mine increases, these oxidation products are put into solution. The first flush discharge may be highly concentrated and produce slug loads of AMD.

Superimposed on this pattern of seasonal change in base flow and AMD concentration are several relationships between concentration and stream biological damage. First, because first flush discharges may be more concentrated, the assimilative capacity of the stream may be overloaded from these slug loads. Second, the capacity of the receiving stream to assimilate a given volume and concentration of AMD varies with stream discharge and is particularly related to the percentage of base flow represented in the receiving stream, the presence of calcareous rocks, and several other physical factors such as temperature.

The stream's assimilative capacity is based on both alkalinity (for neutralization) and stream discharge volume (for dilution). In Figure 4 the stream's assimilative capacity is equated with stream health (i.e., when assimilative capacity is high all indications of stream health will be correspondingly high.) During the spring, volume of AMD is usually high, but dilution is increased by higher discharge in the receiving stream, which improves assimilative capacity. As spring high flows recede, AMD volumes may remain high, thus increasing the impact of the discharge. During the late spring and summer months when AMD volume is low, normal stream discharge is also low. This may result in extended periods of low water quality and corresponding high stress. Although impact may be severe in some areas of Indian Creek, the contribution water with somewhat higher alkalinity improves the assimilative capacity of the stream reach immediately downstream. Based on this pattern, although conditions may improve downstream due to the influence of tributaries, the midportion of Indian Creek (Stations 4 and 5) was under more severe stress, and Station 7 showed decreased water quality, due probably to poor mixing (similar conditions noted at Station 4). In late summer volumes of AMD remain low (in part due to high transpiration rates of vegetative cover which further reduces infiltration). As stream discharge increases in late summer and early fall due to greater rainfall, stream conditions improve (see Fig. 4) because the dilution capacity of the receiving stream is increased. During the winter the cycle begins again. High infiltration rates affect groundwater conditions by increasing mine drainage volumes, but volumes of stream discharge during these periods are generally high, and the assimilative capacity is correspondingly good.

Temperature and seasonal climatic conditions affect AMD in other ways. The mine drainage from underground sources during the summer months is usually poorly oxidized because flow volumes are less, reducing the total oxygen available from further oxidative reactions (Schumate and Smith, 1972). The oxidation of this mine drainage may occur in the receiving stream, placing a severe oxygen stress on the receiving water. Thus a secondary stress occurs due to the high oxygen demand of the AMD. This occurs when stream water temperatures are generally high and dissolved oxygen is low.

A second seasonally related AMD discharge problem occurs from surface sources such as gob piles. Pyritic materials on gob piles are well oxidized. During the winter months the reduced surface temperature reduces the oxidation rates, and temperatures below freezing prevent runoff from gob piles. Initial melt carries the oxidation products into the receiving stream, but high assimilative capacity due to high stream discharge reduces its impact. Rates of chemical reactions on the gob piles are increased during the warm summer months. Rainfall during this period usually occurs as high-intensity storms which flush unvegetated areas rapidly. The accumulation of oxidation end products of pyrite makes initial runoff highly concentrated, and AMD slugs precede increased streamflow while assimilative capacity is low. Major damage may occur under these conditions.

Biological Studies

Biological sampling efforts centered on methods to describe the recovery of Indian Creek as related to both seasonal differences in stream conditions and the distance from the AMD source area where healthy communities were reestablished. Macrobenthic communities were selected as the best indicators of overall biotic effect from AMD. The integration of observed physical and chemical conditions and hydrologic simulations with the results of biological sampling yielded information which assisted in identifying the causes of damage and the nature of its effects on stream benthic communities (seasonally related changes in stream conditions; see Fig. 4).

Analytical Procedures

Sampling was systematized to assure that equal sampling effort was made at each station on each sampling date. Both qualitative (kick-net) and quantitative (Surber) samples were collected. The details of sampling procedure and sampling handling are available (Herricks and Cairns, 1974). Biological samples were collected over a two-year period with major sampling effort occurring in March through November.

A number of techniques were used to analyze the data from the benthic collections. Species lists were compiled, and rank abundance was tabulated. Diversity was analyzed according to equations proposed by Shannon (Shannon and Weaver, 1949) and Brillouin (1962). The results of these analyses were given by Herricks and Cairns (1974). To gain an appreciation of alterations in community structure due to AMD, several other analytical techniques were applied to data from Indian Creek.

Cluster analysis was made using both presence-absence data and taxa abundance data. The analysis using presence-absence data was performed according to procedures outlined by Cairns and Kaesler (1969). The presence-absence data matrix from all samples collected during one year was used to generate coefficients of similarity based on the formula proposed by Jaccard (1908). The computed coefficients of association were then clustered, using the unweighted pair group method (Sokal and Sneath, 1963; Figs. 5-6). The distortion introduced during the clustering procedure was assessed by the cophenetic correlation coefficient (Sokal and Rholf, 1962).

The second clustering procedure included use of taxa abundance data. Each station was related by distance values similar to the taxonomic distance described by Sokal and Sneath (1963). Details of the procedures and results of this analysis were given by Herricks and Cairns (1974). The distance matrix derived from this analysis was also used to ordinate the Indian Creek data. Rohlf (1968) pointed out that this analysis may help clarify confusing points which may arise from cluster analysis. The ordination technique chosen was proposed by Gower (1966, 1972). The method, principal coordinate analysis, had been successfully used in numerical taxonomy (Rohlf, 1968, 1972; Sokal and Rohlf, 1970) and has been applied to ecological problems by Cairns et al. (1974). The generalized procedure includes computation of a distance matrix (Sokal and Sneath, 1963) between all pairs of stations in the study based on their contained faunas. This distance matrix is then transformed so that principal components (normalized eigenvectors) extracted from the matrix can be used to ordinate the stations in a reduced species space.

The data from Indian Creek included 122 taxa (rows) distributed among 122 samples (columns). The data were strongly postively skewed, and a drastic transformation following Ebeling et al. (1970) was applied before standardization by rows (species): $x' = \log_{10} (x + 1)$. Following standardization, a matrix of Euclidean distances was computed in the Q-mode,

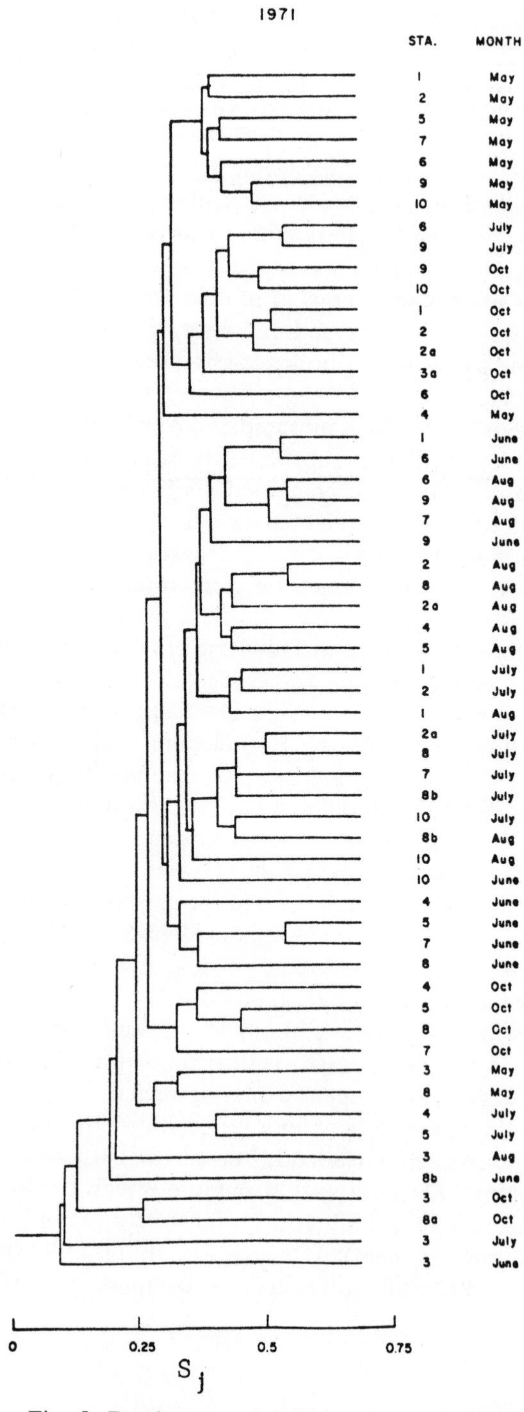

Fig. 5. Dendogram of 1971 biological collections

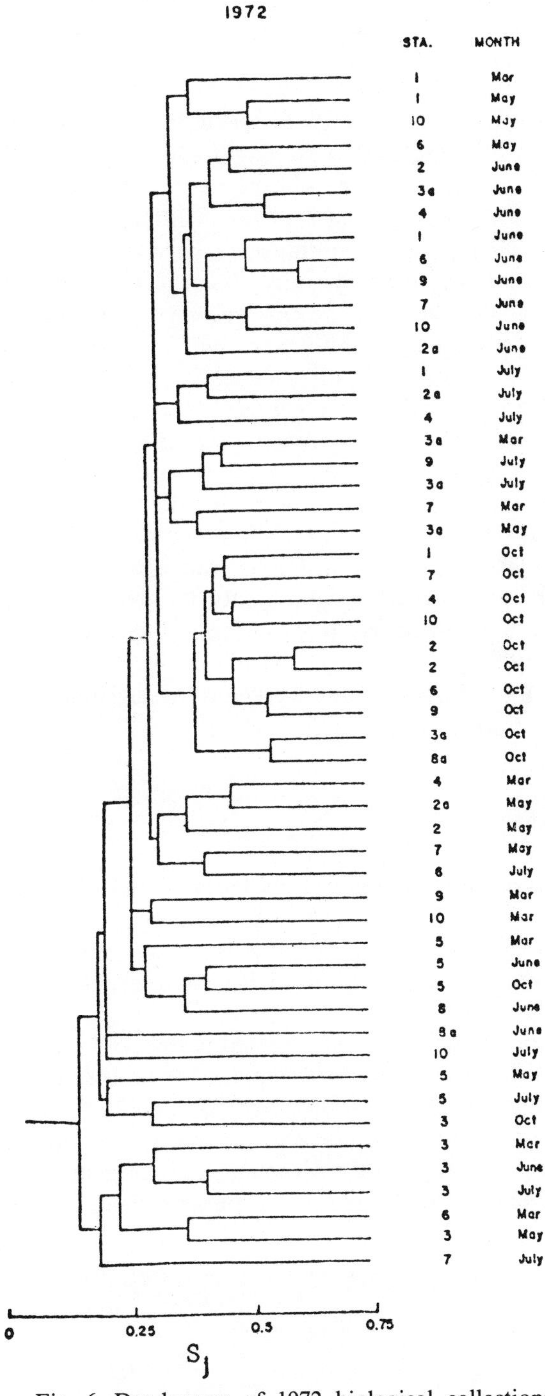

Fig. 6. Dendogram of 1972 biological collections

i.e., among samples. The Gower transformation was then applied, and principal components were extracted, the first two of which were used as principal coordinates in Figures 7-8.

Because the information derived from both cluster analyses and ordination by principal coordinates is useful in interpreting changes in community structure and observing trends due to seasonal or other influences, results of both will be discussed in some detail. Each analytical technique provides slightly different information. Cluster analysis is very sensitive to similarity between pairs of samples. The lower the similarity between pairs of samples, the lower the sensitivity becomes if the samples lie in different clusters. The ordination technique does not have this sensitivity for pairwise comparisons. The similarity of two ordinal points depends on loadings ascribed to that component; thus two data points close to the ordinate on the diagrams may be quite dissimilar on a third, fourth, or even fifth factor, so that they are not, in fact, so closely similar as they appear. Thus the ordination techniques are very useful for comparing the similarities of large groups of data that are very difficult to relate in pairwise fashion. The discussions related to the influence of AMD on Indian Creek using these two analytical procedures overlap significantly, but detail in the cluster analysis provides a basis for direct comparison of presence-absence data from station to station, while principal coordinate analysis illustrates the significant trends which may be observed from station to station and month to month.

Cluster Analysis

The results of cluster analysis of the Indian Creek data for 1971 and 1972 are contained in Figures 5 and 6. The cophenetic correlation coefficient calculated for the 1971 dendrogram was 0.848, and 0.831 for the 1972 dendrogram, indicating only minor distortion introduced by the clustering procedure.

The dendrograms can be interpreted in light of both AMD effects and seasonal differences in the occurrence of certain taxa. For example, the dendrograms for samples from both 1971 and 1972 show the impact of AMD by clustering stations which were affected by AMD together (Stations 3, 4, 5, 8a, and 8). Although Stations 3 and 8a were the only stations which were located on acid tributaries, Station 5, and occasionally Stations 4, 7, and 8, were included in this cluster grouping, due probably to seasonal differences (e.g., low flow and dominance of tolerant taxa).

The stations grouped with the highest degree of similarity were the reference stations (1, 2, 2a), tributary stations (3a, 6, 9), and the station farthest downstream (10). All these stations had similar fauna due to minimal or no influence of AMD.

A detailed analysis of these dendrograms can aid in understanding physical-chemical changes due to AMD and fluctuations of stream discharge and changes in the occurrence of certain taxa (which are also seasonally related). Reference will be made in this discussion to rank-abundance data contained in Herricks and Cairns (1974).

The first cluster of the 1971 dendrograms included only samples collected in May 1971, a period of high discharge. Although samples from Stations 3 and 8 were not included in this cluster, samples from all remaining stations were clustered together. The similarity was probably caused by conditions very suitable for insect drift, in effect homogenizing stream biota. The second cluster contained samples either from upstream reference stations or from tributary stations collected during July and October. During these months, *Hydropsyche* sp. and Chironomidae were present in large numbers throughout the watershed. The fact that non-tolerant organisms occurred with these more tolerant forms made healthy, unstressed stations very similar. The third cluster contained samples from June, July, and August where again seasonal differences were revealed. One grouping in this cluster contained upstream reference stations and healthy tributaries from June and August. A second grouping contained samples from mainstream stations downstream from Station 2a, and a third grouping included samples from upstream reference stations collected in July and August. Stream discharge during this period was low. The emergence of Ephemeroptera in the early spring and the dominance of *Hydropsyche* sp. and Chironomidae contributed to the similarity of collections. The grouping together of samples from mainstem stations collected in August may be attributed in part to dominance of tolerant taxa, and additionally to increased effect of AMD from discharges near Station 2a which stressed the fauna at this station. Cluster four contained samples collected at Stations 2a, 8, and 10 in July and August. The final cluster was made up to 19 samples separated at a low limit of similarity. Samples from all months were represented in this cluster, in which the apparent basis of similarity was AMD influence. Samples from reference, tributary, and recovery stations were not contained in this cluster. The effect of low discharge grouped together samples from Stations 4 and 5 collected during July. The samples from the stations most severely damaged by the September 14 flood were all clustered together.

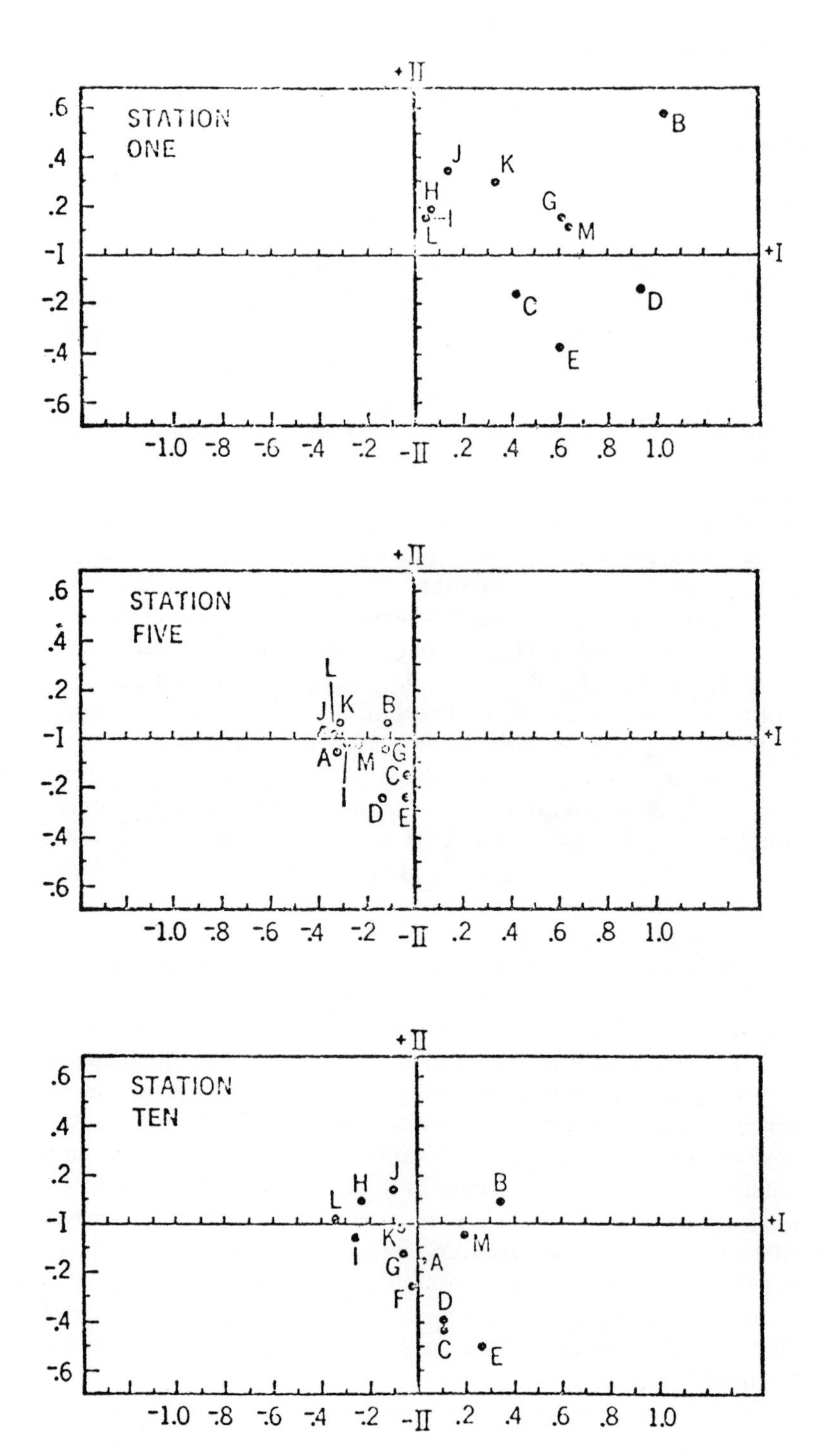

Fig. 7. Principal coordinate analysis of Indian Creek data, all collections grouped by station

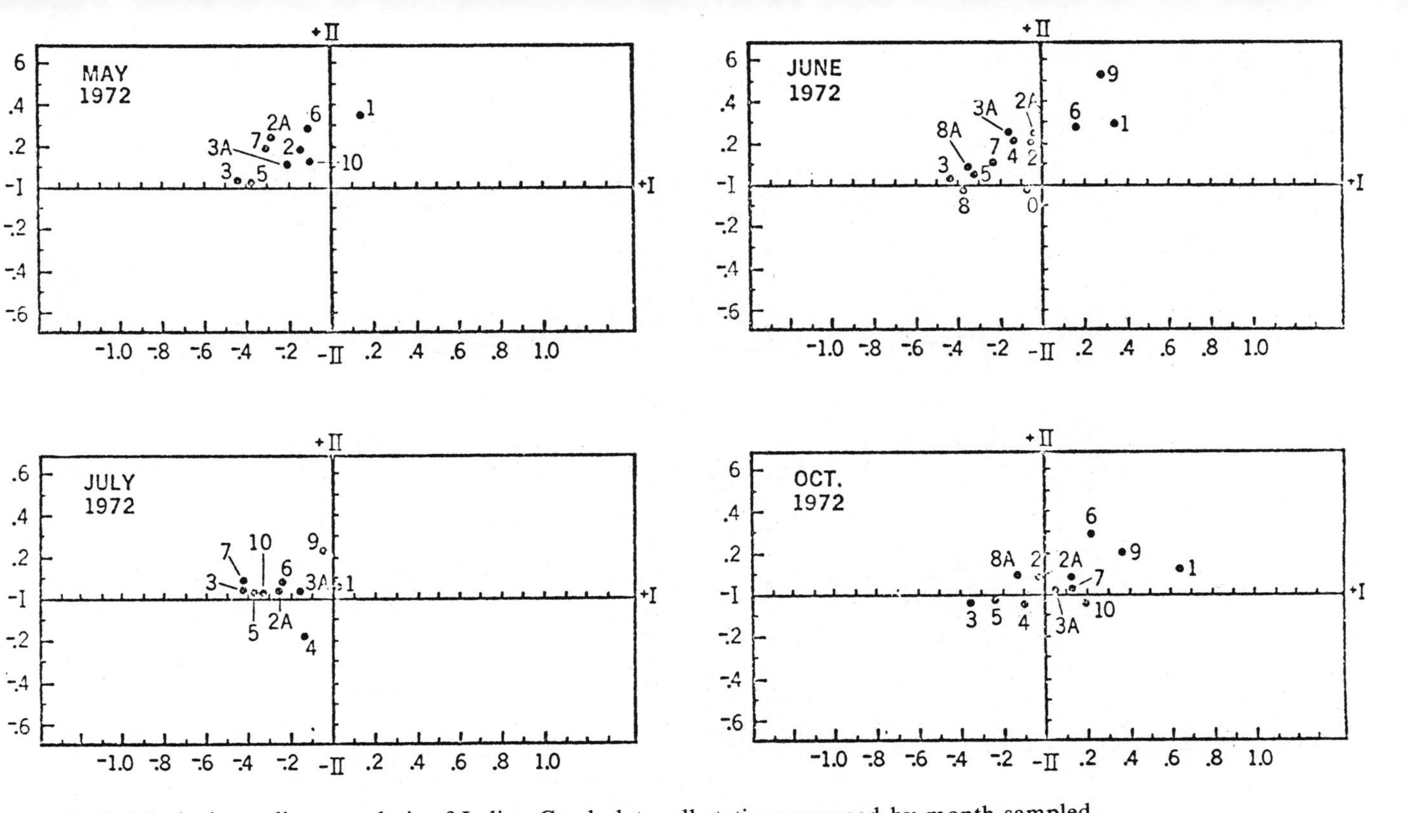

Fig. 8. Principal coordinate analysis of Indian Creek data, all stations grouped by month sampled

The dendrogram produced for 1972 sampling showed similar results. Samples from upstream reference stations and unpolluted tributaries were grouped together in the first cluster. In addition, samples from Stations 1 and 10 and the samples collected in May (a high discharge period) were clustered together. Tributary and mainstem stations from June were grouped together in this same cluster, further revealing the separation of high-discharge periods from other samples.

The second cluster showed some effects of flooding in late June. Samples from Stations 1, 2a, and 4 were clustered together. The third cluster contained samples from healthy tributary stations (6 and 9) from March, July, and May. The fourth cluster was made up of 10 samples from October, but samples from both Champion Run (Station 3) and Station 5 were missing from the cluster. This cluster indicates recovery from the effects of flooding of late June and the normalization of conditions in the stream. Samples from Station 1 were grouped with Stations 4, 5, and 10, indicating similar conditions throughout the mainstem of Indian Creek with the exception of Champion Run (Station 3) and Station 5, where AMD stress effects were still pronounced. The fifth cluster was made up of samples from all months. Stations with high AMD loading can be noted (Stations 3 and 5 were clustered together). Similarly, samples from mainstem stations during high-discharge periods were grouped together, as were samples from stations on tributaries from March and July. Both of these periods were preceded by high stream discharge, and the March fauna still showed effects of the September 1971 flood.

Summarizing stream conditions subjectively as stressed, good, or unstressed, Figure 4 illustrates some of the seasonally biological changes which can be related with stream discharge. High assimilative capacity (e.g., increased dilution or neutralization capacity) produces generally good stream biological conditions; when high spring discharges are combined with significant drift of aquatic insects (especially mayflies), stream conditions appear unstressed. As discharge decreases, stress increases, and biological conditions appear stressed. This seasonal cycle is repeated as stream flow increases in the late fall.

Principal Coordinate Analysis

Figures 7 and 8 illustrate the results of principal coordinate analysis for data from Indian Creek. Each of the strata diagrams (all stations collected in one month, on all monthly samples collected at one station) was computed from the same large matrix; they have been separated for clarity. Since all of the analyses were computed as a part of the same analysis, comparisons of scale can be made between all scatter diagrams to determine similarity of position. Unfortunately, the analytical technique does not identify the component represented on the absissa or ordinate, although familiarity with the data may allow identification of factors which are probable causes of the variability. Even though components are not identified, the positions of the samples indicate the importance of that component in accounting for the sample. Similarly, the position of the sample in the positive or negative quadrants is not so meaningful as the association that sample has in relation to the other samples.

Only the first two components (e.g., the two which accounted for the greatest variability in the data) were used in these displays. The I-axis is probably representative of a factor which is related to general sample size and accounted for 11.34% of the variability in the data. The II-axis appears to reflect some of the seasonal variability in the data and accounted for 5.71% of the total variability. The remaining components each accounted less than 3% of the variability, and have not been plotted.

Figure 7 illustrates the results of the principal coordinate analysis for samples collected during the two-year sampling period from Stations 1, 5, and 10 (e.g., reference, severe stress, recovery stations). Benthic collections for each month during the study have been keyed in sequence alphabetically; that key is contained in Table 2. The general trends indicated by this analysis are apparent when the groups of coordinate locations are compared on the same scale. At Station 1, all samples occur in the positive quadrants of I-axis. If axis I is indicative of general abundance, Station 1 regularly produced samples with the greatest abundance of organisms. Station 5 monthly coordinates were all in the negative I-axis. Since this station was the most severely affected by AMD and had very limited fauna, the general trend of abundance relationships appears to be valid. The monthly samples at Station 10 showed a much greater scatter, with overlap between samples from both Station 1 and Station 5. The scatter pattern indicates a significant influence of axis II at Stations 1 and 10. This pattern is not evident at Station 5, leading to the conclusion that even though seasonal variability is indicated at Station 5, it is not so great as at Stations 1 and 10. Further evidence of seasonal influence on stream conditions (whether physical-chemical or biological) can be observed by noting coordinate similarities for certain months. June, July, and August 1971 (*C, D,* and *E*) were all located closely together, and often stood apart from other monthly coordinates (e.g., Station 1 and 10). From water

Table 2. Benthic collections for months, keyed in sequence alphabetically

A	October 1970
B	May 1971
C	June 1971
D	July 1971
E	August 1971
F	September 1971
G	October 1971
H	December 1971
I	March 1972
J	May 1972
K	June 1972
L	July 1972
M	October 1972

quality data for these months, it is apparent that AMD effect was somewhat greater due to decreased streamflows. Similarly, other close groupings (e.g., *G* and *M,* the October samples from 1971 and 1972) also show close coordinate location for Stations 1 and 5. Again a seasonal interpretation can be made;increased streamflow, as well as seasonal dominance of *Hydropsyche* and Chironomidae, closely link Stations 1 and 5.

This seasonal interpretation of axis II can be verified in more detail in Figure 8, while gaining a further understanding of AMD and tributary influence. Figure 8 illustrates the results of principal coordinate analysis for each of the stations sampled in May, June, July, and October 1972. The stations collected in May have a lower loading on axis I, indicating a lower abundance than is evident in June or October. This is probably due to high seasonal streamflow as well as seasonal occurrence of certain groups of taxa, which are known to drift. The combination of transport capacity (high streamflow) and drifting biota tends to homogenize the biota of the stream. Habitats which are normally unavailable for colonization during periods of lower discharge (i.e., less dilution) are marginally acceptable for drifting organisms; thus benthic communities throughout the stream have similar taxa in approximately equal abundance. These conditions change rapidly, as noted in the June scatter diagram. The upstream and tributary stations (Stations 1, 6, and 9) all plotted in the positive portion of axis I, while the coordinates for the remainder of the stations sampled were grouped similarly to the stations from May. As in May, Champion Run (Station 3) and other stations significantly affected by AMD (Stations 5, 8, and 8a) all had the most negative position on axis I.

The stations collected in July indicated a significant shift from June samples. This was probably due to severe disruption of benthic communities by flooding caused by Hurricane Agnes in early June. Even with this severe disruption, Station 1 was still positive on axis I, even though it was very close to the origin (indicating another factor-controlled variability). The locations of stations sampled in October may be considered in two major groups: those significantly affected by AMD (Stations 3, 4, 5, and 8a) and the upstream or tributary stations (including mainstem stations immediately downstream from tributaries). This indicates that not only the AMD influence, but also the influence of unpolluted tributaries (interpreted in light of recovery) were significant.

Discussion

The general model of reduction of AMD stress through distance proposed in Figure 2 was intended to provide a basis for further discussion. It should be apparent that recovery and restoration of streams from long-duration stress due to AMD theoretically can be equated with the generalized response patterns which occur when any equilibrium or steady state conditions are altered. Unfortunately, the complexity of the stream system makes it particularly difficult to define the steady state or equilibrium condition which existed before AMD was introduced to the stream, and similarly it is almost impossible to define a return to "normal" or "acceptable" conditions which are the requisite of restoration and recovery.

It is possible that the problems of definition (e.g., for both normal and recovery conditions) are particular to stream systems and especially long-duration stresses such as AMD which have such a significant impact on the physical, chemical, and biological environment. In this regard it seems appropriate to quote Annie Dillard from *Pilgrim at Tinker Creek*: "The creeks–are an active mystery, fresh every minute. Theirs is the mystery of continuous creation and all that providence implies: the uncertainty of the present, the intricacy of beauty, the pressure of fecundity, the elusiveness of the free, and the flawed nature of perfection." It is this implied diversity and vitality that links the physical, chemical, and biological systems of a stream together, although some elements of each of these systems cannot be measured on the same scale. For example, many of the physical and chemical elements of the stream system's steady state must be measured in different terms than the biological elements. If we follow William Morris Davis's theories of the slow, progressive aging of rivers and landscapes, then the equilibrium relationship in the physical system must be scaled to millenia rather than hours or days. More recent work in fluvial systems has demonstrated rapid response to changes in physical conditions, but the gross morphology of a river or stream is still incompletely explained. Thus, as observed in Indian Creek, the occurrence of high flows, whether they recur every 1.2 or 1000 years, is the basis of the steady state or equilibrium conditions of the stream channel, substrate, and in part, the water quality. For the most part the biological system is subject to equilibrium states measured in terms of minutes or hours. The integration of the complex interactions which occur when time scales of such vastly different proportions are used leads to many philosophical questions as well as practical conclusions.

With regards to the cause of damage, in Indian Creek the most significant cause of damage to benthic communities downstream from Champion Run was AMD. The additional stress caused by addition of AMD altered the structure of the benthic communities, in general reducing total numbers of organisms present and significantly reducing diversity. The results of cluster analysis, principal coordinate analysis, and changes in water quality demonstrate the overall cause of damage. The specific causes of damage or mechanism of action are discussed in Herricks and Cairns (1974).

The nature of the effect of AMD on Indian Creek can be discussed in both descriptive and philosophical terms. The description of the effect is obvious: damage, with all that the word *damage* implies. In philosophical terms the nature of the effect of AMD (dependent on the time scale adapted) may be relatively minor. For example, effects of AMD in a stream or river will continue until the pyritic or sulfide minerals are exhausted. We may assume that this will occur after several hundred years, a geologic instant, and a much shorter time period than is necessary for most evolutionary damage. The question can then be asked if damage has occurred. From a practical point of view, this question must be answered in terms which accurately reflect the time spans we perceive as important (e.g., a reasonable estimation would be 60 to 100 years, a typical life span, although relatively short individual and societal attention spans lead to even shorter "critical" time periods, as evidenced by the environmental movement).

The mechanisms of recovery are affected by similar arguments. It is first necessary to determine if damage will be defined on a grand scale or on one which has present "environmental" interest. Restoration in 100 or 1000 years may be acceptable. In Indian Creek there is clearly a yearly restoration of benthic communities in the areas having the greatest AMD stress. This restoration may not be recovery, but certainly it can be considered accommodation to conditions by certain groups of organisms during certain seasonal periods. The most significant factor which enters this discussion is the long-duration nature of the stress due to AMD. This constant input of materials which have the potential for causing damage may actually preclude recovery until no more AMD is discharged in the stream. The mechanisms which function downstream (i.e., dilution and chemical reaction) to maintain a steady state can do so if the assimilative capacity, as judged in both distance and beneficial input, is great enough.

Summary

The recovery of streams from chronic pollutional stress caused by AMD can occur if certain conditions are met. First, seasonal variability inherent in any stream system must at all times produce conditions which are suitable for maintenance of the structure and function of stream biological communities. Second, there must be both a means of reducing the stress and one of restoring damaged habitat. Finally there must be a means of restoring biological system structure and function.

These three criteria can be met by naturally occurring stream conditions, as evidenced in Indian Creek. Discharges of AMD did not exceed the assimilative capacity for the total stream system. Tributaries provided good-quality water which both diluted and chemically modified AMD discharges. These unpolluted tributaries also provided source areas for organisms which recolonized damaged sections of the stream.

Acknowledgments

The author wishes to acknowledge the assistance of Dr. Roger L. Kaesler in the preparation of the cluster analysis and principal coordinate analysis sections of this chapter, and his careful review of the manuscript. Computation of cluster analysis and principal component analysis was performed at the University of Kansas Computation Center, using the NT-SYS system of computer programs written by F. James Rohlf and his associates. Special thanks are extended to Dr. Gerald F. Doebbler and George E. Cella for their assistance in preparation of the final manuscript. The quotation from Annie Dillard's *Pilgrim at Tinker Creek* (1974, Harper & Row, Publishers, Inc.) is printed by permission of the publishers.

Literature Cited

Appalachian Regional Commission. 1969. Acid Mine Drainage in Appalachia. Appalachian Regional Commission, Washington, D.C.

Brillouin, L. 1962. Science and Information Theory, 2d. ed. Academic Press, New York.

Cairns, J., Jr. and R. L. Kaesler. 1969. Cluster analysis of Potomac River survey stations based on protozoan presence-absence data. Hydrobiologia 34(3-4): 414-432.

Cairns, J., Jr., R. L. Kaesler, and J. A. Ruthven. 1974. Protozoan distribution in a small stream. Am. Midl. Nat. 92:406-414.

Ebeling, A. W., R. M. Ibara, R. J. Lavenberg, and F. J. Rohlf. 1970. Ecological groups of deep-sea animals off southern California. Los Angeles County Mus. Contrib. Sci. 6:1-43.

Gower, J. C. 1966. Some distance properties of latent root and vector methods used in multivariate analysis. Biometrika 53:325-338.

———. 1972. Statistical methods of comparing different multivariate analyses of the same data. Pages 138-149 *in* F. R. Hodson, D. G. Kendall, and P. Tantu, eds. Mathematics in the Archaeological and Historical Sciences. Edinburgh Univ. Press, Scotland.

Herricks, E. E., and J. Cairns, Jr. 1974. Rehabilitation of Streams Receiving Acid Mine Drainage. Water Resour. Res. Center Bull. 66, Blacksburg. 284 pp.

Herricks, E. E., V. O. Shanholtz, and D. N. Contractor. 1975. Models to predict the environmental impact of mine drainage on streams. Trans. Am. Soc. Agr. Eng. 18(4):657-663, 667.

Jaccard, P. 1908. Nouvelles recherches sur la distribution florale. Bull. Soc. Vaudoise Sci. Nat. 44:223-270.

Pennsylvania Department of Mines and Mineral Industries. 1969. Youghiogheny River Basin Mine Drainage Pollution Abatement Project, Report of Phase I and Phase II. Commonwealth of Pennsylvania, Harrisburg.

Rohlf, F. J. 1968. Stereograms in numerical taxonomy. Syst. Zool. 17:244-255.

———. 1972. An empirical comparison of three ordination techniques in numerical taxonomy. Syst. Zool. 21:271-280.

Schumate, K. S., E. E. Smith, and A. H. Morth. 1972. Pyritic Systems: A Mathematical Model. Office of Research and Monitoring, U.S. Environmental Protection Agency, Washington, D.C. EPA R2-72-002.

Shannon, C. E., and W. Weaver. 1949. The Mathematical Theory of Communication. Univ. of Illinois Press, Urbana.
Sokal, R. R., and F. J. Rohlf. 1962. The comparison of dendrograms by objective methods. Taxon. 11:33-40.
_____. 1970. The intelligent ignoramus, an experiment in numerical taxonomy. Taxon. 19:305-319.
Sokal, R. R., and P. H. A. Sneath. 1963. Principles of Numerical Taxonomy. Freeman, San Francisco. 359 pp.

Restoration and Recovery of the Thames Estuary

A. L. H. Gameson and Alwyne Wheeler

Abstract

The Thames Estuary has a long history of pollution culminating some 20-25 years ago in public nuisance from hydrogen sulfide. A detailed investigation was made by the Water Pollution Research Laboratory from 1949 to 1964. Since 1955 the pollution load from London's sewage discharges has been progressively reduced and there has been no sulfide in the water during the past 10 years.

The Thames as a fishery started to decline about 150 years ago, and, despite occasional partial recovery, by the 1950s the only fish able to survive in the most polluted reaches were eels. The recent restoration of the water quality has been accompanied by a remarkable ecological recovery, fish returning to the estuary in ever-increasing numbers and varieties.

Introduction

The deliberate restoration of the Thames Estuary was solely for the benefit of part of the external ecosystem, namely man. The need for it arose mainly from the offensive smell of hydrogen sulfide emanating from the estuary during the summer months some 20 to 25 years ago. This restoration has brought in its wake a remarkable recovery of the internal ecosystem, with fish returning to the estuary in ever-increasing numbers. The estuary is bounded on its landward side by Teddington Weir (Fig. 1), and for the purpose of this paper its length will be considered as 100 km.

Early History *

Pollution of the tidal Thames has been a matter of concern for hundreds of years. As early as 1620 the bishop of London in a sermon

* See Acknowledgments, p. 97.

expressed the hope that "the cleaning of the river . . . will follow in good time." A century and a half later, Tobias Smollett wrote in *Humphry Clinker*: "If I would drink water, I must quaff the maukish contents of an open aqueduct, exposed to all manner of defilement; or swallow that which comes from the river Thames, impregnated with all the filth of London and Westminster–Human excrement is the least offensive part of the concrete, which is composed of all the drugs, minerals, and poisons, used in mechanics and manufacture, enriched with the putrefying carcasses of beasts and men; and mixed with the scouring of all the wash-tubs, kennels, and common sewers, within the bills of mortality." Although he may have exaggerated for the sake of literary effect, Smollett was an outspoken surgeon with plenty of opportunity for observing the Thames, and this quotation certainly suggests that in the latter half of the eighteenth century the reaches of the estuary within the cities of London and Westminster were grossly polluted by discharges of crude sewage and industrial wastes. Nevertheless, the Thames was still a good fishing river; large numbers of salmon could still be caught during the eighteenth century, 130 being sent to market on a single day in 1766.

It was in the first half of the nineteenth century that serious deterioration set in. During this time the population of what was later to become the County of London rose from less than one million to more than two million, and this must have been one cause of the deterioration. Probably of greater importance, however, was the increased proportion of excrement reaching the estuary. The water closet was introduced toward the end of the eighteenth century when domestic wastes normally were collected in cesspools and removed at intervals for manure. As the use of water closets increased–a rapid process after about 1830–overflows from cesspools had to be constructed, and sewage began to reach the Thames at many points. During this period the numbers of fish declined, although 12 freshwater and 6 saltwater species were listed as occurring in the estuary in 1819, and 400 fishermen earned a living from the river between Deptford and London as late as 1828. So far as is known the last salmon caught was in 1833, and by 1850 all commercial fishing had ceased.

In 1848 the first Metropolitan Commission of Sewers was appointed. This resulted in abolition of all remaining cesspools, improvement of house drainage, and consequent increased pollution of the Thames. In the following year a second commission considered large-scale measures to reduce this pollution; many plans were considered by them and by the series of commissions which succeeded them during the following seven years. No agreement was reached; no action was taken, and the Thames continued to worsen. In 1856 the Metropolitan Board of Works was established by statute and charged with the duty of "preventing all or any part of the sewage within the Metropolis from flowing into the River Thames in or near the Metropolis." This was interpreted to mean that the outfalls had to be downstream of the limits of the Metropolitan District, and it was on these grounds that some early proposals were rejected.

Meanwhile the condition of the Thames in central London had become so foul that sheets soaked in disinfectant were hung in the Houses of Parliament in an attempt to counteract the stench. This culminated in 1858 when the smell at Westminster became so overpowering that its control was of great personal interest to members of Parliament. In the same year, work started on the construction of intercepting sewers to carry sewage from central London to Barking on the north and to Crossness on the south side of the estuary (see Fig. 1)–the outfall sites eventually agreed upon by the board.

In 1864 both outfalls were in use and it was estimated that about a third of the sewage had then been diverted, but the whole scheme was not completed until 1875. No treatment was provided, but the sewage was stored in large reservoirs and discharged only during the first few hours of the ebb tide. This general displacement of the center of pollution, by some 20 km to where the estuary is much wider, gave a correspondingly greater dilution of the sewage, and thereby alleviated the foul condition of the Thames in central London. However, as early as 1868 the vicar and other inhabitants of Barking addressed a memorial to the Home Office in which it was stated that the condition of the estuary was "dangerous alike to navigation and to the health of the inhabitants of the parish of Barking and of all the populous and industrial towns below London." The specific complaints included reference to formation of deposits of foul mud on the foreshore and to the absence of fish from Barking Creek and its neighborhood. At a subsequent inquiry, many witnesses confirmed that fish were no longer able to exist in the Thames near Barking.

During the following decade a protracted argument developed between the Metropolitan Board of Works and the Thames Conservancy about the extent and causes of mud banks in the estuary, and there were many complaints of nuisance from those who lived or worked on or near the estuary in the vicinity of Barking. In 1878 the steamship *Princess Alice* was sunk near Barking with heavy loss of life, and it was alleged that the death toll had been increased by the polluting matter in the water.

In 1882 a royal commission was appointed to "inquire into and report upon the system under which sewage is discharged into the Thames by the Metropolitan Board of Works, whether any evil effects result therefrom, and in that case what measures can be applied for remedying or preventing the same." In their first report, published in 1884, the commissioners summarized the "popular" and "scientific" evidence concerning the effects of the discharges to the estuary. The popular evidence consisted of statements from pilots, police officers, and others alleging that the water and mud gave off foul odors which caused headaches and nausea, that the mud was black and sticky, that bubbles of foul-smelling gas were given off, that fish had disappeared from parts of the estuary where they were formerly plentiful, and that it was no longer possible to bring fish up to London in live wells of boats, as they died while passing through the polluted reaches. The scientific evidence included values for the concentration of dissolved oxygen, over a distance of nearly 100 km, on two days in August and September 1882—thus providing the data for what may well be the earliest oxygen "sag curve" for any river or estuary in the world. At each end of the estuary the oxygen content exceeded 90% of the saturation value, whereas the minimum value (off Crossness) was only 22%. It is interesting to note—for reasons which will become apparent later in the chapter—that one witness stated that nitrate was always present even in the most polluted reaches. Among the general conclusions of the commissioners concerning the "evil effects" of the discharge of sewage were that it did not appear to have had any seriously prejudicial effect on health (but that in hot, dry weather it caused serious nuisance and inconvenience), that fish had disappeared for a distance of some 15 miles below the outfalls and for a considerable distance above them, and that conditions were likely to worsen as the population increased. In their second report (also published in 1884) the commissioners concluded that it was neither justifiable nor necessary to discharge crude sewage to the Thames and that as an immediate measure some process of precipitation should be used to separate the solids from the sewage at the existing outfalls, the liquid then being discharged during the ebb tide and the sludge burned, applied to land, or dumped at sea. They did not consider, however, that the discharge of settled sewage through the existing outfalls would be satisfactory as a permanent measure.

During the next few years, methods of chemical treatment of the sewage were examined and (between 1887 and 1891) precipitation works were constructed at Beckton and Crossness, the sewage being treated with lime and ferrous sulfate. By 1891 the Metropolitan Board of Works had ceased to exist and responsibility for sewage disposal had been taken over by the newly constituted London County Council. Since the construction of the outfalls, the population of the County of London had risen from about three million to over four million.

Installation of the precipitation works did not immediately restore the estuary to an inoffensive condition; thus, in 1891 it was reported that "at certain times, such as during dry summer weather and in particular places, the stream is still apt to become very discoloured, and occasionally to emit offensive odours." However, during the last decade of the nineteenth century fish began to return to the parts of the estuary from which they had long been absent. Whitebait reappeared at Gravesend in 1892 and Greenwich in 1895, and in the latter year flounders were caught in the upper reaches for the first time in 12 years. In 1900 smelts passed right up the estuary to Teddington. Simultaneously with the passage of saltwater fish up the estuary, freshwater fish began to move downstream; large shoals of dace, bleak, and roach appeared throughout the uppermost 20 km of the estuary after 1890, whereas a few years earlier hardly any had been seen more than 10 km below Teddington. From 1861 until at least the end of the century, repeated attempts were made to reestablish a salmon fishery by releasing salmon fry in the upper reaches of the Thames and Lee, but without success–though adult salmon caught occasionally in the most seaward reaches of the estuary may have been derived from these fry.

Twentieth-Century Changes in Pollution

During the first decade of the present century, further intercepting sewers were laid, and the quantity of storm sewage discharged to the estuary was thereby reduced. In 1915 the regular chemical treatment of sewage at Beckton and Crossness was discontinued.

Between 1931 and 1946 paddle-aeration activated-sludge plant and diffused-air reaeration channels were installed at Beckton to give partial treatment to about a quarter of the total flow, and in 1936 a new works was commissioned at Mogden (near the head of the estuary; see Fig. 1) to replace 28 small works, most of which formerly discharged into streams in the area. These two measures were insufficient to counteract the gradually increasing volume of sewage, and there was a progressive deterioration in the quality of the estuary between about 1930 and 1950.

The Second World War and its aftermath delayed, for more than a decade, extensive improvements in sewage treatment proposed by the London County Council. The first of these improvements came in June

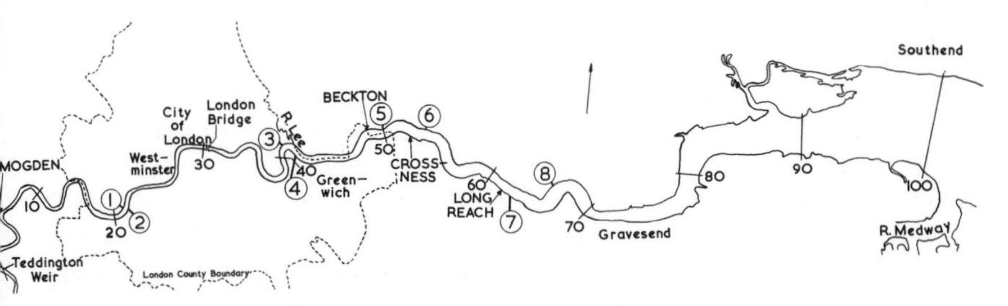

Fig. 1. Sketch map of Thames Estuary. Sewage works mentioned in text are in capitals. Ringed numbers refer to the following power-generating stations: 1. Fulham; 2. Lombard Road; 3. Brunswick Wharf; 4. Blackwall Point; 5. Barking; 6. Ford Motor Co.; 7. Littlebrook; 8. West Thurrock. Other numbers are distances (km) seawards of Teddington Weir.

1955 with the commissioning of a new and more efficient primary settlement plant at Beckton. This was followed, between October 1959 and February 1960, by additional activated-sludge plants, thus bringing the proportion of the total flow receiving biological treatment to nearly 50%. A further large reduction in polluting load was made between June 1963 and January 1964 by reconstruction of the Crossness works to give full treatment for flows up to twice the dry-weather value.

In the course of an investigation made by the Water Pollution Research Laboratory in 1949-64 (Department of Scientific and Industrial Research [DSIR], 1964), it was concluded that a reasonable estimate of polluting strength was given by

$$E^* = \frac{3}{2}(B + 3N) \tag{1}$$

where B is the five-day biochemical oxygen demand and N the content of oxidizable nitrogen; this was a simplified form of an expression developed for what was termed the *effective oxygen demand*, E. Values for the loads (i.e., the products of E^* with rates of discharge) are shown for the four largest sewage discharges, during selected periods, in Table 1.

The Mogden works has generally provided full biological treatment for the whole flow, but from 1956 to 1962 (when extensions to the works were brought into use) part of the flow was discharged without secondary treatment. For a number of years up to the end of 1965 surplus activated sludge was discharged to the estuary, thereby roughly doubling the polluting load from the works. Subsequently there have been further improvements, and in 1973 the final effluent was almost fully nitrified and the BOD had been reduced to 9 mg/l.

Similarly at Beckton, further improvements in treatment continue to be made, the effectiveness of which is evident from Table 1. By the end of 1975 flows up to rather more than the average will receive full biological treatment and the E^* load is expected to fall to about 35 tonnes/day. Table 1 also shows that large reductions in polluting load have been achieved at Crossness since completion of the new works in 1964.

In 1950-53 the discharge from the Long Reach (West Kent) works was of relatively small importance, the load being only about a twelfth of that from Beckton and Crossness, and its effect being far smaller than that of Mogden because of the greater dilution afforded by virtue of the greater width of the estuary at the point of discharge (see Fig. 1). However, the flow arriving at the Long Reach works has risen progressively over the years, and by 1973 the load discharged was over half of the total from Beckton and Crossness. It is understood that by 1978 the works will have been rebuilt and the load reduced to about 32 t/d. By that time the total load from these four works should be only about 115 t/d–or roughly a third of that in 1971-73.

Nothing has yet been said about other sources of pollution. In 1950-53 the four works listed in Table 1 accounted for rather more than 70% of the total effective oxygen demand load entering the tidal Thames. Other sources of pollution included 17 other sewage works, at least 26 storm-sewage overflows, about 30 direct industrial discharges, and 13 freshwater discharges–including, of course, the River Thames entering the estuary over Teddington Weir. The polluting loads from many of these sources have since been reduced, largely as a result of action taken by the Port of London Authority (which was responsible for pollution control in the tidal Thames from 1902 until these duties were taken over by the newly constituted Thames Water Authority on April 1, 1974) and by the Greater London Council (which superseded the London County Council in 1964 and whose responsibilities for many of the tributaries, as well as for Mogden, Beckton, Crossness, and two smaller works, were also taken over by the Thames Water Authority).

Despite its considerable length, this historical prologue cannot do full justice to the efforts of the numerous organizations and individuals who have contributed to the restoration of the Thames Estuary. Attention will now be turned to the effects produced on the water quality by the pollution of the estuary.

Dissolved Oxygen

In 1949 the Water Pollution Research Laboratory of the Department of Scientific and Industrial Research started a 15-year investigation of the Thames Estuary. During the past few years there had been many complaints of the general insanitary condition of the estuary water, particularly during dry summer months. The foul-smelling gas hydrogen sulfide was being evolved to an extent sufficient to constitute a public nuisance–it could be smelled not only by those working on the river but also those living some distance away, and it was also reported to be causing rapid tarnishing of brass and discoloration of lead-based paints.

It was evident, from the results of the detailed surveys made by the London County Council, that sulfide was present in the water only when anaerobic conditions had been established. Consequently, the investigation was directed largely to the study of the factors affecting the distribution of dissolved oxygen. The work started on three independent lines: the practical–regular sampling and chemical analysis of the water from Teddington to Southend; the statistical–examination of the vast accumulation of the chemical records of the condition of the estuary obtained by the London County Council over more than half a century; and the mathematical–development of methods by which the movements of an effluent discharged to the estuary could be predicted. Later, when the mathematical methods had been developed, they were validated by comparison of the observed distributions of dissolved oxygen with those calculated from knowledge of the imposed conditions. Finally, the aim was achieved of predicting the future state of the estuary if certain changes were made in the imposed conditions–such as the amount of pollution discharged, the points of entry, the rate of entry of fresh water from the Upper Thames at Teddington, and the heat discharged from power stations. The predictive model has subsequently been used by the Port of London Authority and the Thames Water Authority as an aid in managing the estuary.

Water movements in the estuary are dominated (in the upper reaches) by the flow from the Upper Thames and (except in the uppermost reaches at times of high flow) by the effects of tides. Both these factors are of importance when considering the distribution of dissolved oxygen.

The flow of the Thames has been gauged at Teddington Weir since 1883 by the Thames Conservancy (now superseded by the Thames Water Authority). In the 50 years 1925-74, the yearly and daily averages varied over the ranges 20-123 m^3/s and 0.9-709 m^3/s, respectively; and the overall average was 69 m^3/s. Water is abstracted by the Thames Water Authority (formerly by the Metropolitan Water Board) a few kilometers upstream of Teddington. The average rate of abstraction has increased almost linearly from 4.3 m^3/s in 1885, through 12.5 m^3/s in 1955, to 17.0 m^3/s in 1970-74. By the Thames Conservancy Act of 1932, water may not be abstracted at such a rate as to reduce the flow at Teddington Weir below 170 million Imperial gallons per day (8.94 m^3/s), but in an emergency certain ministers may make an order varying this provision. However, despite the construction of new reservoirs to meet the rising demand for water, the statutory minimum flow was not maintained, on average, on 24 days/year during the 42 years 1933-74, which may be compared with 15 days/year during the 42 years 1890-1931. It may be noted that the rate of abstraction is now nearly double the statutory minimum flow.

The average tidal range increases from about 2 m at Teddington to 6 m at London Bridge, from where it decreases gradually to 4 m at Southend. The tide carries the water up and down the estuary, giving an average excursion of 11-14 km in those reaches more than about 10 km below Teddington, and polluting matter entering the estuary will be dispersed upstream and downstream by considerably greater distances during successive tidal cycles. This is illustrated by the distribution of chloride, since seawater may be considered as a conservative pollutant entering at the mouth of the estuary: the point where the chloride content is half that of seawater varies between about 40 and 90 km below Teddington Weir (depending on freshwater flow, tidal state, and tidal range), and even at London Bridge the chloride content varies with tidal state.

The dissolved-oxygen contents of two samples taken at the same point in the estuary but at different tidal states are likely to differ appreciably since the samples will have been taken from different bodies of water. When comparing the results for samples taken at different tidal states, it is therefore preferable to reduce the data to a common tidal state. The method adopted (DSIR, 1964) was to replace the true sampling position by that where the water would be at half tide–defined as the instant when the volume upstream (to Teddington Weir) is the mean value for the average tidal cycle. Low water samples are thus, in effect, moved upstream (and high water ones downstream) by about half the tidal excursion.

It had long been recognized that worst conditions were associated with hot, dry summers; but at the start of the laboratory's investigation, it was not clear whether the freshwater flow or the temperature was the dominant factor. The relative unimportance of temperature near the point of lowest oxygen content is shown by Figure 2, where the four-weekly averages of the oxygen content and temperature of the water off Beckton and the flow at Teddington are plotted—the temperature scale has been inverted for convenience of comparison with the other factors. In the first and last years of the period it is not clear whether the oxygen content is following more closely the changes in flow or in temperature, but during the exceptionally dry winter of 1933-34 both the flow and oxygen values are seen to rise little above the summer levels, while the temperature curve is much the same as for the other years.

Average distributions of dissolved oxygen at half tide have been calculated for each quarter year from 1920 to 1974 as well as

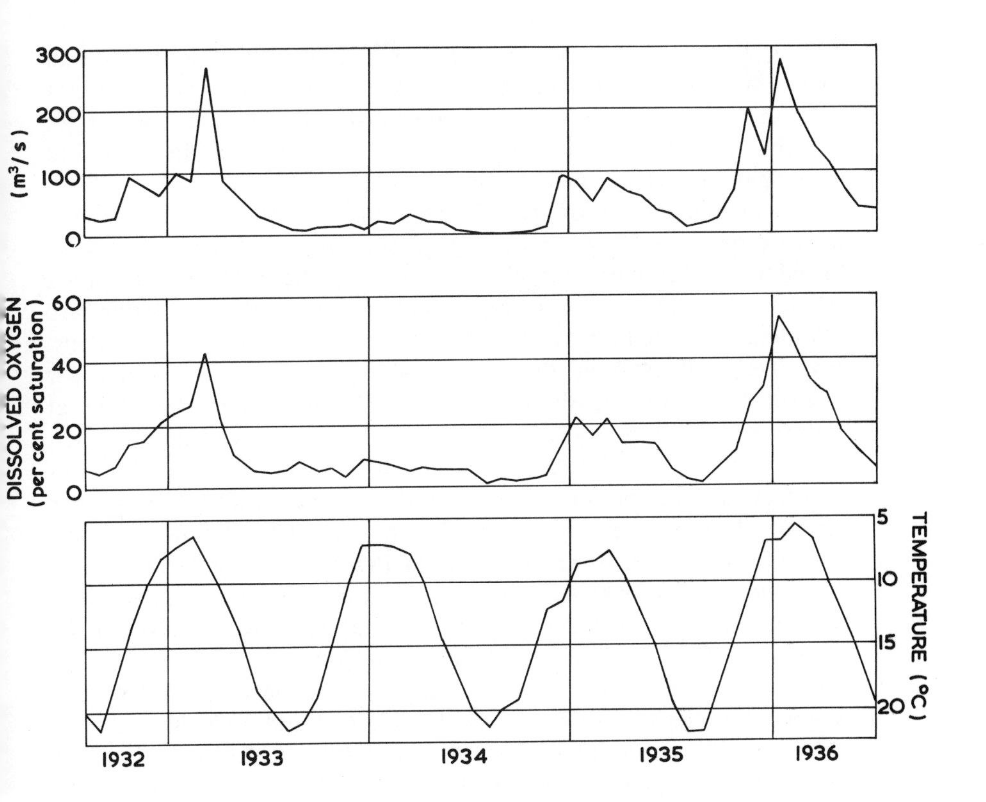

Fig. 2. Four-weekly averages of flow at Teddington and of dissolved-oxygen content and temperature of water off Beckton, July 1932 to July 1936

for occasional earlier years. The data for 1920-52 were subjected to partial regression analysis (DSIR, 1964), and it was found that except when the quarterly averages exceeded about 80% or fell below 10% of the saturation value, there was a nearly linear relationship between dissolved oxygen and freshwater flow, the effect of flow falling off progressively from 22 km below Teddington Weir (the most upstream point examined in the analysis) to virtually zero by 80 km. At London Bridge (30 km), for instance, an increase of 100 m^3/s was associated with an increase in dissolved oxygen of 26% of the saturation value. The effect of temperature was more difficult to determine, since the largest changes in temperature are seasonal and other factors affecting the oxygen content– such as photosynthesis and, to a lesser extent, the polluting loads–also vary seasonally. The largest apparent effect of temperature was found at London Bridge where a decrease of 13°C would be required to give the increase of 26% in oxygen content mentioned above. It is thus evident that in comparing oxygen distributions for different periods it is necessary not only that the data should be reduced to a common tidal state but also that they should refer to the same season of the year and to comparable values of the freshwater flow.

For comparison of oxygen distributions–or "sag curves"–in the summer quarters (July-September) of successive decades, a flow of 13 m^3/s has been chosen; about a fifth of the third-quarter flows are less than this, so it is representative of conditions which are considerably worse than average. The values, at 5-mile (8-km) intervals, have been taken from the individual sag curves, and those for particular decades (or shorter periods) plotted against the flow at Teddington. After interpolating to 13 m^3/s, the data have been replotted against distance to give the curves shown in Figure 3. This illustrates the progressive deterioration from 1893 (the earliest detailed third-quarter survey, when the flow was 12.5 m^3/s) to the 1950s, followed by the extensive restoration of water quality in 1970-74. It may be mentioned that conditions in 1900 and 1905 were comparable with those in 1893, that the curve for the 1930s lies roughly midway between those for the 1920s and 1940s, and that in 1960-62 the upstream limb of the sag curve was similar to that in the 1950s whereas the downstream limb was roughly midway between those for 1950-59 and 1970-74.

It might be considered that on the basis of the reductions in polluting load shown in Table 1, the recent improvements in the water quality should have been much greater than shown by Figure 3. This apparent discrepancy is attributed to the effects of low oxygen

Table 1. Effective oxygen demand loads (tonnes/day), as derived from Equation 1, for four major sewage discharges to Thames Estuary

Period	Mogden	Beckton	Crossness	Long Reach	Total
1950-53	77	504	212	59	852
1956-58	83*	366	222	58	729*
1960-62	75*	283	251	70	679*
1966-67	24	263	117	77	481
1968-70	28	250	60	91	429
1971	16	203	50	97	366
1972	15	170	29	91	305
1973	12	153	34	105	304

* Excluding uncertain load from surplus activated sludge.

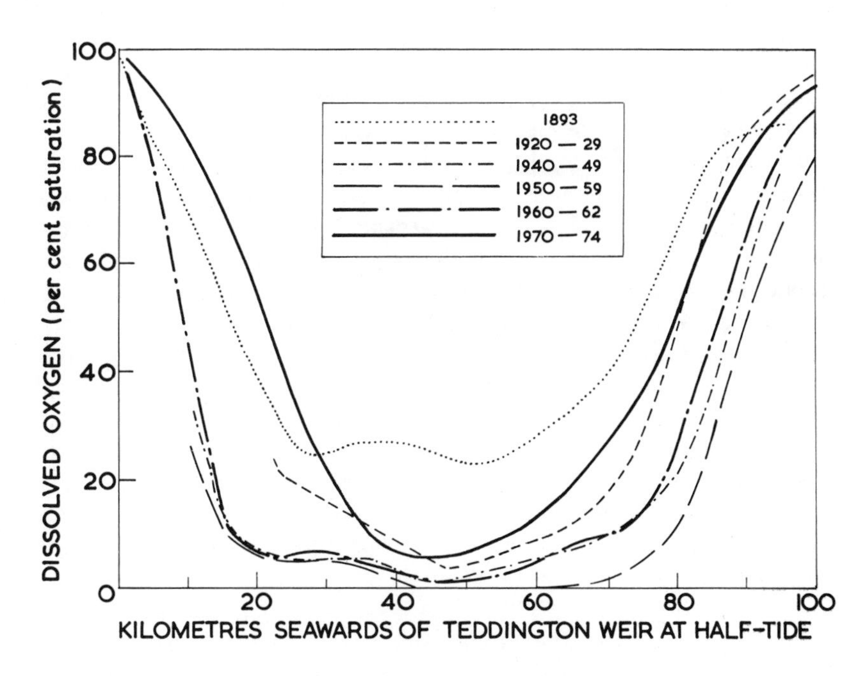

Fig. 3. Oxygen sag curves for July-September quarters with flows about 13 m^3/s at Teddington Weir

concentrations on the rate of oxidation (DSIR, 1964). It appears that when the oxygen concentration exceeds about 10% of the saturation value, the rates of oxidation of carbon and nitrogen are virtually independent of the oxygen level. However, when it falls from 10% to zero, the rate of nitrification decreases and eventually ceases, and denitrification occurs, the end product being mainly molecular nitrogen and the oxygen made available being used in the oxidation of organic carbon. Finally, when fully anaerobic conditions have been established, and the reserves of nitrate exhausted, sulfates (of which there is a plentiful supply in estuary water) are reduced—thus leading to the offensive conditions referred to earlier in this chapter. Consequently, when the sag curve minimum is below 10% saturation, oxidation is retarded, most of the accumulated oxygen demand being later satisfied farther downstream.

In a preliminary stage of the laboratory's prediction of oxygen distribution in the estuary (DSIR, 1964), the rate of oxidation was assumed not to be limited by the rate of supply of oxygen, and the oxygen concentration thus calculated was termed the *nominal oxygen content.* If the nominal oxygen content nowhere fell below 10%, it was then everywhere identical with the finally predicted oxygen content; but when anaerobic conditions were being predicted, the nominal oxygen content was often strongly negative. For instance, it was estimated that for a flow of 9 m^3/s at Teddington, the minimum nominal oxygen content during the summer in the early 1950s was about -240% saturation (or a nominal oxygen deficit of 340%). It was later calculated that for a flow of 13 m^3/s in 1970 the minimum has risen to -50% (or a deficit of 150%). In the third quarter of 1974, though with the higher flow of 27 m^3/s, the observed minimum oxygen content was 12% (or a deficit of only 88%). Although these improvements are somewhat exaggerated by effect of freshwater flow, it may be seen by comparison with Table 1 that the maximum nominal oxygen deficit reflects the reductions in the polluting loads. The point has now been reached where, with high summer flows, further reductions in load may be expected to be followed by proportionate reductions in the oxygen deficit, so that the results of future improvements in the quality of the effluents discharged will appear to be far more rewarding, in terms of the oxygen distribution, than has so far been the case.

By inspection of Figure 3, and from what has been said about the effects of restriction on nitrification, it is evident that the position of the steep part of the recovery (downstream) limb of the sag curve is a sensitive measure of changes in the condition of the estuary, and since this part is virtually unaffected by the flow at Teddington, it can be used for demonstrating the changes from year to year. The third-quarter averages for two selected points—30 and 40 miles below London Bridge (the traditional units and reference point in the estuary) or 79 and 95 km below Teddington Weir—are plotted in Figure 4,

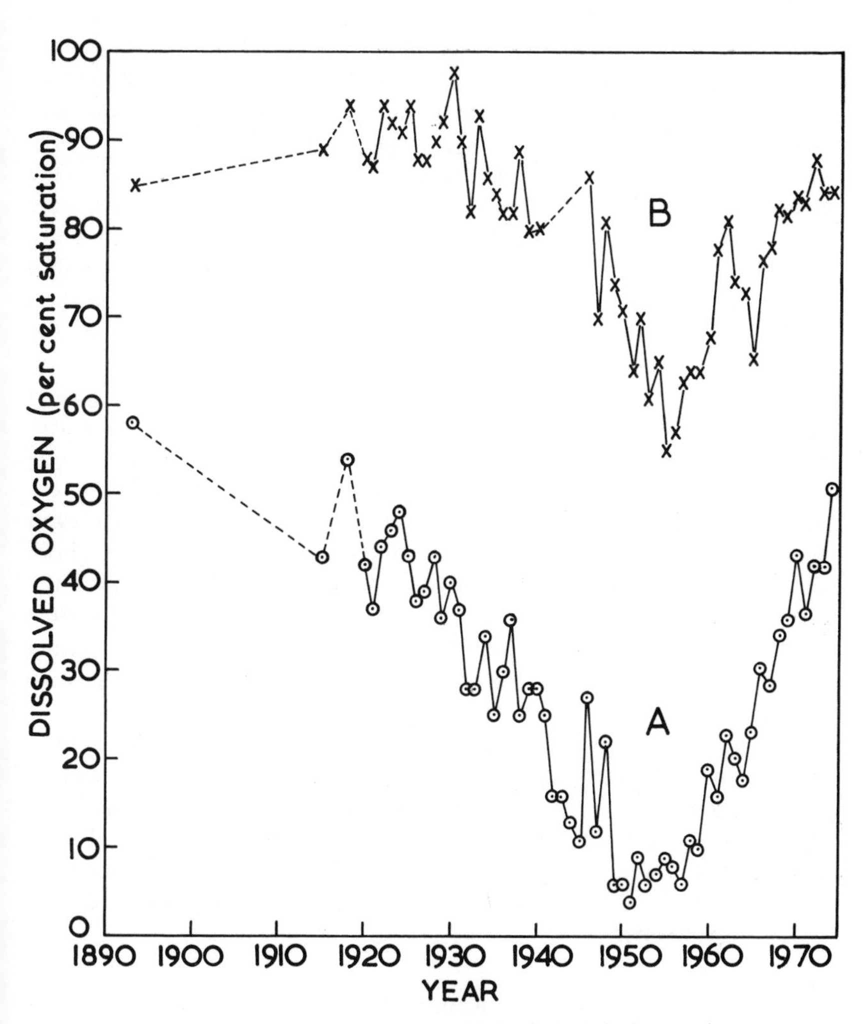

Fig. 4. Average dissolved-oxygen content at half tide in July-September quarter: *A,* 79 km below Teddington Weir (30 miles below London Bridge); *B,* 95 km below Teddington Weir (40 miles below London Bridge)

which shows clearly the progressive deterioration and subsequent restoration of oxygen levels.

Some 25 years ago the major concern was with the elimination of anaerobic conditions, and it is therefore of particular interest to examine how the length of the anaerobic reach has changed in recent years. The approximate length of this reach was found by plotting the individual oxygen data for each third day throughout 1945-65 and drawing contours bounding the anaerobic areas. Since the length that is anaerobic is markedly dependent on freshwater flow, it was necessary to select a particular value of the flow before comparing one year with another; that chosen was 25 m^3/s, which was well within the range of flows occurring in each of the years examined. The lengths were plotted against the corresponding daily flows, and regression lines were fitted. The values given by the regression equations at 25 m^3/s are shown in Figure 5–inspection of the data for 1966-74 showed that no anaerobic reach existed in these years. The marked deterioration in 1949-50 (which may also be seen in Fig. 4) is attributed to the widespread introduction of packaged synthetic anionic detergents which were shown (DSIR, 1964) to lower the rate of reaeration. The lack of improvement between 1955 and 1960, so far as Figure 5 is concerned, may have been due to improvement in the Beckton effluent being offset by the effects of surplus activated sludge discharged from Mogden; in the lower reaches, however, there was some increase in oxygen levels (see Fig. 4). Following the further improvements at Beckton in 1960, both diagrams show corresponding improvements in the condition of the estuary; and on completion of the new works at Crossness in 1964 the anaerobic reach disappeared. The subsequent reductions in polluting loads shown in Table 1 have been accompanied by a progressive increase in dissolved oxygen (see Figs. 3 and 4).

If by 1979 the 304 t/d given in the last line of Table 1 has been reduced to 115 t/d, then–even after allowing for other sources of pollution which become progressively more important as their proportion of the total load increases–it would seem reasonable to expect that curve *A* of Fig. 4 will have passed 70% saturation, and that the sag curve minimum will exceed 40% except, perhaps, during a very dry summer.

The Return of Fish

The introductory section on the early history of pollution in the tidal Thames indicated that before the water became grossly polluted there

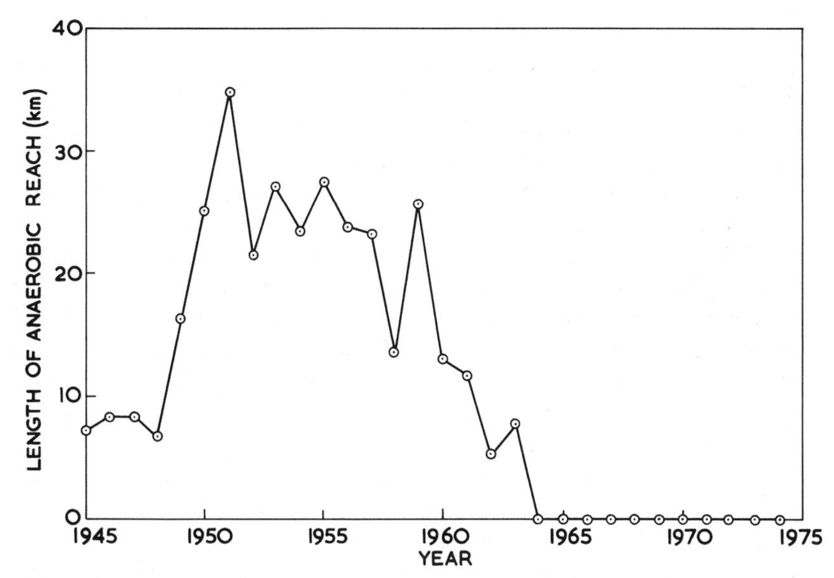

Fig. 5. Average lengths of reaches devoid of dissolved oxygen when flow at Teddington Weir was 25 m^3/s

were substantial populations of fish in the river which provided employment and food for the population of London and the riverside communities. As pollution increased, the number of fish and the diversity of the fauna decreased, most notably affecting the migratory species such as the salmonids. Unfortunately, no monitoring of the changes in the fauna was attempted until recent years and information from the period 1850-1950 is sparse and mostly culled from incidental observations.

However, in the period 1957-58 Wheeler (1958) attempted to establish the status of fish in the river and came to the conclusion that from the region of Gravesend upstream for some 68 km there was no evidence of fish life with the single exception of eels, which were found in the upper reaches of this area. In this period, which coincided with extremely low dissolved-oxygen levels and at times anaerobic conditions (Fig. 3), there were no commercial fisheries in the river, and sportfishing was practiced only at the extreme upper limit of the estuary between Richmond and Teddington Weir. There was also a lack of information in the literature on aquatic organisms other than birds in the upper reaches.

During 1964 and 1965 reports of fish captured on the cooling-water intake screens at the newly commissioned electricity-generating station at West Thurrock (well within the presumed "fishless" zone of 1958) suggested that fish might be returning to the river, following the improvements noted in the level of dissolved oxygen in 1964 (p. 85). It also suggested a method of capturing fish in this large and busy river where conventional fishing methods were precluded. With the agreement of the Central Electricity Generating Board, arrangements were made for fish caught on cooling-water intake screens to be preserved by staff at the generating stations for later study. Commencing in September 1967 the following generating stations were involved: Fulham, Brunswick Wharf, Blackwall Point, Barking, and West Thurrock (see Fig. 1). In February 1969 Lombard Road station was added to the list; Littlebrook generating station participated from January 1970, and in May 1972 the generating station at the Ford Motor Company's plant at Dagenham was included. The staff at all these generating stations participated in the survey until its completion in December 1973 (except that the Lombard Road station closed in April 1972). The data derived from the generating station catches have been supplemented by seine netting in the upstream reaches (the upper 15 km), and in the mouth of the estuary by trawling with a beam trawl from FRV *Tellina* (between 80 and 100 km from Teddington Weir). These collection methods have produced much information on the fish living in the tidal Thames from the upstream freshwater conditions, through the zone of least dissolved oxygen, to nearly marine conditions at the mouth. None of these methods of capture (as employed during the survey) produce data which can be used in a statistical analysis. Nevertheless the results of these collections can be analyzed qualitatively to produce a list of species found in the Thames, and with caution it is possible to make suggestions as to comparative abundance of certain species and to analyze the components of the fauna.

Between 1967 and December 1973 a total of 68 species of fish were captured at the generating stations listed above, and a further 5 species were caught by other means within the 100 km of estuary. Of these, 18 are freshwater fish, 43 are of marine origin, and 11 are euryhaline species, 6 of which are migratory. Two of the freshwater species and 9 of the marine species have been captured once only. The residual total of fish species from all sources between 1967 and 1973 which are represented by more than a single capture therefore amounts to 62 species. Isolated captures since December 1973 have increased the grand total to the 80 species listed in the Appendix. From examination of the capture during each year of the survey, there is no evidence of a significant change in the number of species present, thus suggesting that the recovery had already occurred by 1967.

This recovery of the fish fauna in a formerly grossly polluted river, which has taken place within a relatively short time of the river's becoming inhabitable, is remarkable. With the observed increase in the aquatic invertebrate populations and wildfowl along the lower reaches of the river, it is a graphic illustration of the speed with which a damaged ecosystem will be rehabilitated.

Distribution of Fish

For the purposes of this present contribution, the catches from the various stations are best analyzed region by region and the generating stations can conveniently be grouped in pairs. Analysis of the captures is presented graphically in Figure 6. As would be expected, the catch at the most

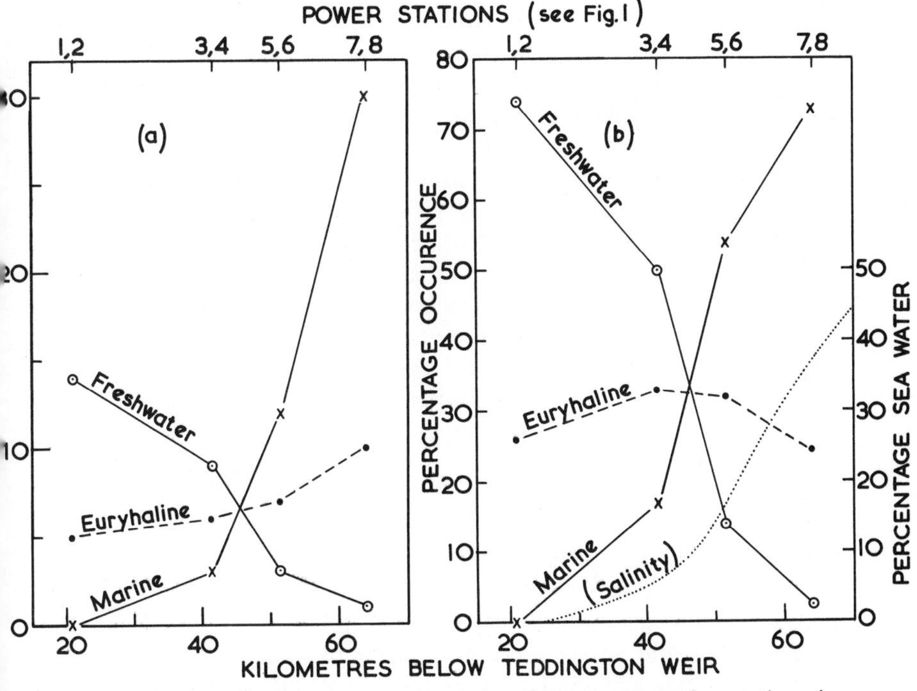

Fig. 6. Analysis of the fish fauna at four pairs of power-generating stations in 1967-73, showing (*a*) numbers and (*b*) proportions of freshwater, euryhaline, and marine elements. Also shown in *b* is approximate half-tide salinity distribution (as percentage seawater) for average flow conditions in 1967-73.

upstream pair of stations (Fulham and Lombard Road) is composed mostly of freshwater fish species (74% of the species caught here more than once can be classed as freshwater fish), the remainder being euryhaline species. Farther downstream, at Brunswick Wharf and Blackwall Point, the fauna is composed of 50% freshwater fish and 17% marine species. Downstream farther, at Barking and Dagenham, the freshwater fish comprise 14% of the total, and marine species 54%, while in the mouth of the estuary (West Thurrock and Littlebrook) the freshwater component amounts to only 2.5% and the marine to 73%. Throughout all station pairs the euryhaline component is virtually constant and varies only between 24.5% and 33%.

To what extent this represents a normal condition of distribution in response to changes in salinity is difficult to judge, mainly because of the sparsity of comparative data on the distribution of fish in an unpolluted estuary. Nor is it possible to make direct comparison with observed salinity in the Thames as the data on fish are a total of captures between 1967 and 1973 and salinity varies with freshwater flow and tidal conditions–the approximate salinity distribution at half tide for the average freshwater flow during the survey is included in Figure 6*b*. However, the total picture that emerges is what might be expected of a fish population in a tidal river, with freshwater fish being confined to the upper reaches, while marine fishes penetrate in decreasing numbers upstream to the middle reaches (about 50 km from Teddington). All the freshwater fish which are indigenous to the Thames system, most of the euryhaline species, and an estimated 90% of the species prevalent in the southern North Sea have now been captured in the tidal Thames.

There are reasons to suggest, however, that the fish in the tidal Thames are not all entirely resident in the areas in which they are captured by generating stations. Marine fish become distinctly less common in the lower reaches of the river in the second and third quarters of the year owing to lowering of the dissolved-oxygen level with rise in temperature and decreased freshwater flow, to normal seasonal migration of the individual species, or to both factors. In the first and fourth quarters, however, marine fish penetrate farther up the river, and are caught in greater numbers in the lowest reaches. High spring tides, especially when backed with a strong northeasterly wind, result in more marine fish in the lower reaches and stronger penetration by them upriver. Exceptional conditions of tide and wind occurred just before the capture of haddock at Barking generating station in January 1970, although the presence of this species in the Thames estuary and southern North Sea in 1969 and 1970 (as also in 1966) was due to other factors.

To some extent, therefore, many of the marine fish, and probably such invertebrates as the brown shrimp and occasional swimming crabs, are carried passively in the tidal excursion of the river. In this way these organisms would avoid many of the physiological stresses due to changes in salinity, temperature, and dissolved oxygen inherent in living in a polluted estuary.

The freshwater fish likewise must be seen as inhabiting a water mass rather than a given reach of river. Again there is evidence that in times of high freshwater flow more fish occur in the region of Brunswick Wharf and Blackwall Point than under normal flow conditions, especially on the late ebb and first half of the flood tide. However, with the average tidal excursion of 11-14 km it is clear that for fish to occur in numbers in this region they must be living within this distance of it continually, which implies that populations are established in the vicinity of London Bridge if not farther downstream. The effective limit to the range of the freshwater species is the vicinity of Barking and Dagenham, for although some have been captured farther downstream, they are few in number and species, and their occurrence could be accounted for by the downstream carriage of impaired or sickly fish by the seaward flow of the river.

The greatest interest, however, centers on the third element of the fish fauna, the presence of euryhaline, and especially the migratory, species. As might be expected, they are approximately equally represented at each of the paired collecting sites.

Some nonmigratory species, such as the three-spined stickleback, have been found at all stations, and there is little doubt that this species in particular now lives throughout the tidal Thames. Others, such as the ten-spined stickleback and the common goby, have occurred only in the Barking-Dagenham area, and it is possible that they are locally established in the adjacent confluence of the tributary River Beam and the Thames (53 km from Teddington). Both species occur in great numbers well within tidal influence in another small tributary, the Mardyke, only 1-2 km from its confluence with the Thames (at 60 km).

Of the migratory species the most abundant and most widely distributed is the freshwater eel. Eels have been captured at all the generating stations involved in the 1967-73 survey; they are also frequently caught by anglers along the entire length of the tidal Thames and have been caught in seine nets in the upper reaches of the estuary. The eel seems to have been found even in the days when the river was badly polluted (Wheeler, 1958), but it was not known whether this population was recruited by a run of elvers up the river, or whether they came from other rivers through the canal system (or even to the headwaters of the Thames system overland), or were the result of stocking with elvers. However, in April and May 1968 elvers were caught in numbers at West Thurrock generating station; in April 1972 a number of elvers were reported (D. Solomon, pers. comm.) at Hammersmith; in May-June 1972 six elvers (65-80 mm long) were caught at Ford's works, Dagenham; and in 1972 and 1973 many elvers were found at low tide in the Richmond area under stones, on the shore. Also in 1972 and 1973 transparent "glass eels," the newly metamorphosed elvers, were captured in the lower reaches of the Mardyke. These observations all tend to show that the eel population is now maintained by direct migration of elvers from the sea.

The flounder is the only flatfish to enter fresh water in Europe, and the Thames was famous for the wealth of flounders it contained until the nineteenth century. There are reports of its presence in the river in the Chiswick region in the 1920s (see Wheeler, 1969*a*), but these are unconfirmed. Flounders were captured in the West Thurrock–Littlebrook region in both 1967 and 1968, and in the second quarter of 1969 one was captured at Fulham. Since that date they have been captured at all the collecting sites between West Thurrock and Fulham, and by netting up to the head of the estuary at Teddington. The flounder is now extremely abundant in the upper reaches of the tidal Thames, as many as 63 being caught between September and November 1972 at Fulham generating station alone, while 31 were caught in a single sweep of a seine net at Richmond in July 1973. The particular interest of the flounder is that it breeds in the sea at depths of 27-55 m, mainly in March to April (Wheeler, 1969*b*) and as it is doubtful whether it spawns within the 100 km of the Thames Estuary as here defined, all the fish captured have thus migrated presumably as young fish from the sea, some for distances of 100 km or more.

An anadromous migratory fish, the smelt, has also reestablished itself in the river, if anything sooner than the eel and the flounder, but not in so great numbers. The smelt enters rivers in spring to spawn in fresh water within tidal influence. The first captures in the tidal Thames were made late in 1967 at West Thurrock and in the second quarter of 1968 at Fulham and Blackwell Point generating stations. Since then this fish has occurred each year in the river and from 1967 to 1970 in increasing numbers at the upstream collecting sites (i.e., in the upper 50 km of the estuary) as follows: 1968—2 specimens, 1969—4, 1970—11. Two smelt were captured in 1971, 7 in 1972, 3 in 1973. All, except one specimen captured in the upper 50 km, were sexually mature (20-25 cm long) with well-developed gonads often in spawning condition. The exception was a specimen of 7-cm body length caught at Fulham in the first quarter of 1972. Its size would suggest that it was the progeny of breeding in the Thames, but no other young smelt or eggs have been found in the river. However, this species is known to breed in the estuaries of the Rivers Crouch, Roach, and Blackwater on the Essex coast, all associated with the outer Thames Estuary, and other immature smelt (9-10-cm body length) were caught at West Thurrock in June 1972.

Among other euryhaline fish in the tidal Thames, considerable interest centers on the salmonids, both trout and salmon. During the period of the 1967-73 survey, specimens of migratory (sea) trout were captured at Deptford (1971), Tilbury (1971, 1972, 1973), and Northfleet (1972), respectively 38, 70, and 73 km below Teddington. Another was captured by an angler at Teddington (1971). Since then several sea trout have been captured by anglers in the mouth of the river or caught at Tilbury generating station (1974, 1975). The origin of these migratory trout is obscure, but it might be suggested that they came from other rivers in southeast England where sea trout are indigenous. Accordingly, no great significance can be placed on their occurrence other than that being alive when captured it is obvious they were not deterred by the quality of the water from entering the Thames. The capture of a salmon in November 1974 at West Thurrock generating station has similar significance in that it shows that such species which normally require water of good condition can now survive in the Thames.

The collections of fish made during the 1967-73 survey show that the fish fauna as a whole is still affected by the polluting discharges to the river. During the second and third quarters of each year the total catches of freshwater fish show a considerable decline in the vicinity of Brunswick Wharf, Blackwell Point, Barking, and Dagenham. This was particularly noted during the warm summer of 1972. The effect of warm, dry summers on the dissolved oxygen in the river has been noted already (p. 79). Those fish which do occur in the vicinity of the oxygen sag during these quarters are mainly the hardy, adaptable species physiologically tolerant of low dissolved oxygen levels, or able to adapt their behavior to survival in poor conditions. Of the former group the freshwater eel, roach, carp, and goldfish have all occurred widely in the middle reaches of the river even in conditions of low dissolved oxygen. Other fish, such as the three-spined stickleback and bleak, although not tolerant of low dissolved-oxygen levels, are able to survive by swimming at the surface, taking advantage of the well-oxygenated layer at the air/water interface.

Other Aquatic Organisms

The improved quality of water in the tidal Thames also has resulted in the appearance of a considerable invertebrate fauna in the lower reaches. Unfortunately, no comparative data are available to indicate the poverty of the fauna at a time when the river was anaerobic, and it can only be inferred from that fact and the lack of any information in the literature that (other than tubificids) invertebrates were, like fish, absent.

Captures of the brown shrimp (*Crangon crangon*) at West Thurrock generating station were noted from early 1967 and through 1968-69 formed the basis of a study by Huddart and Arthur (1971). These authors found that this shrimp was absent during the early summer months. More recent observations by one of us (A. W.) show that this shrimp is present in small numbers throughout the summer months at West Thurrock and occurs in great abundance during the remainder of the year. Shrimps were captured at Barking and Dagenham in late 1972, and over 1000 specimens were captured at Dagenham in September-October 1973. Whereas Huddart and Arthur found only one berried female in the total of 42,370 shrimps they examined in 1967-69, in the December 1972 and January 1973 samples from West Thurrock berried females were numerous (a total of 153 in a random sample of 700 in the latter month). This suggests that berried females have become more numerous in recent years. Other decapod crustaceans, for example the shore crab (*Carcinus maenas*) and swimming crab (*Portunus* sp.), have occurred at West Thurrock, the former being very numerous there and also occurring at Dagenham.

In addition to shrimps, Huddart and Arthur reported the jellyfishes *Aurelia aurita* and *Chrysaora isosceles* to be present at West Thurrock during 1967-69, the former being abundant. *Aurelia* occurred in some numbers in the vicinity of Brunswick Wharf in June and July 1972 and 1973. The ctenophore *Pleurobrachia pileus* also was recorded at Dagenham on a number of occasions in the spring and summer of 1973. Such planktonic species might be supposed to occur in the river to some extent owing to passive drift in tidal inflow; it is evident, however, that they can tolerate the condition of the tidal Thames in its present state.

The dramatic, and welcome, return of wildfowl to the lower reaches of the tidal river has been commented upon by Grant et al. (1973). It closely parallels that observed for the fish fauna. The lower reaches of the river were not especially well known as a haunt of ducks or waders, except that in exceptionally severe winters flocks of up to 500 ducks could be seen (as in 1962-63). Within recent years the number of ducks seen has increased, so that peak counts in the London-Tilbury region in 1971-72 showed 1174 mallard, 730 teal, 4000 pochard, and 1349 shelduck, with lesser numbers of other species. Some analysis of gut contents of a number of species showed that most had been feeding on tubificid worms, which are abundant in the middle reaches of the Thames. The reason for the increase in the numbers of birds is not, however, due solely to a new or increased food source in the supply of tubificid worms, for these have been present in the benthos for many years. Although it has occurred during the period of the decrease of pollution in the river, the association is probably more complex than is the direct relationship involving fish and the quality of the water.

It is ironic in a sense that the improved water quality which has resulted in the restoration of an aquatic fauna has also brought problems to industrial water users in the shape of fouling organisms. Hydroids, such as *Sertularia argentea* and *Eudendrium* sp., have been found growing in intake culverts in the Gravesend and West Thurrock regions in such quantities as to cause local blockage on intake screens. As a result it has proved necessary to chlorinate the water at certain intakes.

Despite these minor disadvantages the return of fish, invertebrates, and wildfowl to the tidal Thames has been a major advance in the restoration of an aquatic ecosystem. It illustrates that a grossly polluted river can be restored sufficiently well to support diverse fauna, to conserve wildlife in urban areas, and to provide recreational potential for human inhabitants.

Other British Estuaries

It may be of interest to conclude by referring to the condition of other British estuaries, which have been reviewed by Gameson (1973). The definition of an estuary is necessarily arbitrary, and for the purpose of the review an *estuary* was defined as "a body of water (at least 5 km long) where there is a discernible effect of the sea resulting in a rise and fall in water level under average tidal conditions of river flow and tide, and where there is a discernible effect of one or more rivers resulting in salinity variations over the average tidal cycle." The seaward limit is often difficult to define.

There were found to be 66 estuarine systems in England, 16 in Scotland (excluding the west coast north of the Clyde), 18 in Wales, and 4 in Northern Ireland. The excluded Scottish coast includes some 30 sea lochs which may or may not be estuaries as defined above, but there was too little information readily available on which to decide. The physical characteristics of these sea lochs tend to be very different from those generally associated with British estuaries, so that it seemed both reasonable and convenient to exclude them; their total length exceeds 400 km and they are nearly all virtually free of pollution. Some of the estuarine systems considered have tidal tributaries which may be classed as estuaries in their own right, and their inclusion raises the total number of estuaries from 104 to 134.

The total length of these 104 systems was estimated to be 3170 km, and the median 22 km. Only four have lengths exceeding 100 km, by far the largest being the Humber system with some 340 km of tidal waters and a tidal run of 147 km in the Trent-Humber. The length of an estuary depends, of course, to some extent on the tidal range. Average ranges vary widely in Britain–from 0.5 m on that part of the Scottish coast which is nearest Ireland to 9.6 m in the Severn Estuary. Average summer freshwater flows vary from less than 0.1 m^3s to about 100 m^3/s, with a total of about 1100 m^3/s and a median about 3.5 m^3/s.

In 1971 (the year to which the review applies) chemical surveys were made in only 76 of the 134 estuaries, though earlier surveys had been made in 35 others. For some of the smaller estuaries information was sparse, but 43 estuaries were reported unequivocally as "clean," 27 as "slightly polluted," 16 as "moderately polluted," 16 as "grossly polluted," and more than one classification was given for the remaining 32. It is thought that anaerobic conditions may have existed at some time during the year in 20 estuaries (or 15% of the total).

Among the questions asked of the River Authorities and other organizations supplying information was how the general condition of each estuary compared with that which had existed 5 and 20 years previously; the replies, summarized in Table 2, were distinctly encouraging.

All but 20 of the estuaries were reported to contain established fish populations: 19 contained freshwater species (predominantly roach and dace), 70 contained salmonids (43 with salmon and 35 with sea trout),

Table 2. Condition of 134 British estuaries in 1971 as compared with those in two earlier years (in percentages of total)

Earlier year	Better	Similar	Worse	Not known
1966	19	74	5	2
1951	33	47	9	11

and 69 contained marine species, 42 of these including flatfish (predominantly flounder) and 44 including roundfish (predominantly mullet and bass).

Acknowledgments

Much of the material for this chapter has been taken from earlier publications. The section headed "Early History" is an abridged and rearranged version of Chapter 5 in Department of Scientific and Industrial Research (1964), in which the references to the original publications will be found. This chapter was written by the late Nora H. Johnson, who, with Gameson, edited the whole volume. The permission of Her Majesty's Stationery Office to reproduce this material is gratefully acknowledged.

The section on Twentieth-Century Changes in Pollution is based on DSIR (1964) and Gameson et al. (1973), supplemented by information published in Greater London Council (1972, 1974a, and 1974b) and further details kindly supplied by the Thames Water Authority and the West Kent Main Sewerage Board. The section on Dissolved Oxygen was derived largely from DSIR (1964), Gameson (1964), Gameson et al. (1965), and Gameson et al. (1973), additional data again being made available by the Thames Water Authority.

The section on fish and other aquatic animals is based largely on the work carried out (by A. W.) with the cooperation of the Central Electricity Generating Board, many of the data of which are as yet unpublished. Acknowledgment is due to the station superintendents and staffs of the generating stations concerned, and to the manager and staff of the power station at the Ford Motor Company's Works, Dagenham.

Literature Cited

Department of Scientific and Industrial Research. 1964. Effects of Polluting Discharges on the Thames Estuary. Water Pollut. Res. Tech. Paper, No. 11, H. M. Stationery Office, London. xxviii + 609 pp.

Gameson, A. L. H. 1964. Pollution of London's river. New Sci. 22:295-298.

_____. 1973. Estuaries of the United Kingdom. Pages 3-13 *in* A. L. H. Gameson, ed. Mathematical and Hydraulic Modelling of Estuarine Pollution. Water Pollut. Res. Tech. Paper, No. 13, H. M. Stationery Office, London.

Gameson, A. L. H., M. J. Barrett, and W. S. Preddy. 1965. Predicting the condition of a polluted estuary. Pages 167-177 *in* Proc. 2nd Int. Conf. Water Pollut. Res., Tokyo, 1964, and Int. J. Air Water Pollut. 9:655-664.
Gameson, A. L. H., M. J. Barrett, and J. S. Shewbridge. 1973. The aerobic Thames Estuary. Pages 843-851 *in* Proc. 6th Int. Conf. Water Pollut. Res., Jerusalem, 1972.
Grant, P., J. Swift, and J. Harrison. 1973. The return of wildfowl to the inner Thames. Pages 44-52 *in* Annual Report of the Wildfowlers' Association of Great Britain and Ireland, 1972-1973.
Greater London Council. 1972. Pages 20-21 *in* Annual Report of the Scientific Adviser 1971.
_____. 1974*a*. Pages 36-37 *in* Annual Report of the Scientific Adviser 1972.
_____. 1974*b*. Pages 34-35 *in* Annual Report of the Scientific Adviser 1973.
Huddart, R. and D. R. Arthur. 1971. Shrimps and whitebait in the polluted Thames Estuary. Int. J. Environ. Stud. 2:21-34.
Wheeler, A. 1958. The fishes of the London area. London Nat., pp. 80-101.
_____. 1969*a*. Fish-life and pollution in the Lower Thames: a review and preliminary report. Biol. Conserv. 2:25-30.
_____. 1969*b*. The Fishes of the British Isles and North West Europe. Macmillan, London. 613 pp.

Appendix

Total List of Fish Recorded from Tidal Thames in 1967-74

Nomenclature and systematic order follow Wheeler (1969b) and fish captured at generating stations:

col. 1,2 Fulham (Sept. 1967–Dec. 1973) and Lombard Road (Feb. 1969–April 1972)
col. 3,4 Brunswick Wharf and Blackwall Point (Sept. 1967–Dec. 1973)
col. 5,6 Barking (Sept. 1967–Dec. 1973) and Dagenham (May 1972–Dec. 1973)
col. 7,8 West Thurrock (Sept. 1967–Dec. 1973) and Littlebrook (Jan. 1970–Dec. 1973)

Numbers in columns indicate

0 - none caught
1 - a single catch
2 - two or more catches

Indentation of common name of species indicates the three elements:
Freshwater
Euryhaline
Marine

Fish		1,2	3,4	5,6	7,8
Lamprey	*Petromyzon marinus* L., 1758	0	0	2	2
Lampern	*Lampetra fluviatilis* (L., 1758)	0	0	0	2
Smooth Hound	*Mustelus asterias* Cloquet, 1819	0	0	0	1
Sting Ray	*Dasyatis pastinaca* (L., 1758)	0	0	0	1
Anchovy	*Engraulis encrasicolus* (L., 1758)	0	0	1	2
Twaite Shad	*Alosa fallax* (Lacépède, 1803)	0	0	0	2
Pilchard	*Sardina pilchardus* (Walbaum, 1792)	0	0	0	2
Sprat	*Sprattus sprattus* (L., 1758)	0	2	2	2
Herring	*Clupea harengus* L., 1758	0	0	2	2
Salmon	*Salmo salar* L., 1758	0	0	0	0
Trout	*Salmo trutta* L., 1758	2	2	0	0
Smelt	*Osmerus eperlanus* (L., 1758)	2	2	2	2
Pike	*Esox lucius* L., 1758	2	2	1	0
Carp	*Cyprinus carpio* L., 1758	2	1	0	0
Barbel	*Barbus barbus* (L., 1758)	2	0	0	0
Gudgeon	*Gobio gobio* (L., 1758)	2	1	1	1
Tench	*Tinca tinca* (L., 1758)	2	2	1	0
Crucian Carp	*Carassius carassius* (L., 1758)	0	1	0	0
Goldfish	*Carassius auratus* (L., 1758)	2	2	1	0
Bleak	*Alburnus alburnus* (L., 1758)	2	2	0	0
Bream	*Albramis brama* (L., 1758)	2	2	2	1
Minnow	*Phoxinus phoxinus* (L., 1758)	0	0	0	0
Rudd	*Scardinius erythrophthalmus* (L., 1758)	2	0	0	0
Roach	*Rutilus rutilus* (L., 1758)	2	2	2	2
Chub	*Leuciscus cephalus* (L., 1758)	2	1	0	0
Dace	*Leuciscus leuciscus* (L., 1758)	2	2	1	0
Loach	*Noemacheilus barbatulus* (L., 1758)	0	0	0	0
Eel	*Anguilla anguilla* (L., 1758)	2	2	2	2
Conger Eel	*Conger conger* (L., 1758)	0	0	0	2
Skipper	*Scomberesox saurus* (Walbaum, 1792)	0	0	0	1
Garfish	*Belone belone* (L., 1758)	0	0	0	0
Broad-nosed Pipefish	*Syngnathus typhle* L., 1758	0	0	0	0
Greater Pipefish	*Syngnathus acus* L., 1758	0	0	2	2
Nilsson's Pipefish	*Syngnathus rostellatus* Nilsson, 1855	0	0	2	2

Fish		1,2	3,4	5,6	7,8
Whiting	*Merlangius merlangus* (L., 1758)	0	0	0	2
Bib	*Trisopterus luscus* (L., 1758)	0	0	2	2
Poor Cod	*Trisopterus minutus* (L., 1758)	0	0	0	2
Cod	*Gadus morhua* L., 1758	0	0	0	2
Haddock	*Melanogrammus aeglefinus* (L., 1758)	0	0	2	2
Pollack	*Pollachius pollachius* (L., 1758)	0	0	0	1
Tadpole-fish	*Raniceps raninus* (L., 1758)	0	1	0	2
Four-bearded Rockling	*Rhinonemus cimbrius* (L., 1766)	0	0	0	0
Five-bearded Rockling	*Ciliata mustela* (L., 1758)	0	0	0	0
Northern Rockling	*Ciliata septentrionalis* (Collett, 1875)	0	0	0	0
John Dory	*Zeus faber* L., 1758	0	0	0	2
Bass	*Dicentrarchus labrax* (L., 1758)	0	2	2	2
Ruffe	*Gymnocephalus cernua* (L., 1758)	2	2	0	0
Perch	*Perca fluviatilis* L., 1758	2	2	2	1
Scad	*Trachurus trachurus* (L., 1758)	0	0	2	2
Red Mullet	*Mullus surmuletus* L., 1758	0	0	0	2
Ballan Wrasse	*Labrus bergylta* Ascanius, 1767	0	0	0	0
Lesser Sand-eel	*Ammodytes tobianus* L., 1758	0	2	2	2
Raitt's Sand-eel	*Ammodytes marinus* Raitt, 1934	0	0	2	2
Greater Sand-eel	*Hyperoplus lanceolatus* (Lesauvage, 1824)	0	0	0	2
Mackerel	*Scomber scombrus* L., 1758	0	0	0	0
Transparent Goby	*Aphia minuta* (Risso, 1810)	0	0	1	2
Common Goby	*Pomatoschistus microps* (Krøyer, 1840)	0	0	2	0
Sand Goby	*Pomatoschistus minutus* (Pallas, 1770)	1	2	2	2
Black Goby	*Gobius niger* L., 1758	0	0	0	1
Dragonet	*Callionymus lyra* L., 1758	0	0	0	2
Viviparous Blenny	*Zoarces viviparus* (L., 1758)	0	0	0	0
Thick-lipped Grey Mullet	*Chelon labrosus* (Risso, 1826)	0	0	0	2
Thin-lipped Grey Mullet	*Liza ramada* (Risso, 1826)	0	1	1	2
Grey Gurnard	*Eutrigla gurnardus* (L., 1758)	0	0	0	2
Red Gurnard	*Aspitrigla cuculus* (L., 1758)	0	0	0	2
Tub Gurnard	*Trigla lucerna* L., 1758	0	1	2	2
Bullhead	*Cottus gobio* L., 1758	1	0	0	0
Hooknose	*Agonus cataphractus* (L., 1758)	0	0	0	2
Lumpsucker	*Cyclopterus lumpus* L., 1758	0	0	0	1
Sea Snail	*Liparis liparis* (L., 1766)	0	0	0	2

Fish		1,2	3,4	5,6	7,8
Three-spined Stickleback	*Gasterosteus aculeatus* L., 1758	2	2	2	2
Ten-spined Stickleback	*Pungitius pungitius* (L., 1758)	0	0	2	0
Brill	*Scophthalmus rhombus* (L., 1758)	0	0	0	2
Dab	*Limanda limanda* (L., 1758)	0	0	2	0
Plaice	*Pleuronectes platessa* L., 1758	0	0	0	2
Flounder	*Platichthys flesus* (L., 1758)	2	2	2	2
Lemon Sole	*Microstomus kitt* (Walbaum, 1792)	0	0	0	1
Sole	*Solea solea* (L., 1758)	0	0	1	2
Trigger-fish	*Balistes carolinensis* (Gmelin, 1789)	0	0	0	0
Angler	*Lophius piscatorius* L., 1758	0	0	0	0

Recovery of Lake Washington from Eutrophication

W. T. Edmondson

Abstract

Changes in the condition of Lake Washington have promptly accompanied changes in the input of nutrients. In the period 1941-63 the lake received increasing amounts of treated sewage effluent. The abundance of algae increased and the population changed character, most notably with the appearance of *Oscillatoria rubescens* in 1955. Sewage was diverted during the period 1963-68, during which time the proportion of blue-green algae began to decrease. By 1974 the lake seemed to be finishing its response to the diversion of sewage.

Discussion

Lake Washington has been through two episodes of pollution, one of which has been described in some detail (Edmondson, 1972*a*). In the early part of this century, the lake received raw sewage from the city of Seattle, the maximum being from about 50,000 people. This sewage was diverted to Puget Sound by a project starting in the late 1920s and ending in 1936. There appears to be no limnological record of the conditions of the lake during this phase. Apparently much of the motivation for the diversion was based on health considerations, with BOD and the *E. coli* count being the main criteria of condition.

By 1933 the lake appeared to be in acceptable condition (Scheffer and Robinson, 1939). *Anabaena* and other blue-green algae were present, but the authors described the lake as oligotrophic and did not mention nuisance algal conditions. High concentrations of oxygen prevailed in the hypolimnion, and the maximum concentration of total phosphorus was 26 μg/1 with 8 μg/1 of dissolved inorganic phosphate phosphorus in the winter. Unfortunately they did not measure the transparency of the lake by Secchi disc or with any other instrument.

The second episode of pollution began in 1941 with the construction of a secondary sewage treatment plant. By 1954 ten secondary treatment plants had been built. Another limnological study in 1950 showed that while the lake was in a publicly acceptable condition, it was showing distinct signs of eutrophication, relative to 1933 (Comita and Anderson, 1959). The winter phosphate concentration was about twice as high, and the oxygen deficit distinctly higher (Edmondson, 1966). The Secchi transparency ranged between 3.2 and 4.0 m in summer (Edmondson, 1972*b*).

The progressive deterioration of the lake that could be predicted as the result of these observations was confirmed in 1955 by the appearance of *Oscillatoria rubescens.* By 1963 the Secchi transparency was about one meter during the summer, and the abundance of phytoplankton was about 12 times as much as in 1950.

Fortunately, this deterioration had been foreseen and action had been taken to forestall it. A public vote in 1958 provided for construction of a sewerage system that would divert all effluent from Lake Washington to Puget Sound without causing the same kind of problem there. Another part of the project was to collect the raw sewage that was entering Puget Sound at sea level at 70 million gallons per day and put it through one of the treatment plants that was also receiving effluent diverted from the lake. The process of building the trunk sewers and progressively connecting the treatment plants took five years, from March 1963 to 1968, during which time the input of effluent was decreased progressively.

These changes in sewage caused large changes in the nutrient income. In 1957 the total input of phosphorus to the lake was between about 100,000 and 108,000 kg, of which 42,000 kg were contained in sewage effluent and 43,000 kg were carried in streams. The rest was in industrial wastes, combined storm sewer overflow, and septic tank drainage, much of which has now been eliminated. By 1963, the beginning of diversion, the sewage input had decreased because of increases in population. The total input of phosphorus in 1962 was estimated to be about 231,000 kg, of which about 150,000 kg, or 65%, were attributed to sewage sources (Municipality of Metropolitan Seattle, unpublished data). In 1974, with sewage eliminated, the streams brought in about 41,000 kg of phosphorus.

The lake responded promptly to the decrease in nutrient income (Fig. 1). With the first diversion of about one third of the effluent, deterioration was halted, as measured by summer Secchi transparency (see Fig. 3 of Edmondson, 1972*b*). The phosphorus content of the lake decreased sharply from 1963 through 1969, and has shown relatively little change since then. Winter dissolved phosphate showed a distinct minimum in 1972, a most unusual year in that heavy rains caused landslides along the Cedar River which brought much silt into the lake (Fig. 2). Presumably

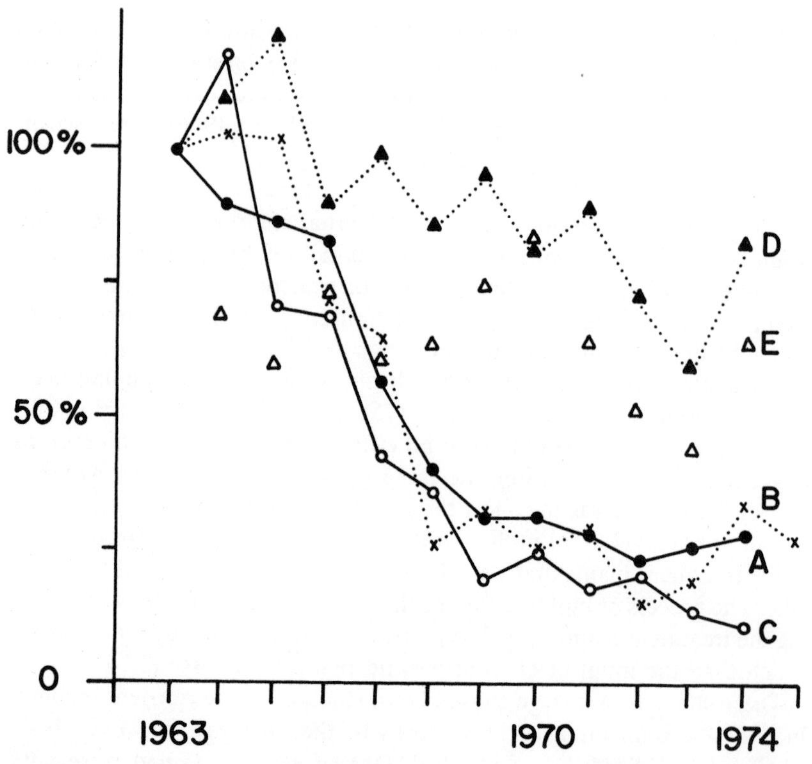

Fig. 1. Recovery of Lake Washington. Mean values for top ten meters. The value for 1963 (in parentheses) is shown as 100%. A = total phosphorus for whole year (perchloric acid digestion; 65.7 μg/l). B = dissolved inorganic phosphate phosphorus, January-March (55.3 μg/l). C = chlorophyll, July-August (34.8 μg/l). D = Nitrate nitrogen, January-March (423 μg/l). E = Carbon dioxide, January-March (4.05 mg/l). Note that the winter values are for a slightly different time from those published earlier (Edmondson, 1970). Sewage diversion started in 1963 and ended in 1968. Winter phosphate phosphorus can exceed the annual mean for total phosphorus when the latter decreases during the summer as in 1974 (see Fig. 2).

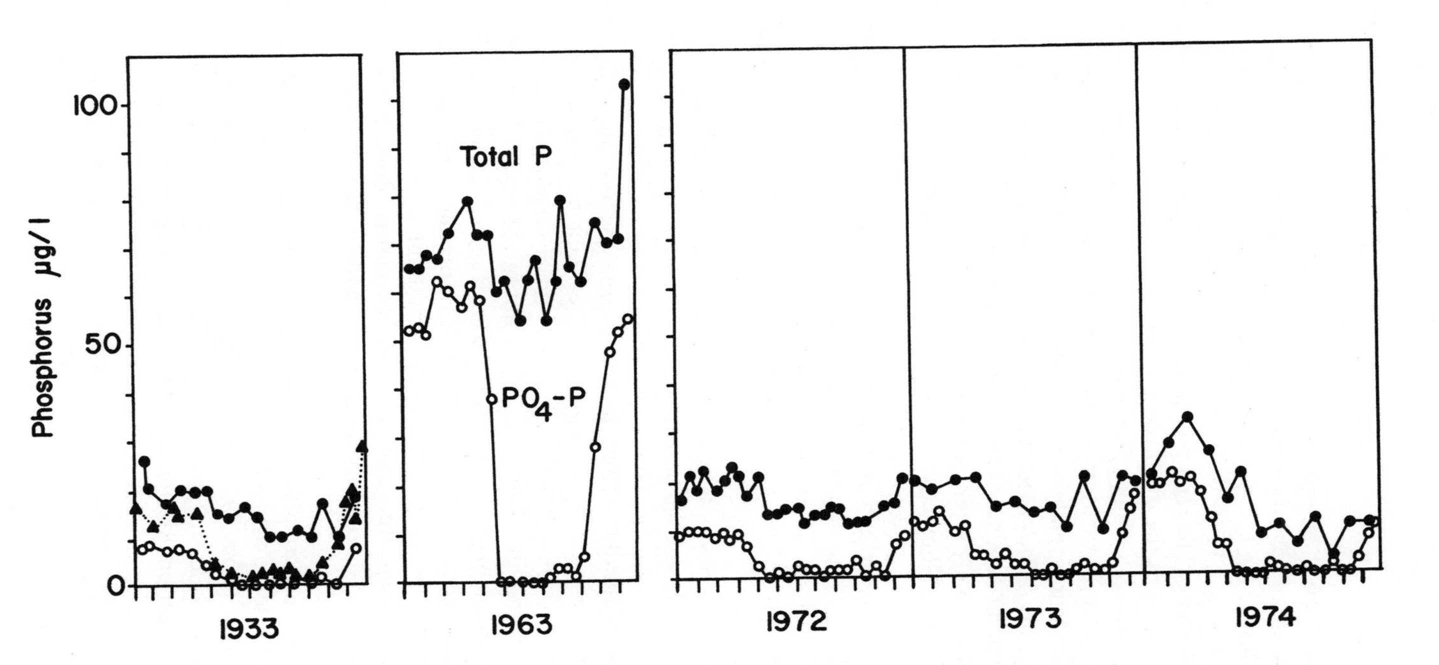

Fig. 2. Concentration of total phosphorus and dissolved phosphate phosphorus in surface water of Lake Washington. Triangles show phosphate for 1950.

the clays adsorbed phosphate and settled. The Secchi transparency in January-March 1972 varied between 2.1 and 5.2 m in contrast to 5.7-7.5 m in 1975. Despite the heavy local rainfall, the total inflow into the lake was not uniquely heavy; so the change in phosphate cannot be attributed solely to flushing. The total phosphorus averaged for the whole year was not as different as the dissolved phosphate.

The abundance of phytoplankton in summer decreased in close proportion to phosphorus during the period of recovery, and has shown a slightly decreasing trend since 1969 (see Fig. 1). Several important changes in the character of the phytoplankton have occurred. *Oscillatoria rubescens* and *O. agardhii* dominated the summer plankton in the middle 1960s but have been seen only rarely and in small quantities since 1972. At the time of maximum enrichment, as much as 98% of the volume of phytoplankton was made up of colonial blue-green algae, largely *Oscillatoria*. By 1971 the percentage had dropped to 64%, but it was up to 89% in 1972. It decreased sharply to 44% in both 1973 and 1974, the rest being mostly a variety of small species of green algae. In the recent years, the filamentous blue-green algae are mostly slender species of *Lyngbya* and *Pseudanabaena*.

In contrast to phytoplankton, the species list of zooplankton had no major changes during the period of eutrophication and recovery. Some of the species showed major changes in abundance over the years, but not in a way systematically related to the degree of eutrophication. More recently, an important change appears to be taking place in that the genus *Daphnia* has reappeared in the lake. *Daphnia* was reported to be common in 1933 (Scheffer and Robinson, 1939), but in the period 1949-71 it was essentially absent from the open water of the lake, only isolated individuals being found occasionally. In 1972 a few specimens of *Daphnia* were seen, and the genus became more abundant in the two following years, being present from June through November. Both *D. thorata* and *D. schoedleri* have occurred consistently, with occasional specimens of *D. ambigua.*

Changes have taken place in the fish population, the most striking being a large increase in the abundance of sockeye salmon *(Onchorhynchus nerka)* starting about 1964 (Ames, 1970). It is difficult to connect the changes in *Daphnia* and salmon directly and simply with each other and with eutrophication and recovery of the lake. The *Daphnia* disappeared before the salmon became abundant and before the lake was strongly eutrophicated. The increase of the salmon did not follow a major increase in abundance of zooplankton. Evaluation of these questions will depend upon future changes in the population and upon experimental and paleolimnological work now in progress.

The question of the relative importance of different nutrients has been given much attention in the literature. Considerations especially relevant to Lake Washington have been given by Edmondson (1970, 1972*a*, 1974*a*). In 1933 dissolved phosphate was removed from the water, leaving an excess of nitrate, but when the lake was receiving much sewage, phosphate was in excess (see Fig. 2 of Edmondson, 1970).

Much concern has been expressed about sediments as a source of nutrients. Sediments may well be a problem in shallow lakes enriched for a long time with a low rate of flushing. But the hypolimnetic phosphorus concentrations in Lake Washington took only about two years to return to those prevailing in 1933 and 1950 (Fig. 3).

It appears that the lake is no longer responding to the diversion of sewage and is now showing changes in response to year-to-year variations in other properties that affect it, such as solar radiation and flushing rate. We intend to study the properties that were used to evaluate the condition of the lake during the period of enrichment long enough that we can see how they respond and to define the endpoint of the Lake Washington experiment.

It is not clear by what standard the condition of the lake should be judged. In previous publications I have used the conditions in 1933 as the standard simply because it was the earliest time of a detailed study of a lake, and it appeared to be in acceptable condition. Yet the lake was still ending its first period of pollution then. A single sampling in 1913 did not provide definite data, although *Anabaena* and *Aphanizomenon* were present (Kemmerer et al., 1924). Presumably the study being made of the lake now will provide the best standard. We are seeing how the lake will perform in an urban setting with storm runoff but deprived of its concentrated sewage effluent. This condition must be different from that prevailing before the arrival of European settlers. Possibly the paleolimnological studies now under way will provide a description of that condition (Edmondson, 1974*b*).

Acknowledgments

The work reported here has been supported by the National Science Foundation and in part by the Environmental Protection Agency since 1973.

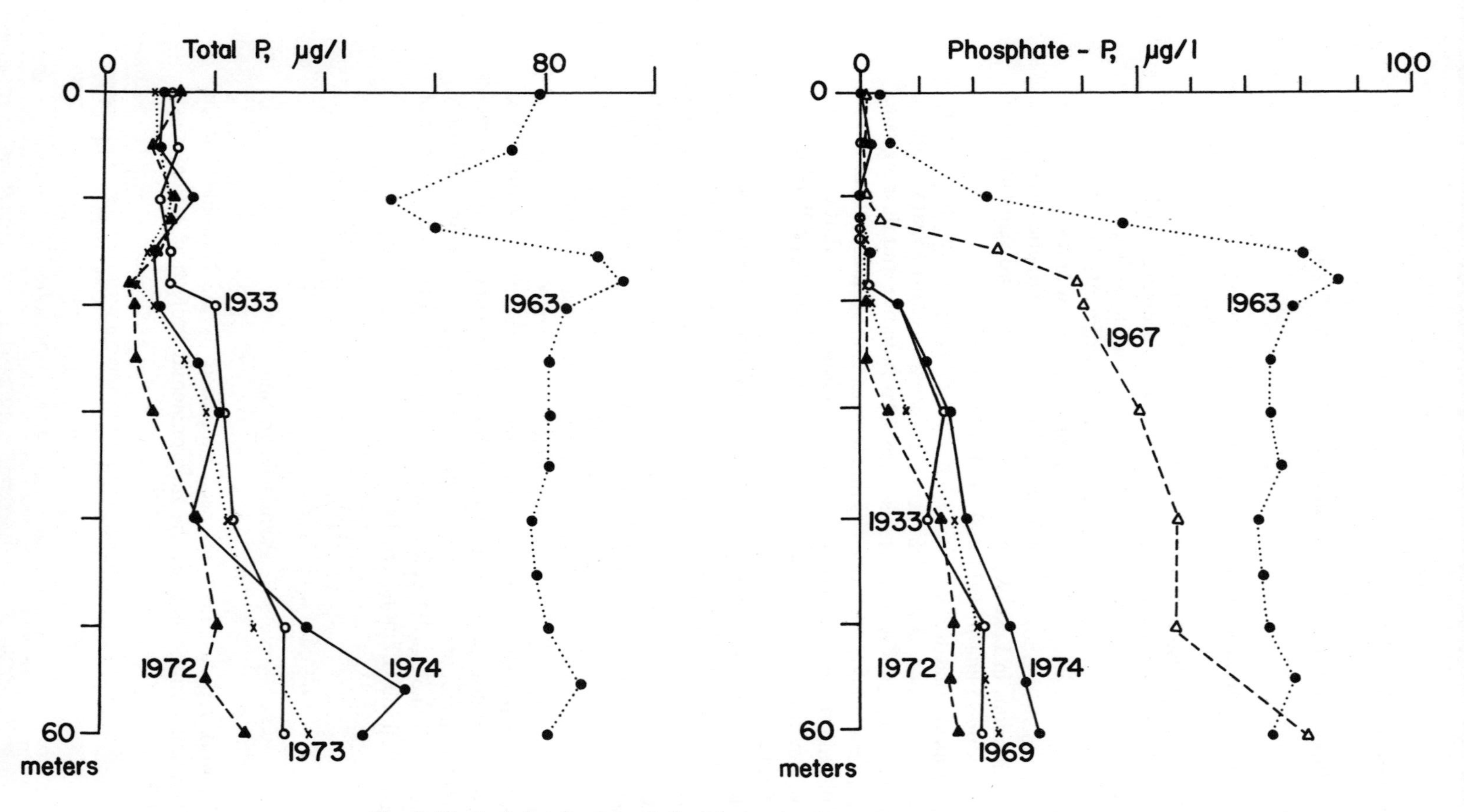

Fig. 3. Vertical distribution of phosphorus in mid-September of several years

Literature Cited

Ames, J. 1970. Lake Washington sockeye salmon. Washington Dept. Game Ann. Rep. 1970. pp. 67-68.

Comita, G. W., and G. C. Anderson. 1959. The seasonal development of a population of *Diaptomus ashlandi* Marsh, and related phytoplankton cycles in Lake Washington. Limnol. Oceanogr. 4:37-52.

Edmondson, W. T. 1966. Changes in the oxygen deficit of Lake Washington. Verh. Int. Verein. Limnol. 16:153-158.

_____. 1970. Phosphorus, nitrogen and algae in Lake Washington after diversion of sewage. Science 196:690-691.

_____. 1972*a*. Nutrients and phytoplankton in Lake Washington. Pages 172-173 *in* G. Likens, ed. Nutrients and Eutrophication, Special Symposia No. 1. Am. Soc. of Limnol. and Oceanogr., Lawrence, Kansas.

_____. 1972*b*. The present condition of Lake Washington. Verh. Int. Verein. Limnol. 18:284-291.

_____. 1974*a*. Review of The Environmental Phosporus Handbook. Limnol. Oceanogr. 19:369-375.

_____. 1974*b*. The sedimentary record of the eutrophication of Lake Washington. Proc. Natl. Acad. Sci. 71:5093-5095.

Kemmerer, G., J. F. Bovard, and W. R. Boorman. 1924. Northwestern lakes of the United States: biological and chemical studies with reference to possibilities in production of fish. Bull. U.S. Bureau of Fish. 39:51-40.

Scheffer, V. B., and R. J. Robinson. 1939. A limnological study of Lake Washington. Ecol. Monogr. 9:95-143.

Additional Literature

Edmondson, W. T. 1961. Changes in Lake Washington following an increase in the nutrient income. Verh. Int. Verein. Limnol. 14:167-175.

_____. 1963. Pacific Coast and Great Basin. Pages 371-392 *in* D. G. Frey, ed. Limnology in North America. Univ. of Wisconsin Press, Madison.

_____. 1968. Water quality management and lake eutrophication: the Lake Washington case. Pages 139-178 *in* T. H. Campbell and R. O. Sylvester, eds. Water Resources Management and Public Policy. Univ. of Washington Press, Seattle.

_____. 1973. Lake Washington. Pages 281-298 *in* C. R. Goldman, James McEvoy III, and Peter J. Richerson, eds. Environmental Quality and Water Development. Freeman, San Francisco. (Originally published as a report to the National Water Commission).

Shapiro, J., W. T. Edmondson, and D. E. Allison. 1971. Changes in the chemical composition of sediments of Lake Washington, 1958-1970. Limnol. Oceanogr. 16:437-452.

Thut, R. 1969. A study of the profundal bottom fauna of Lake Washington. Ecol. Monogr. 39:79-100.

Recovery and Restoration of Damaged Lakes in Sweden

Sven Björk

Lakes Damaged by Water Level Lowering— The Lake Hornborga Project

During the nineteenth and the beginning of the twentieth century, the water levels of most lakes in southern Sweden were lowered to obtain arable land. In northern Sweden the water levels have been either raised or regulated in the recent exploitation of many waters for hydroelectric power. The damage caused by the water level changes is quite serious within certain areas of the country. However, no efforts comparable to those made for improvement of the conditions in polluted lakes have yet been begun for restoring the several damaged lowered lakes. Until now only a few experimental studies for restoring lowered lakes have been made. The most well-known of these is the Lake Hornborga project.

Lake Hornborga is situated on a plain, composed in part of silurian deposits (limestones, etc.), between the great lakes Vänern and Vättern (Fig. 1). Until the beginning of the nineteenth century the lake area was about 30 km^2 and the maximum depth about 3 m. The lake was surrounded by vast wetlands. The catchment area is 616 km^2, and the water renewal time on an annual basis could be calculated to two months. However, at normal high water flow it was 22 days and at normal low water flow 560 days. In spite of its shallowness and its status as a highly productive water, Lake Hornborga remained in a stable condition. The growth rate of the sediment was kept low, as there was an effective transport from the lake of organic matter produced within the lake. The main inlet discharged water on the eastern long side of the lake, and the outlet was on the opposite side, giving strong water movements across the lake.

As the lake has its greatest length in the same direction as the prevailing winds, the conditions necessary for strong water movements along the long axis were also favorable. In winter the ice pushed against the shores, and in spring the ice floes piled up in the littoral zone. Thus, the bulldozer effect of the ice kept this zone free from macrophyte vegetation within the exposed reaches. The emergent vegetation developed along shores sheltered from winds, waves, and ice.

Since 1802 the water level of Lake Hornborga has been lowered five times. The most severe drainings were carried out in 1904-11 and 1932-33. After the latest one the lake was laid dry in summer. The entire draining project was an economic failure as it was impossible to cultivate the calcareous sediments, and the lake area is easily flooded.

Until the early twentieth century, Lake Hornborga was considered one of the most famous waterfowl lakes in northwestern Europe. It was also highly productive with respect to fish and crayfish. The diversified and smoothly functioning ecosystem of the undamaged lake totally collapsed due to the lowerings (Fig. 2). The whole lake was invaded by emergent vegetation: by sedges (above all *Carex acuta*), willow bushes, and the common reed (*Phragmites communis*). In 1968 about 12 km^2 were covered by reed monoculture (see Fig. 8).

As the incoming water was conducted directly to the outlet through canals (see Fig. 2), the water movement and transport system was rendered inoperable. Coarse organic detritus–impregnated with silica–produced in the macrophyte areas was deposited there. The growth rate of the unconsolidated sediment was roughly on the order of magnitude of 10 mm/year over large parts of the lake area. Of course, fishery was spoiled and the value of the lake as a nesting site and a resting place for migratory birds was rapidly declining. The Swedish government investigated the value of Lake Hornborga with respect to nature conservancy. It was found that "a restoration of Lake Hornborga is one of the most urgent conservancy projects at present in our country–at least as far as the safeguarding of a special nature conservancy subject is concerned" (National Swedish Nature Conservancy Board, 1967). After being ordered by the government to investigate the possibilities of restoring Lake Hornborga to the status of a waterfowl lake, the National Swedish Environment Protection Board organized a broad study of the complex of man-made problems (National Swedish Environment Protection Board, 1973).

As a result of lowering the water level, the lake ecosystem had been irreversibly damaged. Simply raising the water level would not have given satisfactory results because of the severely damaged lake bottom. This had been interwoven by roots of, above all, sedge and common reed. After a raising of the water level, the root felt within the sedge-covered and part of the felt within the reed-covered areas would have been lifted to the water surface. The cause of this phenomenon was gas (methane) formation within and beneath the felt. Furthermore, the thick layer of coarse detritus within the reed monoculture would have constituted an unsuitable substrate for the development of submerged vegetation and bottom fauna. Another negative factor was that the oxygen consumption of the accumulated and voluminous detritus masses would have brought about oxygen deficiency during periods with ice cover.

The goal of a restoration is to recover a waterfowl lake. Large-scale field experiments made it possible to work out methods to solve the ecological problems in an economically favorable way. In Lake Hornborga the root felt of the sedge-covered areas was impossible to remove. However, the reed root felt could be destroyed, and thus it was possible to expose a clean bottom of consolidated calcareous mud.

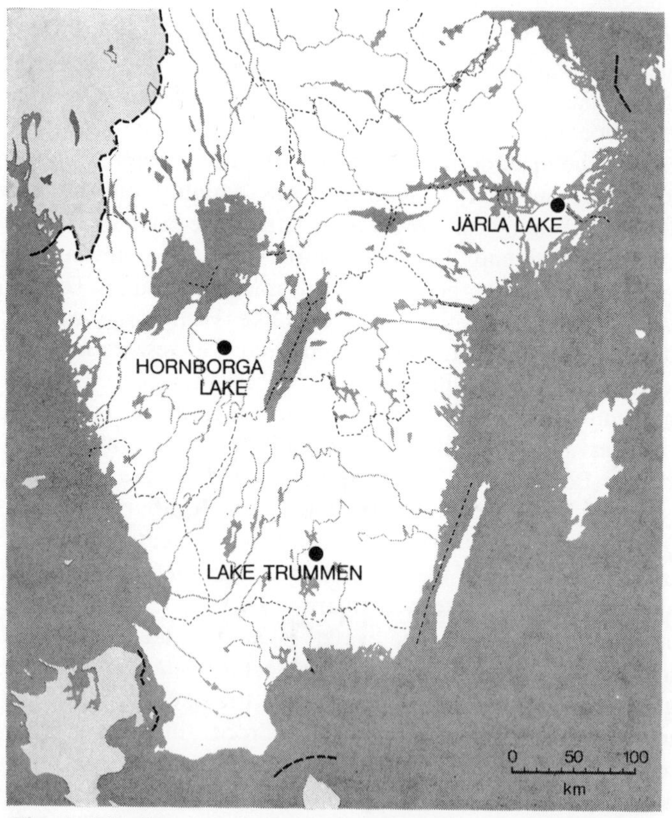

Fig. 1. The location in South Sweden of Lake Hornborga, Lake Trummen, and Lake Järla, the subjects of a Swedish Lake restoration research program. (From Björk, 1972)

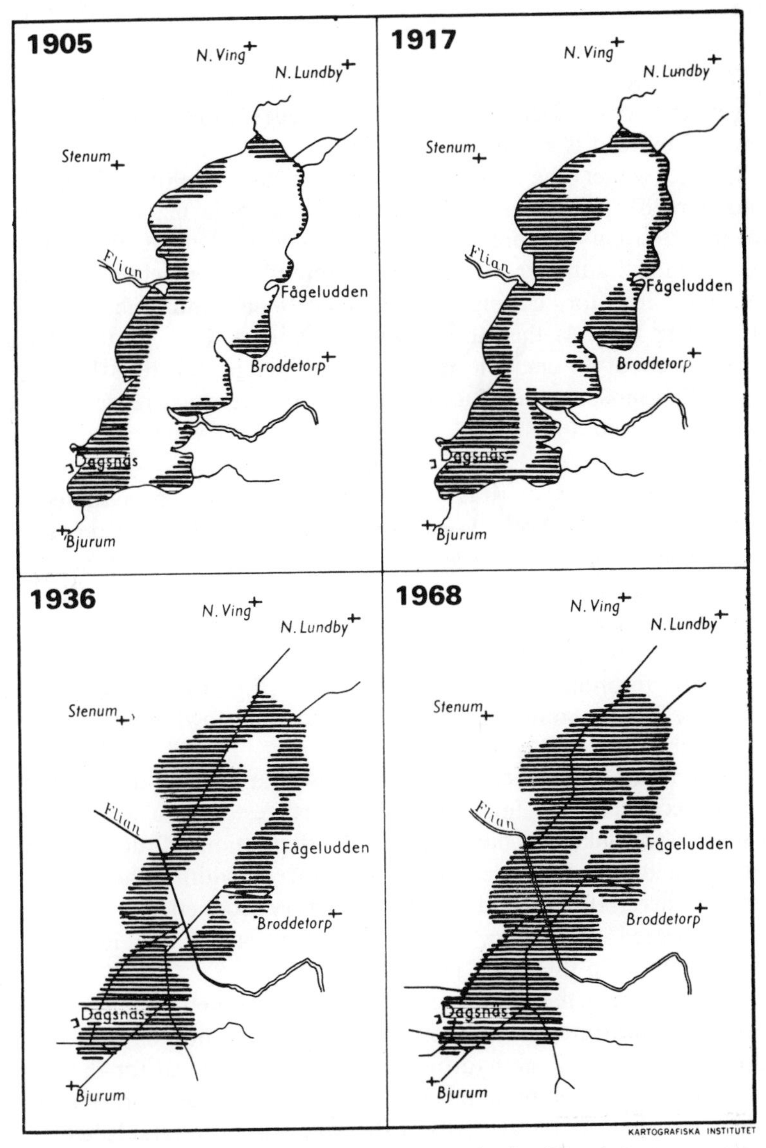

Fig. 2. Lake Hornborga (area about 30 km²). The last two water level lowerings (1904-11 and 1932-33) have caused an almost complete overgrowth by emergent macrophyte vegetation. At the beginning of the restoration investigation (1968) the emergent vegetation consisted primarily of stands of common reed (about 12 km²) and areas of sedge or sedge mixed with stands of willow bushes. The small open waters were choked with charophytes. (From Björk, 1970)

The procedure for achieving the change from production of emergent vegetation to production of submerged vegetation in the areas now covered by reed is as follows (Fig. 3).

In the winter the dry stems are cut by amphibious harvesters and burned on the ice. In the spring the stubble mats are shortened to about 40 cm by pontoon-equipped mowers. During the low-water period in the summer and autumn, amphibious machines are again utilized, first for cutting the green shoots and then for rotor cultivating (Fig. 4) the stubble mats and root felts, the thickness of which is about 40 cm. The requisite time per hectare is 8-10 machine hours. Enormous masses of accumulated coarse detritus are loosened from the bottom and transported by the spring high water to the shores, where they are burned in the summer.

With the coarse detritus out of the way, the consolidated mud again becomes the bottom, and the reed monoculture is replaced by underwater vegetation (Fig. 5) with charophytes, *Myriophyllum, Potamogeton,* and *Utricularia.* The microbenthos and periphyton communities become rich in species and quantitatively well developed. The same trend toward higher diversity and greater quantities has been recorded in the bottom fauna (Fig. 6). In the prepared areas the bottom animals are available to waterfowl, which is not the case in the untreated reed stands.

As a result of the structural changes that can be brought about in the ecosystem, the function will also be restored to working order. The lowered lake is characterized by deposition of detritus, which means that the organic matter produced within the lake area remains there. After the redirection of primary production from emergent to submerged vegetation, a decomposition of organic matter takes place again as in the intact lake (see Fig. 6).

After repairing the damaged bottom, it is planned to raise the water level to a maximum depth of 2.4 m (summer normal water level). Upon restoration of the hydrological conditions typical for the original Lake Hornborga, the prerequisites for an effective transport of dissolved and suspended matter away from the lake will be reestablished.

The large-scale field experiments have shown what great gains can be made, from a nature management point of view, when the highly simplified, damaged ecosystem of the lowered lake is replaced by the structurally complex and functioning system of a waterfowl lake. After preparation of just 1 km^2 previously covered by reed, there was a marked recolonization by waterfowl when suitable precipitation conditions supplied the necessary water to the lake (see

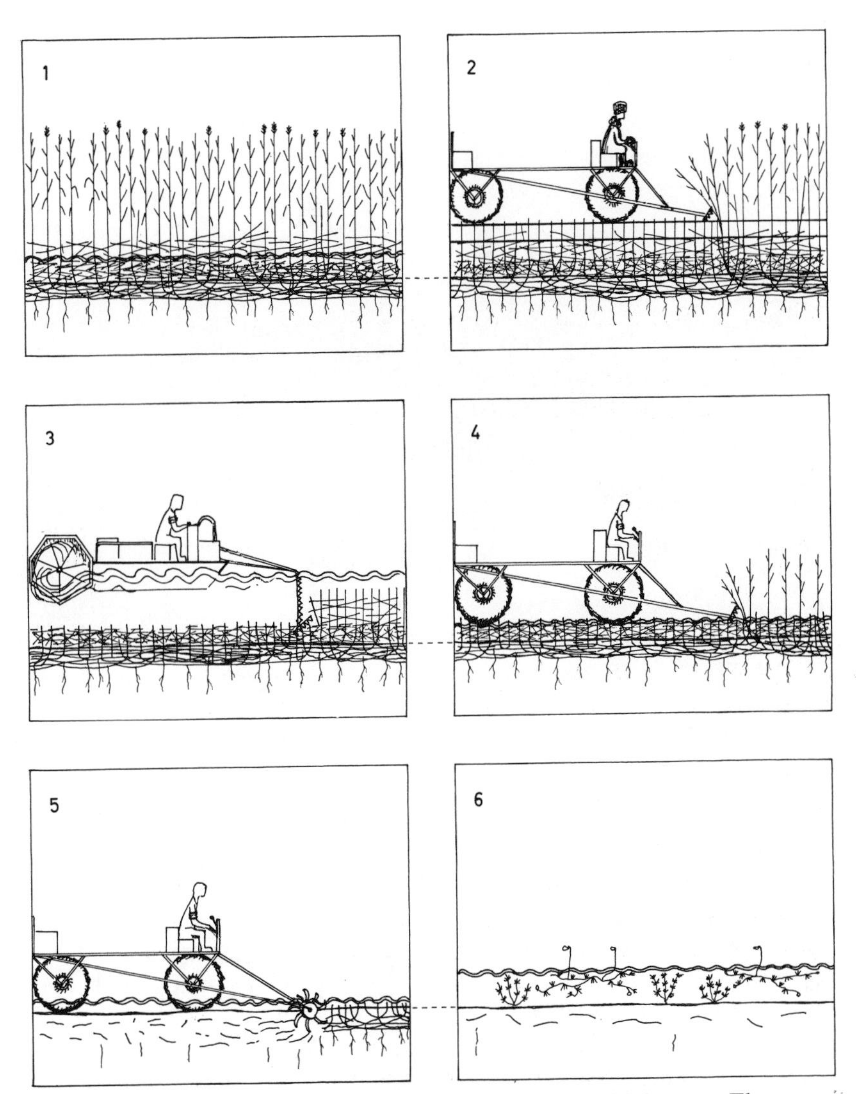

Fig. 3. The Lake Hornborga restoration project. 1. Initial state. The area overgrown by common reed, the consolidated mud covered by coarse detritus, and a dense root felt developed in the top layer of the mud. 2. The work starts with cutting during the winter. Capacity 2 hectares per hour. The reed material is being burnt. 3. During the spring high-water period, pontoon-equipped mowing machines are used for shortening the stubble and clearing the bottom from the layer of horizontal reed stems. 4. At low water in the summer the green shoots are cut. 5. Final preparation of the bottom by a rotor cultivator. 6. Emergent vegetation is replaced by submerged plants, and bottom fauna communities rich in species and individuals are developed (From Björk, 1971)

Fig. 4. The Lake Hornborga restoration project. Amphibious machine rotor cultivating reed stubble mats and root felts. Rotor cultivators are also used on pontoon-borne machines (Photo Björk, 1974.10.03)

Fig. 6). Altogether, 74% of the nesting species increased in number. The tufted duck increased from 10 to 40 pairs, the pochard from 20 to 110, the crested grebe from 3 to 52, and the horned grebe from 15 to 40 pairs.

In order to make the sedge-covered areas attractive to birds, amphibious excavators are used to create mosaics of small open waters and nesting islands (Fig. 7). The excavators make holes in the root felt, and the dug-up material is simply piled at the side. After the water level increases, the root felt will float to the surface. It will again be covered by vegetation, surrounding the still fairly deep holes.

The exposed shores can regain, to a considerable extent, the characteristics of suitable biotopes for waders. This, too, has been proved during the experimental period 1968-71.

The cost of investigations of the possibilities of restoring Lake Hornborga amounted to about $600,000 (U.S.) by 1974. This sum includes all studies made by ornithologists, limnologists, technologists, economists, and agriculturists. It also includes field experiments, purchase and construction of technical equipment, and so on. Within the near future

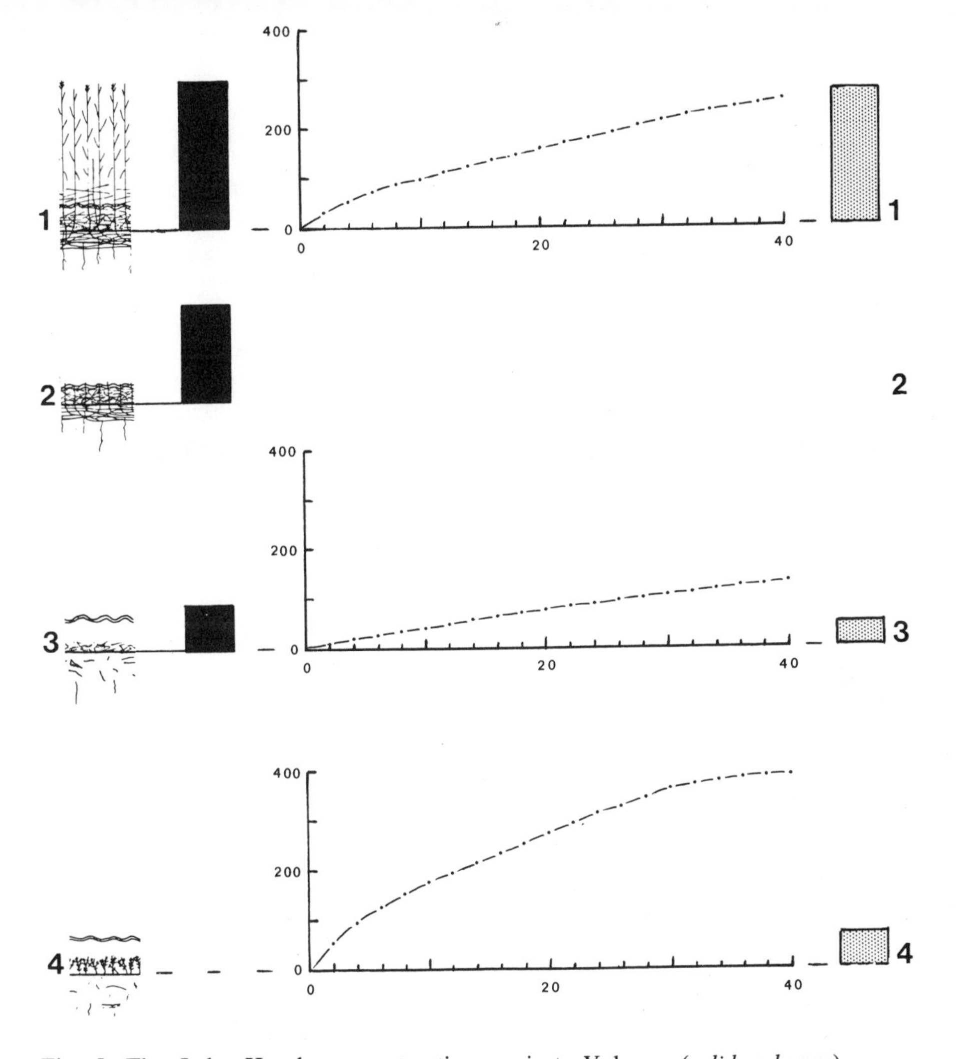

Fig. 5. The Lake Hornborga restoration project. Volume (*solid columns*) and oxygen consumption (*curves and dotted columns*) of coarse detritus (of *Phragmites*) and of charophytes (*Chara tomentosa*) per m^2 within intact and treated areas. The area of the columns proportional to detritus volume and oxygen consumption. 1=Intact stand of *Phragmites communis*. 2=Stubble mat after reed cutting. 3=Rotor-cultivated area (rhizome pieces and other freely floating plant material removed by water currents). 4=Flushed calcareous mud bottom with charophytes. Volume of coarse *Phragmites* detritus per m^2 (*solid columns*): 1=120, 2=82, 3=37, and 4=0 liter. The curves denote oxygen consumption (20°C) in mg/g dry matter of detritus and charophytes respectively (*vertical axis*) during 40 days (*horizontal axis*). Oxygen consumption per m^2 at 20°C during 40 days (*dotted columns*): 1=1100, 3=190, and 4=270 gram O_2. The vertical, dry shoots of *Phragmites* are not included in these determinations. Sampling 1971.11.02. (From Nat. Swed. Env. Prot. Board, 1973, PM 280:2)

the Swedish government will decide on the final restoration. The cost for the remaining limnological restoration of Lake Hornborga, i.e., the measures for directing the development of the vegetation, and so on, is estimated to be another $600,000. The cost for the total project is of the order of magnitude of $4,000,000 (Fig. 8).

In South Sweden the lakes are as a rule shallow, and in the southeastern part of the country at least 90% of the lakes are suffering from damage due to water level lowering. Today there is a common wish to restore the lowered lakes for nature conservancy reasons, water supply, aesthetic purposes, and fishing and other forms of recreation. Therefore, countywide investigations have been started in order to rank the lakes with respect to the need for restoration (Kronoberg County Administration, 1973). For making realistic evaluations of the possibilities of restoring lakes and wetlands

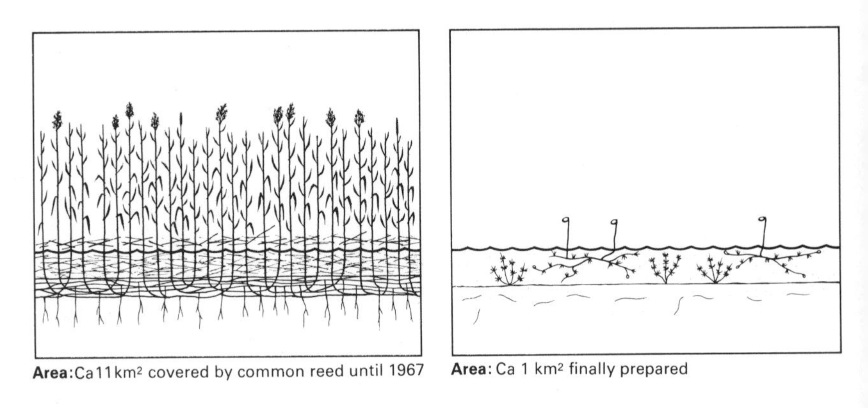

Fig. 6. Lake Hornborga before and after experiments for directing the primary production from emergent to submerged vegetation. Water level not yet raised. Comparison between conditions in 1965 and 1971. Data from H. Berggren and P. O. Swanberg. (From Björk, 1972)

Fig. 7. The Lake Hornborga restoration project. Amphibious excavators making a mosaic of small open waters in sedge-covered areas. (Photo Björk, 1971.07.23)

and for estimating the costs, the results from the Lake Hornborga demonstration project are of great value.

In Lake Hornborga, sedge (*Carex*) and the common reed (*Phragmites communis*) are replaced by underwater vegetation. In other lakes comparable procedures are utilized to remove bulrush (*Scirpus lacustris*) in favor of submerged plants.

The National Swedish Environment Protection Board presents special management instructions for waterfowl habitat improvement in connection with lake restoration (Nilsson, 1974). The birds could be looked upon as sensitive control instruments, indicating if the ecosystems are in balanced function.

Recovery and Restoration of Recipients

By January 1, 1975, 586 sewage treatment plants in Sweden included chemical or biological-chemical treatment, i.e., phosphate precipitation. This means that over 57% of the population in clusters with at least 200 inhabitants (i.e., approximately 83% of the total

Fig. 8. The Lake Hornborga restoration project. Experimental area at the start of treatment (*above*) and after reed harvesting and treatment of stubble mat and root felt. (Photo Björk, August 1968 and April 1970)

population) have effective sewage treatment. By the same date, 6 plants included a fourth step (as a rule, sand filters). Only 0.4% of the population in population clusters with at least 200 inhabitants are still without sewage treatment. Collection of the sewage from large areas to big treatment plants also means the diversion of sewage from several lakes.

Thus, after several decades of discharge of inadequately treated sewage, the nutrient loading on recipients is decreasing. At the same time there is, at last, a trend toward decreasing sewage production, which means that people have started to conserve water (the consumption per capita is decreasing).

Studies of the effects on lakes of chemical sewage treatment and diversion of sewage effluent have been carried out since 1972 under a research program of the National Swedish Environment Protection Board (cf. Forsberg et al., 1975). The program included 18 lakes and 15 sewage treatment plants distributed over the country. The objectives of the program (Forsberg et al., 1975) are to:

a) Record the municipal wasteload before and after the introduction of chemical wastewater treatment or more total diversion of sewage effluent.

b) Estimate the percentage of municipal load of phosphorus, nitrogen and organic matter in relation to the corresponding load from other sources.

c) Calculate nutrient budgets for the recipient lakes and study the role of the sediments in the recovery process.

d) Note the changes of the composition of phytoplankton; study the algal growth potential, the specific growth rate and the growth limiting factors in the lake water by using algal assay procedures.

e) Elucidate for each lake the time required 1) until the concentration of nutrients will decrease, 2) until the biomass will decrease and the transparency increase, 3) until a new balance between nutrient loading and the concentrations of nutrients in the lake will be achieved.

f) Elucidate the influence of nutrient loading on trophic level and how changed loading conditions will influence the water quality.

g) Give a base for estimating if and under what conditions nitrogen removal from sewage effluent will be necessary for combating eutrophication.

Because the program was started as late as 1972, it is not yet possible to exemplify any significant changes or improvements of conditions in the investigated lakes. The "between-seasons" and "between-years" changes make evaluation of the results from short-term studies very difficult.

Studies like these are, of course, absolutely necessary for increased information concerning the effects of diversion and advanced wastewater treatment methods, including the effects in limnologically different recipients of the addition of chemicals such as aluminum sulfate.

Without any doubt, the results from these investigations will show that some recipient lakes react and recover more or less immediately, while others need decades to attain a tolerable status. A third group of lakes could, in the time perspective of man, be looked upon as irreversibly damaged.

Lake Lyngby Sø

Lake Lyngby Sø (0.6 km^2, 2.8 m), close to Copenhagen, has been studied by Mathiesen (1971) for several years. Until 1959 the lake was heavily polluted, resulting in very high primary productivity and blooms of blue-green algae. After sewage diversion, the production immediately decreased. However, as the quality of the water from Lake Furesø, discharging into Lake Lyngby Sø, became eutrophic, the productivity of Lake Lyngby Sø again increased. A restoration project for Lyngby Sø is under discussion, with phosphorus precipitation by aluminum sulfate of the water from Furesø as a first step.

Lake Norrviken

Lake Norrviken (2.7 km^2, 12.5 m), situated immediately to the north of Stockholm, became highly eutrophicated by sewage (Ahlgren, et al., 1975). For several decades the lake was characterized by heavy water blooms in summer and oxygen deficiency during the winter. At the beginning of 1970 all sewage effluents were diverted from Lake Norrviken. Extensive studies on the ecosystem have been carried out by Ahlgren and coworkers (1975), who focused their interest on nutrient dynamics and phytoplankton. At the end of 1974 (i.e., five years after the diversion), the total phosphorus concentration was about 30% of the concentration at the end of 1969 (i.e., just before the diversion). The concentration of total nitrogen has been about constant since 1970, but in the summer the concentrations of the inorganic fractions reach such low values that nitrogen might be a limiting factor for the development of phytoplankton.

Since the diversion the phytoplankton diversity has increased, but the annual production is still high. From the wealth of material on phytoplankton and zooplankton over several years, Ahlgren et al. (1975) conclude that species composition in the summer phytoplankton is probably dependent on spring conditions. Important prerequisites for an early bloom of blue-greens are probably an early thermal stratification and temperatures above 10°C. If conditions are suitable for an early development of blue-greens immediately following the spring outburst of diatoms, then a "bad" year will follow in Lake Norrviken. The Secchi disc transparency will be low as there is little or no grazing by zooplankton. It is evident that in the case of Lake Norrviken, factors other than simple nutrient supply are to be taken into consideration when judging the figures recorded for biomass and primary productivity of phytoplankton. Light and temperature regimes as well as zooplankton grazing have been mentioned. According to the nutrient budget, the production of phytoplankton seems at present to be most dependent on the internal loading provided by nutrient release from the sediments. In comparison with Lake Lyngby Sø (before the eutrophication of Lake Furesø), the recovery of Lake Norrviken is considerably slower.

Lake Trummen

Lake Trummen and other polluted lakes in the town of Växjö (Figs. 7 and 9) in the classical oligotrophic area of central South Sweden have been studied by limnologists since the 1920s. The lake, which has an area of 1 km² and, until 1969, a maximum depth of 2 m, was used for water supply until about 1920 and for swimming until the middle of the 1930s. From 1936 to 1958 Lake Trummen was exploited as a receiver of biologically treated sewage from a maximum of about 1,500 toilets. During the period 1941-57 waste water from a flax factory was also discharged into the lake.

The lake ecosystem responded to the sewage input in the usual way. Blooms of blue-greens (*Microcystis* as a rule dominating) depressed the Secchi disc transparency to 15-20 cm, and winter oxygen deficiency resulted in fish kills.

Eleven years after sewage diversion the situation was still the same. Preservation of the bad conditions for such a long period makes it correct to talk about irreversible damage. Thus, observations like those from Lake Trummen made it quite clear that diversion or advanced sewage treatment would not always bring about a recovery of the recipient. Methods are therefore needed for recipient management and for restora-

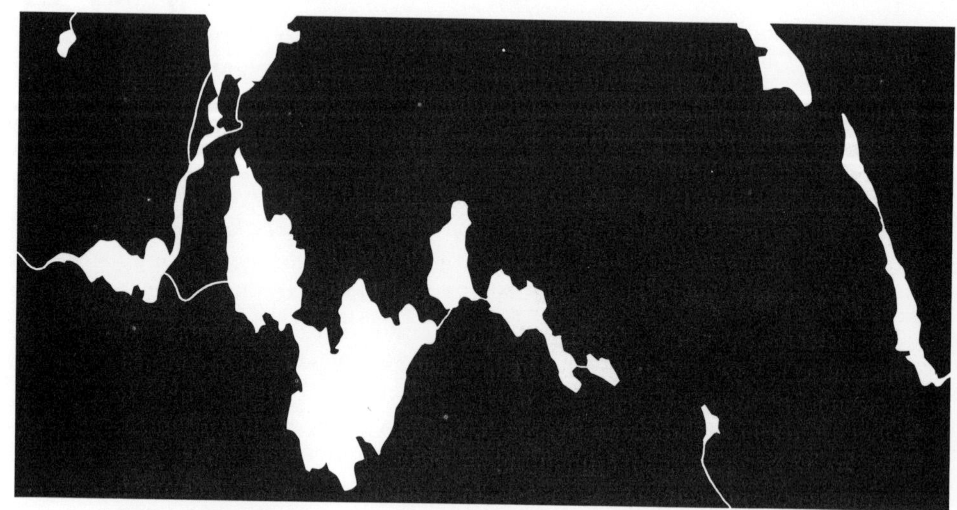

Fig. 9. The location of Lake Trummen in the town of Växjö. (From Björk, 1972)

tion, for reduction or elimination of the internal loading remaining from the period with insufficient sewage treatment.

Thorough paleolimnological investigations (Digerfeldt, 1972) showed that the sediment growth rate in the oligotrophic Lake Trummen was 0.2 mm/year just before the recipient period. During that period it increased to 8 mm/year. The sediment deposited during the last decades (20 to 50 cm) was loose and black with iron sulfide and released large amounts of phosphorus and nitrogen under anoxic conditions. In contrast, the underlying brown, well-consolidated sediment from the intact lake adsorbed phosphate and released no nitrogen even under anoxic conditions.

Smooth cooperation among limnologists and other ecologists, administrators, and technologists resulted in the Lake Trummen restoration project. The restoration research project started in 1968 and will continue until 1980. It is a combination of research and training of limnologists in theoretical and applied limnology.

A suction dredge with a nozzle specially designed according to the limnologists' wishes was used. This equipment made it possible to pump sediment from the lake up on land without making the lake water turbid (turbidity is an indication of supply of sediment-interstitial water to the lake) and with a minimum of lake water intermixture with the pumped sediment.

In 1970, 0.5 m (theoretically) of sediment was removed from the main part of the lake, and in 1971 another 0.5 m (Fig. 10). Altogether about 400,000 m^3 of sediment were pumped to simple settling ponds. From these ponds the run-off water was conducted to a simple treatment plant for phosphate precipitation with aluminum sulfate. The water returning to the lake had a total phosphorus content of about 30μg/l.

The expanding macrophyte vegetation was removed by means of a dragline. Part of this vegetation was floating due to the lively methane formation in and beneath the root felt.

After sewage diversion but before the restoration, the structure and function of the ecosystem of the damaged lake were directed by the internal circulation of nutrients. During this period more phosphorus, nitrogen, etc., were transported from than to the lake. However, the losses were small in comparison to the store accumulated during the recipient period. After the restoration the lake again acts as a trap for nutrients, just as the unpolluted lakes in the vicinity. As the development of the ecosystem is now dependent on the long-term supply of nutrients and other substances, the water of the brooks discharging into Lake Trummen from the urbanized surrounding is continuously kept under observation. Oil traps are installed upstream from the inlets, and there are arrangements for phosphate precipitation if this should be necessary.

The restoration caused dramatic changes in the Lake Trummen ecosystem with respect to structure as well as function. The phosphorus and nitrogen concentrations decreased (Bengtsson et al., 1974; Fig. 11). The *Microcystis* blooms disappeared and were replaced by plankton communities with higher diversity (Fig. 12). Of course, the phytoplankton primary productivity decreased, but the proportion of production by small species of algae increased (Cronberg et al., 1974).

The bottom in the restored lake has an age of about 2000 years. The sediment has good phosphate-adsorbing capacity, and the leakage of phosphorus and nitrogen during anoxic conditions is negligible. After the restoration the oxygen content is high under ice cover. The improved light conditions and the bottom of consolidated mud now permit the development of submerged macrophytes such as *Nitella* and *Potamogeton* species.

In summary, the restoration resulted in an ecosystem with good balance between production and destruction (Fig. 13). Situated within an urbanized area, the lake is now available for fishing, swimming, and other forms of recreation (Fig. 14). Continuous cooperation between the town authorities and ecologists will certainly bring about a

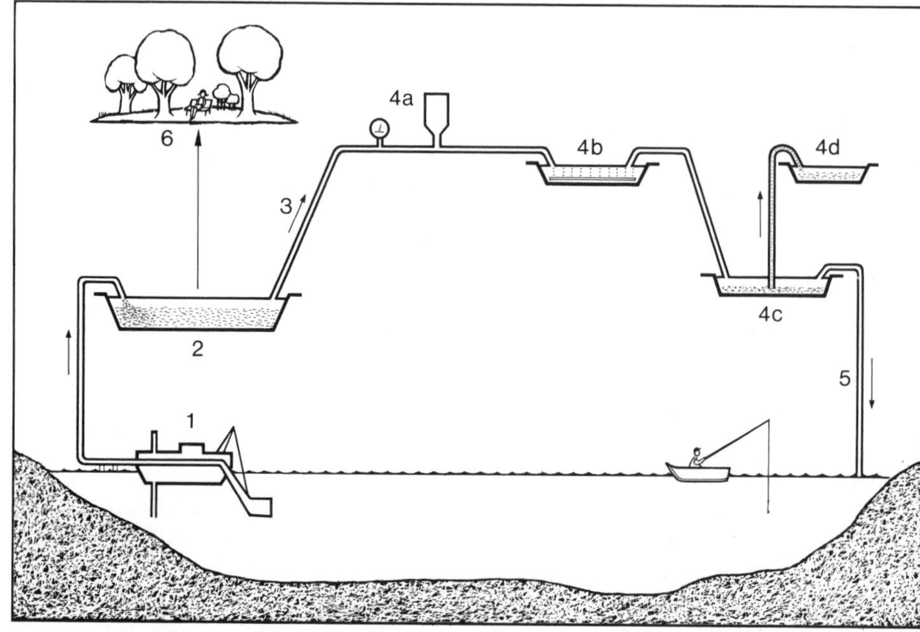

Fig. 10. The restoration scheme as practiced for Lake Trummen. 1. Suction dredger. 2. Settling pond. 3. Run-off water. 4. Precipitation with aluminum sulfate (4*a*: Automatic dosage; 4*b*: mixing through aeration; 4*c*: sedimentation; 4*d*: sludge pond). 5. Clarified run-off water. 6. The dried sediment is used as fertilizer for lawns and parks. The area utilized for sediment deposition will become recreation grounds. (From Björk, 1971)

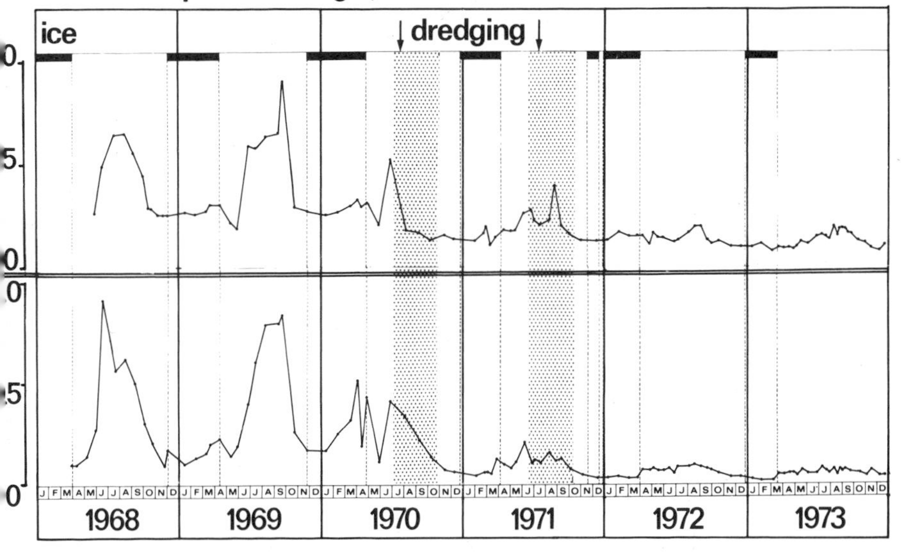

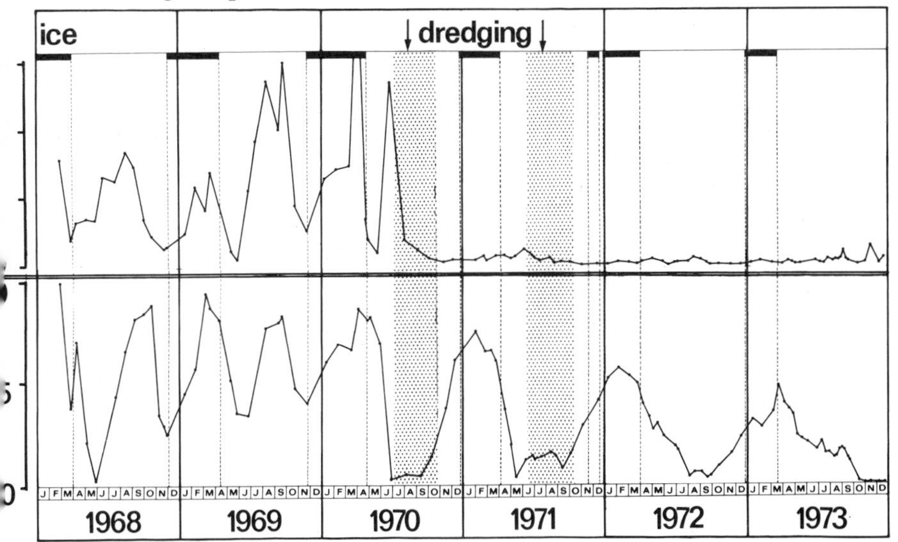

Fig. 11. Kjeldahl-nitrogen, total phosphorus, phosphate phosphorus, and silica in Lake Trummen (0.2 m) before, during, and after restoration. (From Bengtsson et al., 1974)

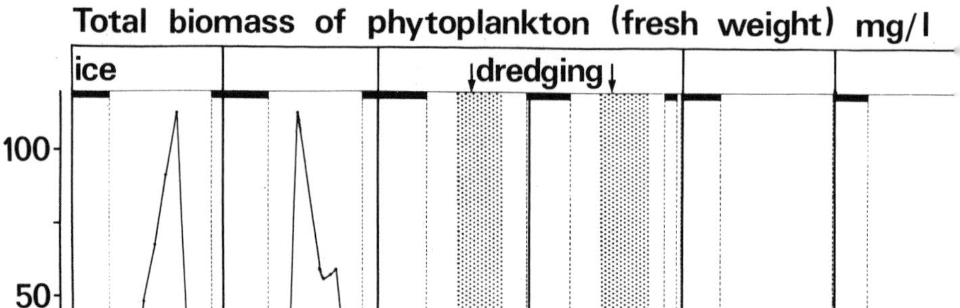

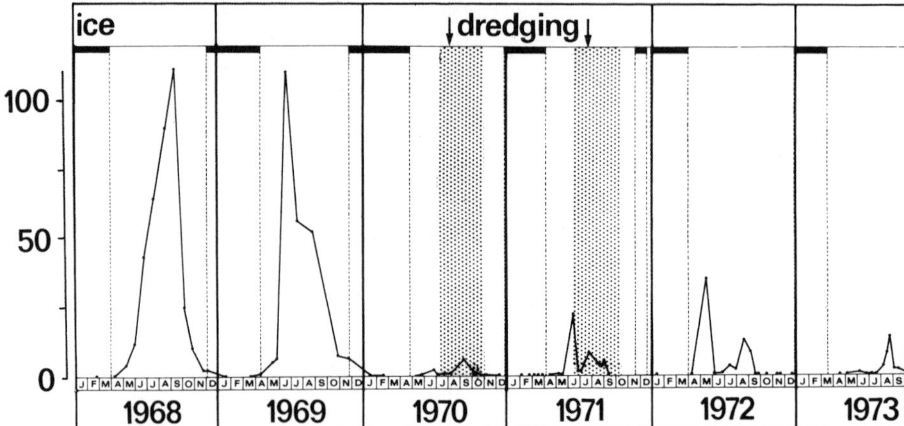

Fig. 12. Phytoplankton and blue-green algae biomass (fresh weight) at 0.2 m depth in Lake Trummen before, during, and after restoration. (From Cronberg et al., 1974)

good natural planning for the lake and its surroundings, making it attractive especially for children. Artificial arrangements and disturbing activities must, of course, be avoided. The unit of land and water should be restored to a natural state dictated by local characteristics. This should be a generally accepted rule.

The total cost of the Lake Trummen restoration amounted to about $500,000 (U.S.). The dry sediment is now being sold as soil fertilizer for parks, gardens, roadsides, and so on. The price is about $2 per m^3.

Hypolimnetic Aeration—the Lake Järla Project

Lake Järla (1 km^2, 23 m) in Stockholm is a recipient from which the sewage has been diverted (see Fig. 7). It is thermally stratified and oxygen deficiency and development of hydrogen sulfide occur in the hypolimnion. As in the unrestored Lake Trummen, the superficial sediment layer has a high oxygen demand, and a strong leakage of nutrients takes place under anoxic conditions.

The goal for a joint limnological/technical three-year research program was to increase the hypolimnion oxygen content without disturbing the thermal stratification and to study the limnological consequences. Engineers from Atlas Copco Ltd. constructed a hypolimnion aerator, a modified design which later received the trade name Limno.

In Lake Järla the hypolimnetic oxygen content was increased to 7 mg/l during the experimental period. As a result, a thin, oxidized sediment surface layer developed. The leakage of phosphate was reduced and the ammonium concentration decreased, while the concentration of nitrate increased. In this experiment aeration was performed for four months only, and no permanent changes in the ecosystem could be recorded afterwards. From other European long-term aeration projects, the positive effects of, for example, controlling the redox conditions at the sediment/water interface are described (Mercier and Gay, 1954; Bernhardt, 1974). Hypolimnetic aeration has also been combined with addition of phosphate-adsorbing material (Ohle, 1972).

Planning for Lake Restoration

In Sweden, plans have been made for the restoration of lakes within, for example, the city of Stockholm (Stockholm City Environment Board,

Fig. 13. The Lake Trummen restoration project. The lake just before (*above*) and at the end of restoration. (Photo Björk, 1969.08.12 and 1971.08.04)

Fig. 14. Lake Trummen after restoration. The former recipient, situated within the town of Växjö, is now available for fishing, swimming, etc. (Photo Björk, 1972.08.16)

1973) and Stockholm County areas. The planning has been preceded by a systematic inventory of the status of the lakes, the recreational interests estimated to be focused on them after restoration, and so forth. Thereafter, the lakes have been ranked in order of priority and will be subjected to detailed limnological studies before tailor-made restoration plans are developed.

Thus, it is possible to restore damaged lake ecosystems. However, it takes some time to start people thinking of restoration and management, of active therapy of natural waters. For such a long time they have been programmed to exploitation and building of artificial things. Nearly two generations became accustomed to chlorinated swimming pools instead of natural lakes in good health. Sooner or later, votes will be given to those politicians who have acquired knowledge of the necessity to take care of nature, because it is a sign of common sense.

Literature Cited

Ahlgren, I., G. Ahlgren, and B. Ulén. 1975. Limnological studies of lakes Norrviken, Edssjön, and Oxundasjön [in Swedish, English summary]. Report 15. Environmental factors, plankton, and nutrient dynamics in 1974. Mimeo. Institute of Limnology, Uppsala.

Bengtsson, L., S. Fleischer, G. Lindmark, and W. Ripl. 1974. The Lake Trummen restoration project. I. Water and sediment chemistry. Abstracts, XIX Congress of International Association Limnology. Mimeo. Institute of Limnology, Lund.

Bernhardt, H. 1974. Die hypolimnische Belüftung als Verfahren zur Steuerung der Redoxvorgänge am Gewässergrund einer eutrophen Talsperre. Abstracts, XIX Congress of International Association Limnology.

Björk, S. 1970. Hornborgasjön till nytt liv. Svenska Turistföreningens årsskrift.

———. 1971. Reversibla skador i sjöekosystem. (Reversible damage in lake ecosystems). Svensk Naturvetenskap, 1971.

———. 1972. Swedish lake restoration program gets results. Ambio 1. With the title "Restoring lakes in Sweden," 1973, Technology Review, No. 76. With the title "Joining forces to save damaged lakes in Sweden and Tunisia," 1974, Landscape Architecture, No. 64.

Cronberg, G., C. Gelin, and K. Larsson. 1974. The Lake Trummen restoration project, II. Bacteria, phytoplankton, and phytoplankton productivity. Abstracts, XIX Congress International Association Limnology. Mimeo. Institute of Limnology, Lund.

Digerfeldt, G. 1972. The post-glacial development of Lake Trummen. Regional vegetation history, water level changes, and palaeolimnology. Folia limnologica scandinavica, No. 16.

Forsberg, C., S. O. Ryding, and A. Claesson. 1975. Recovery of polluted lakes. A Swedish research program on the effects of advanced waste water treatment and sewage diversion. Water Research, No. 9.

The Kronoberg County Administration (Länsstyrelsen i Kronobergs län). 1973. Sänkta och utdikade sjöar i Kronobergs län. I. Ronnebyån och Mieån. (Lowered and drained lakes in Kronoberg County I. The rivers Ronnebyån and Mieån). Mimeo. Växjö. (In Swedish).

Mathiesen, H. 1971. Summer maxima of algae and eutrophication. Mitt. Int. Verein. Limnol. No. 19.

Mercier, P., and S. Gay. 1954. Effets de l'aération artificielle sous-lacustre au lac de Bret. Schweiz. Z. Hydrol. No. 16.

The National Swedish Nature Conservancy Board (Statens Naturvårdsnämnd). 1967. Regarding investigation concerning the preservation of Lake Hornborga from a nature conservancy point of view. (Ang. utredning rörande Hornborgasjöns säkerställande ur naturvårdssynpunkt). Official letter to the Ministry of Agriculture. Dnr. 367/65 (PN 48/64).

The National Swedish Environment Protection Board. 1973. PM 280. The Lake Hornborga investigation. 1. Summary. 2. Björk, S., The limnological investigation. Possibilities, methods, and costs for the restoration of Lake Hornborga. 3. Swanberg, P. O. Lake Hornborga as a waterfowl lake. Ornithological study within the investigation by the National Environment Protection Board on the future of the lake. 4. Technical investigation, estimation of damage, and agricultural economics concerning the restoration of Lake Hornborga. Mimeo. Stockholm. (In Swedish).

Nilsson, L. 1974. Program for ornithological investigation work in connection with lake restorations. (Program för ornitologiskt utredningsarbete i samband med sjörestaurering). Mimeo. The Natural Swedish Environment Protection Board, Stockholm.

Ohle, W. 1972. Zur Seentherapie. Ein Forschungsprojekt am Grebiner See. Schrift des Max-Planck-Instituts für Limnologie anlässlich des Besuchs des Symposium Semisaeculare der SIL in Plön am 4. October 1972.

The Stockholm City Environment Board. The Working Group for Lake Restoration. 1973. Stockholms sjöar. Stockholm. (In Swedish).

Artificial Aeration and Oxygenation of Lakes as a Restoration Technique

Arlo W. Fast

Abstract

Artificial aeration is one of many lake restoration techniques. These techniques operate by modifying either the symptoms or the causes of eutrophication. Those techniques which modify the cause of eutrophication operate by reducing either the inflow of nutrients to the lake (external loading) or the recycling of nutrients already in the lake (internal loading). Artificial aeration may reduce the internal loading rate by maintaining aerobic conditions throughout the lake, but the full consequences of this modification on internal loadings are still unknown. Artificial aeration can be accomplished either through the total mixing of the lake (destratification) or through aeration of the hypolimnion only without thermal destratification (hypolimnetic aeration). Each method of aeration has its unique attributes and beneficial impacts on the symptoms of eutrophication irrespective of their effects on nutrient loadings.

Introduction

Although the theme of this book is the recovery of damaged ecosystems, I must confess that I do not think of eutrophic lakes as damaged. I think of them as different; not good, not bad, but simply different. Eutrophic lakes, here taken to include culturally eutrophic lakes, differ in many respects from other kinds of lakes. They are characterized by greater quantities of biomass, rapid and large changes in the biomass, and in general, extremes of conditions with respect to both time and space. Eutrophic lakes are not the "sky waters" of Thoreau, nor are they the aseptic aqueous pools so prized by "overcivilized" man. Conditions associated with eutrophic lakes often conflict with man's intended usage and mental concepts of what a lake should be.

Having said all that, I must confess that I thoroughly enjoy investigating eutrophic lakes with the forethought of how to improve them—improvement in the general sense of restoration, that is, "the manipulation of a lake ecosystem to effect an in-lake improvement in degraded, or undesirable conditions" (Dunst et al., 1974). Unfortunately, our present ability to restore or improve most lakes is vestigial at best. Sufficient interest and financial resources have only recently been focused on lake restoration. The restoration efforts have followed two approaches:

1. The basic approach, where the object is to discover the underlying principles of lake ecosystems and thereby learn to manipulate lakes to greatest advantage.
2. The applied approach, where a reasoned manipulation is made even though we do not fully understand its implications or consequences. After the fact, we then attempt to deduce what went right, or wrong, and in so doing, we often learn much about the underlying principles of how lakes work.

Both approaches are necessary for the advancement of lake management, just as they were for the advancement of medicine. The only difference is that lake medicine is still in the fifteenth century. Lake doctors with their bags of toxicants, dredges, coagulants, and other devices are still in the medical equivalent of the bloodletting stage.

Causes of Eutrophication

Eutrophication is caused by excess nutrients such as nitrogen and phosphorus. Certain levels of nutrients are necessary to maintain life, but excess amounts, like excess human sustenance, lead to undesirable changes. In lakes, these changes include nuisance plant growth, undesirable changes in species composition (both plants and animals), odors, oxygen depletion, insect infestations, fish kills, and a general deterioration of water quality. There are lake restoration techniques which attack both the causes of excess nutrients as well as the symptoms of eutrophication. Before we consider these restoration techniques, we will review further some emerging principles of eutrophication.

Principle: Lakes are nutrient traps. A considerable amount of the nutrient load entering a lake is tied up in the lake's sediments and is either recycled back to the water or removed from the cycle due to additional sedimentation (Fig. 1). In most cases, the amount of nutrients leaving the lake is relatively small in comparison to the inflow of nutrients. The concentration of nutrients in the water is therefore a function of the inflow of nutrients (external loading) and the recycling of nutrients from the sediments (internal loading).

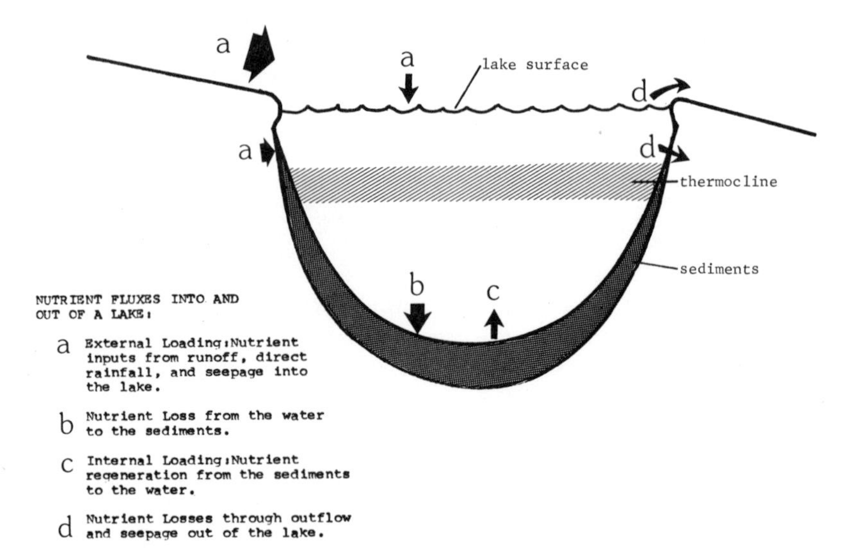

Fig. 1. A lake as a nutrient trap, illustrating nutrient fluxes into and out of the lake. Sizes of the arrows do not necessarily represent relative magnitude of fluxes.

Principle: Lakes which have low nutrient loading to begin with are thought to undergo succession from an organically unproductive (oligotrophic) condition to a productive (eutrophic) condition (Lindeman, 1942). This may occur due to increases in external and/or internal loading. Internal loading may increase with time due to the accumulation of nutrients in the sediments and the recycling of these nutrients back into the water. External loading is effected by changes in the watershed and atmosphere. Man's impact on the watershed is of paramount importance.

Principle: The amount of organic production is related to the nutrient concentrations of the water. Sawyer (1947) suggested that nuisance plant growth would result if inorganic phosphorus and nitrogen levels exceed 0.01 and 0.3 mg/1 respectively. Vollenweider (1968) later confirmed these results elsewhere and at the same time developed criteria for permissible and dangerous loading rates (Table 1).

Principle: The rate of internal loading is greatly accelerated by hypolimnetic oxygen depletion (in thermally stratified lakes). If we begin with an oligotrophic, thermally stratified lake, most of the nutrients which enter the lake are thought to be either tightly bound up in the aerobic sediments or flushed from the lake. As the lake ages, nutrients accumulate

Table 1. Specific nutrient loading levels for lakes (expressed as total nitrogen and total phosphorous in g/m^2/yr)

Mean depth up to (meters)	Permissible loading, up to		Dangerous loading, in excess of	
	N	P	N	P
5	1.0	0.07	2.0	0.13
10	1.5	0.10	3.0	0.20
50	4.0	0.25	8.0	0.50
100	6.0	0.40	12.0	0.80
150	7.5	0.50	15.0	1.00
200	9.0	0.60	18.0	1.20

Source: Vollenweider (1968).

and organic production gradually increases. Concurrent with increased organic production, hypolimnetic oxygen is depleted (Fig. 2). The point in time at which the oxygen is exhausted during the annual cycle of thermal stratification is seen as an important point because when the hypolimnion becomes anaerobic, the solubility of nutrients such as phosphorus are greatly increased. Nutrients which were formerly bound in the sediments are brought back into solution, where they presumably will be available for additional plant growth. Hence, a feedback mechanism is established whereby the eutrophication process may be accelerated. As will be discussed later, artificial aeration may be one means of preventing this nutrient regeneration and, therefore, reducing the internal loading rate. Mortimer (1941, 1942) clearly documented the increased solubility of phosphorus from sediments during anaerobic conditions, but no one has yet established the general relative importance of this source of phosphorus. Most limnologists feel that anaerobic regeneration is a significant source of nutrient, but under some circumstances it may not be of consequence (Schindler, 1974).

These principles have led to the conclusions that eutrophication is potentially controllable and that lakes can be restored by reducing the external and/or internal nutrient loading rates. Restoration techniques based on this conclusion include:

1. External load reduction by nutrient diversion, wastewater treatment, watershed management, treatment of inflow, or product modification. The Lake Washington story is a classic example of the success of nutrient diversion (Edmondson, 1961, 1966, 1970, 1972*a*, 1972*b*).

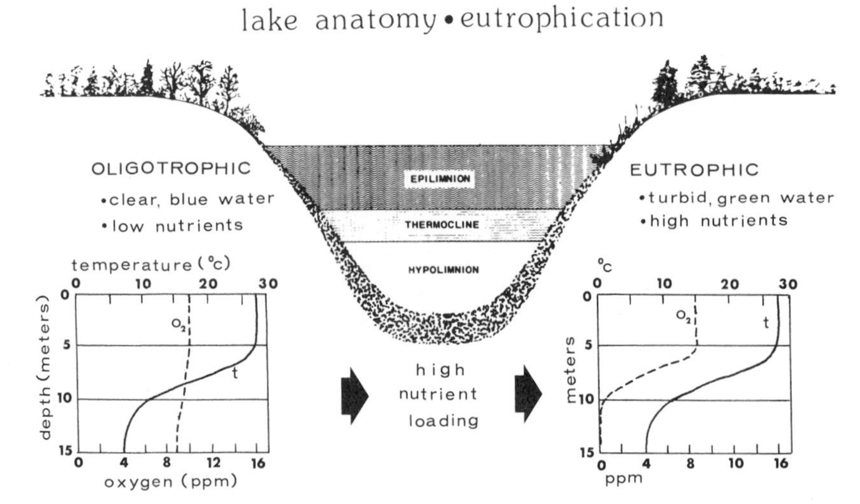

Fig. 2. Cross-sectional anatomy of a lake showing three depth zones. Typical oxygen-temperature profiles of oligotrophic and eutrophic lakes are also shown. (Fast, 1974).

However, nutrient diversion alone may not be enough. Nearby Lake Sammamish, Washington, did not respond significantly after nutrients were likewise diverted. It appears that about one-third of the total nutrient load came from external sources, whereas two-thirds came from the sediments through internal loading (Rock, 1974). Lake Trummen (Björk et al., 1972) is another example of a lake where the eutrophic state was not materially diminished even though most of the external nutrients were diverted. Lake Trummen recovered only after dredging.

2. Internal load reduction by dredging. Sediment removal has been used successfully, especially if nutrient inputs are reduced and if a low nutrient layer of sediment underlies the nutrient-rich sediments which are removed. If the accumulation of nutrient-rich sediments is very deep, then dredging will only expose additional nutrient-rich deposits. Removal of all nutrient-rich deposits under these conditions may be economically prohibitive.

3. Internal load reduction by sediment layering. Researchers have used alum (Peterson et al., 1973), fly ash (Tenney and Echelberger, 1970), bentonite clay (Ohle, 1972), and various other materials (Gahler, 1969) to settle nutrients from the water column, form a layer overlying the sediments, and/or reduce the release of nutrients back into the water. This line of investigation is promising, but it needs much more evaluation.

4. Internal load reduction by artificial aeration. Artificial aeration using compressed air, pure oxygen, or mechanical pumping can create aerobic conditions throughout the lake. This may reduce nutrient regeneration from the sediments and thereby reduce internal nutrient loading.

The remainder of this chapter will center on artificial aeration, as a means of reducing both nutrient regeneration and certain symptoms of eutrophication irrespective of nutrient consideration. Dunst et al. (1974) review in greater detail lake restoration state-of-the-art and on-going restoration research.

Artificial Aeration of Lakes

Much has been written about the benefits of artificial aeration, but the underlying impact of aeration on nutrient cycling is far from clear. The efficacy of aeration to cause certain desirable changes in lakes is without question, providing that the aeration is properly applied. However, most of the proved benefits represent changes in the symptomatic characteristics of eutrophication rather than in the prime cause, nutrient loading. This state of affairs reflects directly on our inability to accurately and simply measure sediment-water exchange rates under various conditions. If we wish to assess the impact of aeration on internal nutrient loading, then we must develop this capability.

This introduction to the subject may sound overly pessimistic, or negative, but it should not be so construed. On the contrary, I am optimistic about the use of aeration as a restoration technique, but at the same time anxious to more fully understand its ramifications. I am also concerned that aeration, like other restoration techniques, be appropriately applied. Artificial aeration, probably more than any other restoration process, directly affects almost all aspects of the lake ecosystem–nutrient cycling, water chemistry, the heat budget, bacteria, phytoplankton, zooplankton, zoobenthos, and the fish. It directly affects distribution of these things as well as their rates of changes.

Artificial aeration of lakes refers to processes whereby air or oxygen is directly added to the water and/or the water is circulated. There are a variety of aeration techniques, including systems for ponds, lakes, and streams. We will consider only lake aeration systems, of which there are two types: *destratification* and *hypolimnetic aeration.*

Destratification is presently the most widely used means of aerating a thermally stratified, eutrophic lake. (Oligotrophic lakes can also be destratified using air injection, but the object is not to add oxygen; see Fast, 1971). After destratification is complete, the lake will be nearly isothermal, and oxygen will be distributed throughout. Likewise, many of the chemical parameters will be more uniform, and barriers to the distribution of fish and other biota will be minimized. Aeration by destratification was reportedly first used by Scott and Foley (1919). It has been widely used within the past ten years for domestic water supply management. The most common destratification method is diffuse aeration using a perforated air line (Fig. 3). Other techniques include mechanically pumping hypolimnetic water to the surface (Hooper et al., 1952; Ridley et al., 1966; Symons et al., 1967; Ditmars, 1970) and pumping epilimnetic to the bottom (Quintero and Garton, 1973). Ridley et al., (1966) destratified Queen Elizabeth II Reservoir using angled jets of water located at the bottom of this aboveground reservoir. Modifications of the air lift tube include the Aerohydraulic Cannons (Bryan, 1964), Helixor (Wirth, 1970), and simple diffuse air injection in submerged pipes (Müller, 1963; Johnson, 1969).

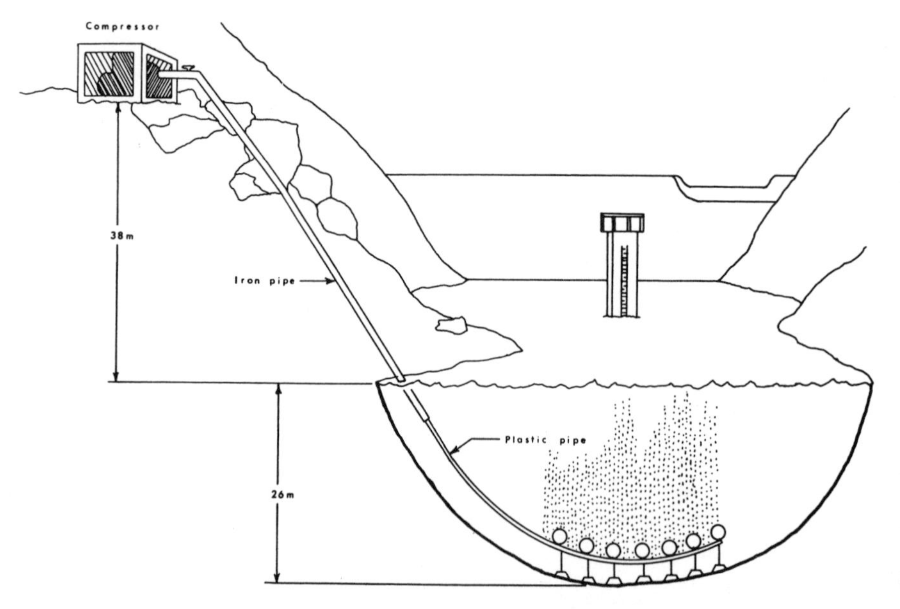

Fig. 3. A typical lake destratification system (Fast, 1968). The shore-based compressor delivers compressor air through a single air line. The air is released from the perforated section of pipe, 100 feet or more in length.

Criteria for sizing destratification systems are not well developed. Although some attempts have been made to compare the efficiencies of the various destratification systems (Symons, et al., 1967; Steichen et al., 1974), these comparisons are inconclusive. Most comparisons are based on in-lake tests and changes in the thermal stability of the lake. However, the validity of this approach is suspect.

Destratification systems are also widely used in northern states to prevent winterkills (Wirth, 1970). The results have been mixed, if you will pardon the pun, but generally successful when sufficient mixing is attained.

Hypolimnetic aeration is a more recent development. Here the object is to maintain thermal stratification but at the same time oxygenate the hypolimnion. In terms of oxygen and temperature conditions, hypolimnetic aeration creates oligotrophic conditions in a eutrophic lake. Without aeration, most of the fish and other biota are excluded from the hypolimnion of the eutrophic lake by hypolimnetic oxygen depletion. Many of these species are cold-water, stenothermic forms which may perish in the shallow warm epilimnion. Hypolimnetic aeration may create suitable yearlong conditions for these species and lead to greater species diversity.

Hypolimnetic aeration was reportedly first used by Mercier and Perret (1949) to aerate Lake Bret. Their system mechanically pumped water from the hypolimnion, discharged the water into a splash basin on shore (where it was aerated), and allowed the water to return by gravity flow through a return pipe to the hypolimnion. Probably the next most significant breakthrough in hypolimnetic aerator design was an injection system developed by Heinz Bernhardt (1967). This was followed by other compressed air designs by Fast (1971; Fig. 4); Bernhardt (1974; Fig. 5); Atlas Copco AB (Fast et al. 1975*a*; Fig. 6). Union Carbide Corporation later adapted a stream oxygenation system to hypolimnetic oxygenation. This system uses liquid oxygen and a mechanical water pump rather than compressed air (Fast et al., 1975*b*; Fig. 7). Speece (1971) proposed the injection of very fine pure oxygen bubbles at great depth as a means of hypolimnion aeration. Presumably the bubbles would be totally absorbed before breaking the thermocline. Speece et al., (1971) and Speece (1969) proposed the downflow bubble contractor and Speece's U-tube aerator (Fast, 1968) for hypolimnetic aeration/ oxygenation. As far as I know, these last three systems have not been tested full-scale in a lake.

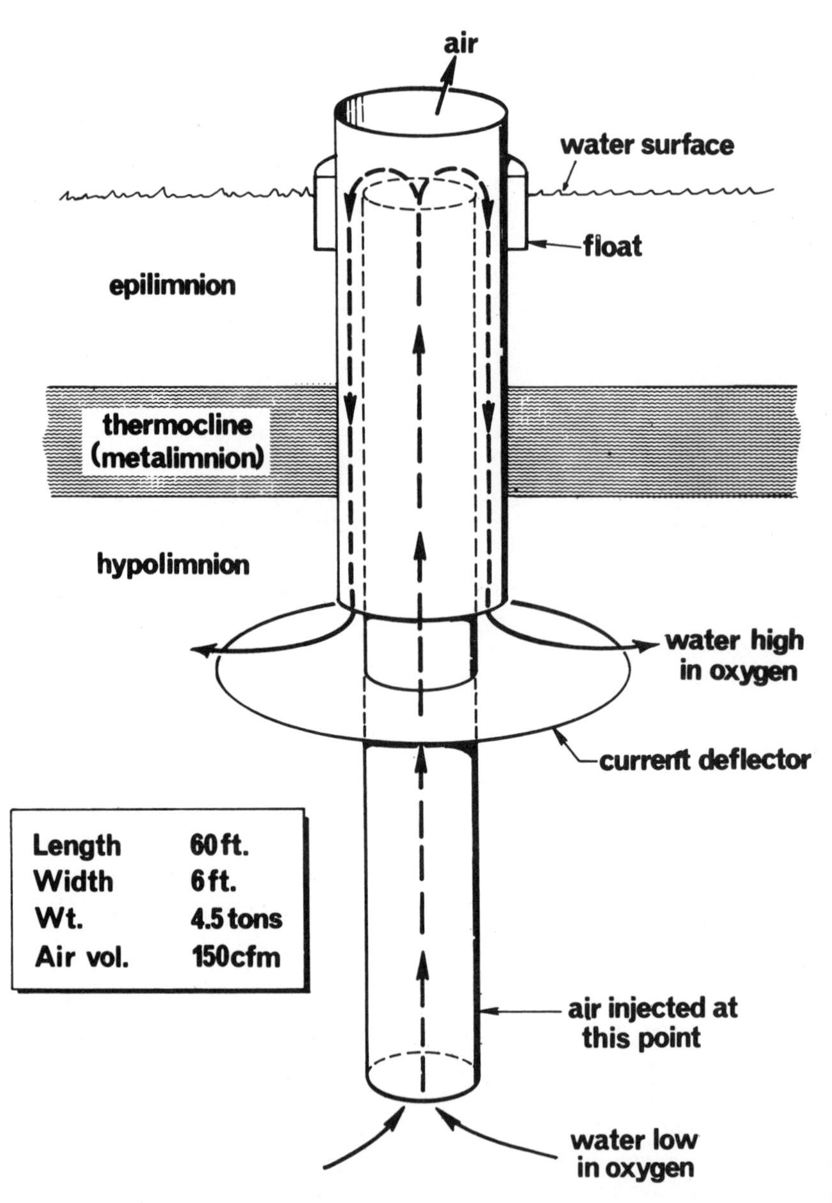

Fig. 4. The first hypolimnetic aerator used in North America (Fast, 1971). Air was injected at the base of the aeration tower from a shore-based compressor. The tower floated about 10 feet off the bottom.

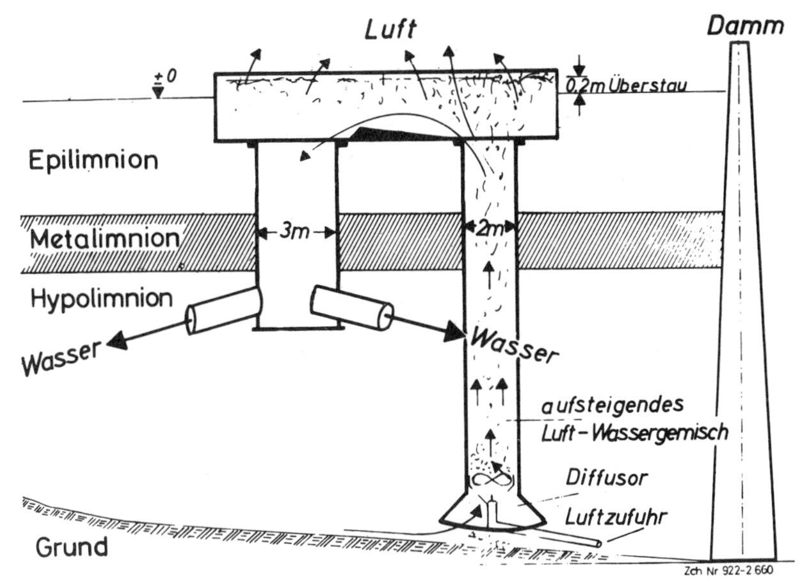

Fig. 5. A hypolimnetic aerator designed by Dr. Heinz Bernhardt of West Germany. This aerator floats above the bottom, and is one of several designed by Dr. Bernhardt (H. Bernhardt, pers. comm., 1971).

Design capacities for most hypolimnetic aeration systems are better defined than for destratification systems, but most of this information is not publicly available because of its proprietary nature. Sizing a given system to a lake also involves estimates of the oxygen consumption rates within the hypolimnion, but methods of estimating these rates are not well developed.

Reviews of artificial aeration of lakes, ponds, and streams have been made by Symons (1969), Toetz et al. (1972), King (1970), Smith et al. (in prep.), and Hogan et al. (1970).

I will now discuss in more detail the rationale for artificial aeration and present some examples of both destratification and hypolimnetic aeration.

Destratification

Destratification is generally very effective in improving drinking water quality, particularly when hydrogen sulfide, iron, or manganese are a problem; but it has been notably less effective in reducing

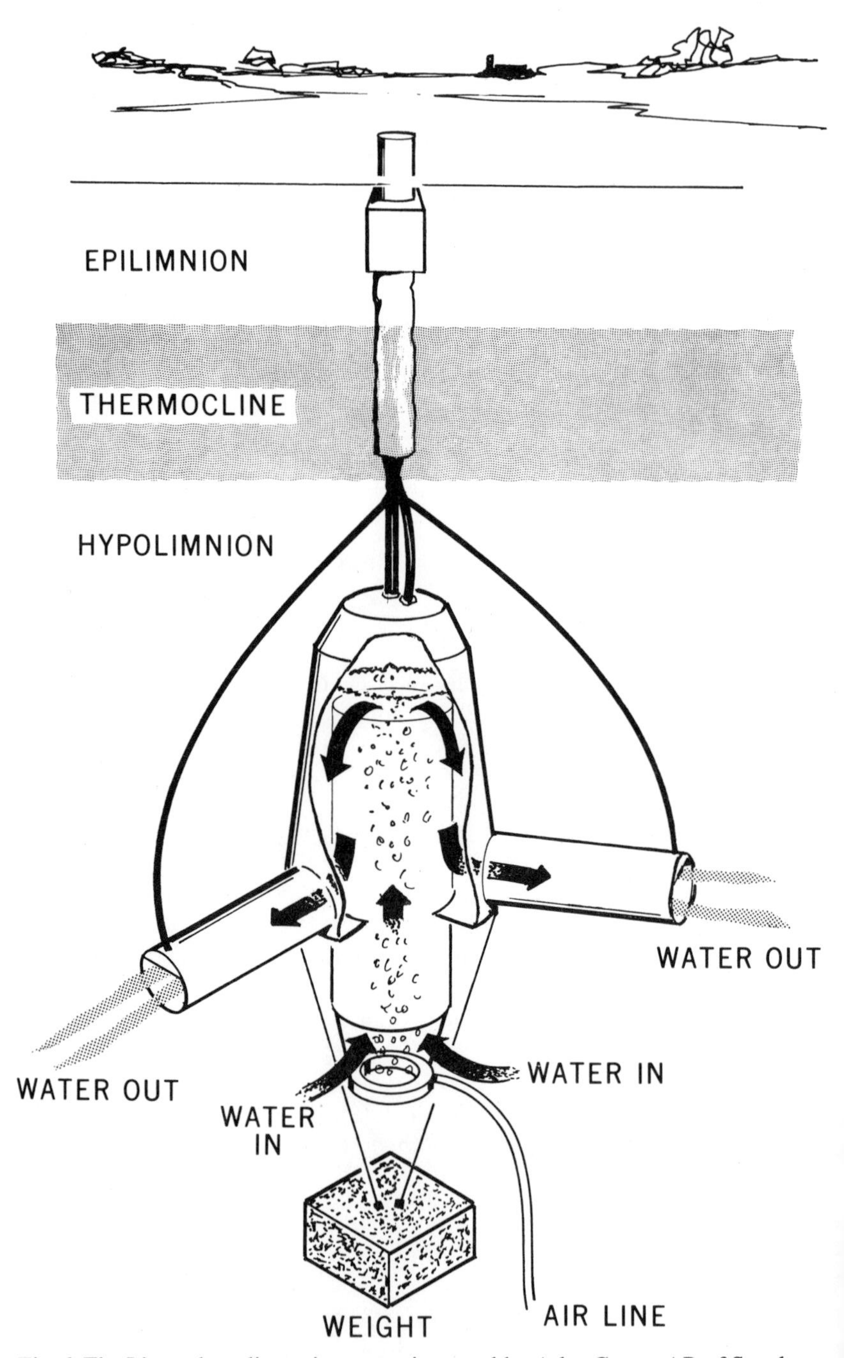

Fig. 6. The Limno hypolimnetic aerator invented by Atlas Copco AB of Sweden. This aerator is anchored to the bottom of the lake. Unlike most other aerators, the water does not rise to the surface of the lake during the aeration process (Fast, Dorr, and Rosen, 1975).

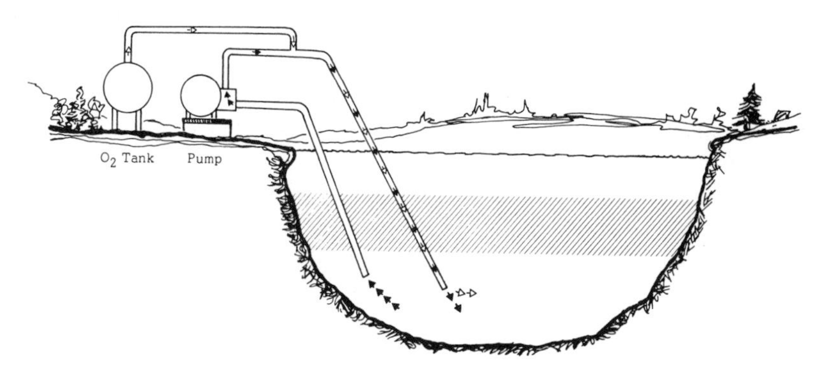

Fig. 7. The Side Stream Pumping system of hypolimnetic oxygenation. This unique system uses liquid oxygen and a mechanical water pump, rather than compressed air and a special aeration tower (Fast, Overholtz, and Tubb, 1975).

algal densities and primary production. At one time, destratification was thought to be a good method of reducing algal growth by one or more of three influences:

1. Preventing nutrient regeneration during anaerobic conditions and thereby reducing internal loading
2. Increasing the mixed depth of the algae and thereby reducing algal growth due to light limitation
3. Subjecting the algae to turbulence and rapid changes in hydrostatic pressure as they are swept through a large vertical distance of the water column

Although these influences may be operative in specific instances, in general they do not seem to be manifestly effective in reducing algal densities. Possibly we can see why if we further evaluate these influences.

First, in regard to the influences of destratification on nutrient regeneration, especially phosphorous, as discussed earlier, it is generally accepted that phosphorous regeneration from the sediments to the water is greatly increased when the hypolimnion becomes anaerobic. The two important questions then, are: (1) Is this phosphorous of consequence in the annual phosphorous budget of a lake? and (2) If it is important, does artificial aeration decrease the total amount regenerated? Although I cannot definitively prove the answers to either question, I believe the answers are, respectively: yes in most cases, and probably not in most cases. One of the most significant unanswered questions of limnology is how important is the phosphorous which is regenerated under anaerobic conditions? While I obviously cannot answer this point here, I believe most limnologists agree that it is important in cases where the internal loading is large relative to the external loading and where algal growth is usually limited by phosphorous. How often is this the case? I do not know.

Regarding the second question, phosphorous concentrations in deep water almost always decrease during destratification. However, this does not mean that the total amount of phosphorous exchanged with the sediments is less; that is the important point. Even though the phosphorous concentration of the water decreases, the rate of exchange could increase. If this happens, the total availability of phosphorous to the plants may actually be greater during destratification. It is my intuitive feeling that destratification does increase the exchange rates between sediments and water since destratification greatly increases the sediment temperature and the flow of water over the sediments. Profundal sediments which may never exceed 10°C to 15°C under normal conditions may be heated to 27°C or warmer by destratification (Fast, 1968, 1971; Fast and St. Amant, 1971; Fig. 8). Furthermore, benthic organisms will reinvade the profundal zone during destratification (Fast, 1973*a*). These burrowing organisms vertically mix the sediments (Davis, 1974) and circulate water through the sediments (Lee, 1970; Brinkhurst, 1972) and thereby undoubtedly affect the nutrient exchange rates. In any event, there is no clear evidence of reduced nutrient concentrations in the shallow, euphotic zone during destratification. In the final analysis, these concentrations and exchanges are what will count.

If a thorough destratification and mix of the lake can be achieved, then algal growth may be limited. Several models predict this result (Murphy, 1962; Bella, 1970; Oskam, 1971; Lorenzen, 1972). A thorough mix would result in algae spending greater time in the dimly lit depths of the lake than when the lake was normally stratified (Fig. 9*A,B*). This would result in less time for photosynthesis and a net reduction in algal growth, or possibly a shift toward "shade" species. However, there is evidence that a thorough mix is difficult to achieve since the closer the lake comes to isothermy, the less efficient the destratification process becomes (Fast, 1973*a*). More thermal energy may be absorbed at the lake's surface than the destratification system is able to redistribute. This is especially true for low-energy (relative to water volume) destratification systems. Unfortunately, most aeration systems are low energy. The results is microthermal stratification of 2°C to 3°C or more near the lake's surface (Fig. 10). This shallow

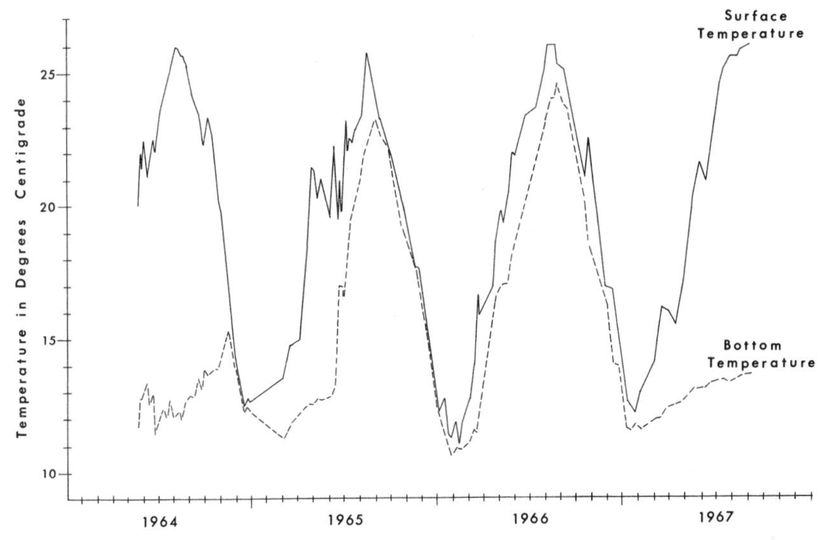

Fig. 8. El Capitan Reservoir surface and bottom temperatures from 1964 to 1967. The reservoir was artificially destratified during 1965 and 1966 (Fast, 1968).

zone of warm water may in effect keep the algae in an even shallower depth than before aeration began (Fig. 9*C*). Within limits, this reduction in mixed depth of the algae should increase the algal growth rates. Consequently, an inadequate destratification system may result in greater algal densities than if the lake were not aerated at all.

There is some evidence that destratification turbulence will reduce certain algae (Bernhardt, 1967; Robinson et al., 1969; Malueg, 1971). Blue-green species which have gas vacuoles which are sensitive to rupture by rapid pressure changes and species which have specific light and temperature optima may be most sensitive to turbulence. However, there is also evidence that destratification may increase the total amount of blue-green algae (Lackey, 1973; Haynes, 1971). This question is largely unresolved, and may be directly related to the degree of mixing as already discussed. An inadequate mix may allow the algae to remain in shallow water, and if other conditions are suitable, blue-green algae may prosper.

A 1970 survey by the American Water Works Association (1971) was most revealing regarding the effects of destratification on algal densities. The survey polled 26 water suppliers who used artificial destratification, and 7% said algal blooms were decreased by destratifi-

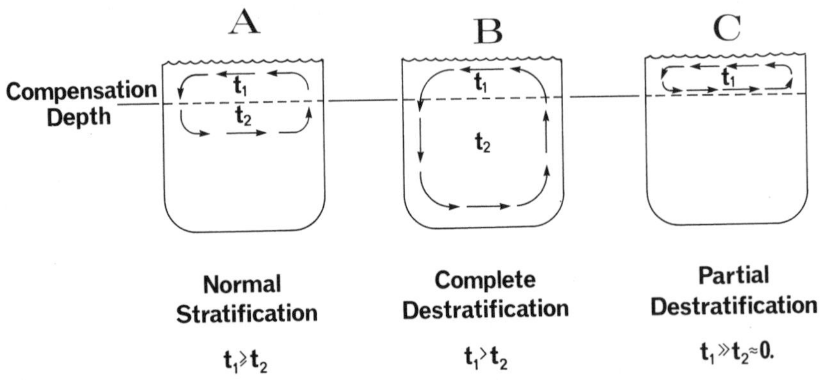

Fig. 9. Schematic of algal mixed depths during (*A*) normal stratification, (*B*) complete destratification, and (*C*) partial destratification. If destratification is not thorough, mixed depth conditions may be created which favor increased algal growth.

cation, 12% said algal blooms were increased by destratification, and 81% observed no change due to destratification. These destratification systems varied greatly in their energy input: water volume ratio, so that the results are not surprising. What was surprising is that 90% of the suppliers were satisfied with their destratification systems. The reason for this apparent disparity is that algae are only one of many factors affecting water quality. Although algal growth may have increased, improvements in other conditions were compensatory.

Destratification is generally considered beneficial to fish, but in some cases it may be undesirable. Since destratification eliminates the cold water in the lake (see Fig. 8), it can eliminate certain cold-water fish species or preclude their establishment (Fast, 1968; Fast and St. Amant, 1971). However, these species may not be jeopardized in lakes where surface temperatures normally are not lethal (Fast and Miller, 1974).

Hypolimnetic Aeration

Hypolimnetic aeration is an effective means of oxygenating the hypolimnion, eliminating hydrogen sulfide, iron, and manganese and otherwise creating suitable habitat for cold-water biota while at the same time maintaining the cold hypolimnetic temperatures. However, as yet we do not know conclusively if hypolimnetic aeration

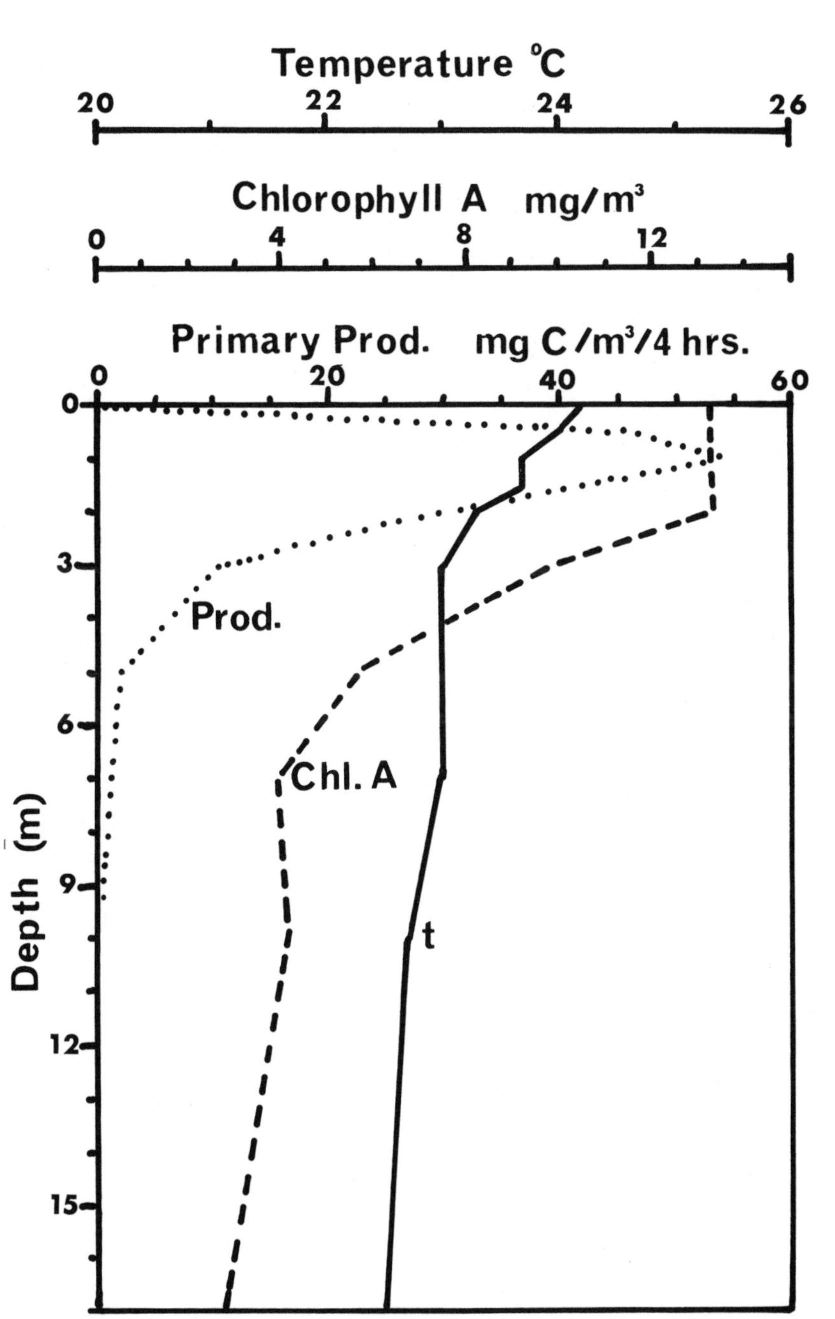

Fig. 10. Phytoplankton net primary production, total chlorophyll, and temperature depth profiles in El Capitan Reservoir during July 1966. The reservoir was artificially mixed at the time, but note the microthermal stratification between 0 and 3 meters (Fast, 1973).

will substantially reduce eutrophication by reducing internal nutrient loading.

Hypolimnetic aeration may reduce internal nutrient loading by creating aerobic conditions and preventing anaerobic regeneration of phosphorous. The best evidence for this hypothesis comes from Lakes Waccabuc and Oscaleta. These neighboring lakes are located about 80 kilometers north of New York City (Fig. 11). They are moderately eutrophic with hypolimnetic oxygen depletion typically occurring by June. A Limno hypolimnetic aeration system (Figs. 6 and 12) was installed in Lake Waccabuc during June 1973 and began operation during the first week of July 1973 (Fast et al., 1975*a*). Lake Oscaleta served as the control lake. Lake Waccabuc stratified thermally during May 1973, and hypolimnetic oxygen was 0.0 by June (Figs. 13 and 14). Concurrent with this oxygen depletion, hypolimnetic phosphorous increased dramatically from less than 300 μg/l during April to more than 1200 μg/l by the end of June (Fig. 15). After aeration began, Waccabuc's hypolimnetic oxygen increased to more than 4 mg/l throughout most of the hypolimnion, and to about 2 mg/l one meter off the bottom. Concurrent with this oxygen increase, hypolimnetic phosphorous concentrations decreased sharply from more than 1200 μg/l to less than 500 μg/l. Phosphorous further decreased to less than 300 μg/l at fall overturn during the late October. The oxygen and phosphorous changes in Lake Oscaleta differed markedly from Waccabuc's during 1973. The period of thermal stratification was nearly identical to Waccabuc's, but hypolimnetic oxygen was 0.0 mg/l from May 1973 through October 1973 (Figs. 16 and 17). Oscaleta's hypolimnetic phosphorous did not show a dramatic increase following anaerobiosis. Instead, there was a gradual but steady increase in the phosphorous from less than 200 μg/l in April to more than 300 μg/l in October (Fig. 18). These data indicate that hypolimnetic oxygenation in Lake Waccabuc has reduced the internal phosphorous load by about 100 kg, or about 30% of the total phosphorous content of the hypolimnion (John Confer, pers. comm.). They also indicate that Waccabuc has a higher phosphorous loading rate than Oscaleta.

The changes in hypolimnetic phosphorous during 1974 are more perplexing. Lakes Waccabuc and Oscaleta stratified thermally during April and destratified thermally during October. Oscaleta's hypolimnetic oxygen profile is nearly identical to its 1973 profile. Hypolimnetic aeration began in Lake Waccabuc soon after thermal stratification developed, and anaerobiosis was avoided until October. The hypolimnion became

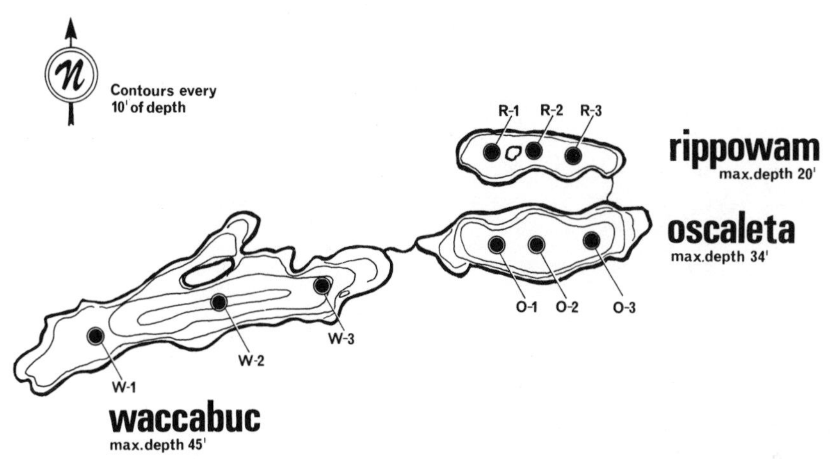

Fig. 11. Map of Lakes Waccabuc, Oscaleta, and Rippowam. Water flows from Rippowam to Oscaleta to Waccabuc. Lake Waccabuc was artificially aerated during 1973 and 1974, using the Limno aerator (see Fig. 6). Oscaleta served as the control lake (Fast, Dorr, and Rosen, 1975).

anaerobic during October, because the compressor failed. Although oxygen concentrations at one meter off the bottom were never less than 1 mg/l during most of the summer, and typically more than 4 mg/l, the hypolimnetic phosphorous concentration increased from less than 400 μg/l during April to more than 1200 μg/l during June 1974 (see Fig. 15). It then decreased slightly during August. By itself, this increase in phosphorous in Waccabuc is perplexing and suggests that aeration did not reduce the rate of internal loading. However, Lake Oscaleta's hypolimnetic phosphorous concentrations showed an even more dramatic increase during this period. Oscaleta's hypolimnetic phosphorous increased from less than 200 μg/l during April to more than 700 μg/l during August (see Fig. 18). These data indicate that some event caused substantial increases in both lakes irrespective of aeration. I believe that the external loading rate increased and overrode any decrease in internal loading which may have occurred in Lake Waccabuc due to the aeration. Unfortunately, we have no data on external loads, although they could be substantial. The surrounding terrain is glacial with overlying granite bedrock. More than 300 houses are located close to these lakes, and all houses are on

Fig. 12. Helicopter lowering a Limno aerator into Lake Waccabuc in June 1973

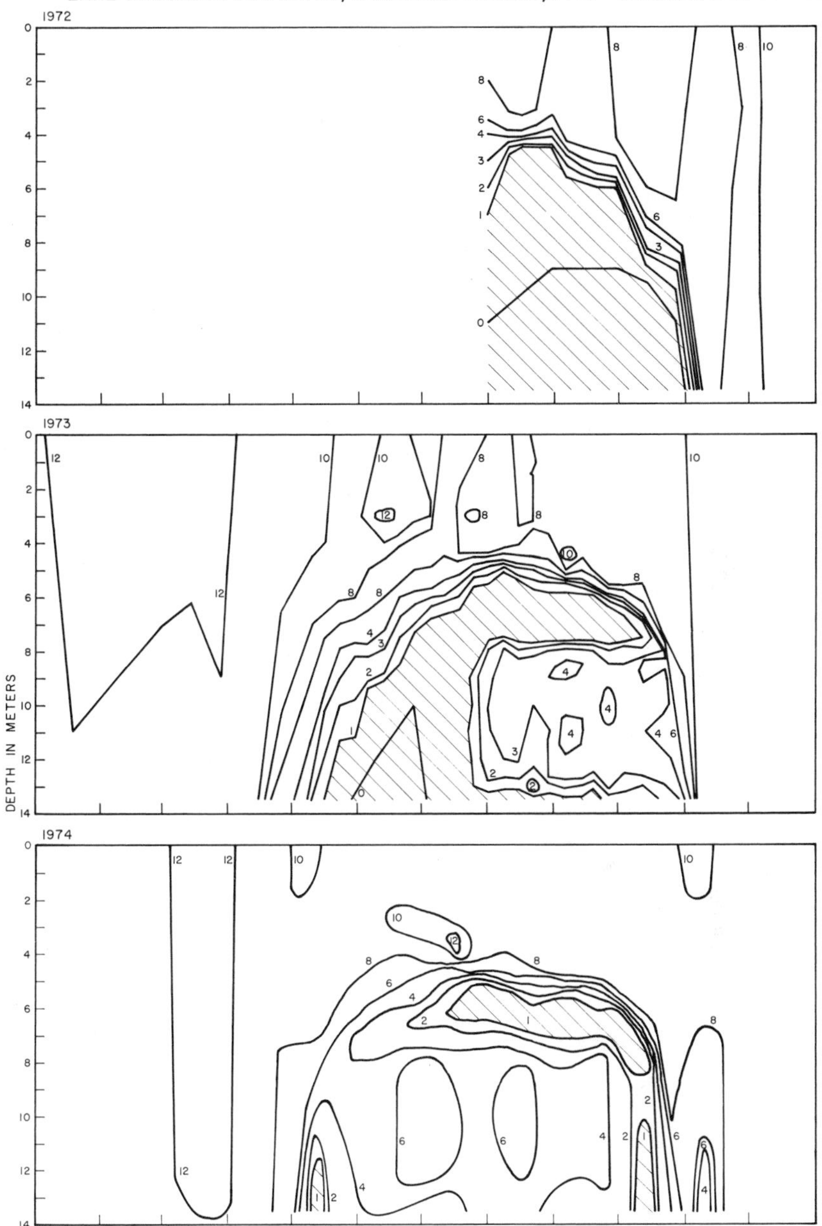

Fig. 13. Lake Waccabuc oxygen isopleths (mg/l) during 1972, 1973, and 1974. Measurements were made at W-2 (see Fig. 11). The lake was aerated during July-October 1973 and May-October 1974.

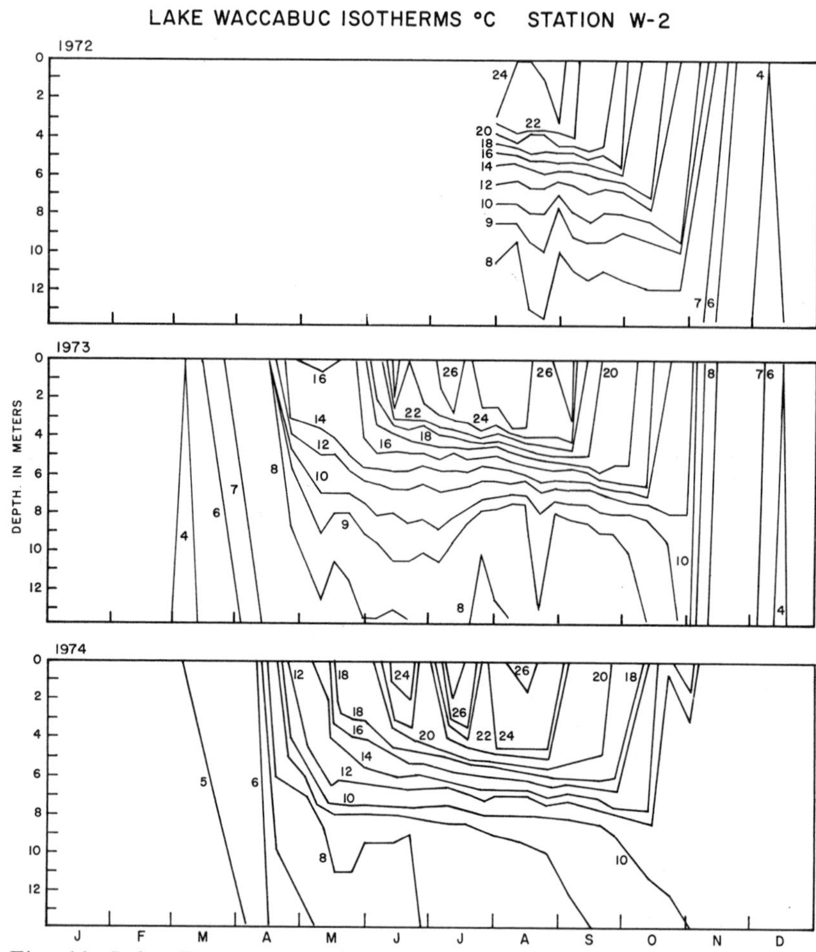

Fig. 14. Lake Waccabuc isotherms (°C) during 1972, 1973, and 1974

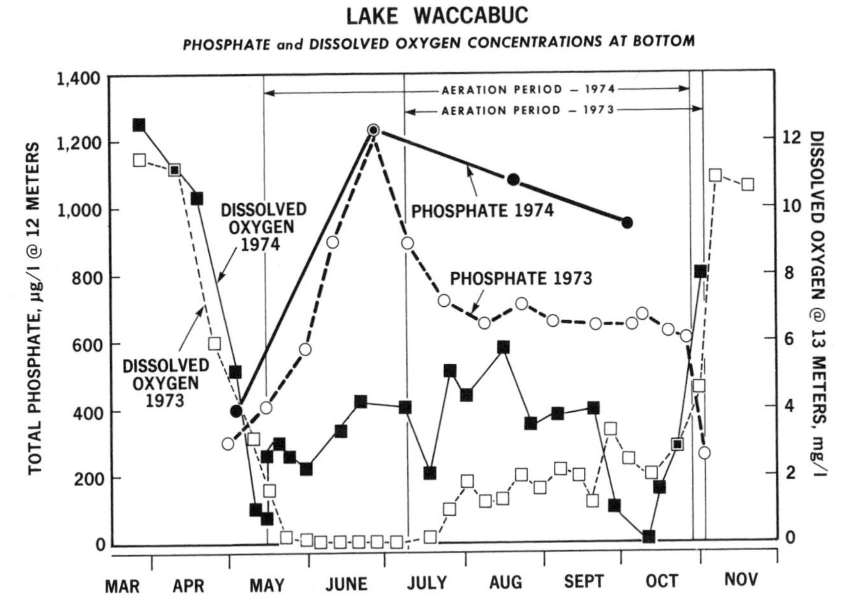

Fig. 15. Lake Waccabuc total phosphate at 12-meters depth during 1973 and 1974. Oxygen concentrations less than one meter off the bottom and the period of aeration each year also shown. Measurements were made at W-2 (see Fig. 11).

septic tanks. There was more rain during the spring and summer of 1974, compared to 1973, and this situation could have washed in greater quantities of septic tank seepage and lawn fertilizer. Even under the best of conditions, it could be virtually impossible to accurately measure this diffuse and subsurface seepage of nutrients into the lake. Further, the movement into the hypolimnion could take several paths. The changes in phosphorous and nitrogen concentrations are being analyzed more thoroughly by Garrell et al. (in prep.).

Circumstantial evidence that the external loading rate increased in both lakes during 1974, relative to 1973, is that the average transparency decreased in both lakes during 1974 (Fig. 19). Since both lakes have a very limited surface inflow and sandy bottoms in shallow water, this turbidity is mainly due to algae. Both lakes were clearest during mid-May 1974 and showed much variability every year.

The use of hypolimnetic aeration for fisheries management represents an improvement in the symptoms of eutrophication,

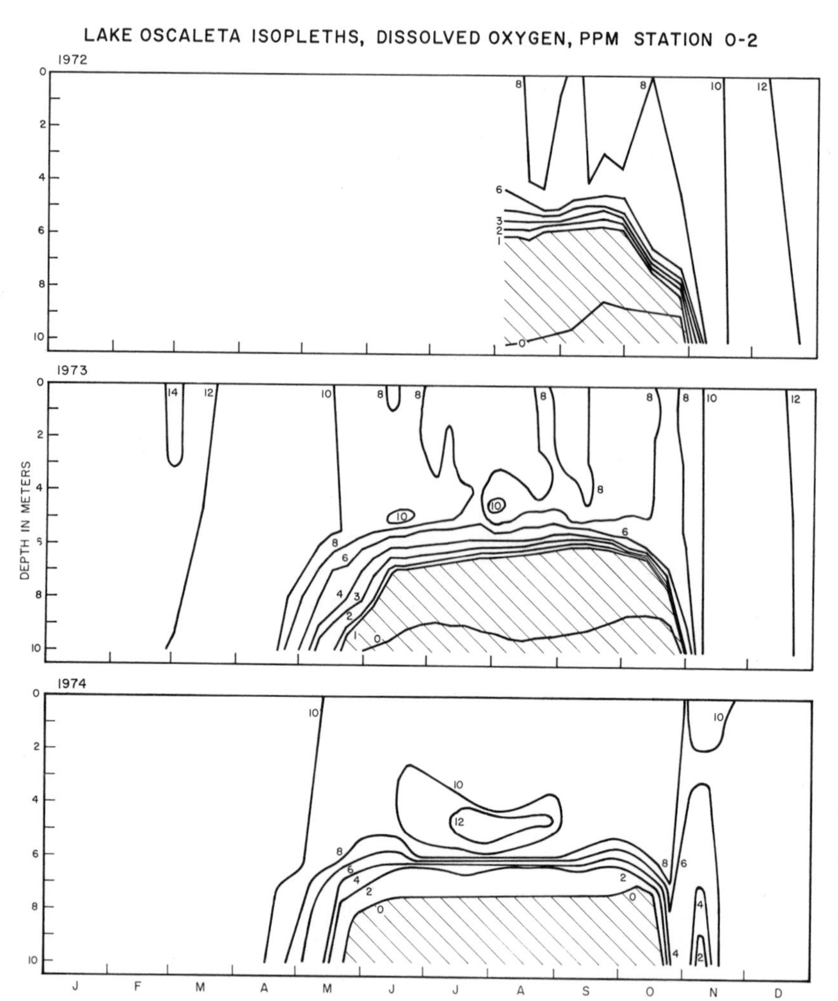

Fig. 16. Lake Oscaleta oxygen isopleths (mg/l) during 1972, 1973, and 1974. Measurements were made at 0-2 (see Fig. 11). The lake was not aerated.

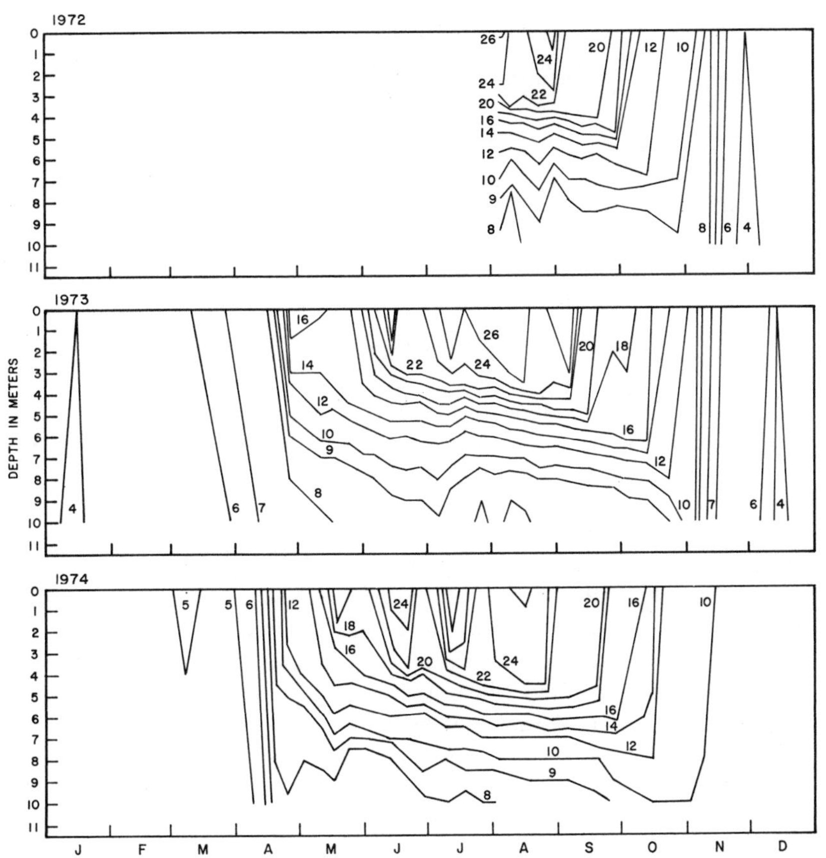

Fig. 17. Lake Oscaleta isotherms (°C) during 1972, 1973, and 1974

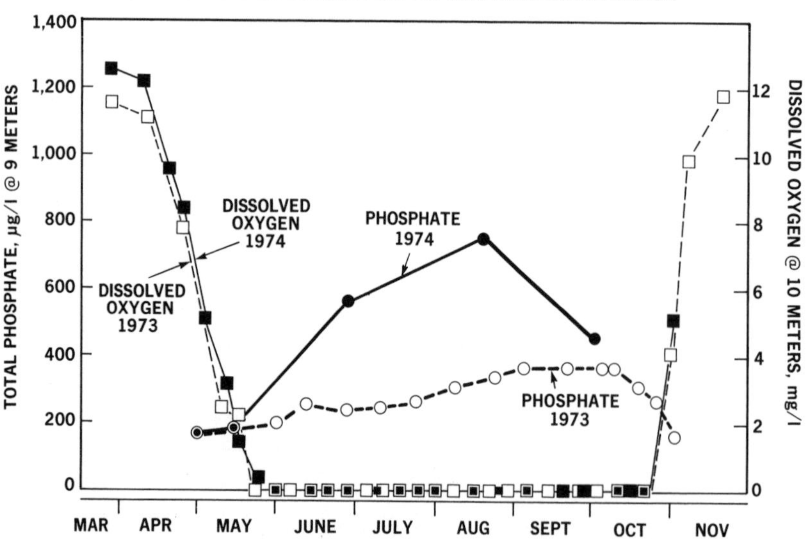

Fig. 18. Lake Oscaleta total phosphate at 9-meters depth during 1973 and 1974. Oxygen concentrations less than one meter off the bottom are also shown. Measurements were made at 0-2 (see Fig. 11). The lake was not aerated.

rather than an improvement in the causes. Without hypolimnetic aeration many eutrophic lakes will not support cold-water species, due to hypolimnetic oxygen depletion and warm surface waters. The use of hypolimnetic aeration to create a suitable yearlong habitat for trout is well documented (Fast, 1971, 1973*b*, 1974; Overholtz, 1975). In most instances it is the only known means of creating a suitable habitat. Lake Waccabuc is a good example. According to the local residents, trout had never been caught in either Lake Waccabuc or Lake Oscaleta. The summertime oxygen and temperature profiles indicate that little or no suitable water existed. The surface water was too warm, and the bottom waters were anaerobic. However, during August 1973, after one month of artificial aeration, we stocked 2100 rainbow trout (*Salmo gairdnerii*), brook trout (*Salvelinus fontinalis*), and brown trout (*Salmo trutta*) in the lake. Surface water temperatures exceeded 27°C on that date, while the fish were delivered to the lake in water of less than 10°C. In order to avoid thermal shock, we "piped" the fish into the hypolimnion using a system similar to one described by Overholtz (1974). We observed

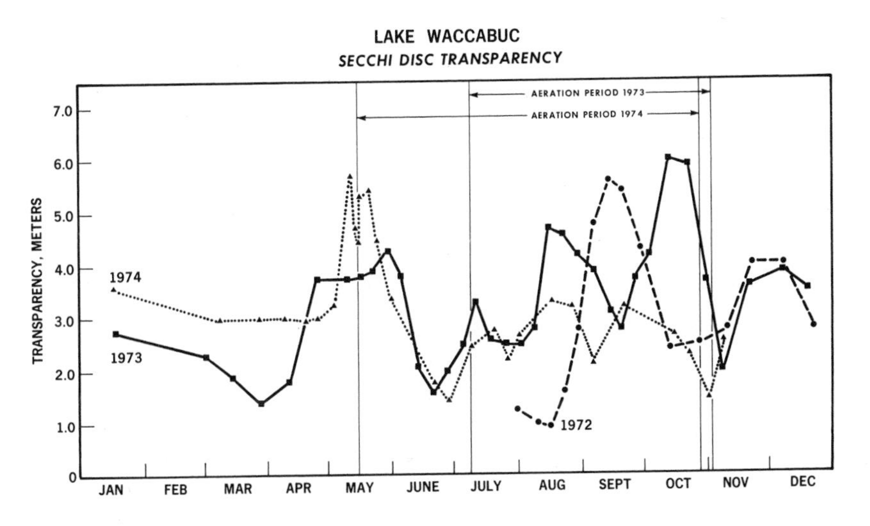

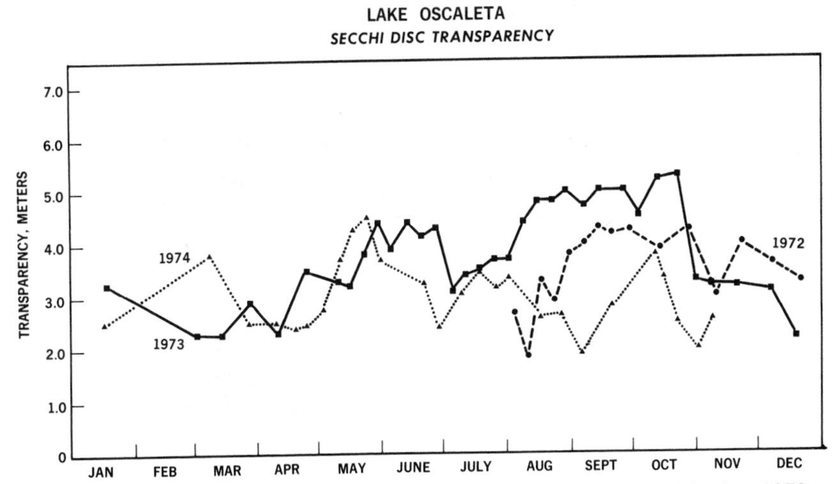

Fig. 19. Lakes Waccabuc and Oscaleta Secchi disk transparency during 1972, 1973, and 1974. Lake Waccabuc was aerated as shown, but Lake Oscaleta was not aerated.

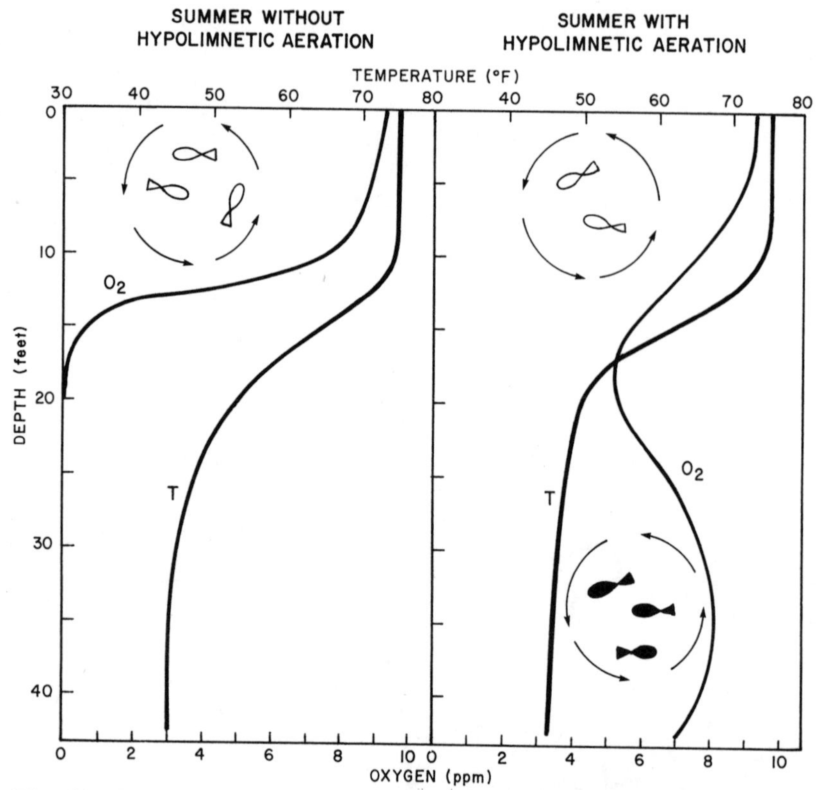

Fig. 20. Schematic of Lake Waccabuc oxygen and temperature depth distributions before and during hypolimnetic aeration. Before aeration only warmwater species of fish existed in the shallow waters. During aeration, a two-story fishery existed, with warmwater fish in the shallow water and trout in the deep water.

no mortality due to this stocking technique. We observed the fish's depth distribution using sonar and vertical gill nets. They initially remained in the thermocline but later spread throughout the hypolimnion. In effect, we created a classical "two-story" fishery with warmwater species in shallow water and cold-water species in deep water (Fig. 20). We did not measure their survival or growth rates, but some fish were caught during the summer of 1974, one year after stocking. Growth rates should be excellent, since the lake is still eutrophic.

Acknowledgments

I particularly wish to acknowledge the help of Drs. John Confer and Martin Garrell, who collected and analyzed most of the Lakes Waccabuc and Oscaleta data. Richard L. Miller and Robert Rosen assisted with the preparation of this report. Union Carbide Corporation partially supported the preparation and presentation of this paper and kindly allowed me the use of their data.

Literature Cited

American Water Works Association. 1971. Artificial destratification in reservoirs. J. Am. Water Works Assoc. 63:597-604.

Bella, D. A. 1970. Simulating the effects of sinking and vertical mixing on algal population dynamics. J. Water Pollut. Control Fed. 42:R140-R152.

Bernhardt, H. 1967. Aeration of Wahnback Reservoir without changing the temperature profile. J. Am. Water Works Assoc. 63:943-963.

_____. 1974. Ten years' experience of reservoir aeration. Paper presented at International Association Water Pollution Conference, Paris, France.

Björk, S., L. Bengtsson, H. Berggren, G. Cronberg, G. Digerfeldt, S. Fleischer, C. Gelin, G. Lindmark, N. Malmer, F. Plejmark, W. Ripl, and P. O. Swanberg. 1972. Ecosystem studies in connection with the restoration of lakes. Verh. Int. Verein. Limnol. 18:379-387.

Brinkhurst, R. O. 1972. The Role of Sludge Worms in Eutrophication. Ecol. Res. Series, U.S. Environmental Protection Agency, Washington, D.C.

Bryan, J. G. 1964. Physical control of water quality. British Water Works Assoc. J. 64(395):546.

Davis, R. B. 1974. Stratigraphic effects of tubificids in profundal lake sediments. Limnol. Oceanogr. 19(3):466-488.

Ditmars, J. D. 1970. Mixing of density-stratified impoundments with buoyant jets. W. M. Keck Laboratory. Hydraulics and Water Research, California Institute of Technology, Pasadena. Report No. KH-R-22.

Dunst, R. C., S. M. Born, P. D. Uttormark, S. A. Smith, S. A. Nichols, J. O. Peterson, D. R. Knauer, S. L. Serns, D. R. Winter, and T. L. Wirth. 1974. Survey of Lake Restoration Techniques and Experiences. Wis. Dep. of Nat. Resour. Tech. Bull. 75. 179 pp.

Edmondson, W. T. 1961. Changes in Lake Washington following an increase in the nutrient income. Verh. Int. Verein. Limnol. 14:167-175.

———. 1966. Changes in the oxygen deficit of Lake Washington. Verh. Int. Verein. Limnol. 16:153-158.

———. 1969. Eutrophication in North America. Pages 124-149 *in* Eutrophication: Causes, Consequences, Correctives. Natl. Acad. Sci. Washington, D.C.

———. 1970. Phosphorous, nitrogen and algae in Lake Washington after diversion of sewage. Science 169(3946):690-691.

———. 1972*a*. Nutrients and phytoplankton in Lake Washington. Special Symposia on Nutrients and Eutrophication. Am. Soc. of Limnol. and Oceanogr., Inc. 1:172-193.

———. 1972*b*. The present condition of Lake Washington. Verh. Int. Verein. Limnol. 18:284-291.

Fast, A. W. 1968. Artificial Destratification of El Capitan Reservoir by Aeration. Part 1. Effects on Chemical and Physical Parameters. Calif. Dep. of Fish & Game Fish. Bull. 141.

———. 1971. The Effects of Artificial Aeration on Lake Ecology. Environmental Protection Agency, Water Pollut. Control Res. Series. 16010 EXE 12/71. 470 pp.

———. 1973*a*. Effects of artificial aeration on primary production and zoobenthos of El Capitan Reservoir, California. Water Resour. Res. 9(3):607-623.

———. 1973*b*. Effects of artificial hypolimnion aeration on rainbow trout (*Salmo gairdnerii* Richardson) depth distribution. Trans. Am. Fish. Soc. 102(4):715-722.

———. 1974. Restoration of eutrophic lakes by artificial hypolimnetic oxygenation. Pages 21-34 *in* Proceedings Human-Accelerated Eutrophication of Fresh-Water Lakes, Dec., 1973. Teatown Lake Reservation, Brooklyn Botanical Garden, Ossining, New York.

Fast, A. W., and J. A. St. Amant. 1971. Nighttime artificial aeration of Puddingston Reservoir, Los Angeles County, California. Calif. Fish & Game 57(3):213-216.

Fast, A. W., and L. W. Miller. 1974. Effects of Artificial Destratification on Rainbow Trout (*Salmo gairdnerii* Richardson) Depth Distribution and Growth in a Northern Michigan Lake. Mich. Dep. Nat. Res., Fish. Res. Rep. No. 1814. 15 pp.

Fast, A. W., V. A. Dorr, and R. J. Rosen. 1975*a*. A submerged hypolimnion aerator. Water Resour. Res. 11(2):287-293.

Fast, A. W., W. J. Overholtz, and R. A. Tubb. 1975*b*. Hypolimnetic oxygenation using liquid oxygen. Water Resour. Res. 11(2):294-299.

Gahler, A. R. 1969. Sediment-water nutrient interchange. Pages 253-257 *in* Proceedings of the Eutrophication Biostimulation Assessment Workshop. University of California, Berkeley.

Garrell, M., J. Confer, and A. W. Fast. In prep. The effects of hypolimnetic aeration on nutrient cycling and water chemistry.

Haynes, R. C. 1971. Some ecological effects of artificial circulation on a small eutrophic New Hampshire lake. Ph.D. diss. University of New Hampshire, Durham. 166 pp.

Hogan, W. F., F. E. Reed, and A. W. Starbird. 1970. Optimum mechanical aeration systems for rivers and ponds. U.S. Government Printing Office, Washington, D.C.

Hooper, F. F., R. C. Ball, and H. A. Tanner. 1952. An experiment in the artificial circulation of a small Michigan lake. Trans. Am. Fish. Soc. 82:222-241.

Johnson, R. S. 1966. The effects of artificial circulation on production of a thermally stratified lake. Washington Dep. Fisheries, Fish. Res. Paper 2(4):5-15.

King, D. L. 1970. Reaeration of streams and reservoirs: analysis and bibliography. Bureau of Reclamation, Denver, Colorado. REC-OCE 70-55 (NTIS PB 197 954).

Lackey, R. T. 1973. Artificial reservoir destratification effects on phytoplankton. J. Water Pollut. Control Fed. 45(4):668-673.

Lee, G. F. 1970. Factors affecting the transfer of materials between the water and sediments. University of Wisconsin Water Resources Center, Eutroph. Inf. Prog., Lit. Rev. No. 1, Madison, Wis.

Lindeman, R. L. 1942. The trophic-dynamic aspect of ecology. Ecology 23:399-418.

Lorenzen, M. W. 1972. The role of artificial mixing in eutrophication control. Ph.D. diss. Harvard University, Cambridge, Mass.

Malueg, K. W., J. R. Tilstra, D. W. Shults, and C. F. Powers. 1971. The effects of induced aeration upon stratification and eutrophication processes in an Oregon farm pond. Paper presented at International Symposium on Man-Made Lakes, Knoxville, Tenn. May 3-7.

Mercier, P., and J. Perret. 1949. Aeration station of Lake Bret. Monastbaull, Schweiz, Ver. Gas. u. Wasserfachm. 29:25.

Mortimer, C. H. 1941. The exchange of dissolved substances between lake mud and water in lakes. J. Ecol. 29(2):280-329.

_____. 1942. The exchange of dissolved substances between lake mud and water in lakes. J. Ecol. 30(1):147-201.

Muller, G. 1973. Kann unseren seen geholfen werden? Schweizer Verkehrs- und Industrierevue, Juliheft 5.

Murphy, G. I. 1962. Effects of mixing depth and turbidity on the productivity of fresh-water impoundments. J. Am. Fish. Soc. 91:69-76.

Ohle, W. 1972. Zur Seentherapie. Ein. Forschungsprojekt am Grebiner See. Schrift des Max-Planck-Institus für Limnologie, Symposium Semisaeculare Societas Internationalis Limnologiae, 42-52. Plön, West Germany.

Oskam, G. 1971. A kinetic model of phytoplankton growth, and its use in algal control by reservoir mixing. Paper presented at International Symposium on Man-Made Lakes, Knoxville, Tenn. May 3-7.

Overholtz, W. J. 1974. A method for stocking trout in the hypolimnion of a lake. Prog. Fish-Cult. 36(13):175-176.

_____. 1975. An ecological evaluation of hypolimnetic oxygenation by the side stream pumping process on Ottoville Quarry, Ottoville, Ohio. M.S. thesis, Ohio State University, Columbus. 100 pp.

Peterson, J. O., J. P. Wall, T. L. Wirth, and S. M. Born. 1973. Eutrophication Control: Nutrient Inactivation by Chemical Precipitation at Horseshoe Lake, Wisconsin. Wis. Dep. of Nat. Resour. Tech. Bull. 62. 20 pp.

Quintero, J. E., and J. E. Garton. 1973. A low energy lake destratifier. Trans. ASAE. 16(5):973-978.

Ridley, J. E., P. Cooley, and J. A. P. Steel. 1966. Control of thermal stratification in Thames Valley reservoirs. Proc. Soc. Water Treatment and Exam. 15:225-244.

Robinson, E. L., W. H. Irwin, and J. M. Symons. 1969. Influence of artificial destratification on plankton and populations in impoundments. Pages 429-448 *in* J. M. Symons, ed. Water Quality Behavior in Reservoirs. Public Health Service Publ. No. 1930. Cincinnati, Ohio.

Rock, C. A. 1974. Problems in restoring a mesotrophic lake using nutrient diversion. Paper presented at 37th Annual Meeting of Am. Soc. of Limnol. Oceanogr., June 23-28, 1975.

Sawyer, C. N. 1947. Fertilization of lakes by agricultural and urban drainage. J. N. Engl. Waste Works Assoc. 51:109-127.

Schindler, D. W. 1974. Eutrophication and recovery in experimental lakes: implications for lake management. Science 184:897-898.

Scott, W., and A. L. Foley. 1919. A method of direct aeration of stored waters. Pages 71-73 *in* Proc. Ind. Acad. Sci.

Smith, S. A., D. R. Knauer, and T. L. Wirth. In prep. Aeration as a lake management technique. Wis. Dep. Nat. Res., Madison.

Speece, R. E. 1969. The use of pure oxygen in river and impoundment aeration. Paper presented at the 24th Ind. Waste Conf., Purdue, Ind.

_____. 1971. Hypolimnion aeration. J. Am. Water Works Assoc. 64(1):6-9.

Speece, R. E., M. Madrid, and K. Needham. 1971. Downflow bubble contract aeration. J. San. Eng. Div. Proc. Am. Soc. Civil Eng. 97:SA4, 433.

Steichen, J. M., J. E. Garton, and C. E. Rice. 1974. The effects of 'lake destratification on water quality parameters. Paper presented at Annual Meeting Am. Soc. Agric. Eng., Oklahoma State Univ., June 23-26.

Symons, J. M., W. H. Irwin, E. L. Robinson, and G. G. Robeck. 1967. Impoundment destratification for raw water quality control using either mechanical or diffused-air pumping. J. Am. Water Works Assoc. 59(10):1268-1291.

Symons, J. M., ed. 1969. Water Quality Behavior in Reservoirs: A Compilation of Published Research Papers. U.S. Government Printing Office, Washington, D.C. 616 pp.

Tenney, M. W., and W. F. Echelberger, Jr. 1970. Fly ash utilization in the treatment of polluted water. Reprinted from Bureau of Mines Inf. Circ. 8488, Ash Util. Proc., pp. 237-265.

Toetz, D., J. Wilhm, and R. Summerfelt. 1972. Biological Effects of Artificial Destratification and Aeration in Lakes and Reservoirs: Analysis and Bibliography. U.S. Dep. of Inter., Bureau of Reclamation, Rep. No. 12-D-2121 and -2122 (REC-ERC-72-33). Denver, Col. 117 pp.

Vollenweider, R. A. 1968. Scientific fundamentals of the eutrophication of lakes and flowing waters, with particular reference to nitrogen and phosphorous as factors in eutrophication. Organ. for Econ. Cooperation and Dev. Dir. for Sci. Aff. (Reference DAS/CSI/68.27/Bibliography). Paris, France.

Wirth, T. L. 1970. Mixing and aeration systems in Wisconsin lakes. Pages 31-46 *in* E. Schneberger, ed. A Symposium on the Management of Midwestern Winterkill Lakes. Am. Fish. Soc. North Central Div., Spec. Publ.

The Santa Barbara Oil Spill: An Ecological Disaster?

Michael S. Foster and Robert W. Holmes

Abstract

In its initial stages, the 1969 Santa Barbara oil spill released over 70,000 barrels of crude oil into the Santa Barbara Channel. This oil eventually polluted the entire channel and over 230 km of mainland and Channel Islands shore. The greatest known damage occurred in surfgrass communities and barnacle and bird populations; an estimated 9,000,000 barnacles and 9000 birds were killed, and over 14 tonnes of surfgrass blades and associated organisms were lost. The oil also affected other populations, but, in general, total damage cannot be determined. Cleanup procedures resulted in additional damage on both rocky shores and sandy beaches. Although some populations have recovered, the lack of prespill and postspill information makes it impossible to determine adequately the long-term effects. In addition, the size of the spill, diversity of habitats affected, and lack of ecological information combine to make an overall "disaster" assessment impossible. The inherent difficulties in assessing biological damage and recovery resulting from such a pollution incident suggest that every effort must be made to prevent and contain future spills.

Introduction

At 10:45 A.M., January 28, 1969, a flow of mud from the drilling pipe occurred as it was being withdrawn from well No. 21 on Union Oil's Platform A in the Santa Barbara Channel (Federal Lease OSC-P-0241, shared between Union, Texaco, Gulf, and Mobil; Fig. 1). After an abortive effort to install a valve on the drill stem, the drill pipe was dropped into the hole and the blind rams on the blowout preventers closed. Very soon thereafter oil and gas began to appear at the sea surface adjacent to Platform A. This flow remained unchecked until about midnight on February 7, when efforts to reduce the uncontrolled flow were partially successful. On February 8 Union Oil announced publicly that the well had been successfully plugged (Holmes, 1969).

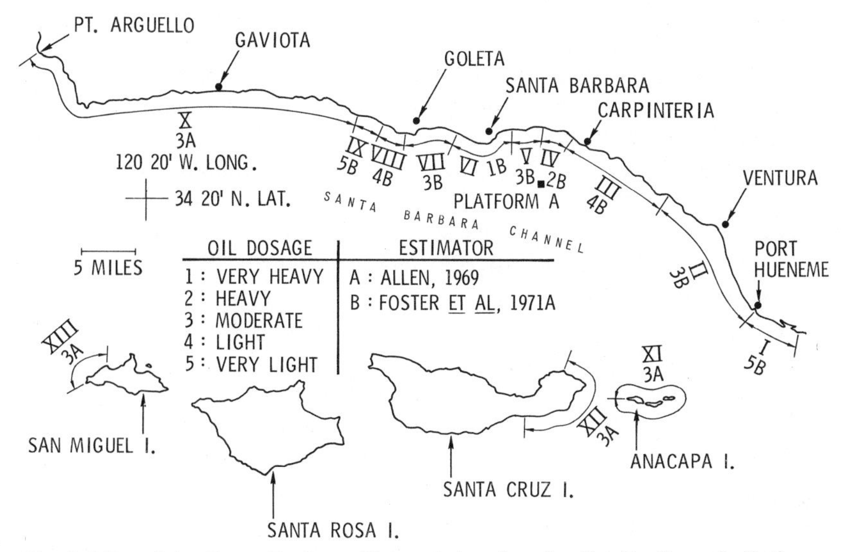

Fig. 1. Map of the Santa Barbara Channel showing the distribution of oil along mainland and Channel Island shores. Roman numerals refer to shore sections sampled by Foster et al. (1971*a*) and Allen (1969). The oil dosages in tonnes per kilometer corresponding to Very Heavy, Heavy, Moderate, Light, and Very Light are: greater than 100, 70-100, 30-70, 3-30, and less than 3, respectively.

The amount of oil released into the Santa Barbara Channel during this initial 10½-day period is a matter of some uncertainty (Table 1). Kolpack et al. (1971) state that the U.S. Geological Survey estimate is too low but do not give their own estimate.

Oil continued to leak into the channel in the vicinity of Platform A after Union Oil's February 8 announcement. The data in Table 1 give the available estimates for the period February 8 to March 5 and for the total amount spilled through May 31. According to Allen (pers. comm.), the well continues to leak at a rate of about 10 barrels per day (see Appendix I for conversion factors).

Foster et al. (1971*a*) measured the amount of oil found in sandy beach deposits and on solid substrates at nine locations on the mainland beaches between El Capitan and Port Hueneme (see Fig. 1) on February 8, 1969. Coupling these observations (supplemented by additional observations on February 9, 10, 12, and 13, 1969) with interpretation of aerial photographs, they calculated that a total of 4,508 tonnes of oil had reached the mainland beaches between El

Capitan and Port Hueneme by February 8. The estimated average dosage for this section was 51.4 tonnes per kilometer with a range of 2.7 to 118 t/km. Calculations by Foster et al. (1971*a*) of the amount of oil that could have reached the mainland beaches—assuming a circular pattern of dispersal from the spill site, ignoring the effects of currents, winds, and the amount retained in the kelp beds, and using Union Oil's and Allen's flow estimates—were also made (see Table 1). Allen's maximum estimate of 16,000 barrels per day (corrected for 25% evaporative losses) would have led to an average beach dosage of 115 kg/km.

Prespill concentrations of oil in beach sands for Southern California beaches were obtained by Mertz (1959). The Mertz concentrations may be converted to values comparable to those given in legend of Figure 1 by assuming an average beach width of 50 meters. Such a calculation yields an average value for Southern California beaches of 0.03 t/km. The mean value for Coal Oil Point, which has natural oil seeps immediately seaward and had the highest oil content of all beach samples, was 10.5 t/km, with a maximum value of 47 t/km. Since the flow rates from the Coal Oil Point seeps are quite variable, the amount of oil in adjacent beach sands must vary considerably. Nevertheless, the oil content of Santa Barbara beaches after the spill almost certainly increased some four orders of magnitude in many locations.

Smith (1968) gives estimates of oil dosage along the French and Cornish coasts following the *Torrey Canyon* wreck. The dosage on the French coast was about 169 t/km and on the Cornish coast 60 t/km (Foster et al., 1971*a*).

This brief review of the conflicting estimates of the amount of oil released into the marine environment of the channel illustrates the first of many problems we faced in our attempt to assess the biological impact of the spill as related to oil pollution dosage. It is our opinion that Allen's estimate of at least 5,000 barrels per day for the first 10½ days of the spill is the most reliable, since this estimate agrees reasonably with the independent estimates of Foster et al. (1971*a*), which have not been challenged in the scientific literature.

Effects of the Oil Spill on the Marine Ecosystem

It is possible to make an assessment, albeit rough, of the immediate biological effects of the oil spill, in some instances of the effects

Table 1. The estimated daily flow and total spillage (January 29–May 31) of oil into the Santa Barbara Channel resulting from the Union Oil well No. 21 blowout and the calculated concentrations of oil on the mainland beaches as of February 8, 1969

Source of estimate and reference	Flow (barrels per day) Jan. 29–Feb. 8	Feb. 8–Mar. 5	Total spillage (barrels) Jan. 29–May 31	Calculated average concentration of oil on beaches (t/km)
Union Oil (Holmes, 1969)	500	6.9-13.8		4.7*
U.S. Geological Survey (Holmes, 1969)	1,000	3.5-35		
A. Allen (Allen, 1969)	5,000-16,000	500	78,000	48-115*
Foster et al. (1971*a*)				51.4†

* Calculated by Foster et al. (1971*a*) from flow rate data.
† Determined from beach samples.

resulting from cleanup procedures, and finally of some of the long-term effects. For this purpose we have divided the channel area into a number of habitats: shoreline habitats (rocky intertidal and sand beaches); inner (shallow) subtidal habitats (-0.6 to -30 m); outer (deep) subtidal habitats (-30 to -183 m); and the neritic habitat, which refers to the aqueous habitat overlying the bottom down to approximately 183 m depth.

Shoreline

Approximately 13% (24.8 km) of the Santa Barbara mainland coast (191 km) polluted by oil is rocky intertidal (Appendix II). A much greater percentage (81.7%) of the coast on the Channel Islands is rocky intertidal (Emery, 1960).

Organisms which were reported (Foster et al., 1971*b*) to have suffered damage in the rocky intertidal were the high intertidal barnacle *Chthamalus fissus* and the low rocky intertidal surfgrasses, *Phyllospadix* *scouleri* and *P. torreyi* (which have been combined and called surfgrass in the subsequent discussion). In the field *C. fissus* is difficult to distinguish from another high intertidal barnacle, *Balanus glandula*, and the estimates of the number killed almost certainly include members of both species. The immediate damage to barnacles and surfgrass as reported by various authors has been summarized in Table 2, together with oil dosage estimates taken from Figure 1. In Table 3 estimates of the mortality percentages and the number of organisms or living biomass killed as a result of oil pollution and cleanup activities in various habitats have been tabulated. The methodology employed in making these estimates is described in Appendix II. Unfortunately, prespill abundance data were unavailable for most of the species living in the shoreline habitat, precluding any estimate of the effects of oil or cleanup activities.

In addition to the barnacles, over 90 other species commonly inhabit the high intertidal (including the splash zone). These species commonly live in cracks and crevices and subsist on food carried into these locations or forage on the shore at night. These areas were particularly heavily oiled during the spill, and mortalities have been extrapolated (Table 3) from the data of Evans (1970, and pers. comm.). Additional rocky intertidal damage has been reported (Battelle Northwest, 1969; Jones et al., 1969; California Dep. of Fish and Game, 1970; Foster et al., 1971*b*; Nicholson and Cimberg, 1971), but it is impossible to make quantitative estimates of mortalities.

There are indications that the damage to the rocky intertidal may have been far greater than suggested by the few data in Table 3. Straughan (1971*a*) noted a reduction of breeding in three out of six sessile species, two species of mussels, and goose barnacles. The effects of oil pollution on other plants and animals was not documented. Mitchell et al. (1970) reported that in some areas all barnacles, limpets, mussels, and periwinkles were killed and that as much as 1% to 10% of rocky intertidal biomass may have been destroyed on the mainland. California Department of Fish and Game (1969*c*) suggests that 5% to 10% of all intertidal life may have been damaged in polluted areas on the Channel Islands.

Many of the surveys carried out were by their very nature incapable of detecting any but the grossest effects. Foster et al. (1971*b*) and Nicholson and Cimberg (1971) attempted to evaluate species changes, primarily among the algae, along transect lines relative to prespill surveys. They did not investigate changes in abundance systematically. Whether or not the detected effects at the species level were related to oil

Table 2. Barnacle and surfgrass damage and oil dosage estimates (from Fig. 1)

Location	Damage estimate (% killed or % blades lost)	Reference	Oil dosage estimate
Barnacles			
El Capitan Beach	1	Foster et al., 1971*b*	Light
Campus Pt.	20	Foster et al., 1971*b*	Moderate
Santa Barbara Pt.	80-90	Foster et al., 1971*b*	Very heavy
Santa Barbara Pt.	60	Cubit, 1970	Very heavy
E. Cabrillo Beach	100	Nicholson and Cimberg, 1971	Very heavy
Loon Pt.	10	Foster et al., 1971*b*	Heavy
Rincon	<10	Jones et al., 1969	Light
Surfgrass			
Coal Oil Pt.	–†	Nicholson and Cimberg, 1971	Light
Santa Barbara Pt.	50–60	Foster et al., 1971*b*	Very heavy
Carpinteria Reef	30–50	Foster et al., 1971*b*	Light
–*	–†	Cubit, 1970	Light
Anacapa Island	–†	Calif. Fish & Game, 1970	Moderate
Loon Pt.	90–100	Foster et al., 1971*b*	Heavy

* No location given.
† Damage but no % estimate given.

pollution is questionable since no adequate control areas were studied. Furthermore, species may have been present but not observed due to transect line location and/or lack of replication (Connell, 1973). Other studies (California Dep. of Fish and Game, 1969*a*, 1969*b*, 1969*c*, 1970; Mitchell et al., 1970) consisted of little more than reports of damage observed during beach walks or rapidly made survey dives. The inadequacy of such methods is also shown by Chan (1973).

Damage to the rocky intertidal resulting from cleanup activities is difficult to estimate quantitatively since once again very little information is available on the effects of these procedures on marine organisms.

Cleanup activity took place in the rocky intertidal, harbors, and around piers. On solid substrates, it involved sandblasting, steam cleaning, hydroblasting, and spraying with a naptha-talc mixture (Gaines, 1971). Cleaning was initiated in February 1969 and continued through

Table 3. Number of organisms or living biomass killed as a result of oil pollution and cleanup activities

Rocky intertidal	Number killed	Mortality % in oiled areas—mean (range) %
Barnacles (*Chthamalus*)	9,000,000 killed	13 (1–100)
Surfgrass (*Phyllospadix* spp.)	14.6 tonnes of leaves and attached algae and invertebrates killed	27 (0–100)*
Polychaete worms	81,000 killed	27 (0–100)
Limpets (*Acmaea paleacea*)	52,000 killed	27 (0–100)
High intertidal crevice fauna (mostly arthropods)	20,000 killed	—†
Mussels (*Mytilus* spp.)	30,000 killed	60‡
Sandy Beaches		
Sandy beach macrofauna (*Euzonus, Emerita, Orchestoidea*)	15,000,000 removed from beach during cleanup	—†
Deep subtidal zone		
Benthic invertebrates	6,000 tonnes killed	18§
Neritic habitat		
Marine birds	9,000 killed (60% loons and grebes)	3-8″

* Estimate for intertidal populations only. Worms and limpets live on surfgrass.
† Cannot be calculated from available data.
‡ For Santa Barbara Harbor only. From Jones et al., 1969.
§ For 89 km^2 offshore from the mainland near the city of Santa Barbara only.
″ See text.

August 1969 (Gaines, 1971; Lt. G. Brown, U.S. Coast Guard, pers. comm.). Areas reported cleaned included Santa Barbara Harbor and rocks at Bonnymede, Padero Lane, Emma Wood State Beach, Hobson Beach, Sea Cliff, Rincon, and Sandyland (Gaines, 1971; Lt. G. Brown, U.S. Coast Guard, pers. comm.).

No studies were carried out to determine the effects of cleanup procedures. The only "data" consist of comments by individuals who happened to observe effects in areas being cleaned. It is reasonable to assume that the procedures used to remove dried oil from solid substrates will also remove and possibly kill attached organisms (Straughan, 1970; Foster et al., 1971*b*). Foster et al. (1971*b*) suggest that steam cleaning removed an extensive community of limpets, snails, crabs, and algae in the high intertidal. Jones et al. (1969) estimated that 60% of the mussels in Santa Barbara Harbor were killed and believed that this damage resulted from steam cleaning. Thus, in the two cases where the effects of cleanup procedures were observed, damage was probably high.

Using the 60% mortality estimate of Jones et al. (1969), we (see Appendix II) estimate that some 30,000 mussels were killed by cleanup procedures in Santa Barbara Harbor alone.

Quantitative estimates of long-term damage to the rocky intertidal cannot be made due to the lack of data. Foster et al. (1971*b*) predicted that the long-term effects of the spill would be severe in the high rocky intertidal where the oil dried and blanketed rocks and other solid substrates. Sessile organisms do not seem to settle or grow as well on dried oil as on non-oil-contaminated rocks, perhaps because of the higher surface temperatures found with black substrata (Foster et al., 1971*b*), or because of the texture of dried oil.

Straughan (1970) found that barnacles which settled on dried oil-straw substrates are frequently lost due to the instability of this substrate. In addition the oil entered cracks and fissures, dried, and unquestionably reduced the habitat available to the crevice fauna.

Other potential but undocumented long-term effects probably include reduction in breeding success (from sublethal oil contamination) and an indirect reduction in predatory species resulting from the loss of prey (such as a reduction of larval abundance of intertidal species normally eaten by zooplankton, a reduction in predatory snails and starfish which feed upon barnacles and mussels, etc.).

The effects of oil pollution and cleanup activities upon the sandy beaches are even less well documented than in the rocky intertidal. Morin (1971) concluded that oil did not kill the foraminifera and attributed changes in abundance to seasonal changes in the beaches. Straughan (1971*a*) observed breeding and newly settled *Emerita* and breeding of *Euzonus* on previously oiled beaches in June and July 1969. Cubit (1970) found that *Euzonus* avoided contact with oil by altering its normal migratory behavior. Foster et al. (1971*b*) stated that *Orchestoidea* and *Euzonus* appeared undamaged in the vicinity of their rocky intertidal survey sites. Trask (1971) surveyed the sandy beach macrofauna after the initial large spill and found that some organisms were still alive. However, other than this qualitative observation, his methods were such that the data have no relevance on the effects of oil pollution (Connell, 1973).

Westree (unpublished manuscript) has examined the effects of sand removal during the cleanup operations on sandy beaches. Her preliminary estimates suggest that over 15 million organisms were removed from the beaches during the cleanup, including 11 million *Euzonus*, 2 million *Emerita*, and 800,000 *Orchestoidea* (for methods see Appendix II).

Castle (pers. comm.) has found buried oil from the spill in the backshore of some mainland beaches in the vicinity of Santa Barbara. The biological effects of this buried oil are not known at present.

Inner Subtidal

Several qualitative surveys (California Dep. of Fish and Game, 1969*a*, 1969*b*, 1969*c*, 1970; Foster et al., 1971*b*; Nicholson and Cimberg, 1971) revealed no apparent immediate damage to the shallow subtidal biota during the early stages of the spill. During cleanup, oil was physically and chemically sunk in Santa Barbara Harbor (Gaines, 1971), the biological effects of which have not been determined. Long-term damage to the shallow subtidal, except for possible indirect effects of intertidal life on the shallow subtidal, appear minimal.

Deep Subtidal

The effects of the oil spill on the deep subtidal are uncertain. Fauchald (1971) compared the biomass, composition, and distribution of benthic invertebrates in the deep subtidal in the vicinity of Santa Barbara before and after the initial spill. Peak biomass in the prespill survey was 2400 g/m^2. Peak biomass in the postspill survey was 860 g/m^2, a dramatic decline. A substantial decrease in mean biomass also was observed in the 75 samples examined (from 363 g/m^2 to 168 g/m^2). Most of the decline was due to the drop in abundance of the echiuroid worm *Listriolobus pelodes*. Changes in abundance in the less abundant species cannot be determined from the data given (Fauchald, 1971; Connell, 1973), except that 17 species observed in the prespill survey were not recorded in the postspill survey.

The role of oil pollution in this decline cannot be assessed adequately because Fauchald (1971) gives no information on the presence of oil in his samples. He did suggest that increased sewage discharge may be partially responsible for the decline. Juge (1971), however, recorded oil in the sediments in the area and observed that when oil was present, burrowing invertebrates were absent. Thus there is some direct evidence from the Santa Barbara area of a negative correlation between animals and oil in the deep subtidal. Assuming that one-third of the biomass loss estimated by Fauchald (1971) was due to oil, we calculate that some 6000 tons of deep subtidal biomass were lost within the three-mile limit (see Table 3; see also Appendix II).

Dispersing and sinking agents were used during the cleanup over the deep subtidal (Gaines, 1971), and almost certainly contributed to the accumulation of oil on the bottom and to the damage discussed above.

Unfortunately, the long-term effects of oil pollution in this habitat have not been documented. However, the effects may be considerable, and Kolpack et al. (1971) believe that much of the oil from the spill will end up in this habitat.

Neritic

In the neritic habitat, certain marine birds are known to be affected by surface oil (Clark, 1968). The California Department of Fish and Game (1969*b*) reported that 3686 birds were killed by oil between January 28 and May 31, 1969. This estimate was based on birds reported dead on the mainland shores and on birds turned in at bird-cleaning stations on the mainland. This estimate must be minimal since it ignores dead birds which failed to reach the mainland shore or which washed up on the beaches of the Channel Islands (Connell, 1973). Assuming the number of dead birds which came ashore per mile of mainland beach was similar on the mainland and Channel Island beaches, some additional 5000 dead birds may have come ashore on the island beaches, bringing the total to approximately 9000 birds.

Clark (1968) and Nelson-Smith (1970) have observed that shoreline counts of dead seabirds underestimate mortalities. Not all dead birds which come ashore are counted, and many disappear at sea before reaching the shore. However, in Santa Barbara the mainland counts may have been fairly complete, but we have no estimates of birds which disappeared at sea.

Based on a review of California Department of Fish and Game (1969*a*, 1969*b*, 1970) data as discussed by Straughan (1971*c*), Connell (1973) suggests that total mortality to June 1969 was 8000 birds. We believe this is a conservative estimate and that total mortality to June 1969 was probably between 9000 and 15,000 birds.

The impact of these estimated bird mortalities on the bird populations of the mainland, islands, and channel is difficult to estimate. The California Department of Fish and Game (1969*a*) estimated that 12,000 birds were present in the 1075-square-mile (2874-sq-km) area of the channel between February 3 and March 28, 1969—the period of greatest bird mortality. Assuming the mortality of 9000 birds, the total population of the channel at this time would have been 21,000 birds and the mortality would have amounted to 45% of the channel bird population. Connell (1973) suggests that the Fish and Game bird population estimate is too low. This suggestion is substantiated by an analysis of the Christmas bird count (December 28, 1969) by the Audubon Society (Cruickshank, 1969). The total number of common marine birds (loons, western grebes, brown pelicans, double-crested cormorants, and gulls) counted in 1968 was 8758. Assuming that the majority of these birds were found in marine habitats and extrapolating to the entire channel yields a total bird estimate of 322,000—more than an order of magnitude greater than the Fish and Game estimate. This extrapolation of the Audubon data, however, assumes that bird abundance is similar in nearshore and offshore locations and is probably incorrect, leading to an overestimate. If it is assumed that the Audubon abundance data are only typical of the zone within three miles of the coast and that no birds live outside this limit (and ignoring the Channel Islands bird populations), we obtain 116,000 birds present within three miles (4.7 km) of shore. Comparing both of these estimates with a mortality of 9000 birds yields a total mortality percentage of between 2.8% and 7.7%. This figure covers the period January through May 1969.

The oil pollution did not affect all bird species equally. About 60% of the birds reported killed were loons and grebes (Straughan, 1970). Since these species breed in northern freshwater habitats in spring and summer (Peterson, 1961), breeding colonies were not directly affected. An analysis of Audubon Society Christmas counts (Table 4; Cruickshank, 1967, 1968, 1969, 1970, 1971, 1972; Arbib and Heilbrum, 1973; Metcalf, pers. comm.) from Santa Barbara (for loons and western grebes) showed no significant declines (Mann-Whitney U test) between 1966-68 and 1969-73 when San Diego western grebe counts were used as a control.

The initial reports of oil pollution-related deaths of whales, elephant seals, and sea lions have generally been discredited (Brownell, 1971; Brownell and Le Boeuf, 1971; Le Boeuf, 1971). An analysis by Connell (1973) of the data of Brownell and Le Boeuf (1971) concluded that oil may have caused some sea lion pup deaths.

No obvious changes in species composition of marine phytoplankton in Goleta Bay in April and May 1969 were observed relative to prespill observations (Holmes, 1969). Oguri and Kanter (1971) observed a reduc-

Table 4. Audubon Society Christmas bird counts, 1966-73

	1966	1967	1968	1969	1970	1971	1972	1973*
Santa Barbara								
Loons†	40	30	43	53	8	16	51	15
Western grebes	267	414	186	139	172	108	103	192
San Diego								
Western grebes	1508	3861	405	1074	2357	3240	871	

* Data for Santa Barbara from Metcalf (pers. comm.).
† Includes common, arctic, and red-throated loons.

tion in primary production around Platform A in April 1969, but it was not shown that this reduction was due to oil pollution or use of detergents (Connell, 1973). A study of macroplankton by Ebeling et al. (1971) found no postspill changes that could be attributed to oil pollution. The temporary disappearance of a mysid shrimp from the kelp beds may have been caused by oil or reduced surface salinity (Ebeling et al., 1971).

Pelagic fish-spotting records and total commercial catches and commercial sport catches showed no abnormal changes after the spill (California Dep. of Fish and Game, 1970; Straughan, 1971*b*), and bottom fish communities appeared normal (Ebeling et al., 1971).

In preceding pages references have been made to biological damage which could not be quantified in any way. In Table 5 we have summarized this information.

Discussion

It should be apparent from the above summary that too few data are available to assess the overall impact of the spill upon the channel waters and its associated biota. The usefulness of some of the limited number of studies carried out after the spill is impaired by inadequate sampling and the lack of suitable controls. Some of the before and after data do not lead to unequivocal interpretations because the observed changes cannot be shown to be the result of oil pollution or any other specific environmental change. Since the overall impact of the oil spill cannot be adequately evaluated, it is obviously impossible to evaluate overall recovery.

Table 5. List of organisms reported damaged or for which damage was highly probable. There are insufficient data to determine total killed or biomass lost.

	Notes
Rocky intertidal shore	
Kelp crabs (*Pugettia producta*)	
Hermit crabs (*Pagurus* spp.)	In surfgrass
Isopods (*Idothea* spp.)	
Snails (*Lacuna* and *Aceton* [?])	
Barnacles (*Pollicipes*)	Observed damaged by various intertidal investigators
Algae (*Enteromorpha, Ulva, Porphyra, Gigartina, Hesperophycus*)	
All rocky intertidal organisms (including Santa Barbara Harbor)	Affected by cleanup activities
Shallow subtidal zone	
Subtidal benthic organisms in Santa Barbara Harbor	
Neritic habitat	
California sea lions (*Zalopus californianus*)	

Notwithstanding these problems, both short-term and long-term damage to and recovery of portions of the channel ecosystems have been demonstrated. Marine bird abundances are now at levels comparable to prespill levels. Qualitative observations made at several heavily polluted rocky intertidal areas five years after the spill indicate that the surfgrass community and barnacles have reestablished. Without prespill abundance data, it is impossible to determine if this reestablishment has been complete. Although no abundance data are available, the continued presence of dried oil in cracks and crevices in the high intertidal zone suggests that the crevice fauna has not recovered.

We have mentioned on several occasions that it has not always been possible because of inadequate experimental design to assign biological damage or apparent biological change to a particular cause in a given habitat. This is particularly frustrating. The extraordinarily heavy rains and runoff of December 1968 and January and February 1969 caused flooding and introduced large quantities of sediment and freshwater into and onto the nearshore environment of the channel and almost certainly had direct or indirect biological effects. Natural oil seepage, cumulative effect of sewage discharge, introduction of DDT, etc., in channel waters all may have acted singly or in consort with the oil from well No. 21 to cause damage or change. Although the damage estimates made in the present chapter attempt to refer only to damage by oil, some of the damage may have been caused by other agents.

Since the data collected leave so many unanswered questions about the effect of the spill, we might ask whether it is still possible to make observations or studies in the channel which might help determine the biological effects of the spill. It seems to us unlikely that such observations would help resolve the dilemma. The available evidence strongly indicates that all of the surface sediments of the channel have received oil since the spill, leaving only a remote chance that any area might have been left unaffected. Similarly, the location of other areas which could serve as controls in evaluating the effects of the rains and flooding of December 1968, with and without oiled sediments, seems remote. Purposely created spills in carefully selected areas of the channel with adequate biological and physical controls, evaluation of the cleanup procedures employed, and studies on the effects both long- and short-term of oil, if allowable, would yield useful information but would not reveal what happened as a result of the spill of 1969.

If we accept Webster's (*Third International Dictionary, Unabridged,* 1971) definition of a disaster as being "a sudden calamitous event producing great damage and loss," we find ourselves being unable to conclude with certainty whether or not the Santa Barbara spill qualifies. We have attempted to estimate damage and loss. These estimates, although conservative, leave considerable uncertainty as to total damage. Our conclusion is that we do not have and probably never will have an accurate assessment of the initial overall biological effects of the Santa Barbara spill. No really adequate data exist to permit an accurate assessment of these effects or of the recovery of the channel biota.

Our discussion of the Santa Barbara spill points out the great difficulty in assessing biological damage and recovery from such a pollution incident. These studies were carried out on an ad hoc basis and in a crisis atmosphere. Little time was available to assemble research teams and to design adequate sampling procedures and programs. Many competent scientists were unable to participate in the studies, and cooperation and coordination between those who did work on the spill was not great. The initial interest of many academics (faculty and students) in the biological effects of the spill was not sustained, and follow-up studies have been few.

The national and area contingency plans developed by the Coast Guard as a result of the Santa Barbara spill are not really designed to provide answers to the biological effects of a spill, especially the long-term effects. We believe these deficiencies should be rectified immediately by the establishment of national scientific response teams made up of qualified and experienced individuals capable of carrying out or directing studies to assess biological damage and recovery resulting from pollution incidents. We also urge that immediate attention be given to the design of sampling techniques and sampling schemes of sufficient sophistication to greatly increase the probability of assessing damage resulting from a pollution incident.

The inherent difficulty and great expense in assessing biological damage and recovery resulting from oil pollution suggest than an effective program in oil pollution prevention and containment procedures should be maintained.

Acknowledgments

This review represents part of a report prepared for the Department of Justice, State of California, Contract No. 455. We especially thank D. Antenore for his generous assistance. J. H. Connell, R. Castle, B. Westree, W. G. Evans, and A. Allen kindly provided unpublished information and helpful suggestions. D. Pirie provided access to the U.S. Army Corps of Engineers' aerial photographs used in determining the extent of the mainland rocky intertidal.

Literature Cited

Allen, A. 1969. Santa Barbara oil spill. Pages 146-153 *in* Hearings from the Subcommittee on Minerals, Materials, and Fuels of the Committee on Interior and Insular Affairs. Senate, 91st Congress, May 19-20, 1969. U.S. Government Printing Office, Washington, D.C.

Arbib, R., and L. H. Heilbrum, eds. 1973. Seventy-third Christmas bird count. Am. Birds 27:135-540.

Battelle Northwest. 1969. Review of Santa Barbara Channel Oil Pollution Incident. Fed. Water Pollut. Control Admin., Water Pollut. Control Res. Series, DAST 20. U.S. Government Printing Office, Washington, D.C.

Brownell, R. L. 1971. Whales, dolphins and oil pollution. Pages 255-276 *in* D. Straughan, ed. Biological and Oceanographical Survey of the Santa Barbara Channel Oil Spill 1969-1970. Vol. 1. Allan Hancock Foundation, Univ. of Southern California, Los Angeles.

Brownell, R. L., and B. J. LeBoeuf. 1971. California sea lion mortality: natural or artifact? Pages 287-306 *in* D. Straughan, ed. Biological and Oceanographical Survey of the Santa Barbara Channel Oil Spill 1969-1970. Vol. 1. Allan Hancock Foundation, Univ. of Southern California, Los Angeles.

California Department of Fish and Game. 1969*a*. Progress report on wildlife affected by the Santa Barbara Channel oil spill January 28-March 31, 1969. Mimeo. State of California, Sacramento.

_____. 1969*b*. Second progress report on wildlife affected by the Santa Barbara Channel oil spill April 1-May 31, 1969. Mimeo. State of California, Sacramento.

_____. 1969*c*. Cruise report 69A2: Inshore survey of Santa Barbara oil spill. Mimeo. State of California, Sacramento.

_____. 1970. Santa Barbara oil leak iterim report. Mimeo. State of California, Sacramento.

Caplan, R., and R. Boolootian. 1967. Intertidal ecology of San Nicolas Island. Pages 203-217 *in* R. Philbrick, ed. Proceedings of the Symposium on the Biology of the California Islands. Santa Barbara Botanic Garden, Santa Barbara.

Chan, G. L. 1973. A study of the effects of the San Francisco oil spill on marine organisms. Pages 741-759 *in* Proceedings of Joint Conferences on Prevention and Control of Oil Spills, March 13-15, 1973, Washington, D.C. American Petroleum Institute, Washington, D.C.

Clark, R. B. 1968. Oil pollution and the conservation of seabirds. Pages 76-112 *in* Oil Pollution of the Sea. Report of the Proceedings of the International Conference on Oil Pollution of the Sea, Rome.

Connell, J. H. 1973. A review of: Straughan, D. 1971. Biological and Oceanographical Survey of the Santa Barbara Channel Oil Spill. Allan Hancock Foundation, Univ. of Southern California. 426 pp. and A guide to the proper method of investigation of the effects of oil on marine organisms. Submission to California State Land Commission, Sacramento.

Cruickshank, A. D., ed. 1967. Sixty-seventh Christmas bird count. Audubon Field Notes 21:81-384.

_____. 1968. Sixty-eighth Christmas bird count. Audubon Field Notes 22:93-402.

_____. 1969. Sixty-ninth Christmas bird count. Audubon Field Notes 23:113-432.

_____. 1970. Seventieth Christmas bird count. Audubon Field Notes 24:101-464.

_____. 1971. Seventy-first Christmas bird count. Am. Birds 25:131-514.

_____. 1972. Seventy-second Christmas bird count. Am. Birds 26:147-530.

Cubit, J. 1970. The effects of the 1969 Santa Barbara oil spill on marine intertidal invertebrates. Pages 131-136 *in* R. W. Holmes and F. A. DeWitt, eds. Santa Barbara Oil Symposium: Offshore Petroleum Production, an Environmental Inquiry. U.S. Government Printing Office, Washington, D.C. No. 1972 0-463-300.

Ebeling, A. W., W. Werner, F. A. DeWitt, and G. M. Cailliet. 1971. Santa Barbara Oil Spill: Short-Term Analysis of Macroplankton and Fish. Environmental Protection Agency. Water Pollut. Control Res. Series,

15080EAL02/71. U.S. Government Printing Office, Washington, D.C. 68 pp.

Emery, K. O. 1960. The Sea off Southern California. J. Wiley and Sons, New York. 366 pp.

Evans, W. G. 1970. *Thalassotrechus barbarae* (Horn) and the Santa Barbara oil spill. Pan-Pac Entomol. 46:233-237.

Fauchald, K. 1971. The benthic fauna in the Santa Barbara Channel following the January, 1969, oil spill. Pages 61-116 *in* D. Straughan, ed. Biological and Oceanographical Survey of the Santa Barbara Channel Oil Spill 1969-1970. Vol. 1. Allan Hancock Foundation, Univ. of Southern California, Los Angeles.

Foster, M., A. C. Charters, and M. Neushul. 1971*a*. The Santa Barbara oil spill. Part 1: Initial quantities and distribution of pollutant crude oil. Environ. Pollut. 2:97-113.

Foster, M., M. Neushul, and R. Zingmark. 1971*b*. The Santa Barbara oil spill. Part 2: Initial effects on intertidal and kelp bed organisms. Environ. Pollut. 2:115-134.

Gaines, T. H. 1971. Pollution control at a major oil spill. J. Water Pollut. Control Fed. 43:651-667.

Holmes, R. W. 1969. The Santa Barbara oil spill. Pages 15-27 *in* D. P. Hoult., ed. Oil on the Sea. Plenum Press, New York.

Jones, L. G., C. T. Mitchell, E. K. Anderson, and W. J. North. 1969. Just how serious was the Santa Barbara oil spill? Ocean Industry 4:53-56.

Juge, D. M. 1971. A study of the bacterial population of the bottom sediments in the Santa Barbara Channel after the oil spill. Pages 179-222 *in* D. Straughan, ed. Biological and Oceanographical Survey of the Santa Barbara Channel Oil Spill 1969-1970. Vol. 1. Allan Hancock Foundation, Univ. of Southern California, Los Angeles.

Kolpack, R. L., J. S. Mattson, H. B. Mark, and T. C. Yu. 1971. Hydrocarbon content of Santa Barbara Channel sediments. Pages 276-295 *in* R. L. Kolpack, ed. Biological and Oceanographical Survey of the Santa Barbara Channel Oil Spill 1969-1970. Vol. II. Allan Hancock Foundation, Univ. of Southern California, Los Angeles.

Le Boeuf, B. J. 1971. Oil contamination and elephant seal mortality: a "negative" finding. Pages 277-285 *in* D. Straughan, ed. Biological and Oceanographical Survey of the Santa Barbara Channel Oil Spill 1969-1970. Vol. 1. Allan Hancock Foundation, Univ. of Southern California, Los Angeles.

Mertz, R. C. 1959. Determination of the Quantity of Oily Substances on Beaches and in Nearshore Waters. Calif. State Water Pollut. Control Bd., Sacramento. Publ. No. 21. 45 pp.

Mitchell, C. T., E. K. Anderson, L. G. Jones, and W. J. North. 1970. What oil does to ecology. J. Water Pollut. Control Fed. 42:812-818.

Morin, R. W. 1971. Seasonal study of beach foraminiferal populations in the Santa Barbara Channel area. Pages 64-89 *in* R. L. Kolpack, ed.

Biological and Oceanographical Survey of the Santa Barbara Channel Oil Spill 1969-1970. Vol. II. Allan Hancock Foundation, Univ. of Southern California, Los Angeles.

Nelson-Smith, A. 1970. The problem of oil pollution of the sea. Adv. Mar. Biol. 8:215-306.

Nicholson, N. L., and R. L. Cimberg. 1971. The Santa Barbara oil spills of 1969: a post-spill survey of the rocky intertidal. Pages 325-399 *in* D. Straughan, ed. Biological and Oceanographical Survey of the Santa Barbara Channel Oil Spill 1969-1970. Vol. 1. Allan Hancock Foundation, Univ. of Southern California, Los Angeles.

Oguri, M., and R. Kanter. 1971. Primary productivity in the Santa Barbara Channel. Pages 17-48 *in* D. Straughan, ed. Biological and Oceanographical Survey of the Santa Barbara Channel Oil Spill 1969-1970. Vol. 1. Allan Hancock Foundation, Univ. of Southern California, Los Angeles.

Peterson, R. T. 1961. A Field Guide to Western Birds. 2d ed. Houghton Mifflin Co., Boston. 366 pp.

Ricketts, E. F. and J. Calvin. 1962. Between Pacific Tides, 3d ed. revised. Stanford University Press. 516 pp.

Smith, E. J., ed. 1968. Torrey Canyon Pollution and Marine Life. Cambridge University Press. 197 pp.

Straughan, D. 1970. Ecological effects of the Santa Barbara oil spill. Pages 173-182 *in* R. W. Holmes and F. A. DeWitt, eds. Santa Barbara Oil Symposium: Offshore Petroleum Production, an Environmental Inquiry. U.S. Government Printing Office, Washington, D.C. No. 1972 0-463-300.

——. 1971*a*. Breeding and larval settlement of certain intertidal invertebrates in the Santa Barbara Channel following pollution by oil. Pages 223-244 *in* D. Straughan, ed. Biological and Oceanographical Survey of the Santa Barbara Channel Oil Spill 1969-1970. Vol. 1. Allan Hancock Foundation, Univ. of Southern California, Los Angeles.

——. 1971*b*. Oil pollution and fisheries in the Santa Barbara Channel. Pages 245-254 *in* D. Straughan, ed. Biological and Oceanographical Survey of the Santa Barbara Channel Oil Spill 1969-1970. Vol. 1. Allan Hancock Foundation, Univ. of Southern California, Los Angeles.

——. 1971*c*. Oil pollution and sea birds. Pages 307-312 *in* D. Straughan, ed. Biological and Oceanographical Survey of the Santa Barbara Channel Oil Spill 1969-1970. Vol. 1. Allan Hancock Foundation, Univ. of Southern California, Los Angeles.

Trask, T. 1971. A study of three sandy beaches in the Santa Barbara, California, area. Pages 159-177 *in* D. Straughan, ed. Biological and Oceanographical Survey of the Santa Barbara Oil Spill 1969-1970. Vol. 1. Allan Hancock Foundation, Univ. of Southern California, Los Angeles.

Westree, B. Unpublished manuscript. The effects of cleanup procedures on sandy beach organisms.

Appendix I
Conversion Factors

1. California crude oil: specific gravity 0.907
2. 1 barrel ≑ 42 gallons (U.S. liquid)
   1 barrel (321 pounds) ≑ 145.2 kilograms
   6.88 barrels ≑ 1 tonne

Appendix II
Damage Estimates

In the text we attempted, where possible, to estimate the number of organisms killed during the Santa Barbara oil spill. These estimates were made to avoid the often subjective terms such as *severe, significant, moderate*, etc., which are frequently used to describe damage and to provide a basis for an economic evaluation of damage.

The number of barnacles and organisms associated with surfgrass killed and the biomass of surfgrass lost were estimated by combining information on the amount of oil which came ashore along various lengths of rocky intertidal shoreline, the normal abundance of and the reported damage to organisms on the shore, and the linear extent of rocky intertidal shoreline in areas covered by the oil. The amounts of oil which came ashore along various sections of the channel shoreline are shown in Figure 1. To establish the relationship between the percentage of damage and the amount of oil on shore, damage estimates made by various investigators at particular sites were compared with oil dosage estimates at these sites. These data are shown in Table 1. There is a good correlation between barnacle damage and oil dosage. This correlation is generalized into a dose-response relationship in Table A1. The relationship for surfgrass is not so good, but the data suggest a generalized dose-response relationship as shown in Table A1.

Barnacles and surfgrass are ubiquitous members of rocky intertidal communities in the channel. Both organisms were found at 16 out of 18 rocky intertidal mainland stations within the area of pollution surveyed by Nicholson and Cimberg (1971), and at all 10 stations surveyed by Foster et al. (1971*b*). They are also common on the Channel Islands (Caplan and Boolootian, 1967; Nicholson and Cimberg, 1971).

Although the general distribution of common rocky intertidal species is known for most of the coast, few data are available on their abundance. Since abundance estimates were needed to convert mortality percentage to actual numbers in this analysis, we assumed that populations had returned to prespill densities and made counts of barnacles and measured the biomass of surfgrass between Santa Barbara and Carpinteria (see Fig. 1) in April 1974. To estimate the number of barnacles per unit length of shore, individuals were counted within randomly located strips. Five strips, 0.25 m × 2 m, oriented perpendicular to the shore in the high intertidal, were counted, giving values of 25, 75, 76, 154, and 105,000 barnacles per strip. The latter count was derived from counts within subquadrats of 4 cm^2. The variation is extremely high, and it was obvious that estimating the mean abundance would require detailed field surveys beyond the scope of this review. Therefore, to assure a conservative error we used the median abundance of 76 barnacles per strip. In counts at Santa Barbara Point, Cubit (1970) found a normal mortality of 8% in unoiled barnacle populations. Connell (pers. comm.), based on his experience with barnacle populations, suggests 20% as a high estimate of normal mortality. Using this latter figure, the live abundance is estimated at 61 barnacles per strip. This figure is used in the damage estimates below.

To estimate the average width of the rocky high intertidal where barnacle damage occurred, the data on the distribution of organisms along transects at the ten mainland stations surveyed by Foster et al. (1971*b*) were examined. There were 20 surveys taken at these stations which noted the linear extent of barnacle populations. The average width of the barnacle zone was 16.5 m ± 23.2 (one standard deviation). Therefore, the average abundance of barnacles in the rocky high intertidal is estimated to be (61) (16.5/2) = 502/linear 0.25 m of high intertidal, or 2008 barnacles per linear meter.

The slope of the rocky intertidal on the Channel Islands is generally steeper than on the mainland. In two areas on San Nicolas Island, Caplan and Boolootian (1967) found that *Chthamalus* occupied a strip about 1.37 m wide in the high intertidal. Assuming that the abundance per unit area would be proportionally the same relative to the mainland, barnacle abundance for the Channel Islands is estimated at (2008) (1.37/16.5) = 167/linear m.

At the same time barnacle counts were made, surfgrass samples were taken to estimate wet weight per unit area in the intertidal. Weight rather than number of individuals was determined because blades, not individuals, were killed. In addition, it is difficult to determine what an individual is for species with extensive rhizomes. Four 0.25-m^2 quadrats were located in areas of 100% blade cover, and the grass was collected by cutting off the blades just above the substratum. All blades in the samples were green (live). Motile animals were removed, excess water drained off, and the samples weighed. The mean weight was 0.47 kg ± 0.23/0.25 m^2. This weight includes attached plants and animals, primarily *Smithora naiadum* (red alga), *Melobesia mediocris* (red alga), hyroids, and *Lacuna* (snail) eggs. These organisms were probably killed along with the blades. No correction was made for natural mortality. The absence of brown blades in the samples indicates that this mortality is either very low or difficult to detect with the methods used.

The average width of the rocky intertidal occupied by surfgrass was also calculated from transect coverage data discussed by Foster et al. (1971*b*). However, for surfgrass data were available from 36 transects (see Foster et al., 1971*b*, for a discussion of these transects). Along the transects the mean width of the intertidal occupied at least in part by surfgrass was 10.98 m ± 25.8. Unfortunately, there are no cover percentage data available. Based on personal observations at most of these study sites, we conservatively estimate a 10% coverage of live, green blades. Therefore, a mean width of 10.98 m at 10% cover would be equivalent to a 1.098-m width with 100% cover. The estimate of wet weight for mainland surfgrass is then (0.47) (1.098/0.25)=2.05 kg/linear m.

Surfgrass, like barnacles, occupies a narrower band on the Channel Islands due to differences in intertidal slope. On the two transects surveyed by Caplan and Boolootian (1967) surfgrass occurred on one and was found in the lower intertidal along 0.91 m of the survey line. Again based on personal observations on the islands, we estimated an average coverage width of 0.46 m. Assuming 10% cover, the surfgrass estimate for the islands is (2.05) (0.046/1.098)=0.086 kg/linear m. Combining these estimates with the mortality percentages gives the number of barnacles killed and the biomass of surfgrass blades removed per linear meter for the different oil doses (see Table A1).

To complete the barnacle and surfgrass damage estimates, the linear extent of rocky intertidal habitat in the areas affected by the oil was determined. For the mainland coast, the extent of rocky intertidal (including breakwaters, harbors, and piers) was plotted from aerial photographs supplied by the U.S. Army Corps of Engineers. A complete series of photos taken at 3049 m on December 14, 1972, from Point Mugu to Point Arguello was examined, and the linear extent of rocky intertidal plotted on acetate overlays. Sections of the coast were considered rocky intertidal if over 50% of the shore contained exposed

Table A1. Estimates of barnacle and surfgrass damage relative to oil dose on rocky shores

Oil dosage	% barnacles killed	% surfgrass killed	Barnacles killed/ linear m of shore		Surfgrass blade loss in kg/linear m of shore	
			Mainland*	Islands†	Mainland‡	Islands§
Very heavy	60	50	1205	100.2	1.025	0.043
Heavy	50	50	1004	83.5	1.025	0.043
Moderate	10	30	201	16.7	0.615	0.026
Light	1	10	20.1	1.67	0.205	0.007
Very light	1	0	20.1	1.67	0	0

*Estimated normal live abundance = 2008/linear m of shore.
†Estimated normal live abundance = 167/linear m of shore.
‡Estimated normal live abundance = 2.05 kg/linear m of shore.
§Estimated normal live abundance = 0.086 kg/linear m of shore.

rocks. Much of the rocky intertidal on the mainland is composed of large cobbles and boulders surrounded by sand. To correct for the sand in these situations, we measured only half the actual length of the area. The lengths of piers were considered as equivalent lengths of rocky intertidal. The extent of rocky intertidal as measured from the photographs correlates well with our personal observations at selected sites within the photographed area.

After estimating the extent of rocky intertidal from the 3049-m photos, selected areas on 1524-m photos were plotted for comparison. Replicate plots indicated that the 3049-m estimates were consistently about 30% higher than those from 1524 m. The 3049-m estimates were corrected by this amount and are given, arranged according to areas oiled (see Fig. 1), in Table A2.

The total length of mainland rocky intertidal, minus breakwaters, piers, and harbor interiors, is 24,790 m. The total length of the coast is 191 km. Therefore, we estimate that about 13% of the mainland coast polluted by the oil is rocky intertidal (see Table A2). This is in good agreement with Emery's (1960) estimate for the entire southern California mainland coast of 16.5%.

Emery's (1960) estimate of rocky intertidal on the Channel Islands is 81.7%. Since no aerial photographs are available for the islands, the length of rocky intertidal in oiled areas in this region was determined by multiplying the lengths of island shore oiled by this percentage. Allen (1969) shows the sections of island shore polluted (see Fig. 1). His plots were transferred to U.S. Department of Commerce Nautical Chart No. 5202 and the coast lengths measured with a map measure.

These data were combined with the damage per linear meter per unit oil dosage to arrive at total barnacles killed and biomass of surfgrass removed (see Table A2 and Table 3). The lengths of piers and harbor interiors were not used in the calculations for surfgrass since it generally does not grow in these areas.

The limpet, *Acmaea paleacea*, is found exclusively on surfgrass blades (Ricketts and Calvin, 1962), and polychaete worms are found in tubes attached to the blades. These organisms were probably killed along with the blades. To estimate the number killed in Table 3, individuals were counted in the four surfgrass samples discussed above. The variation was large, and, to be conservative, the lowest abundances in the samples (*A. paleacea*, 3.5/kg of blades; polychaete worms, 5.5/kg of blades) were used. These numbers were multiplied by the total surfgrass biomass lost (see Table A1) to derive total individuals killed (see Table 3).

Table A2. Estimate of total damage to barnacles and surfgrass

Section of shore*	(1) Total length of rocky intertidal (m)	(2) Col. 1 minus piers and harbor interiors (m)	(3) Oil dosage*	(4) Barnacles killed/m †	(5) Total barnacles killed: cols. (1) × (4)	(6) Surfgrass blades lost (kg/m)†	(7) Total weight of surfgrass blades lost (kg): cols. (2) × (6)
Mainland							
I	1,582	796	very light	20.1	31,798	0	0
II	4,453	2,788	moderate	201	895,053	0.615	1,715
III	5,447	4,106	light	20.1	109,485	0.205	842
IV	392	392	heavy	1,004	393,568	1.025	402
V	630	490	moderate	201	126,630	0.615	301
VI	3,207	1,945	very heavy	1,205	3,864,435	1.025	1,994
VII	798	798	moderate	201	160,398	0.615	491
VIII	769	281	light	20.1	15,457	0.205	58
IX	281	281	very light	20.1	5,648	0	0
X	13,023	12,913	moderate	201	2,617,623	0.615	7,941
Total mainland	30,582	24,790					
Channel Islands							
XI	14,150	14,150	moderate	16.7	236,305	0.026	368
XII	15,175	15,175	moderate	16.7	253,423	0.026	395
XIII	5,126	5,126	moderate	16.7	85,604	0.026	133
Total islands	34,451	34,451			8,795,427		14,640kg (14.64 t)

*From Figure 1.
†From Table A1.
‡Total mainland coast length from Pt. Arguello to Pt. Mugu = 191 km. Therefore, % rock = 24,790/191,000 = 13%.

To estimate damage to mussels (see Table 3) in Santa Barbara Harbor, we measured normal abundance by counting mussels in 0.25-m-wide strips on three pilings under the Coast Guard pier next to the shore, in one strip on the rocks under the pier, in one strip in an area of rocks near the Breakwater Restaurant, and a strip on the vertical wall on the north side of the harbor in April 1974. Total mussels ranged from 0 to 74 per strip, with a mean of 22.5. The linear extent of rock and cement areas in the harbor interior is about 1524 m as measured from aerial photographs (see above). Using the 60% mortality figure of Jones et al. (1969) and a low estimate of 33 mussels per linear meter, we estimate that some 30,000 ($33 \times 1524 \times 0.6$) mussels were killed.

Westree (unpublished manuscript) examined the effects of cleanup operations on sandy beach organisms. During beach cleanup at least 9826 loads of "debris" were removed from contaminated areas (Gaines, 1971). Assuming these were dump-truck loads of about 7.65 m^3 each and that 25% of the debris was sand, about 18,792 m^3 of sand were removed. Westree estimated the average abundance of sandy beach organisms per unit volume of sand from the data of Trask (1971) for Santa Barbara area beaches in January and April 1970. She found that during cleanup, sand was removed to a depth approximating the depth of the samples taken by Trask (1971). Multiplying the average number of macroorganisms per unit volume by the total volume of sand removed during cleanup gives the estimate in Table 3.

The deep subtidal biomass loss (see Table 3) was calculated using the information derived from Fauchald (1971). There was an average biomass reduction of 195.62 g/m^2 between the prespill and postspill surveys discussed by Fauchald (1971). Most of the samples in these surveys were taken offshore between Santa Barbara and southern Carpinteria (see Fig. 1). The deep subtidal area between these points and seaward a distance of 4.8 km is about 89 km^2. Assuming that one-third of the biomass loss discussed above and in the text is attributable to the oil pollution, then over 5800 metric tons $[(1/3)(89 \times 195.62)]$ of benthic marine life was killed or otherwise removed in this part of the deep subtidal habitat.

The mortality percentages given in Table 3 were derived by dividing the total estimated normal abundance (calculated by multiplying the live abundances in Table A1 and those given for limpets and polychaetes by the total length of rocky intertidal shore given in Table A2) into the numbers killed or biomass lost estimates and multiplying by 100. The ranges given are from Table 2 and Table A1.

Recovery of Some British Rocky Seashores from Oil Spills and Cleanup Operations

Anthony Nelson-Smith

Abstract

The effects of oil spills of varying severity are described. The most striking result is a reduction in the numbers of grazing molluscs, especially limpets, resulting in successive blooms of green and brown algae. The recovery period varied from up to seven years after a major crude-tanker wreck (even longer where toxic cleansers were spilled) to virtually no time after a small spillage of heavier oils. Recovery may be retarded by further spillages at the same site, while chronic pollution by an oily refinery effluent has modified the balance of shore populations so that recovery to the previous state can never occur. However, cleanup procedures need not necessarily inhibit a reasonable recovery. Simple transect surveys can reveal both effects on shore life and the progress of its recovery. Serious oil spills cause about as much ecological disturbance as unusually cold winters but, unlike these, are potentially avoidable.

Discussion

Between October 1946 and March 1947, a strip 10 meters wide down the shore at Port St. Mary in the Isle of Man was systematically cleared both of limpets and the larger algal growths by biologists from the nearby marine station of Liverpool University (Jones, 1948; Lodge, 1948). Twenty years later, the experiment was accidentally repeated on a grand scale along some of the shores at the approaches to the English Channel when the supertanker *Torrey Canyon* broke up on the Seven Stones Reef between the Isles of Scilly and the British mainland (see, e.g., Smith, 1968). She eventually lost her entire cargo of 117,000 tons of crude oil, of which perhaps 40,000 tons were distributed around the coast of Cornwall. In 1967 large numbers of limpets, together with many other organisms common on Cornish seashores, were eliminated partly by immersion in this oil, which came ashore thickly emulsified in seawater, but mostly by relentless efforts at clearing it away using very toxic so-called detergents.

The effect on algal cover was much the same in each case. Fairly exposed rocky seashores around the North Atlantic are dominated by acorn barnacles, limpets, mussels, and two or three species of marine snail. Figure 1 shows the abundance and distribution of common plants and animals on a rocky shore near The Lizard, in southwest Cornwall, which just avoided damage from the spillage and subsequent cleanup operations. Three species of limpet are present, of which *Patella vulgata* is widespread over the middle of the shore, *P. aspera* becomes important toward low water, and *P. depressa* is relatively insignificant. The most important winkles are the tiny *Littorina neritoides* and the small subspecies of *L. saxatilis*, while the most abundant large seaweed is a bladderless form of *Fucus vesiculosus* which is typical of exposed conditions. On this belt transect, stations were surveyed at two-foot vertical intervals, assessing abundance according to standard criteria using methods described by Moyse and Nelson-Smith (1963) and Nelson-Smith (1967, 1970). Limpets and, to a lesser extent, winkles and topshells feed by rasping microalgae or the sporelings of larger plants from the rock surface with a toothed radula. The Port St. Mary experiment showed just how effective these grazing activities are since, within a few weeks of clearance, each section of the strip became covered by a turf of annual green algae (settling from the plankton with much greater success than normal on such shores, from which they are often almost entirely absent). Species of *Enteromorpha* predominated among these, soon joined by *Ulva lactuca* (sea lettuce) and forming a dense growth visible at some distance. Sporelings of the longer-lived brown fucoids then became established in this turf and grew into plants of moderate size which, in turn, provided shelter for a greater diversity of small brown and red algae. However, living or recently overgrown barnacles offer a poor foothold for bushy seaweeds in a wave-washed situation, and full-grown limpets rapidly moved into the strip from either side. Thus the larger plants were torn away during winter storms, while newly settled sporelings became increasingly grazed away as before. The strip reverted to the state of the surrounding rock ledges after just over three years.

Recovery of the worst affected areas in Cornwall reflected the greater extent of damage there. On a number of shores, a careful search

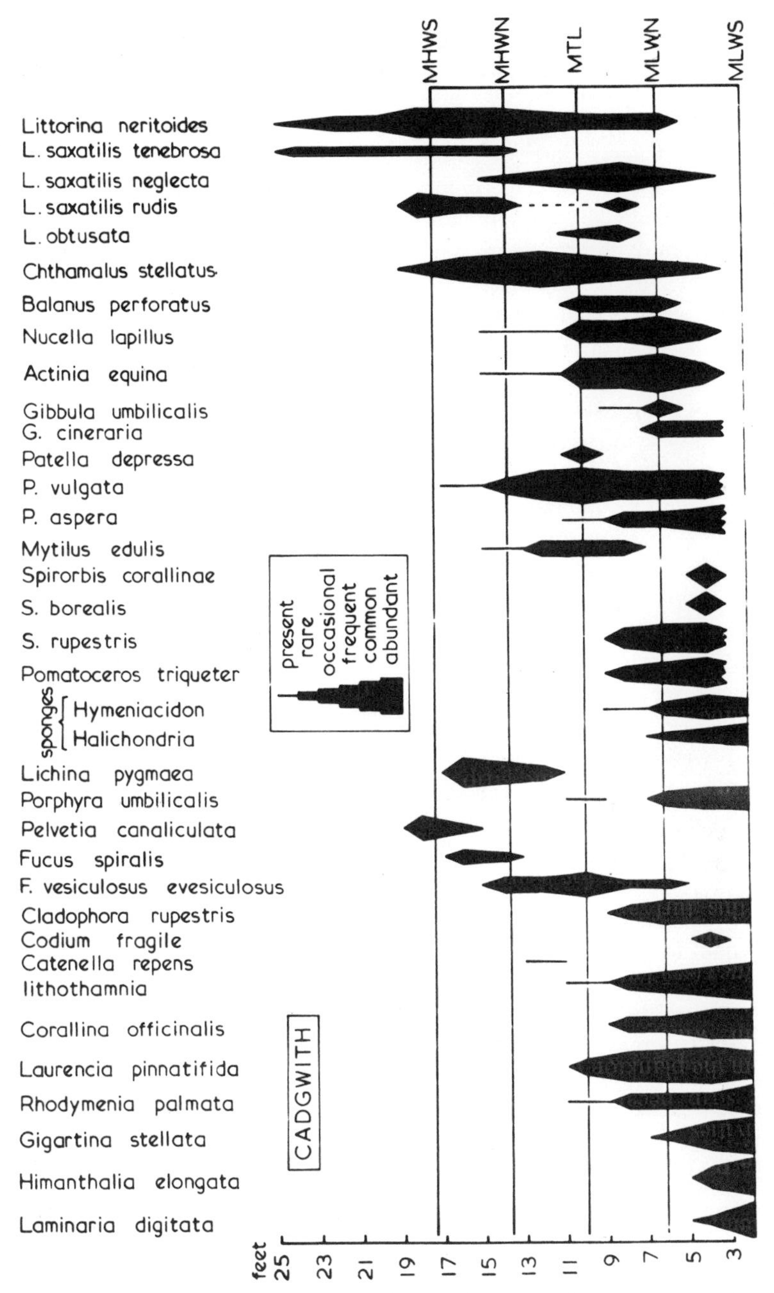

Fig. 1. Zonation of common plants and animals on a typical semiexposed rocky shore in Cornwall. (From Nelson-Smith, 1970)

was necessary to find any sedentary animals or small plants which were still alive (much of Sennen Cove, for example, had been cleared down to a bare rock surface) so that newly settling green algae were able to colonize what was virtually a virgin area. Limpets had been eliminated over a region wide enough to make the intrusion of adults from undamaged shores quite insignificant. The same sequence of events occurred, species of *Enteromorpha* and the oyster thief *Colpomenia peregrina* being the most obvious green weeds, followed, as at Port St. Mary, dominantly by *Fucus vesiculosus*. Figure 2 illustrates the situation at two heavily polluted sites six months after the oil had been cleared. Formal surveys had not been carried out before the oil was stranded there (in fact, oil failed to reach the few shores which were hastily surveyed in order to be able to assess its effects accurately!). Open (white) histograms thus show only the distribution of sedentary organisms found in their normal habitats (but often far from healthy) immediately after cleansing operation ceased in April 1967; many more were either smothered, poisoned, or washed away by high-pressure hosing. The abundance of limpets before pollution was indicated both there and even more clearly on softer rocks at Porthleven Reef by large numbers of the home scars worn at the center of each animal's grazing patch. The prespill condition can be judged more generally by comparison with Figure 1. By September populations had been reduced to the extent shown by the solid (black) histograms. The stippled ones represent new settlement of barnacles and (at Porthleven only) winkles–which did not survive well–and of the green algae which went on to clothe the shore.

On shores to the extreme west (such as the rocks at Sennen), settlement of young limpets from the plankton was poor, but in the south it was more successful. On heavily polluted shores there (like Porthleven Reef), the subsequent growth of these limpets, surrounded by a luxuriance of their food plant, was phenomenal. They attained a shell length of 20-30 mm in six to nine months as against a more normal 10-12 mm in the first year or 16-18 mm by the end of the second year reported by Blackmore (1969). Maturing rapidly, the majority had developed gonads in November 1968. However, the heavy algal cover inhibited the settlement of barnacles in the classical manner, the weeds when stirred by waves sweeping off their cyprid larvae before these could become properly attached; mussels and less common sedentary animals were also unable to penetrate the algae on the worst affected shores for some years. This had an inevitable effect on the welfare of their predators, such as the dog whelk *Nucella lapillus* or the

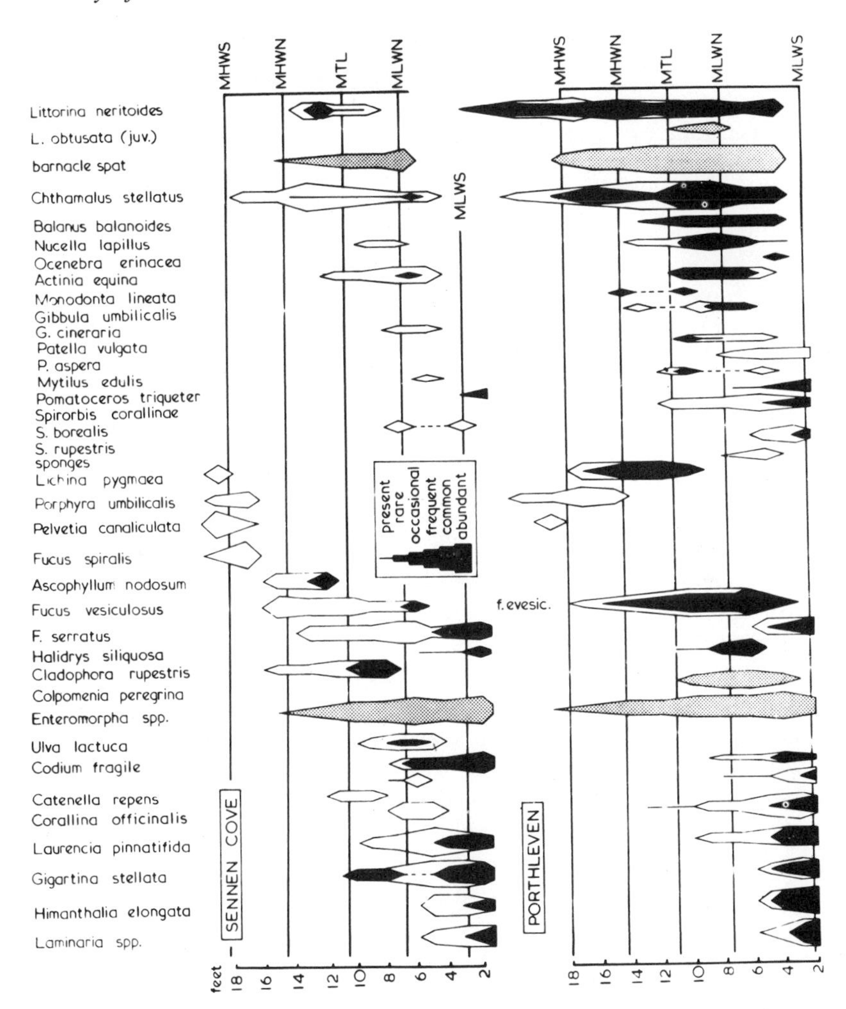

Fig. 2. Distribution of biota on two of the worst affected *Torrey Canyon* shores immediately (*white*) and six months after (*black*) the oil was cleared; new settlement is shown in stippled tone. (From Nelson-Smith, 1968)

polychaeta worm *Eulalia viridis*, although *Nucella* achieved a rapid return as weed clearance proceeded.

These Cornish shores were not resurveyed during the later stages of recovery, but my own rather casual observations have been confirmed and amplified in reports from the Plymouth laboratory (Marine Biological Association, 1969–74). The less seriously disturbed areas regained a normal appearance within 2 to 3 years, although some of their less common inhabitants were still missing or rare. By 1972 algal colonization at the worst sites had been halted, so that although the cover of mature fucoids was still noticeably greater than normal, there were also patches of bare rock from which enormous numbers of limpets were advancing, at times almost touching shell-to-shell. After 6 to 7 years, even those shores were superficially indistinguishable from normal, except for the continuing very high density of limpets. However, a recent paper by Frost (1974) revealed that some cliff top areas, noted for the variety and beauty of their flowering vegetation but badly scarred by the spillage of toxic emulsifiers during cleansing operations, still showed its effects 7 years later. He described the process of recolonization, estimating that the most damaged sites would require another 2 years to achieve complete ground cover and perhaps as many as 15 years altogether before regaining a full diversity of species in their original relative frequencies. Brown (1974) described damage to lichens and their slow recovery at some of these sites, while Tarren and Campbell (1974) reported effects on soil microorganisms there. Paradoxically, these very ill-treated Cornish shores seem nevertheless to have recovered better than the oiled but unsprayed portions of Californian coast facing the Santa Barbara Channel, where crevices are still filled with oily residues and the upper shore is banded with a dried and flaking mixture of oil, straw, and beach material (Holmes, in this volume). Far less initial and residual damage would have resulted had the less toxic dispersants now available been applied in the proper manner—although before this is taken as an unqualified approval of using such treatment on any oiled shore, I should add that Cornwall has the richest coastal flora and fauna in mainland Britain and suffered only a single if catastrophic incident. Its capacity to make a full recovery has thus been maximal.

Milford Haven in southwest Wales is a new oil port which was relatively untouched by industry until the first large-scale installation began operating in 1960. Indeed, most of it lies within the boundary of a national park. It was thus possible to establish baseline surveys on many shores in substantially undisturbed conditions (Moyse and Nelson-Smith, 1963;

Nelson-Smith, 1967). There are now five marine tankship terminals (Fig. 3), four of which serve adjacent refineries or petrochemical plants from which a high proportion of the products are also shipped by sea. During 1974 (the busiest year so far) 53 million tons of oil were moved in and out of the port in 4200 vessels, although most recently there has been a reduction in traffic. During the first ten years, an average of between 0.0001% and 0.0002% of the volume handled has been spilled in small incidents which are quickly detected and efficiently dispersed by a well-organized Harbour Conservancy (see Dudley, 1971). The Haven is a deep inlet which experiences vigorous tidal flushing; so in spite of these regular injections of oils and dispersant, studies from the University College of Swansea and the Oil Pollution Research Unit at Oreilton have revealed only minimal effects on the biota of rocky shores—at about the same level as the effects of long-term climatic changes (Crapp, 1971*a*). Gabriel et al. (1975) similarly failed to find changes in the abundance or composition of the plankton in the first dozen years of major oil movement which could not be regarded as normal long-term changes, in an upward as well as a downward direction (Table 1).

It is impossible to prevent larger spillages from jettyhead mishaps, stranded tankers, or accidents in the refineries themselves from flowing onto adjacent shores. Two in particular, Llanreath on the south and Hazelbeach on the north side, had been studied for several years in order to establish the pattern of normal seasonal changes before the discharge of large quantities of heated cooling water began from the nearby Pembroke Power Station, which was then being built and is now in full operation. Both were heavily oiled by leakage from the hull of a damaged tanker only shortly before the *Torrey Canyon* incident, which of course negated the original purpose of the studies but enabled the effects of the spillage to be recorded rather more accurately than in Cornwall. The pattern of change in populations of some winkles and top shells is shown in Figure 4. These were affected immediately, but it seems that rather than being killed outright, some were washed away—perhaps in a comatose state—but recovered to return later; further, adults probably moved in from the nearest unpolluted areas. Recolonization was thus not entirely dependent on fresh settlement by young animals; so the "green phase" and subsequent fucoid growth were not very marked. However, less than two years later a serious overflow from a refinery tank again covered Hazelbeach. Limpets there well exceeded a density of 100/m^2 even after this second spill, but this declined shortly afterwards to 21/m^2 and further to 5/m^2 a few months later. Four species suffered particularly badly: *Littorina neritoides*, which was eliminated from the upper shore in the first

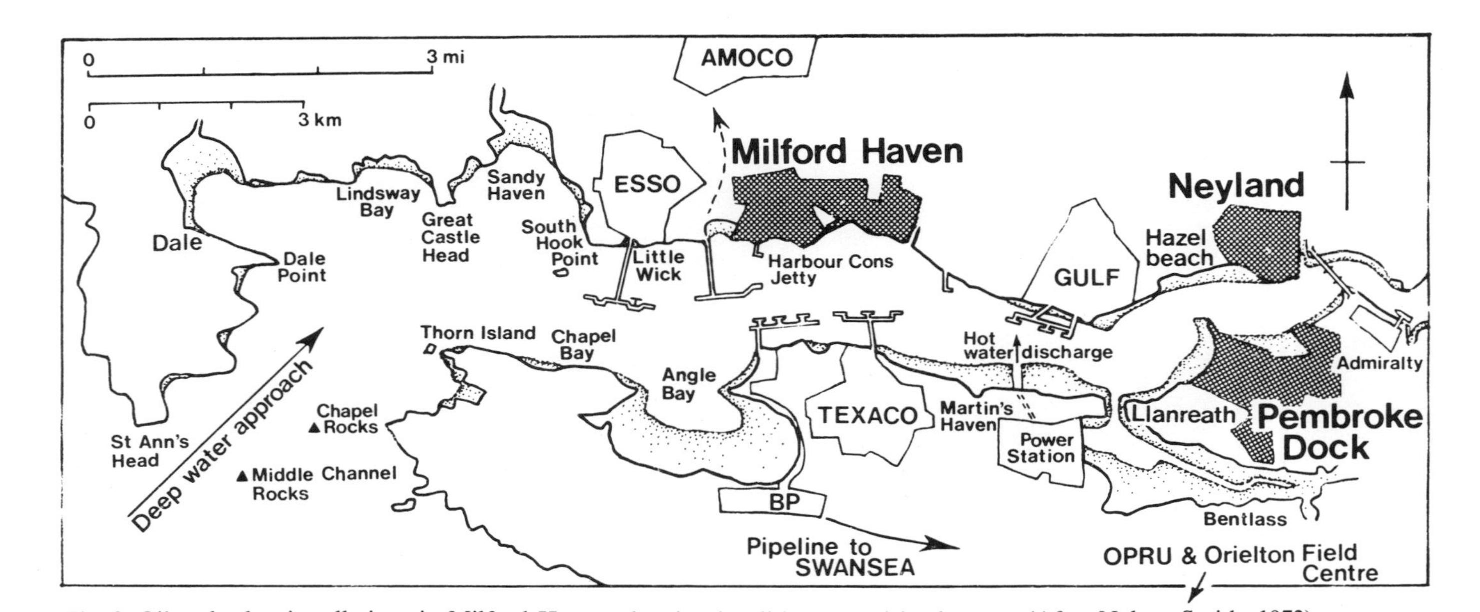

Fig. 3. Oil and other installations in Milford Haven, showing localities named in the text. (After Nelson-Smith, 1972)

Table 1. Monthly standing crop (mean April-Aug.) of plankton in two regions of Milford Haven before and after the establishment of the oil industry there (simplified from Gabriel et al., 1975)

	Upper Haven		Lower Haven	
	1959-60	1971	1959-60	1971
Phytoplankton (per m^3)	25,414	28,728	65,289	51,474
Zooplankton (per 10 m^3)				
Copepods	1,447	2,917	613	3,416
Decapod larvae	867	298	1,605	1,119
Barnacle larvae	27,529	22,200	20,959	21,897
Mysids	29	<10	18	<10
Coelenterates (adult)	29	<10	201	34
Polychaete larvae	3,064	2,212	1,973	1,736
Mollusc larvae	2,545	3,197	890	2,618
Fish eggs + larvae	71	21	117	267

incident, disappeared completely within a year of the second; and *L. obtusata,* a macroherbivore on the lower shore, disappeared for a period after each. Of the top shells, *Gibbula umbilicalis* was never widely distributed but underwent a number of seemingly random changes afterwards, while *Monodonta lineata* showed severe if temporary reductions in density and range. On the Welsh coast, *Monodonta* is near the northern limit of its geographical distribution; during the unusually cold winter of 1962-63, populations in Pembrokeshire were appreciably thinned, while farther up the Bristol Channel to the east, they were virtually killed off (Moyse and Nelson-Smith, 1964). This sensitivity may have enhanced the effect of the spills in Milford Haven, each of which occurred during the winter. By contrast, only about 120 miles farther south on the Cornish coast this gastropod was the only common mollusc surviving *Torrey Canyon* pollution on the rocks around St. Michael's Mount, where it fed well on the bloom of green algae. At Hazelbeach the role of flourishing survivor was taken by the large winkle *Littorina littorea*; after each incident, its range and abundance increased slightly, presumably because its competitors had been more or less inhibited. The situation there slowly returned to normal between 1970 and 1972.

The refineries and associated plant are air-cooled (in contrast to the subject of Dicks's study, in this volume); so refinery effluents consist only of process and storm water which passes through separators before discharge and contains approximately 25 ppm of oil. However, the quantity of oil discharged in these effluents is altogether probably at least half that lost by accidental spillage in a year without a major accident. Outfalls from

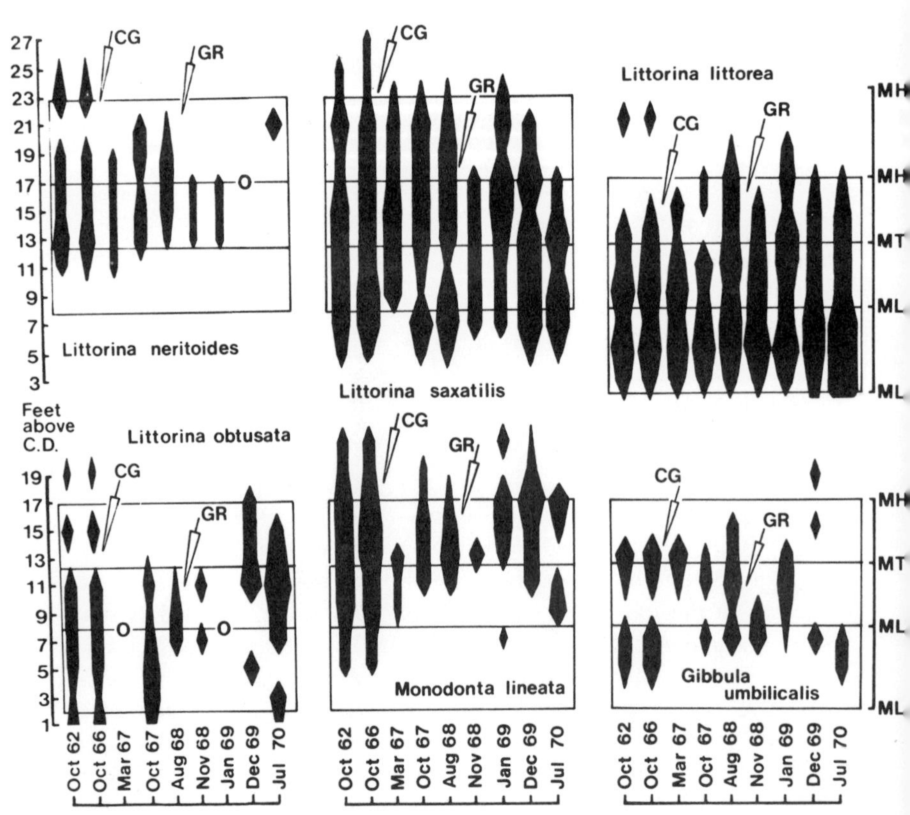

Fig. 4. Effects of successive oil spillages on shore gastropods at Hazelbeach. Crude oil was spilt from the tanker *Chryssi P. Goulandris* in January 1967 (*CG*) and from the Gulf refinery in November 1968 (*GR*). (From Nelson-Smith, 1972)

three of the refineries open at the jettyhead or on a headland and present no local problems, but the fourth discharges across the tidal zone of a small bay, Little Wick. Photographs of this bay, taken before the plant was built, show it to have been dominated by barnacles and limpets, with few large seaweeds. It now has a fairly dense cover of fucoids and, immediately upstream, the bladderless form is intermingled with the typical *Fucus vesiculosus* which one would expect to find at this slightly more sheltered site—a condition rarely seen on normal shores but observed, for example, at Porthleven Reef during recovery from *Torrey Canyon* damage. Surveys of the usual sort, taken down the shore at intervals to each side of the outfall, showed a diminution in the numbers of several gastropods, especially limpets and winkles, as well as among barnacles and other sedentary animals, and confirmed that algal cover is greater in the vicinity of the pipe (Fig. 5). The reduction in limpet density is, in fact, more marked than appears here because the abundance scale originally used is rather too compressed at its upper end. In actual numbers, limpet densities vary from $10/m^2$ at the outfall site to nearly $500/m^2$ at 250 m to either side (Baker, 1973). The shore biota will by this time have reached a new equilibrium; Reynard (1974) has shown that the remaining limpets have much the same reproductive potential as those from nearby but more normal shores. However, no recovery (in the sense of returning to the previous equilibrium) can be expected while the discharge continues.

With the exception of the mixed oils present in this effluent, the incidents described above have involved crude oil in various degrees of freshness. Milford Haven has also experienced two noteworthy spills of products, with diametrically opposite results. In March 1971 the coastal tanker *Thuntank 6*, leaving with a load of light fuel oil, inexplicably turned into the mouth of the Haven too early and struck rocks at Thorn Island, losing about 160 tons from her forward tanks. Most of the oil was dispersed on the water and much of what became stranded was also sprayed away, using the much improved BP1100 dispersant; the rest lay in rock pools from which it was later pumped, or hardened at high-water mark where some still remains. Ottway (1972) found that it had minor effects on some winkles and top shells, which either were weighted down by their covering of oil or could not maintain a foothold on oil-coated rocks: most of these recovered later. Gammarid amphipods and a small proportion of the limpet population were killed, but no long-term effects were observed, so that the shore could be said to have reached a reasonable level of recovery almost as soon as the bulk of the oil had been removed. Laboratory tests showed this oil to have a much lower toxicity than Kuwait crude (the commonest raw material shipped into the Haven) for winkles, although not for limpets (Table 2). It thus differs from the light fuel oil spilled at Buzzard's Bay (West Falmouth, Mass.) in September 1969, which had an aromatic content of over 40% and was therefore much more toxic, as well as very persistent (see, e.g., Blumer et al., 1970), demonstrating that oil products of different formulation or origin may have very different biological effects, even if they bear the same generic name.

In August 1973 a larger tanker, *Dona Marika* dragged anchor in a gale and was stranded on rocks farther around the mouth of Milford Haven at Lindsway Bay. Rather less than half her cargo of high-octane gasoline was eventually pumped to a shallow-draft vessel moored to her seaward side, but about 3000 tons were lost into a fairly small area of heavy surf, so

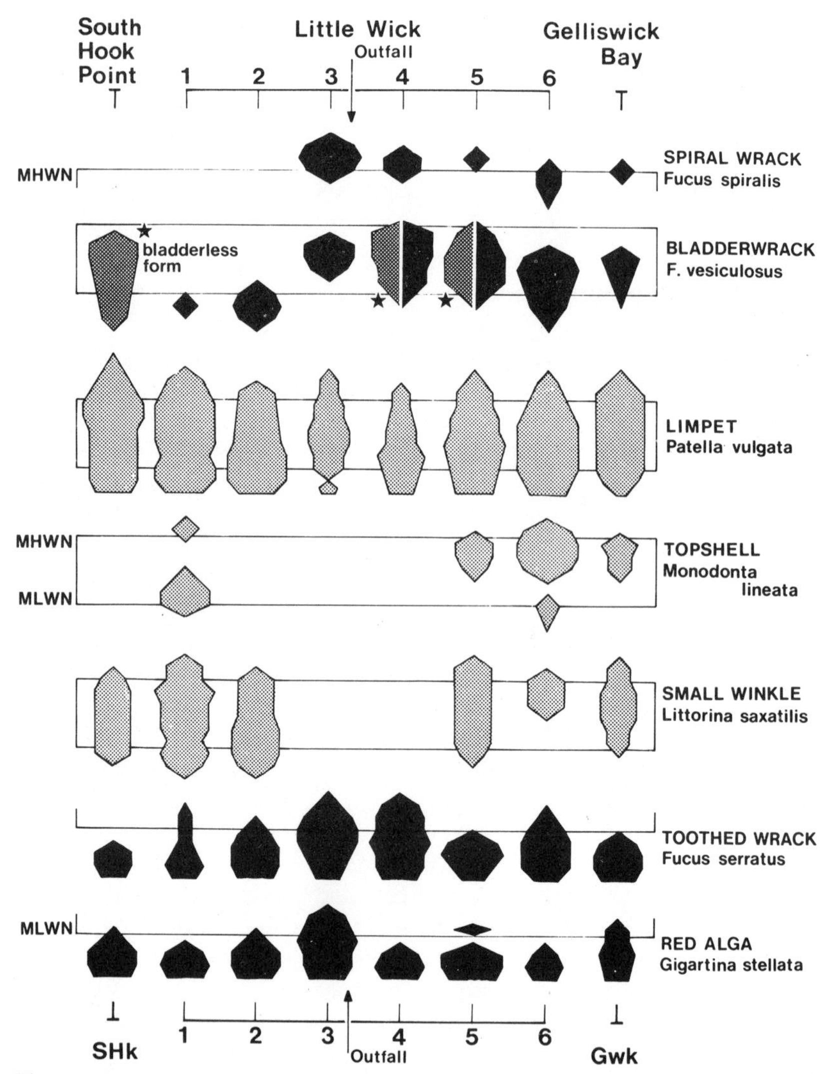

Fig. 5. Distribution of some plants and animals to each side of a refinery effluent in Milford Haven. (From Nelson-Smith, 1975)

Table 2. Effects of light fuel and crude oils on intertidal molluscs. In each test 50 animals of each species were covered with oil for the time shown and allowed to recover in clean seawater for five days; the mortality was calculated from the total number recorded as dead in each group at the end of that period (from Ottway, 1972).

Species	Immersion time (h)	Mortality %		
		Light fuel oil (Gulf)	Kuwait crude oil	Control
Littorina saxatilis (winkle)	1	1	20	0
	6	5	41	4
	21	3	56	0
	24	4	72	0
L. obtusata (winkle)	1	3	36	0
	6	4	42	0
Patella vulgata (limpet)	1	80	98	6
	6	88	92	2
Cardium edule (cockle)	1	7	100	0

that much of it was washed across the shore as an emulsion instead of evaporating almost immediately, as might normally be expected (Blackman et al., 1973). Gasoline is much more toxic to winkles than are crude or fuel oils (Table 3), and in a similar test limpets showed a 100% mortality throughout (Crapp, 1971*b*), but in this case, the first effect was to narcotize them. What actually killed the limpets to each side of Lindsway Bay was being eaten by seagulls while in this state; there is little evidence that the gulls themselves were significantly affected, and dog whelks *Nucella* reappeared quickly enough to suggest that they were merely displaced while comatose. Additional concern was felt because 600 mg/1 of lead antiknock agent had already been added to the gasoline, but analyses provided no evidence of lead accumulation in the sediments or biota around the site. Whatever the cause, the marked reduction in the numbers of limpets and other grazers around the bay gave rise to the familiar "green phase" (Fig. 6) which persisted into 1974. At present, the brown algae which took over are still dominant, but newly settled limpets, although small, are now present at about a prespill density. Early in 1975 a further spillage

Table 3. Effects of various oils on *Littorina littorea*, using 50 animals per test (from Crapp, 1971*b*)

Oil	Mortality % after immersing for				
	6 min	30 min	1 h	3 h	6 h
Gasoline	78	80	76	78	70
Kerosine	10	14	20	50	68
Dieselect	0	0	0	0	0
Fuel (no. 2)	0	0	0	2	2
Fuel (3500)	0	0	0	0	0
Kuwait crude	12	10	28	48	62

of crude oil threatened this site, but only a small patch was stranded in the study area. The rapidity with which this shore can achieve a full recovery depends (as in the case of Hazelbeach) on whether or how frequently it receives further pollution.

On the basis of the still somewhat anecdotal evidence from British rocky seashores, it might tentatively be concluded that the rate of recovery from oil spills (and cleanup treatments which, at least in earlier years, may have been even more damaging) depends very largely on the extent to which the numbers of limpets have been reduced and the speed with which they can reestablish themselves densely enough to control the resulting bloom of large algae. This, in turn, depends not only on the nature of the oil spilled and any cleansing treatments used but also on the extent of damage and the location of the affected shore with respect to sources of adult or larval animals able to recolonize the damaged area. Simple baseline surveys, made before the incident, are of great value in assessing the damage. The time scale of the process depends on how *recovery* is defined: if this is accepted as meaning a reinstatement of the previously dominant community, even if some minor members of the ecosystem may still be absent or fewer, then a single severe oiling incident may be less serious than a climatic aberration of the magnitude of the 1962-63 cold spell. Recovery of intertidal communities from the *Torrey Canyon* incident, even when impeded by destructive cleanup procedures, seems to have taken about 7 years, whereas those animals most affected by the cold spell show little sign of regaining their previous status on the worst-affected shores 12 years later. An oil spillage of *Torrey Canyon* intensity is expected to occur somewhere around the coasts of northwest Europe once every 10 years, on average, whereas unusually cold winters have occurred there recently at approximately 20-year intervals. Their

ROOK'S NEST POINT, Lindsway Bay, Milford Haven

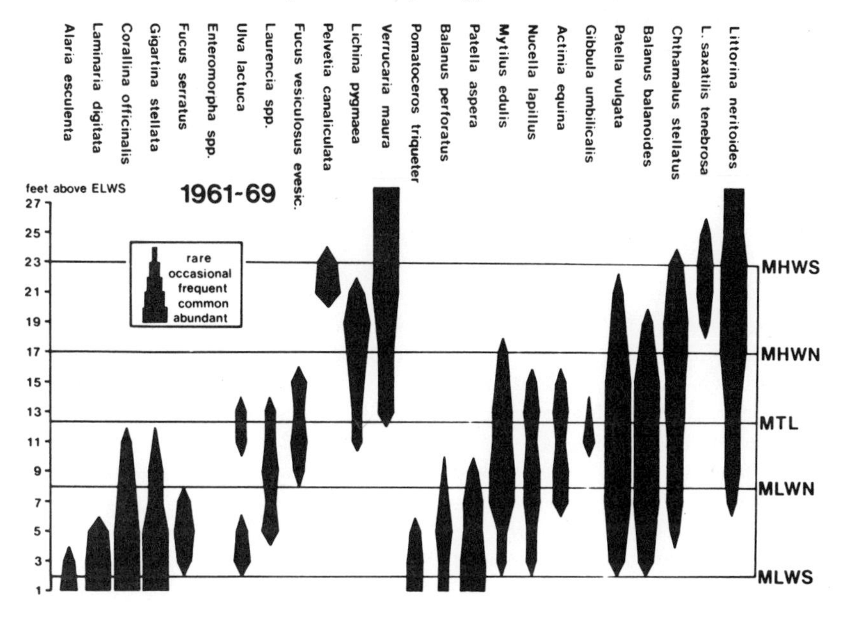

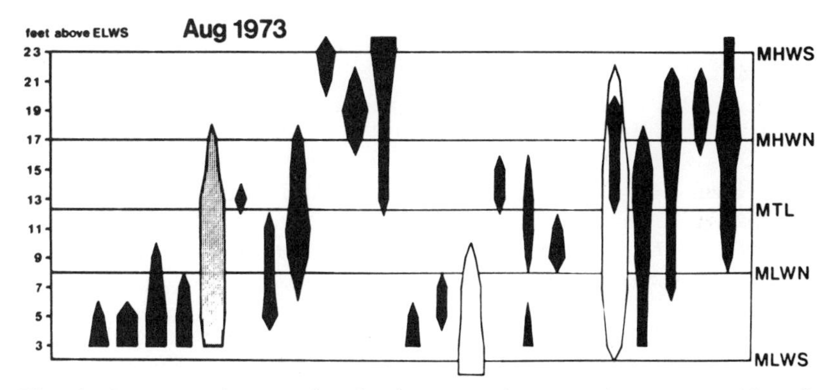

Fig. 6. Common plants and animals on a shore at the eastern side of Lindsway Bay before and after the *Dona Marika* gasoline spill. The normal distribution of the limpets *Patella vulgata* and *P. aspera* has been superimposed in outline on the lower diagram to emphasize the reduction in their numbers and range. The *stippled* histogram indicates new settlement by green seaweeds.

ecological impact on seashores might thus be taken as roughly the same, in the long term. In making such comparisons, however, it must always be borne in mind that little can be done about the weather, whereas oil spills are an additional hazard imposed on the environment by human fallibility and greed; most could be averted if sufficient effort or expenditure were made.

Literature Cited

Baker, Jenifer M. 1973. Biological effects of refinery effluents. Pages 715-723 *in* Proceedings Joint Conference Prevention Control Oil Spills. American Petroleum Institute, Washington, D.C.

Blackman, R. A. A., Jenifer M. Baker, J. Jelly, and Susan Reynard. 1973. The *Dona Marika* oil spill. Mar. Pollut. Bull. 4:181-182.

Blackmore, D. T. 1969. Studies of *Patella vulgata* L. I. Growth, reproduction and zonal distribution. J. Exp. Mar. Biol. Ecol. 3:200-213.

Blumer, M., G. Souza, and J. Sass. 1970. Hydrocarbon Pollution of Edible Shellfish by an Oil Spill. Woods Hole Oceanographic Institution Tech. Rep. 70-1.

Brown, D. H. 1974. Field and laboratory studies on detergent damage to lichens at The Lizard, Cornwall. Cornish Studies 2:33-40.

Crapp, G. B. 1971*a*. Monitoring the rocky shore. Pages 102-113 *in* The Ecological Effects of Oil Pollution on Littoral Communities. Institute of Petroleum, London.

———. 1971*b*. The ecological effects of stranded oil. Pages 181-186 *in* The Ecological Effects of Oil Pollution on Littoral Communities. Institute of Petroleum, London.

Dicks, Brian. 1976. Changes in vegetation of an oiled Southampton water saltmarsh. In this volume.

Dudley, G. 1971. Oil pollution in a major oil port. Pages 5-15 *in* The Ecological Effects of Oil Pollution on Littoral Communities. Institute of Petroleum, London.

Frost, L. C. 1974. *Torrey Canyon* disaster: the persistent toxic effects of detergents on cliff-edge vegetation at The Lizard peninsula, Cornwall. Cornish Studies 2:5-14.

Gabriel, P. L., Nirmala S. Dias, and A. Nelson-Smith. 1975. Temporal changes in the plankton of an industrialized estuary. Estuarine Coastal Mar. Sci. 3: 145-151.

Holmes, R. W. 1976. The Santa Barbara oil spill: an ecological disaster? In this volume.

Jones, N. S. 1948. Observations and experiments on the biology of *Patella vulgata* at Port St. Mary, Isle of Man. Proc. Trans. Liverp. Biol. Soc. 56:60-77.

Lodge, Sheila M. 1948. Algal growth in the absence of *Patella* on an experimental strip of foreshore, Port St. Mary, Isle of Man. Proc. Trans. Liverp. Biol. Soc. 56:78-83.

Marine Biological Association. 1969-74. Reports of the Council. Plymouth, United Kingdom.

Moyse, J., and A. Nelson-Smith. 1963. Zonation of animals and plants on rocky shores around Dale, Pembrokeshire. Fld. Studies 1(5):1-31.

_____. 1964. Effects of the severe cold of 1962-63 upon shore animals in South Wales. J. Anim. Ecol. 33:183-190.

Nelson-Smith, A. 1967. Marine biology of Milford Haven: the distribution of littoral plants and animals. Fld. Studies 2:435-477.

_____. 1968. Biological consequences of oil pollution and shore cleansing. Fld. Studies 2 (suppl.):73-80.

_____. Techniques–surveying rocky shores. Fieldworker 1:50-52.

_____. 1972. Oil Pollution and Marine Ecology. Elek Scientific, London.

_____. 1975. Effects of long term, low level exposure to oil. Proceedings Conference Petroleum and Continent Shelf of North West Europe. Applied Sci. Publ. Barking, U.K.

Ottway, Sheila M. 1972. The "Thuntank 6" oil spill. Pages 29-38 *in* Ann. Rep. Oil Pollut. Res. Unit, Orielton, 1971.

Reynard, Susan. 1974. Reproductive potential of limpets near a refinery effluent in Milford Haven. Pages 44-46 *in* Ann. Rep. Oil Pollut. Res. Unit, Orielton, 1973.

Smith, J. E., ed. 1968. *Torrey Canyon* Pollution and Marine Life. Cambridge University Press, London.

Tarren, Christine, and R. Campbell. 1974. Effects of micro-organisms in the soil of detergents used to combat oil pollution at The Lizard, Cornwall. Cornish Studies 2:23-26.

Changes in the Vegetation of an Oiled Southampton Water Salt Marsh

Brian Dicks

Abstract

This study relates to an area of salt marsh in Southampton Water, Hampshire, England, where the natural vegetation is dominated by *Spartina anglica* C. E. Hubbard.

A new refinery started operation in 1951, and progressive damage to the salt marsh adjacent to the effluent outfalls occurred from 1951 to 1970, by which time an area approximately 1000 m by 600 m had been denuded of vegetation. This appears to have been a result of repeated contamination of the vegetation by oil films. The mechanism of such damage is discussed.

Since 1970 recolonization of some of the previously denuded areas by all the main salt-marsh plants has occurred. This process appears to be associated with the concurrent substantial improvement in effluent quality and absence of serious oiling incidents, although meteorological and other factors may have played a part. As a number of variables may be involved, continued study is necessary, and this report is only of an interim nature.

The pattern of recolonization of damaged areas is discussed. This appears to be similar to the normal successional process of salt-marsh development, but if recovery continues, the final distribution of plants may be different from that of the original salt marsh.

Introduction

Extensive salt marshes dominated by the common cordgrass, *Spartina anglica*, occur in Southampton Water, although some areas are now reclaimed for industrial use. This species of grass, with its rhizomatous root system and extensive aerial shoots, stabilizes soft marine mud and helps trap further silt to raise the marsh level (Hubbard, 1954; Chapman, 1960, 1964). This process, which allows invasion of the area by other less salt-tolerant plants, thus changes the biology of the area over a period of time (a succession or halosere). A typical Southampton Water salt marsh is therefore composed of, near its landward edge, a mixture of less salt-tolerant plant species or those preferring drier conditions, while among the extensive system of creeks and toward the seaward edge more salt-tolerant plants such as *Spartina anglica* and other primary mud colonizers (*Salicornia* spp. and *Suaeda maritima*) occur. Variations in microhabitat throughout the marsh produced by erosion or changes in drainage patterns provide interruptions to this general successional pattern of change in vegetation from seaward to landward edges of the marsh.

One such marsh, on the foreshore at Fawley, has had a refinery effluent discharged into its creek system since 1951. A map of the Fawley saltings, their creek system, and the positions of effluent outfalls is shown in Figure 1.

The marsh is bounded on its southern edge by reclaimed land, on its landward margin by a seawall at the high-tide mark, and on its eastern margin by Southampton Water. The Marsh extends northward for a considerable distance out of the area of influence of the refinery effluent along the shore of Southampton Water.

The common marsh plants found at Fawley are *Salicornia* spp., *Suaeda maritima* (L.) B. C. J. Dumortier, *Spartina anglica* C. E. Hubbard, *Aster tripolium* L., *Halimione portulacoides* (L.) P. Aellen, *Atriplex* sp., and *Juncus maritimus* Lamarck. As *Spartina* is the dominant species, the saltings are at a *Spartinetum* stage in their development. A brief description and illustrations of some of these plants can be found in the Appendix.

The Effects of Oil on Salt-marsh Vegetation

All the above species which occur in the Fawley marsh are sensitive to oil pollution, especially to films of oil on the water surface or to oil slicks. *Saliconia* spp. and *Suaeda maritima*, two types of succulent annual plants, are particularly sensitive, as they have a small underground system from which new growth cannot occur after severe damage to shoots. The reasons for their and other marsh plants' sensitivity have been fully discussed by Baker (1970*a*, 1970*b*, 1970*c*, 1970*d*, 1970*e*, 1970*f*) but can be summarized as follows:

1. There is a high affinity between plant cuticle and oil hydrocarbons. This encourages adhesion of oil to vegetation.

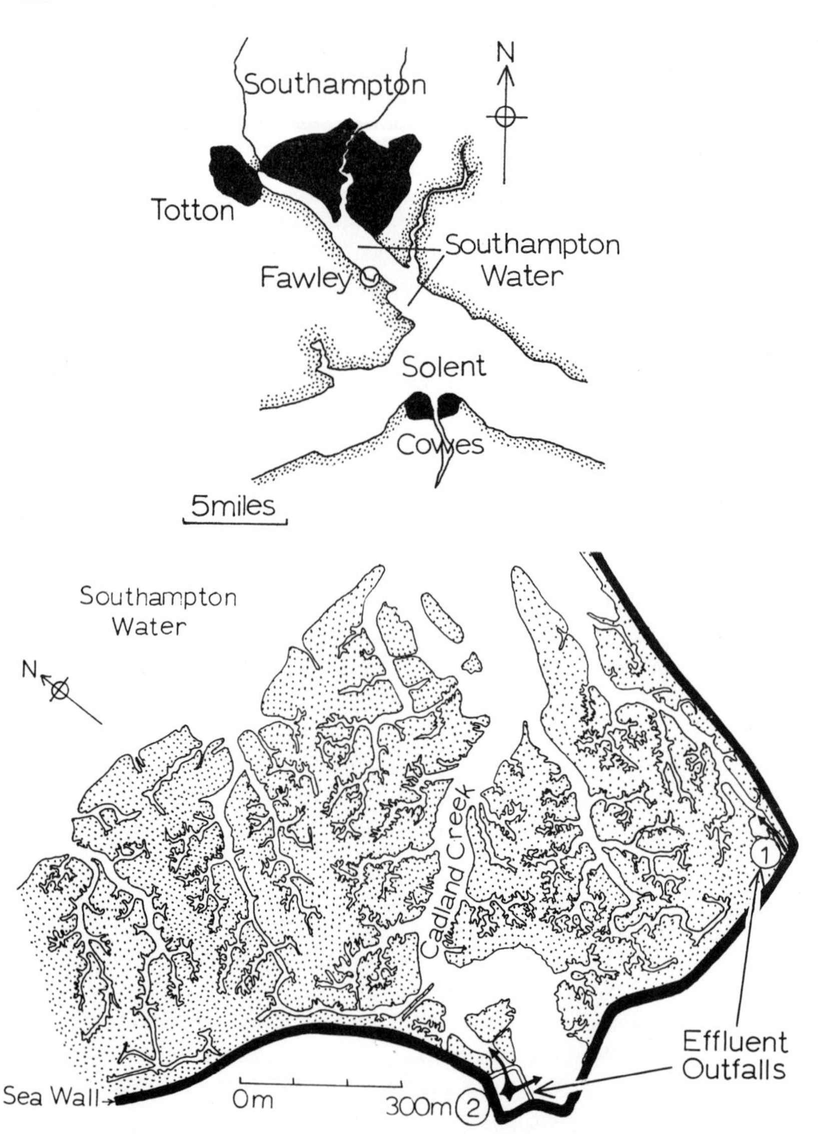

Fig. 1. The saltings and creek system at Fawley, Southampton Water. The location of the saltings in Southampton Water is shown in the upper map, and details of the saltings (*stippled*) and positions of the two effluent outfalls in the lower map.

2. Oil obstructs the leaf pores (stomata) and prevents or restricts gas exchange to root and shoot.

3. Oil films reduce the light available for photosynthesis.

4. Some constituents of crude oils are directly toxic to metabolic processes.

5. Plants with large underground vegetative systems (the perennials) survive oiling better than those without (the annuals), as new growth can occur from these systems after shoots and leaves have been destroyed.

Spartina anglica, which has an extensive underground rhizomatous root system, would be expected to be capable of surviving considerable oiling of its shoots. However, as *Spartina anglica* primarily grows in fine and often reduced muds, its root system is dependent for oxygen on a diffusion down the stems, and oxygen diffuses out of the root system into the mud around its roots (Armstrong, 1967; Baker, 1970*d*). This can be demonstrated using the method described in detail by Baker (1970*d*). This involves measurement of oxygen diffusion rate (ODR) from the *Spartina anglica* root system to the soil using a cylindrical platinum electrode surrounding a root. Under certain conditions, current resulting from electrolytic reduction of oxygen at the platinum electrode is governed solely by the rate at which oxygen diffuses to the electrode surface from the surrounding medium. By embedding the root system of *Spartina anglica* in anaerobic agar jelly, with the root electrode and a control electrode (Fig. 2), ODR could be measured by applying a constant voltage (0.6v) and recording differences in current produced at the root and control electrodes. This method clearly demonstrates that oxygen passes from the roots to the surrounding medium (Fig. 3).

Painting of the leaves of *Spartina anglica* with 90% crude oil residue during this experiment produced rapid falloff in the OD current (see Fig. 3), and had obviously affected diffusion of oxygen to the root system. It is therefore possible that oiling of shoot systems of *Spartina anglica* can produce serious effects on the roots, and perhaps death of both shoot and root if oxygen diffusion is reduced sufficiently. As the root system of many such marsh plants living in reduced muds is dependent on the shoot system for oxygen (Armstrong, 1967), it is not necessarily so extensive a reservoir from which new growth can come after oil damage as might be expected.

The effects of oil on salt-marsh plants are also influenced by the number of times incidents occur, and above a certain number, damage can be extensive (Baker, 1970*a*). The results of successive sprayings of *Spartina anglica* with Kuwait crude oil carried out by Baker (1970*a*) are shown in Figure 4.

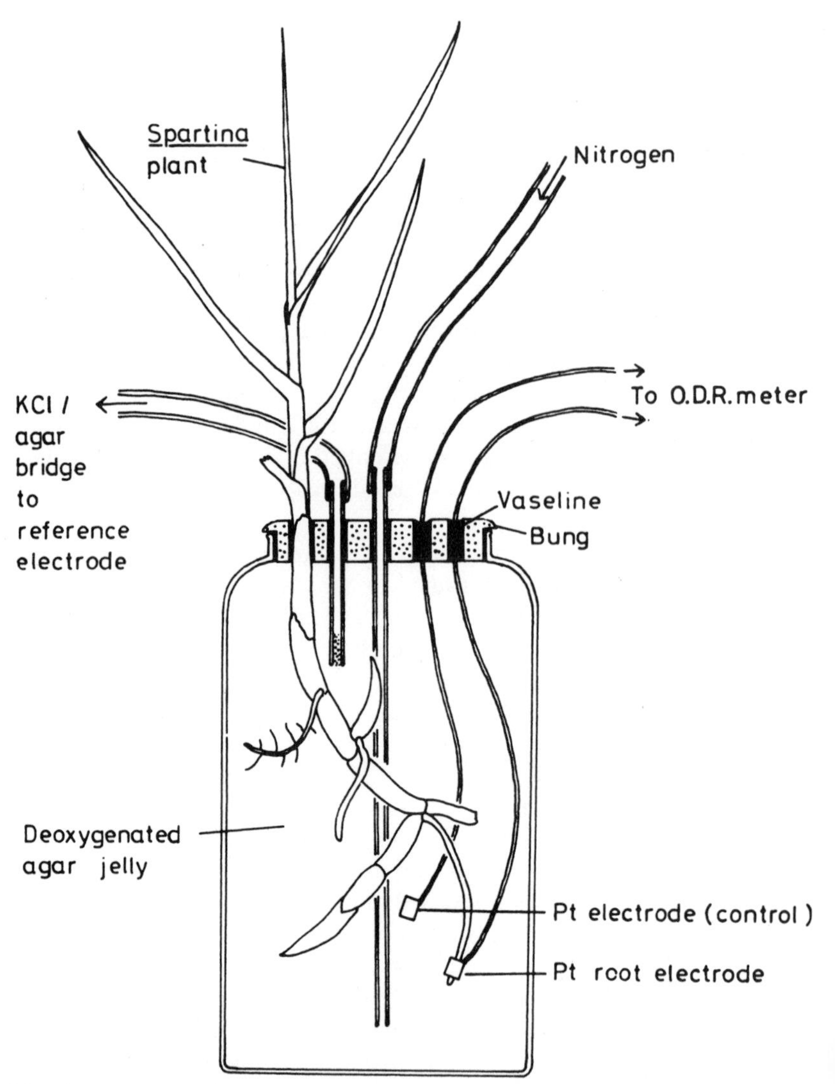

Fig. 2. Apparatus for measurement of ODR from *Spartina anglica* roots

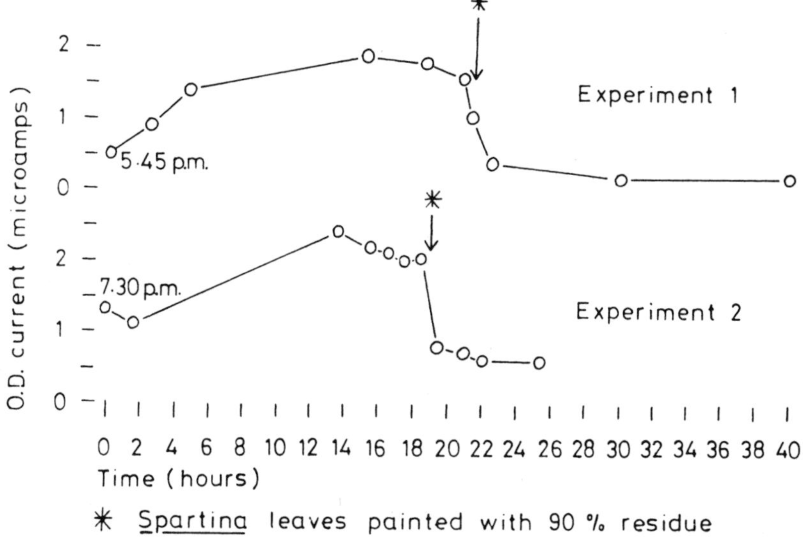

Fig. 3. OD current (microamps) produced by diffusion of oxygen from *Spartina anglica* roots. 90% Kuwait crude oil residue was painted on the leaves at the points marked.

Marked plots of salt-marsh plants were sprayed with Kuwait crude oil (4.5 l of oil per plot 5 m × 2 m) at intervals, and the numbers of healthy tillers (shoots) of *Spartina anglica* counted in ten randomly placed 25 cm × 25 cm quadrats.

The 95% confidence limits of the mean of the tiller counts in each plot were calculated from:

$$\text{Standard error } \sigma_n = \left(\frac{\Sigma(x)^2 - \frac{(\Sigma x)^2}{n}}{n\,(n-1)} \right)^{1/2}$$

and 95% confidence limits (0.05 probability limits) from

$$\bar{x} \pm t\sigma_n$$

where $t =$ probability constant dependent on $n - 1$,
$n =$ number of samples,
$x =$ each sample value,
$\bar{x} =$ mean value of samples.

The results show that the normal dying back of *Spartina anglica* during the winter months (top graph, Fig. 4) is speeded up by more than four successive oilings (at monthly intervals). If oil continues to be applied through the winter months, growth during the following year can be seriously reduced. Continual (chronic) pollution, such as that produced by a refinery effluent, probably has its main effect on salt-marsh plants by the almost continual presence of thin oil films on the surface of the water, which, at high tide, reach the plants.

The Fawley Effluent: Its Characteristics and Discharge

The main pollutants that occur in refinery effluents are oil, dissolved compounds such as sulfides, phenols, and nitrogen compounds, and sludge (Blokker, 1970). Typical refinery waste characteristics of the older type of refinery as produced by the American Petroleum Institute (1963) average around 57 ppm oil. A more recent report (Blokker and Marcinowski, 1970) showed a wide range in effluent quality. The older coastal refineries reported effluent quality to be commonly in the range 11-40 ppm, but approximately 30% reported over 40 ppm. The Fawley effluent, in 1970, averaged 31 ppm oil content with a total discharge volume of 114,000 imp gal/min (136,800 U.S. gal/min) through outfalls Nos. 1 and 2. Recently these figures have been substantially improved with, in 1974, an average oil content of 14 ppm and a total discharge volume of 86,000 imp gal/min (103,200 U.S. gal/min).

The effluent is discharged at the high-tide mark to two of the large creeks in the saltings, as illustrated in Figure 1. Discharge volumes for the two outfalls are roughly equal.

As well as the oil normally dispersed in the effluent, oil from other sources occasionally enters the creek system. These sources include:

1. Accidental spillage within the refinery, which might produce unusually high oil contents in the effluent
2. Oil spillage from the refinery jetty installations and tankers
3. Oil spillages from other jetty installations in the vicinity
4. Oil spillage from the considerable ship traffic using Southampton Water

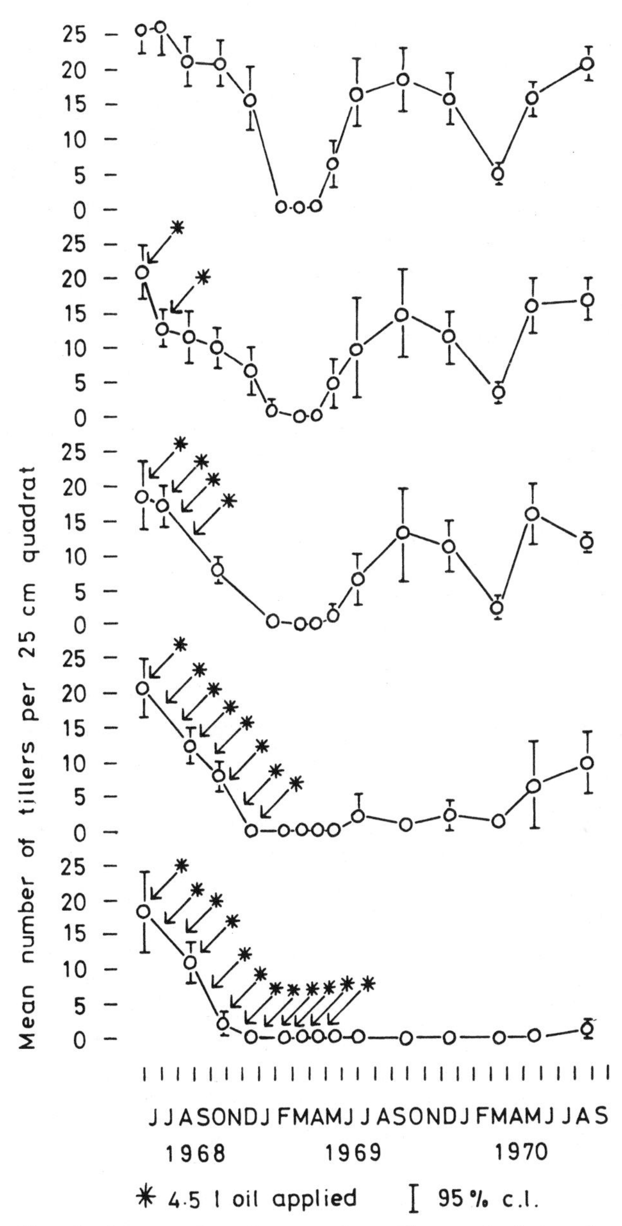

Fig. 4. The results of successive oilings of plots of *Spartina anglica.* Oil was applied at the points marked. Densities of tillers (shoots) in ten quadrats were averaged to produce the points on the curves. 95% confidence limits (0.05 probability limits) are marked.

While thin oil films are commonly present in the area, coherent dark slicks appear to be rare. This has been particularly so since 1970, the number of acute incidents both from refinery operations and elsewhere being small.

Whatever the source of the oil, it first affects vegetation near the creek edges and later, when the initial barrier of the creek edge vegetation has succumbed, affects areas further into the marsh.

A History of the Damage to the Fawley Salt Marsh, 1950–70

Information is available on the condition of the saltings at Fawley from various sources since 1950, although some of the information is rather limited. The history of the marsh up to 1970 can be summarized as follows:

1950 Aerial photograph taken by the RAF (RAF Sortie No. 541/533, print 4005) shows the whole marsh area to be covered by healthy *Spartina anglica* (Fig. 5).

1951 Outfall No. 1 starts operation.

1953 Outfall No. 2 joins No. 1.

1954 Aerial photograph taken by the RAF (RAF Sortie No. 82/895 F21, print 0104) shows what can be interpreted as oily vegetation along the edges of the creek through which the two outfalls pass (illustrated, Fig. 6).

1962 Photographs of the marsh (taken by Dr. D. S. Ranwell, Nature Conservancy Coastal Ecology Research Station) show that large areas of *Spartina anglica* have died and decomposed around the two outfalls.

1966 Ordnance survey map the area and produce sheets Nos. SU4404 and SU4504 with the boundary of the salt marsh as shown in Fig. 7.

1969 A series of transects across the marsh show extensive areas of bare mud with the remains of *Spartina anglica* stems and roots (Baker, 1971). At the edge of the denuded area was a belt of oily vegetation, behind which the marsh was healthy. The mud level in denuded areas was 15-25 cm lower than the healthy marsh, presumably due to erosion.

1970 Resurvey of the transects (Baker, 1970*b*) shows small increases in the amount of oily vegetation at the boundary of the healthy marsh but otherwise no change from 1969 (Fig. 8). Baker concluded that (*a*) the damage to the marsh was due to repeated light oilings of the vegetation and shoots from films of oil stemming partly from

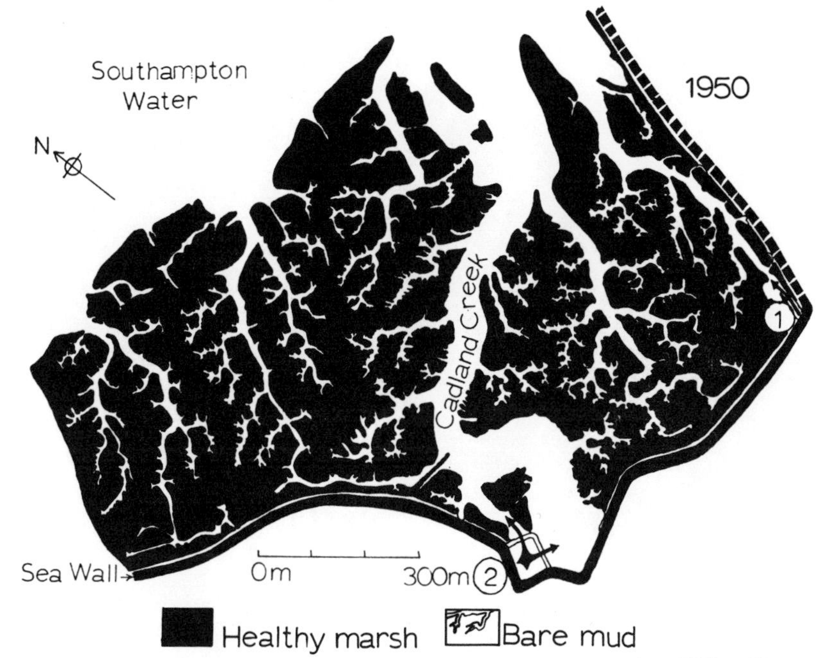

Fig. 5. The extent of salt-marsh vegetation at Fawley in 1950. (Data from RAF Photograph Sortie No. 541/533 print 4005)

the effluent and partly from spillages and (*b*) recolonization would be prevented by the continual presence of these films, although the mud itself, while containing variable amounts of oil, was not toxic and growth of *Spartina anglica* occurred in samples of the mud in the laboratory. The denuded area at this time was approximately 1000 m by 600 m.

Monitoring of the Fawley Salt Marsh, July 1972–July 1974

Methods

Vegetation mapping. Subsequent to the results of the marsh surveys by Baker (1970*b*), a monitoring program was started to assess the condition of the marshes twice a year, once in the summer (July), when the plants grow vigorously, and once during the winter (late December or January), when *Spartina anglica* and the annual plants such as *Salicornia* spp. and

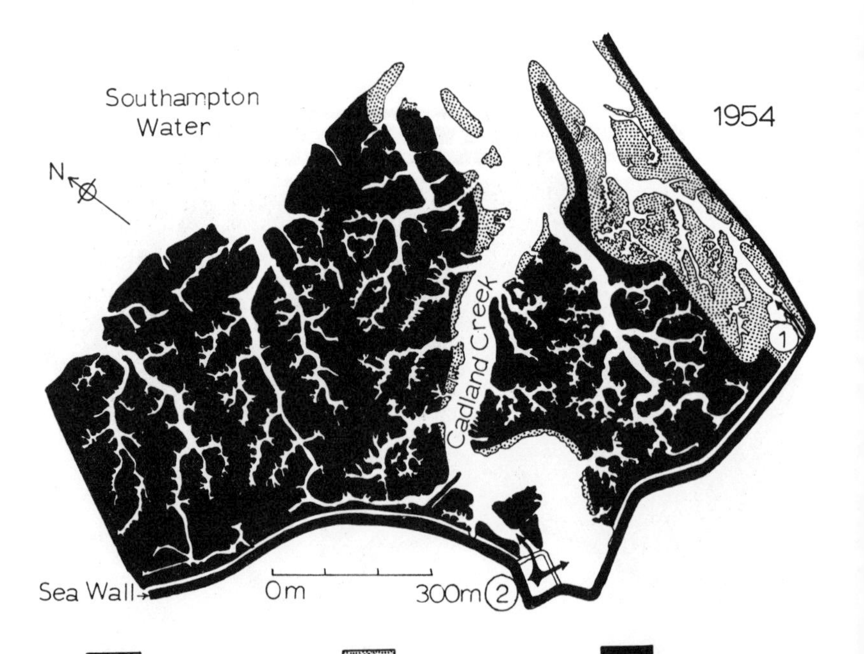

Fig. 6. The extent of salt-marsh vegetation at Fawley in 1954. Two areas of what can be interpreted as oily vegetation are shown around the two effluent-carrying creeks. (Data from RAF Photograph Sortie No. 82, 895F21 print 0104)

Suaeda maritima die back. The objective of the program was to locate changes in the extent of marsh vegetation in association with the effluent improvement program which had been started by the refinery. The series of line transects radiating from the outfall surveyed by Baker (1970*b*) do not give a complete picture of the plant distribution or effluent effects on the marsh due to unevenness in the effluent distribution caused by the complex creek system and the considerable marsh area involved. This method was replaced by a vegetation-mapping technique involving surveying the whole of the marsh area and, using a simple abundance scale based on plant density, roughly quantifying the distribution of all the main marsh plants. Each species was rated as follows:

Abundant (A): The majority of the plants less than 50 cm apart, and often very close to each other.

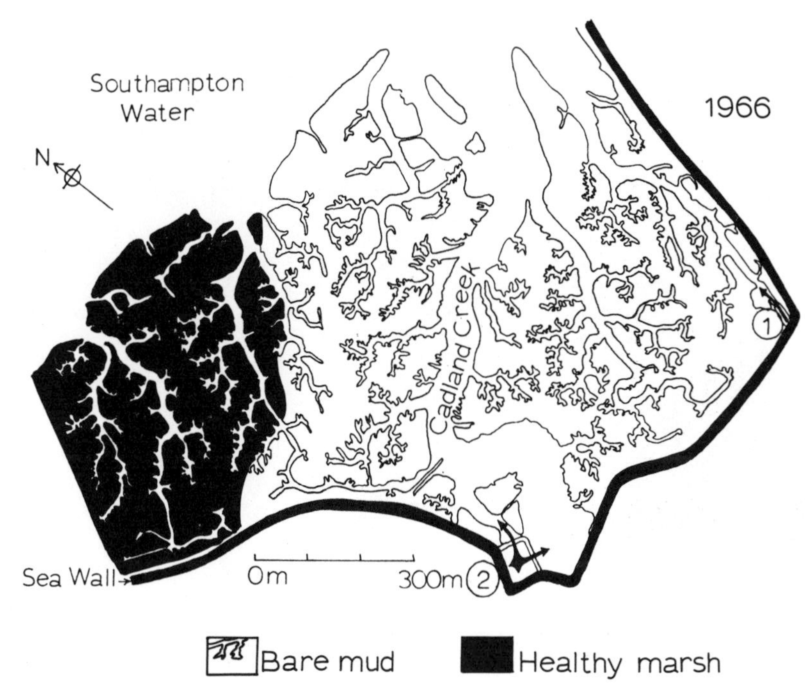

Fig. 7. The extent of salt-marsh vegetation at Fawley in 1966. (From O.S. charts SU 4404 and SU 4504)

Common (C): The majority of individual plants between 50 cm and 1 m apart. There may be small clumps of individuals growing closer together within this category, or small patches of less dense vegetation.

Rare (R): Individual plants more than 1 m apart and may be very widely scattered.

Two further categories of vegetation assessment were utilized: bare mud or absent–recorded where either no plants or none of the species under consideration occurred for 10 m in any direction; "healthy" saltings–the norm for this category was the salt marsh to the north of the outfall area where there were no manifest effects of pollution damage. This category was extended to cover areas previously affected by oil but now containing all the main marsh plants growing vigorously, either commonly or abundantly, but at lower densities or in slightly different proportions to the unaffected area. To fit into this category, vegetational

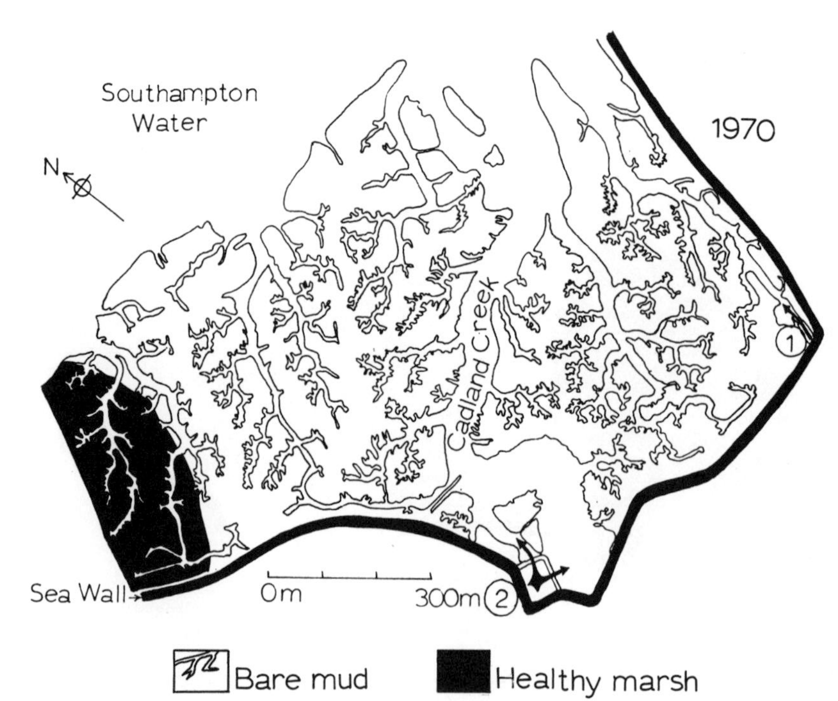

Fig. 8. The extent of salt-marsh vegetation at Fawley in 1969-70. (From Baker, 1970*a*)

cover of the mud should be complete with no sign of oil or pollution damage to the existing population of plants or shoots.

The categories for vegetation classification are arbitrary, and were chosen for speed of use and to fit with the observed plant distributions. They conveniently describe the broad difference in observed density and provide a clear picture of the present plant community.

Quantification of vegetational measurement was not possible because of the large areas of marsh involved and the often patchy distribution of individuals.

In addition to the assessment of distribution of the main marsh plants, the extent of other mud colonizers, e.g., blue-green algae, diatomaceous films, and filamentous green algae, were noted, particularly in the winter months when these organisms grow well due to the wetter conditions. Two further experimental schemes were initiated to monitor vegetational changes in association with mapping.

Spartina clump measurement. Detailed measurement was made of the shape and size of individual clumps of *Spartina anglica* growing in the areas which had been denuded by pollution. It was expected that spread or decline of these clumps would act as a pollution indicator. The location of these clumps is shown in Figure 9. All occur in the area where no living plants were recorded in 1970 (Baker, 1970*b*), and have therefore grown since 1970. Some (13, 14, 15, *A* and *B*) appeared as clumps of shoots (presumably from viable root stock in areas denuded of shoots between the surveys in July and December 1972).

Transplants. As a follow-up to the *Spartina anglica* clump measurements, transplants of groups of shoots and root stock of *Spartina anglica* were made in July 1973 from the healthy marsh to areas near to and farther from Cadland Creek. The progress of these transplants was recorded by the counting of healthy shoots during subsequent surveys. The location of the transplants and the number of healthy shoots at that time are shown in Figure 10.

Results

Change in marsh plant distribution, July 1972–July 1974. The vegetational mapping shows change in the distribution of all the dominant marsh species between July 1972 and July 1974. For comparative purposes, all maps show the boundary of salt-marsh vegetation found by Baker in 1970. This boundary, with the exception of its strandline end, has proved to be the limit of pollution effects on plants to date. The distribution of *Salicornia* spp. is illustrated in July 1972, July 1973, and July 1974 (Fig. 11); of *Spartina anglica* in July 1972, July 1973, and July 1974 (Fig. 12); and of *Halimione portulacoides* in July 1972 and July 1974 (Fig. 13), as examples of change in the extent of salt-marsh vegetation over the monitoring period. *Suaeda maritima* has shown very similar change in distribution to *Salicornia* spp. and *Aster tripolium* to *Halimione portulacoides*. All species have clearly begun recolonization of those areas previously denuded by pollution, with the exception of the area of the strandline. Oil was found only occasionally in small areas of the marsh during this period. In the strandline area (high-tide mark), however, further regression of all species had occurred between 1970 and 1972 (see Figs. 11-13), although between 1972 and 1974 no further damage was noted. Seedlings of *Salicornia* spp. and *Suaeda maritima* established at the edge of this area between 1973 and 1974.

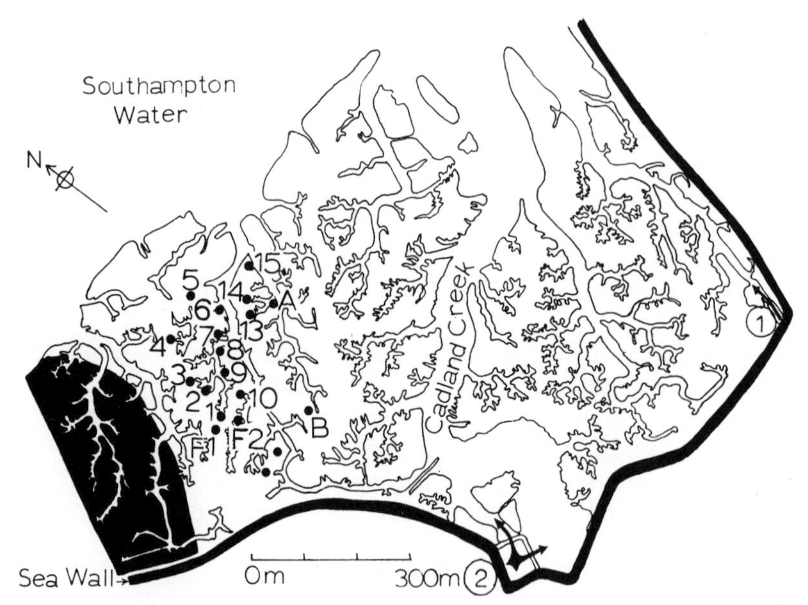

Fig. 9. The location of the measured clumps of *Spartina anglica*, Fawley salt marsh. Measurement started in December 1972 and has continued to the present time. The extent of salt-marsh vegetation in 1970 is shown (*shaded*).

Measured clumps of Spartina anglica. The progress of the measured *Spartina anglica* clumps between December 1972 and July 1974 is illustrated with selected examples of the 19 monitored clumps in Figure 14. Every clump has grown in size over the period monitored, some considerably, although some small parts of some clumps died back, apparently not related to oiling. Some of the clumps are no longer mapped because they have been enveloped by surrounding *Spartina anglica* growth and are no longer distinguishable.

Transplanted Spartina anglica. The numbers of healthy shoots in the transplants were counted at transplantation in 1973 and during the subsequent surveys of January and July 1974. The results are shown in Figure 15. While those transplants made to previously denuded areas but well away from the creeks carrying effluent have all flourished, those transplanted too near these creeks have all died.

Diatomaceous films, blue-green algae, and filamentous green algae. All these types of plant have flourished, particularly during the winter

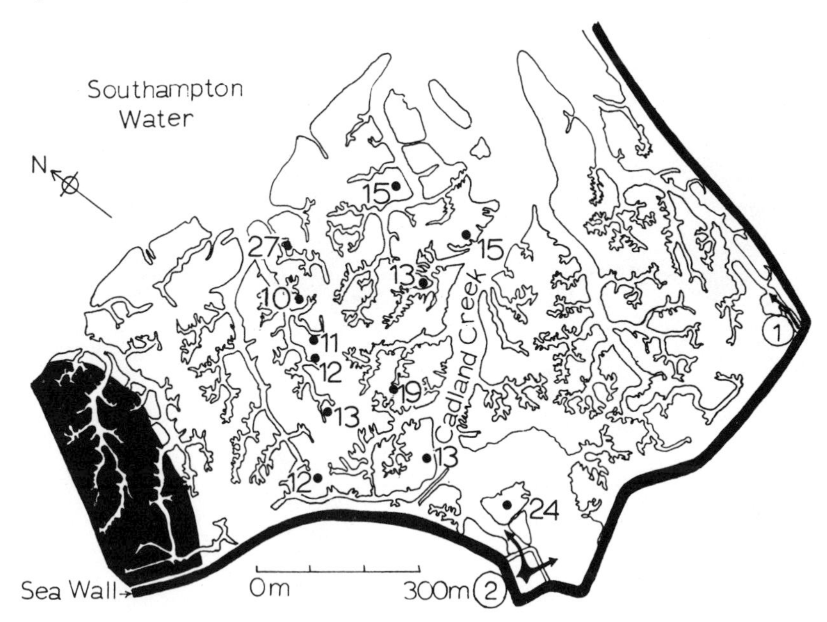

Fig. 10. The location of groups of *Spartina anglica* plants transplanted from the healthy marsh to denuded areas in July 1973 at Fawley. The numbers of shoots in each transplant is shown. The extent of salt-marsh vegetation in 1970 is also shown (*shaded*).

months of the survey. In December 1972 these organisms occurred over most of the bare mud area with the exception of approximately a 50-m bare area around each outfall and along the edges of the creeks taking the main effluent streams. Their distribution has not changed since, but the wet mild winters have produced very dense growths in most areas.

Discussion

The series of marsh vegetation distribution maps (Figs. 5-8) clearly show two phases in the progress of the Fawley marsh. In 1950 the area was a flourishing *Spartinetum*, presumably similar to the present unaffected areas of marsh to the north of the refinery. This was followed by a phase of extensive ecological damage between 1951 and 1970 when large areas of vegetation were killed and decomposed to leave bare mud (see Figs. 6-8). After 1970 the denuded area entered a

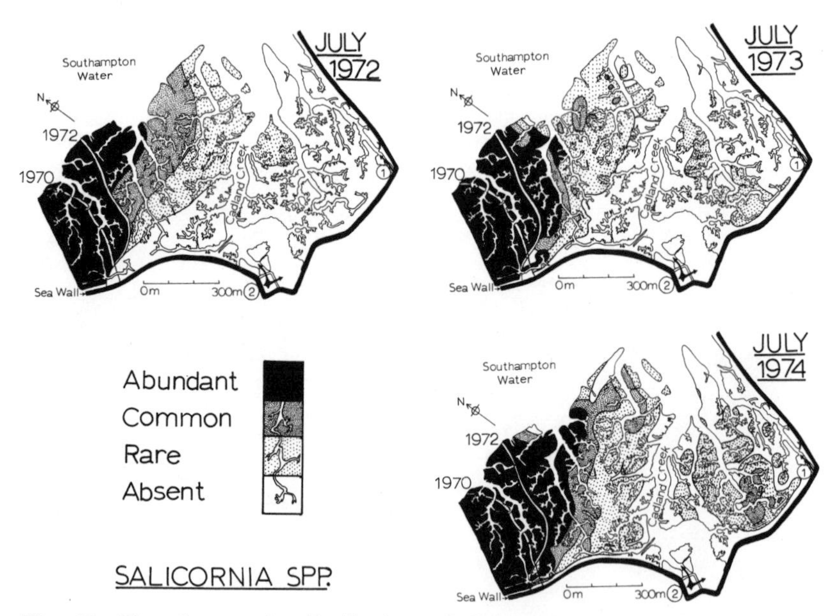

Fig. 11. The changes in distribution of *Salicornia* spp. at Fawley from July 1972 to July 1974. The boundaries of "healthy" saltings in 1970 and 1972 are marked.

recolonization phase when all the main marsh species have reestablished in various areas of the bare mud. The initial colonizers have been *Salicornia* spp. and *Suaeda maritima* followed by *Aster tripolium* and *Halimione portulacoides* and, with the least spread into the denuded areas, by *Spartina anglica*. This process appears similar to a normal successional process of colonization of mud flats in Southampton Water, where *Salicornia* spp. and *Suaeda maritima* are primary colonizers and, when established, start raising the mud level by silt accumulation, forming a *Salicornietum*. This *Salicornietum* stage is normally followed by a *Spartinetum,* when *Spartina anglica* establishes and stabilizes the mud with its extensive root system and accumulates more silt, raising the level even higher. The marsh then becomes better drained and less affected by tides, which allows colonization by *Aster tripolium* and *Halimione portulacoides* and *Juncus maritimus.* This successional sequence is determined by the rising mud level, increased drainage, and reduced exposure to the sea salt (Chapman, 1964).

The observed sequence of plant recolonization between 1970 and 1974 at Fawley, although similar to normal succession, has some important

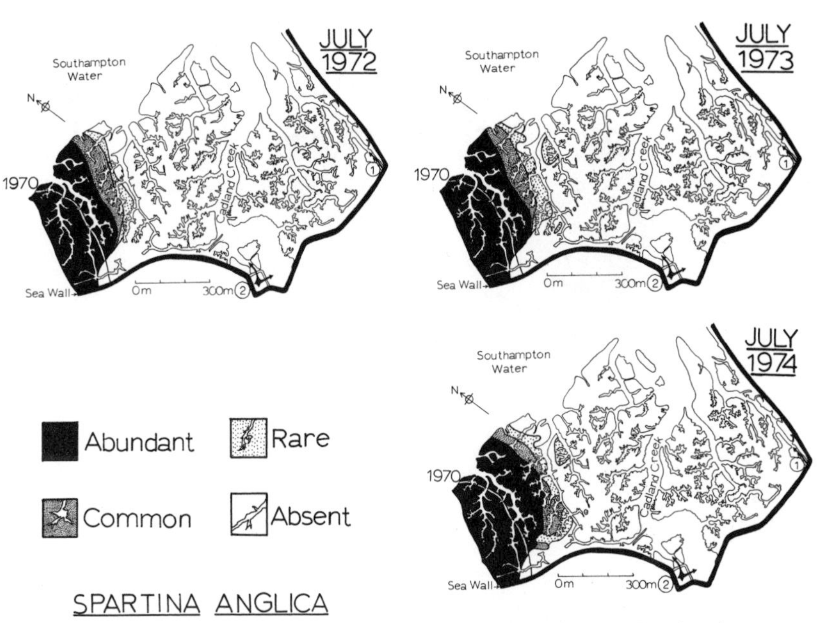

Fig. 12. Changes in the distribution of *Spartina anglica* at Fawley between July 1972 and July 1974. The boundary of the vegetation in 1970 is marked on each map.

differences. For example, *Halimione portulacoides* and *Aster trimpolium* appear out of sequence before *Spartina anglica*, and when *Spartina anglica* is transplanted to some areas where it has not spread naturally, it grows vigorously. Therefore the appearance of *Aster tripolium* and *Halimione portulacoides* before *Spartina anglica* in the recolonization sequence is not determined by a successional factor like attainment of a suitable mud level, but presumably by factors which influence the spreading of the plants. It is likely that mud level is still suitable for growth of most of the marsh species over a large area of the marsh, for the whole area was previously covered by healthy marsh and the mud is still bound and stabilized to some extent by the old *Spartina anglica* root systems and has eroded only a small amount (15-25 cm; Baker, 1970*b*). In some localized areas, erosion may be greater or less than average. Those with greater erosion may only be suitable for recolonization by *Salicornia* spp. The main difference between those plants which spread to bare areas first and those which come later seems to be the success with which the plants

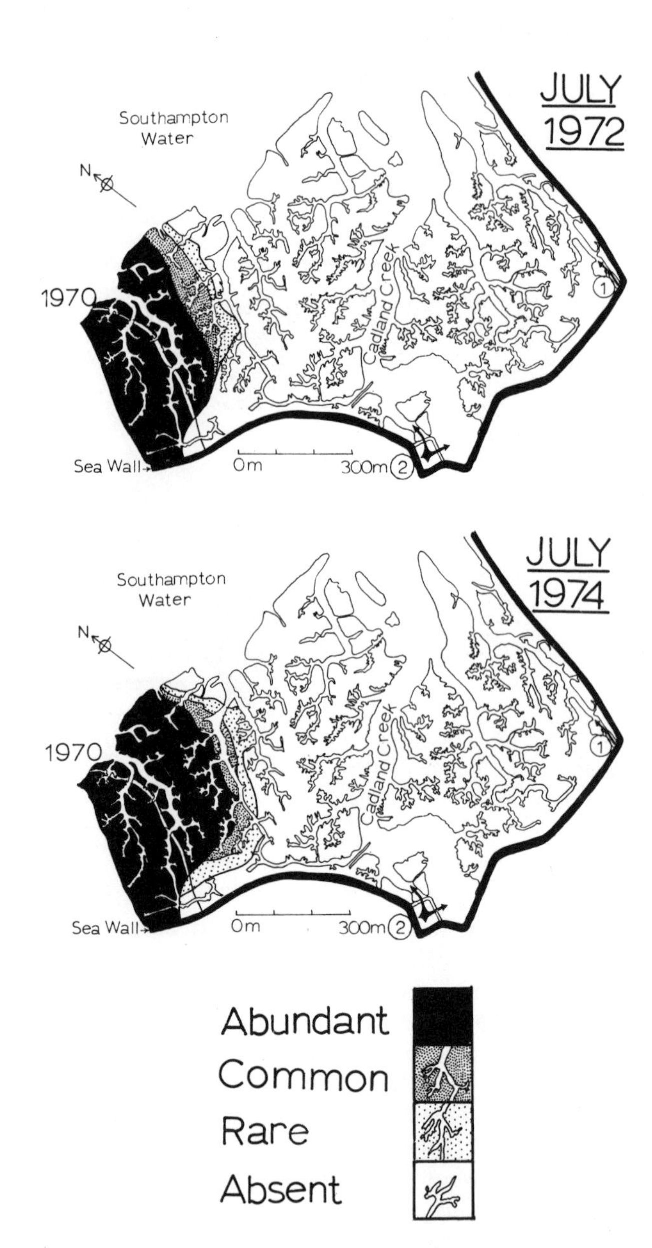

Fig. 13. The distribution of *Halimione portulacoides* at Fawley in July 1972 and July 1974. The boundary of the vegetation in 1970 is also marked.

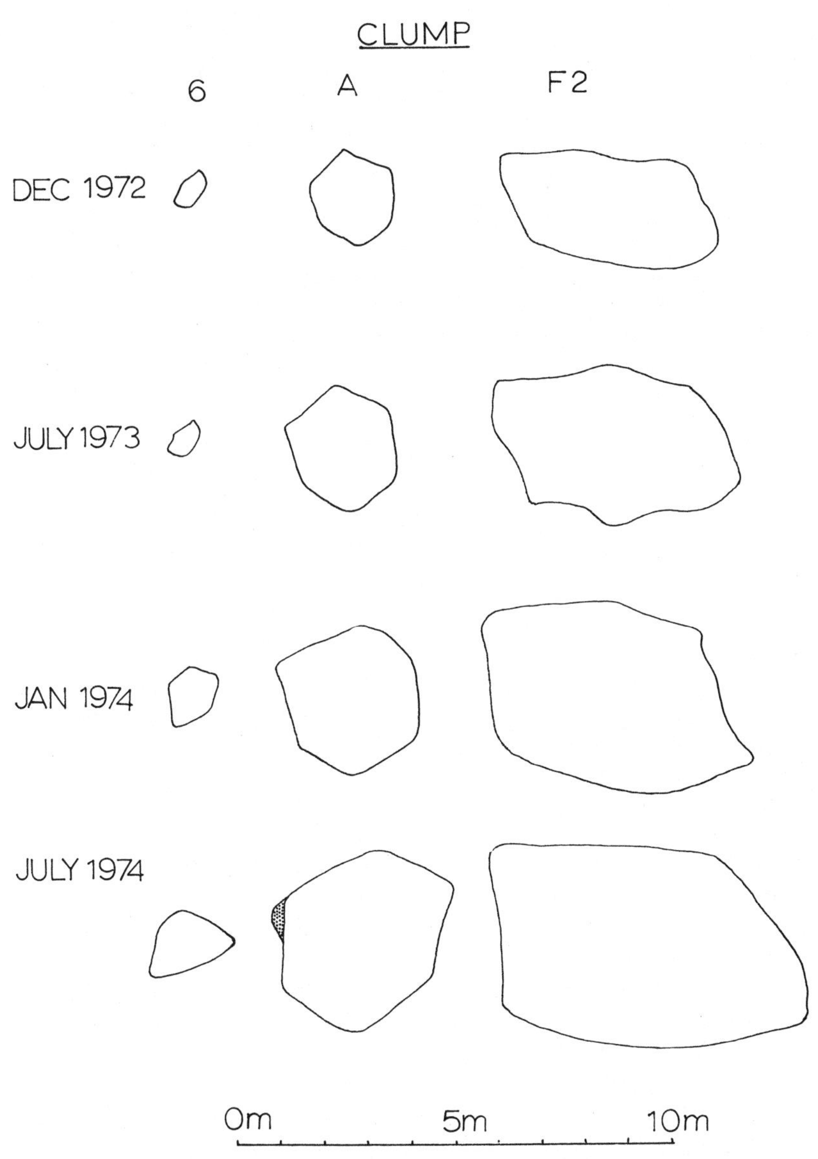

Fig. 14. Change in size of 3 of the 19 measured clumps of *Spartina anglica* between December 1972 and July 1974. Stippled area has died back after growth.

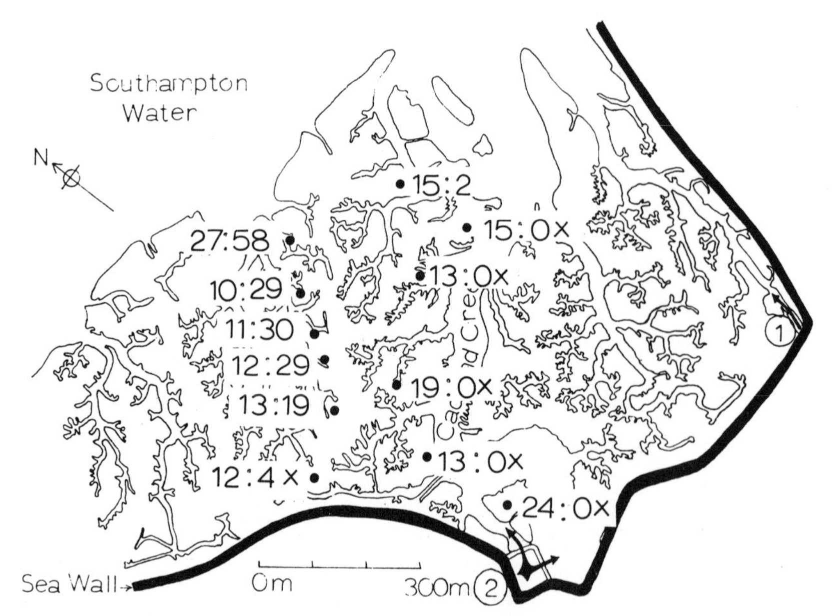

Fig. 15. Changes in the numbers of healthy shoots in the *Spartina anglica* transplants between July 1973 and July 1974 at Fawley. The two figures at each transplant site show numbers of shoots at transplanting in July 1973 (*left-hand figure*) and in July 1974 (*right-hand figure*). Transplants marked *X* were oily on the shoots in January 1973, and the majority of these subsequently died.

spread by seed. *Salicornia* spp., *Suaeda maritima*, and *Aster tripolium* are annual plants reproducing primarily by seed. The abundance of *Salicornia* spp. and *Suaeda maritima* in the healthy marsh means that large numbers of seeds will be available to establish in any new areas where the seeds can germinate and grow. Hence they will be among the first to enter any suitable areas. *Halimione portulacoides* spreads by both seed and vegetative means. *Spartina anglica* however, spreads mainly by vegetative methods and seeds very poorly (Dalby, 1970).

It is feasible, therefore, that if recovery continues, large areas of the denuded mud will become dominated by *Aster tripolium, Halimione portulacoides, Salicornia* spp., and *Suaeda maritima* rather than *Spartina anglica*. In the longer term, however, *Spartina anglica* may once again come to dominate the marshes if it is able to compete successfully against other established species. If this is not the case, the original *Spartinetum* may not return.

Hence the recolonization pattern and its species composition for the present will probably be determined by seed distribution and seedling success. Observation of the marsh plants during surveys supports this. *Salicornia* spp., during their spread from July 1972 to July 1974, first established in the area between Nos. 1 and 2 outfalls all around the heads of the small creeks off the main creeks. This fits well with distribution of the seeds by the rising and falling tides. Around these few established individuals, the following year's seeds greatly increased the density and spread of this species in the area. Similar seedlings spread have been observed in *Suaeda maritima* and *Halimione portulacoides*. The slow spread of *Spartina anglica*, therefore, seems due to its predominantly vegetative reproduction.

As the *Spartinetum* stage may not be the end result of recolonization over the whole marsh (although it is in some of the areas near the 1970 boundary which have already recovered almost completely; see below), the definition given for "healthy" saltings may not be applicable. A revision of the definition to include the return of the marsh to 100% vegetational cover, although not necessarily dominated by *Spartina anglica*, should be made.

In the two areas marked in Figure 16 vigorous growth of all the main marsh species has occurred, and there is no evidence of fresh oil damage. The area nearest the 1970 boundary was in this "healthy" condition in 1972, while the second area reached this state in 1974. The monitored *Spartina anglica* clumps in this area have grown vigorously, and many are no longer measurable as they have been incorporated into extensive spreading *Spartina anglica*. While there are occasional patches of dead oiled vegetation (pre-1970) in these areas, the overall condition of the salting vegetation is healthy and recovered to a *Spartinetum*. Although growth in these areas is not quite so dense as that in the nearby unaffected marsh, they can now be considered as "healthy" saltings. It is probable that the recovery of these areas to a *Spartinetum* was by regrowth of *Spartina anglica* from viable root stock, the repetitive oiling which had killed shoots in the few years previous to 1970 not killing roots as well.

Those areas which have recovered to a *Spartinetum* and where new measured clumps of *Spartina anglica* were found which had grown between 1970 and 1972 were healthy marsh in 1966; so oil has not been affecting them for so long a period as elsewhere in the denuded areas. Hence *Spartina anglica* root survival may have been possible in these areas, and produced the recovery to a *Spartinetum*. Elsewhere, there is no sign of regrowth of *Spartina anglica* from root stock, and seeding will be important in determining the species composition of these areas if recovery continues.

Changes in the extent of the "healthy" saltings between 1950 and July 1974 are summarized in Figure 17, and in the extent of salt-marsh plants in Figure 18. Overall, recolonization has occurred in those areas away from the main effluent-carrying creeks, which are still having a toxic effect on plants, as shown by death of transplanted clumps of *Spartina anglica* to these creek edges. The patterns of recolonization shown in Figures 11, 12, and 13 also indicate that reestablishment of vegetation is in those areas removed from the main effluent creeks.

While recolonization of large areas of denuded mud has occurred, with the exception of the immediate areas of the effluent creeks, there is also an interruption to this pattern in the strandline area (see Figs. 11-13, 17-18). Any new oil usually ends up in this area after deposition by high tides. This has probably been the case since damage to the marsh started, and has resulted in very oily mud in this region (Baker, 1971). The occasional accumulation of new oil in this area and the refloating of old oil at high tides produced further regression of the marsh plants

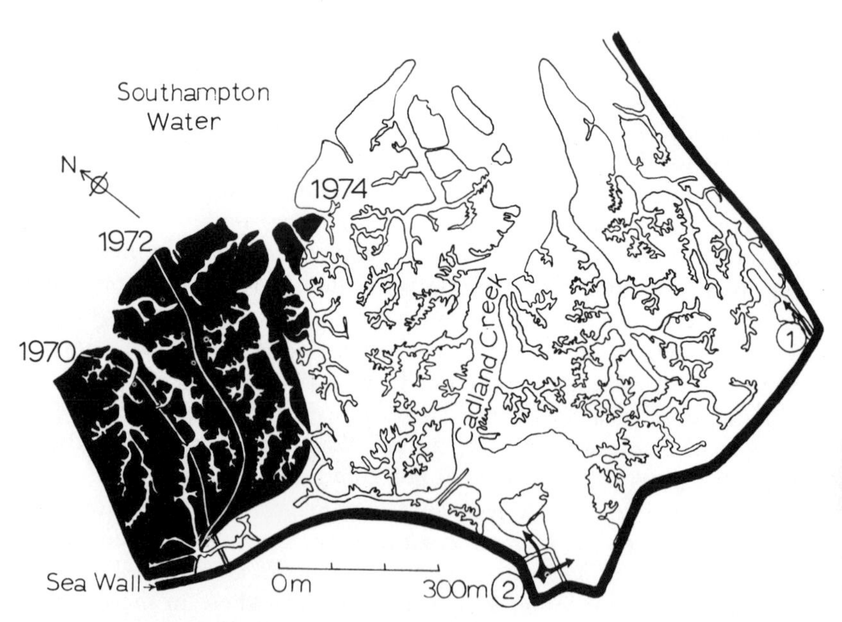

Fig. 16. Extent of "healthy" saltings at Fawley in 1970, 1972, and 1974. The areas marked 1972 and 1974 were determined as "healthy" saltings after the surveys of July 1972 and July 1974.

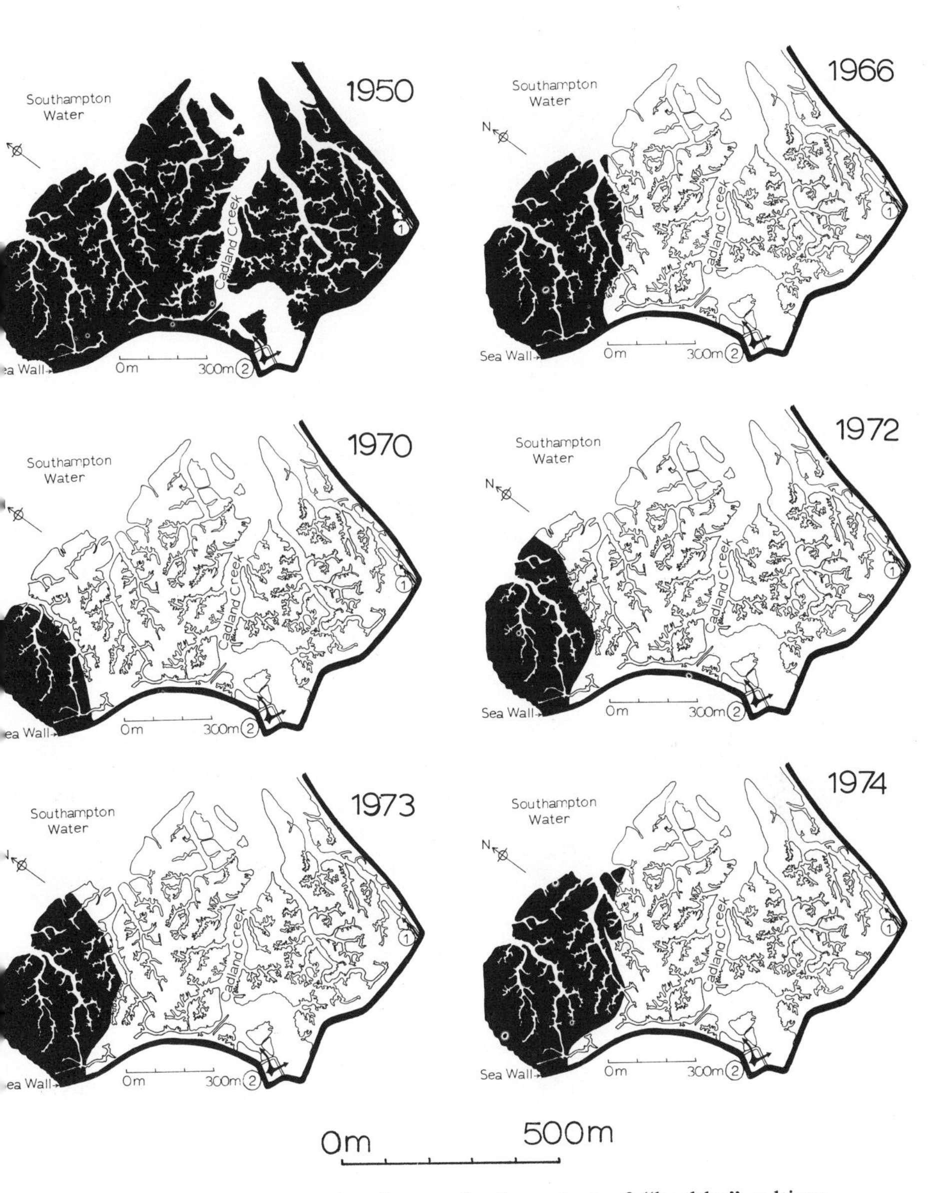

Fig. 17. Maps summarizing the changes in the extent of "healthy" saltings at Fawley from 1950 to July 1974

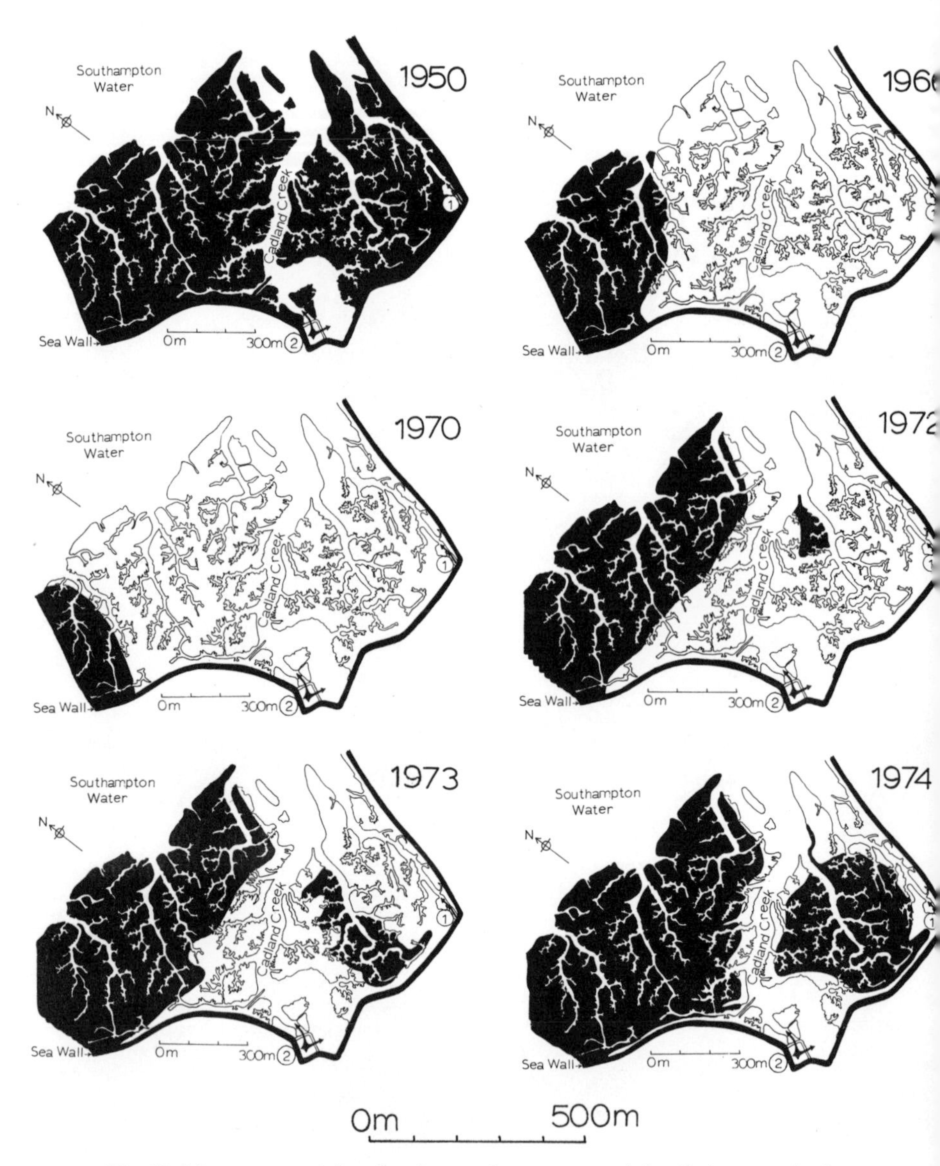

Fig. 18. Maps summarizing the changes in areas containing living marsh plants at Fawley from 1950 to July 1974

between 1970 and 1972, although since then no further damage has been observed. Some seedlings of *Salicornia* spp. and *Suaeda maritima* have established nearby among dead oily vegetation, but none have established in an area between 10 m and 20 m wide immediately below the strand area. This area may well not recover for a considerable period of time, either until all oiling stops or until the oil there is stabilized in the sediment and can no longer float into and onto the vegetation.

An interesting point on recolonization is that neither mud containing old oil (Baker, 1970*a*) nor dead oily *Spartina anglica* seems to prevent growth of seedlings, particularly of *Salicornia* spp. and *Suaeda maritima*. There are areas of dead oily *Spartina anglica* vegetation (oiled before 1970) in the boundary region of the healthy marsh of 1970 which are colonized now by *Salicornia* spp., *Suaeda maritima, Halimione portulacoides, Aster tripolium,* and even by vegetative shoots of *Spartina anglica.*

The reasons for the observed recolonization patterns are not certain at present. It is likely that improvement of effluent quality has played a part. Little fresh oil has been observed on the salt-marsh plants since monitoring started in 1972. However, the mildness of winters over the last three years accompanied by early springs and late autumns may well have influenced marsh plant growth and reproduction, causing early germination and a very long growing season. A severe winter or cold spell after early germination may have considerable deleterious effects. A further point of importance may be the relatively few oil slicks which have affected the marsh since 1970. The sensitivity of *Salicornia* spp. to oil on the water surface and its role as a primary recolonizer may mean that one serious spillage could considerably alter the present recolonization pattern.

For these reasons, and to watch the situation in the strandline area where oil which occasionally occurs on the sea surface usually ends up and around the main effluent-carrying creeks, continued monitoring is necessary. To accurately determine the relative importance of effluent improvement, climatic change, or oil spillage to the recolonization of denuded areas, study of the marsh for some time to come is required, and the interim nature of this report should be stressed.

Summary

1. The effects of oil on salt-marsh vegetation are listed. Experimental evidence indicates that oiling of *Spartina anglica* shoots and leaves can affect underground vegetative systems and that successive oilings of *Spartina anglica* have progressively more serious effects.
2. The salt marsh under study has been progressively damaged by repeated oilings over a period of twenty years from 1951 to 1970, resulting in an area of bare mud approximately 1000 m by 600 m.
3. Recent field surveys (1972-74) show that some previously damaged areas are recovering and being recolonized by most of the original marsh plant species. These areas are removed from the main effluent-carrying creeks.
4. In those areas showing recolonization by salt-marsh plants, the pattern of recolonization appears to depend on the success with which the plants seed, successful seeders reinvading more quickly than poor seeders. This might result in a marsh dominated by species other than the original *Spartinetum anglica*, which reproduces primarily in a vegetative manner and thus spreads only slowly.
5. Some denuded areas of salt marsh, however, have recovered to *Spartinetum*. These areas were only subjected to oiling for a few years, and *Spartina anglica* growth has probably come from viable root stock.
6. Transplanting groups of *Spartina anglica* shoots to denuded areas resulted in some cases in their establishment and growth. Those transplanted near the salt-marsh creeks carrying the effluent streams died, showing the effluent still to be affecting at least part of the denuded area and preventing recolonization.
7. Oil trapped in the strandline area has so far prevented recolonization of this region, and may do so for some time.
8. Whether recolonization of some parts of the previously denuded areas is a result of effluent quality improvement or climatic factors is not yet certain. It is likely that both have played a part.

Acknowledgments

I am grateful to Esso Petroleum Company for sponsoring this work, and to Peter Sutton for his assistance during survey work and in checking the manuscript. I am particularly grateful to Dr. Jenifer Baker for providing helpful suggestions and some of her previous experimental results. Other members of the Oil Pollution Research Unit at Orielton Field Centre, Pembroke, have also willingly assisted in the field, and I would like to thank Andrew Arnold of the Field Centre staff for his help during field surveys.

Literature Cited

American Petroleum Institute. 1963. Manual on the Disposal of Refinery Wastes, Vol. 1. Waste Water Containing Oil, 7th ed. 104 pp.

Armstrong, W., 1967. The oxidising activity of roots in waterlogged soils. Physiologia Plantarum 20:920–926.

Baker, J. M. 1971*a*. Studies on saltmarsh communities–successive spillages. Pages 21–32 *in* E. B. Cowell, ed. The Ecological Effects of Oil Pollution on Littoral Communities. Institute of Petroleum, London.

_____. 1971*b*. Studies on saltmarsh communities–refinery effluent. Pages 33–43 *in* E. B. Cowell, ed. The Ecological Effects of Oil Pollution on Littoral Communities. Institute of Petroleum, London.

_____. 1971*c*. Studies on saltmarsh communities–seasonal effects. Pages 44–51 *in* E. B. Cowell, ed. The Ecological Effects of Oil Pollution on Littoral Communities. Institute of Petroleum, London.

_____. 1971*d*. Studies on saltmarsh communities–oil and saltmarsh soil. Pages 62–71 *in* E. B. Cowell, ed. The Ecological Effects of Oil Pollution on Littoral Communities. Institue of Petroleum, London.

_____. 1971*e*. Studies on saltmarsh communities–comparative toxicities of oils, oil fractions and emulsifiers. Pages 78–87 *in* E. B. Cowell, ed. The Ecological Effects of Oil Pollution of Littoral Communities. Institute of Petroleum, London.

_____. 1971*f*. Studies on saltmarsh communities–the effects of oils on plant physiology. Pages 88–98 *in* E. B. Cowell, ed. The Ecological Effects of Oil Pollution on Littoral Communities. Institute of Petroleum, London.

_____. 1971*g*. The effects of oil pollution and cleaning on the ecology of saltmarshes. Ph.D. diss., Univ. Coll. S. Wales, Swansea.

Blokker, P. C. 1970. Prevention of water pollution from refineries. Paper 3, Seminar on water pollution by oil, Institute of Water Pollution Control, Institute of Petroleum and WHO, May 4–8, 1970.

Blokker, P. C. and H. J. F. Marcinowoski. 1970. Survey on quality of refinery effluents in western Europe. Stichting Concawe. Report 17/17.

Chapman, V. J. 1960. Saltmarshes and Salt Deserts of the World. Leonard Hill, London. 392 pp.

_____. 1964. Coastal Vegetation. Pergamon, London. 245 pp.

Dalby, D. H. 1970. The saltmarshes of Milford Haven, Pembrokeshire. Field Stud. 3(2):297–330.

Hubbard, C. E. 1954. Grasses. Penguin Books, New York. 463 pp.

Appendix
Notes on the Major Species of Plants in the Fawley Salt Marsh

Salicornia spp. (Fig. A1)

Clapham et al. (1962) describe eight annual species of *Salicornia*. Species in this genus can be positively identified only by means of chromosome count, which is considered pointless for the purposes of this survey. *Salicornia* spp. have hairless succulent leaves borne on branched stems. Being annuals, they are shallow-rooted plants with no underground storage organs. Propagation is by seed dispersed by the tide, and most germination and seedling strike occurs in the spring tide zone because seedlings require at least three days of emersion to firmly establish an anchorage. Germination is most pronounced during April and May. Mature plants can grow up to 40 cm in height. *Salicornia* plants are very susceptible to oiling, and due to shallow roots and small food reserves, they show no powers of recovery.

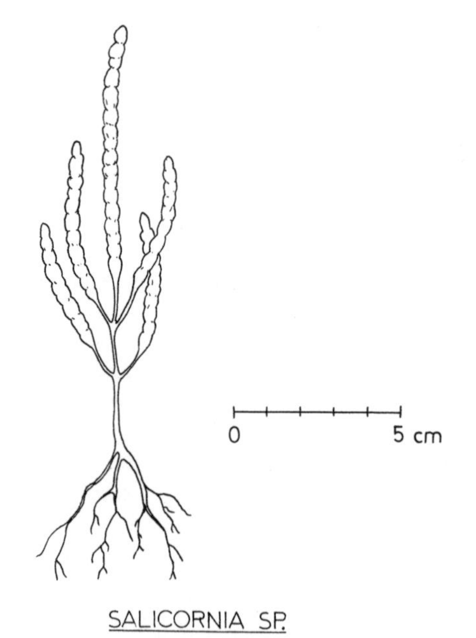

Fig. A1. *Salicornia* spp.

Suaeda maritima (L.) B. C. J. Dumortier (Fig. A2)

Suaeda maritima is a somewhat variable annual plant with glaucous- or red-tinged, fleshy, alternate leaves. It propagates by seed, most germination occurring during April and May. Like *Salicornia*, its seedlings and mature plants are particularly susceptible to oiling.

Halimione portulacoides (L.) P. Aellen (Fig. A3)

Halimione portulacoides is a small woody perennial shrub with a creeping growth habit. The leaves are elliptical in shape with a frosted appearance. The mature plant has a gray-greenish hue and grows up to 80 cm in height. It prefers conditions of good drainage such as the edges of creeks or shingle banks, where it characteristically forms a fringe of vegetation. Oil is readily trapped on the leaves in very fine hairs, and the branch ends of the plant are badly damaged in oiling. Recovery from oiling can occur by the production of new shoots from root stock.

Aster tripolium L. (Fig. A4)

Aster tripolium is a short-lived perennial herb with a short rhizome and stout, erect stems growing to a height of 15-20 cm. The leaves are bright green, elongated ovoid in shape, and fleshy with three faint veins. The flowers are a daisy type with inner (ray) florets white and outer (disk) florets yellow. Seeds are attached to a hairy appendage (pappus) which enables dispersal by wind over a short distance. Seed dispersal is also achieved by tidal movement. The plant employs a typical weedlike strategy, often colonizing bare areas (Ranwell, 1972). The underground rhizome is presumably an asset to the plant in recovery from oiling.

Spartina anglica C. E. Hubbard (Fig. A5)

Spartina anglica is a deep-rooting perennial grass 30-130 cm high with a rhizomatous root system and green to gray-green leaves with overlapping sheaths and small hairy ligules. The species forms extensive clumps or clones growing in concentric rings out from the original stock.

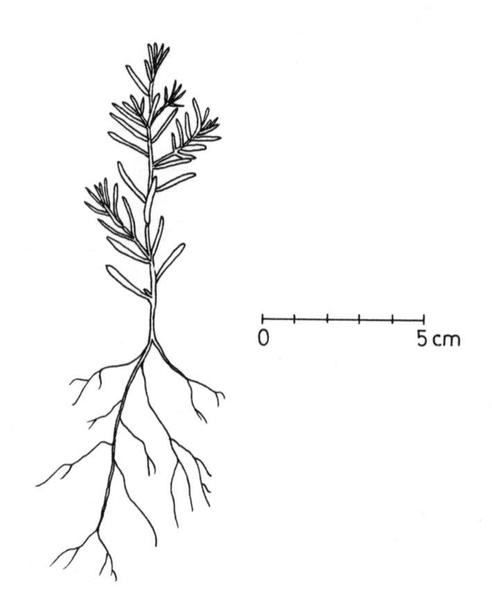

Fig. A2. *Suaeda maritima* (L.) B. C. J. Dumortier

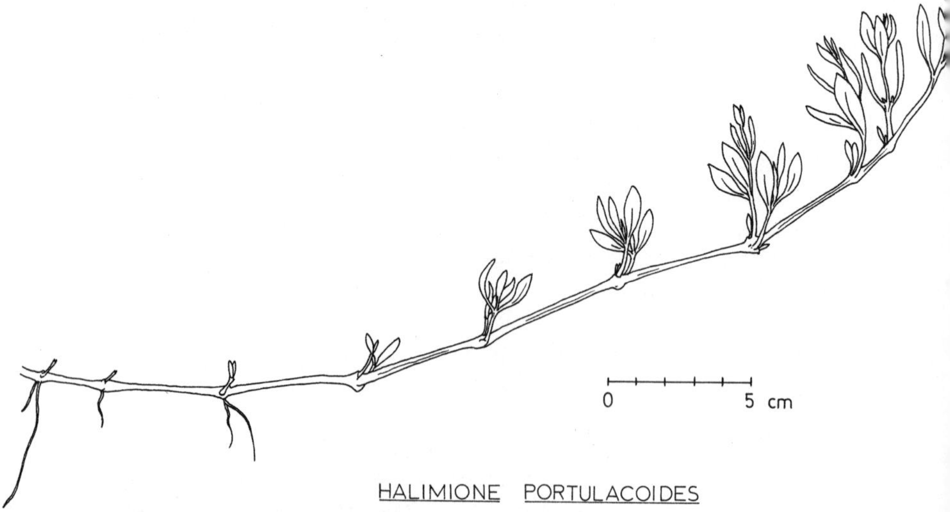

0
5 cm

HALIMIONE PORTULACOIDES

Fig. A3. *Halimione portulacoides* (L.) P. Aellen

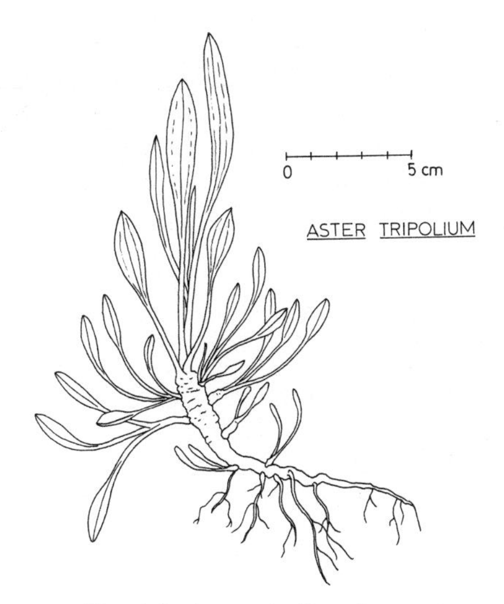

Fig. A4. *Aster tripolium* L.

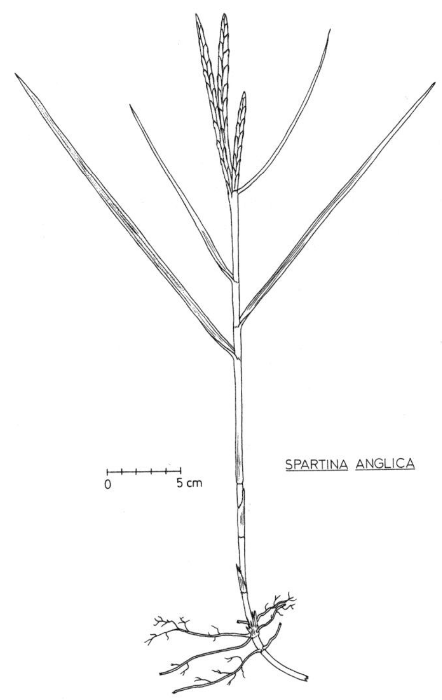

Fig. A5. *Spartina anglica* C. E. Hubbard

Propagation occurs vegetatively by spreading of the clones, from fragments of the plant dispersed by the tide, and occasionally from seed production and dispersal. Oil is readily trapped by furrows in the leaves, causing mechanical blockage and smothering. Recovery from oiling occurs, provided pollution is not chronic, and transplants of healthy shoots are known to grow successfully in soils containing weathered oil residues.

Literature Cited

Clapham, A. R., T. G. Tutin, and E. F. Warburg. 1962. The Flora of the British Isles. Cambridge University Press.

Ranwell, D. S. 1972. Ecology of Salt Marshes and Sand Dunes. Chapman & Hall, London. 258 pp.

Strategies for the Reintroduction of Species into Damaged Ecosystems

Daniel B. Botkin

Abstract

Many introductions, reintroductions, and transplants of species have been tried; most have failed. Success can never be guaranteed, but careful analyses of genetic, physiological, and ecological factors can greatly aid our attempts.

This chapter suggests a methodology for such reintroductions. Relevant ecological concepts and theories are discussed, including the concepts of *r* and *k* selection, niche theory, population interactions, and ecosystem dynamics.

Four case histories, dealing with the Indian lion, bighorn sheep, whooping cranes, and the moose and wolves of Isle Royale, are used to illustrate the application of these concepts and typical problems which must be faced in attempting reintroductions.

Introduction

The reintroduction of a species involves a considerable amount of chance; no matter what we do, there is a finite probability that the species will not persist. Although no one can guarantee the success of a reintroduction, there are a number of factors that can help us to evaluate and to increase the probability of the survival of reintroduced species. The purpose of this chapter is to outline these factors and to provide the basis for a methodology for such reintroductions.

The reintroduction of species into their former habitats is an important task. The purpose of reintroduction may include aesthetic, recreational, ecological, or economic factors. For example, the reintroduction of a plant may help restore soil fertility or prevent erosion; the reintroduction of large mammals, like deer or beaver, may be desired by tourists or hunters. In some cases, such as salmon, reintroductions involve major

A contribution to the Ecosystems Center, Marine Biological Laboratory, Woods Hole, Massachusetts.

sources of food. Reintroduction can play a major role in the preservation of endangered species. Extinction rates are increasing in spite of an increased awareness of the need to preserve natural environments and species. Forty-seven species of wildlife became extinct in the United States between 1700 and 1970, and 25 of these were lost within the last 50 years (Silcock et al., 1973). Ehrenfeld (1972) estimates that the current rate of extinction among most groups of mammals is about a thousand times greater than the "high" rate of extinction that occurred at the end of the last glaciation, when the geological record suggests there were massive extinctions of large birds and mammals.

There are 59 endangered species of artiodactyls (deer, antelope, etc.) that are hunted as game; 45 species of carnivores (cats, bears, weasels, etc.) that are endangered because they are predators which man considers harmful; 35 species of primates, endangered mostly because of their use in medical research; 33 species of marsupials, endangered partly because they are killed in large numbers for hides (Miller and Botkin, 1974). There are approximately 3000 endangered species of plants listed by the Smithsonian Institution for the United States. If man continues to alter more of the earth's environment, more species will join the endangered list, and extinctions will continue to increase. The reintroduction of species into their former habitats is an important part of the strategy to help assist the survival of wild animals and plants.

There are a number of factors which one should take into account to determine whether the probable success of a reintroduction is high. These factors can be classified into (*A*) innate species characteristics, (*B*) interactions between the species and other living organisms, and (*C*) habitat characteristics.

A. Innate Species Characteristics

Innate species characteristics include demographic, genetic, and physiological factors. Questions one must ask include:

1. What is the maximum longevity?
2. What is the maximum reproductive rate?
3. What is the age of first breeding?
4. What is the maximum growth rate?
5. How do maximum longevity, maximum reproductive rate, age of first breeding, and maximum growth rate change as resources change? In particular, how tolerant of scarcity is the species? Can it utilize or store an overabundance of a resource?
6. What population size is needed to avoid serious consequences of inbreeding or chance fluctuations?

B. Biological Interactions

Questions to be asked about biological interactions include:

1. What are the competitors of the species? Which are present in the ecosystem? Do superior competitors already exist or are they likely to enter the habitat?
2. Who are its predators and parasites, and which are present? Which, if not present, could easily immigrate once the prey/host were introduced?
3. Does the species require symbionts? Are they present? Can they persist in the habitat?
4. Are the required prey (if any) available in the habitat?

C. Habitat Characteristics

Determination of habitat characteristics requires that the following questions be asked:

1. Has the habitat been altered to the extent that it is no longer suitable to the species?
2. Are microclimatic and macroclimatic conditions suitable?
3. If shelter is required, is it or the material for it available?
4. Are all necessary mineral nutrients available?
5. Is the habitat sufficiently large?
6. Does the habitat have sufficient variety (or uniformity) for the species?
7. Are there toxic materials that endanger the species?
8. Is more time necessary for the habitat to recover to an extent that it will be suitable?

In the terms of ecological theory, what has been outlined above are requirements of a species in terms of *niche theory*. This theory includes the concepts of (1) a *fundamental niche* and (2) a *realized niche*. As originally formulated by Hutchinson (1957), the fundamental niche of a species corresponds to all the states of the environment which would permit a particular species to exist indefinitely in the absence of adverse interactions with other species. The realized niche is defined as those states of the real environment where the species can persist indefinitely in the presence of actual interspecific interactions. The niche can be represented by a series of graphs, with the abscissa representing an environmental resource and the ordinate the individual or population responses of the species to this resource. Resources axes include light intensity, amount of heat energy, amount of water, concentrations of essential nutrients or of some other measure of dietary resources. The niche represents the totality of responses of the species to all possible resource states important to it. In these terms, an idealized management scheme for reintroduction would be to know the fundamental niche of the species and to know whether the candidate habitat includes the necessary conditions for the realized niche of the species. In this scheme, one would thus know the functional relationships between all resources and the expected response of the species (see Botkin and Miller, 1974; Botkin, 1975 and in press).

The process of reintroduction of a species will include the following considerations:

1. A goal. The reintroduction of the species may be sought (*a*) to return the ecosystem to its predisturbance state for aesthetic reasons; (*b*) to aid in the preservation of a previously endangered species; (*c*) for hunting or food supply or other direct economic gain; (*d*) to affect some ecosystem property (e.g., erosion rates, loss of essential nutrients, to increase or otherwise change productivity or rates and storage of mineral cycling). It is important that the goals be specified; different goals may require different approaches.
2. Careful study of the habitat to determine if the requirements of the species can be met.
3. Monitoring the population over a time period sufficient to determine whether or not the reintroduction has been successful. This may not be a simple matter. The population of a reintroduced species may change temporarily but abruptly. The species may decrease rapidly to some low point, or it may increase abruptly to what seem to be dangerously high levels. In some cases, these temporary fluctuations may be followed by a return to a comparatively stable population. Careful evaluation of the population levels must be made to decide when human intervention may be necessary.

Reintroductions frequently involve small initial populations. Among the difficulties which lead to failure of small populations are:

1. Random fluctuations, even under constant environmental conditions, leading to extinction.
2. An environmental catastrophe which leaves no survivors.
3. A small genetic pool leads to inbreeding, accumulation of pool deleterious genes, and little genetic variability to allow adaptation to a changing habitat.
4. The habitat size is determined for the initial population and is too small for a population capable of long-term persistence.

Reintroductions, like introductions of alien species into new habitats, are subject to two types of potential failure: (1) extinction and (2) epidemic increase to such a level that the species may become detrimental to man's interest or may cause significant ecosystem damage. Most introductions fail. The attempts to reintroduce species into previously occupied habitats share much in common with attempts to introduce alien species into new habitats or preserve species which have been reduced to extremely small populations. Those interested in species reintroduction can learn much from the past history of attempted introduction as well as attempts to preserve populations reduced to low numbers.

Although any species has a unique combination of characteristics, species may be classified into functional groups which share certain general morphological, physiological, or ecological attributes. There are two general types of species. One type is relatively short-lived but has high reproductive and growth rates. This kind tends to be highly productive when resources are in abundant supply. The second has relatively low reproductive and growth rates but is long-lived and tends to be relatively efficient in obtaining resources which are scarce and for which there is considerable competition.

These types are known as r-selected and K-selected species, respectively. These terms arise from the logistic equation which has traditionally been used by ecologists to describe the growth of hypothetical populations:

$$\frac{dN}{dt} = rN\,[(K-N)/K]$$

where N is the number of individuals in the population, r is the intrinsic rate of increase (births minus deaths), and K is the saturation density or the maximum N which can persist in a particular environment. An r-strategist is able to colonize new habitats quickly or to take rapid advantage of changing abundances of resources within habitats, or in transient habitats, such as temporary ponds and pools that may be highly productive for short periods. Many native and introduced pests are r-strategists.

Among K-selected species, less energy is diverted to reproduction and more to individual survivorship. Such species have traded a high rate of increase and the ability to exploit transient environments for the ability to maintain more stable population levels in more stable environments and to make use of resources when they are scarce or where competition for them involves a high density of organisms (Miller and Botkin, 1974).

Each real species has its own unique set of characteristics, but the general idea of *r* and *K* strategies can be helpful in our attempt to reintroduce and to preserve species. In general, it is easier to introduce *r*-selected species and prevent their extinction, but such a species is more likely to become a pest than a *K*-selected one. Many species of economic importance are of the first kind. These include major crops like corn and wheat, and domestic or wild ducks like the mallard. Species of the second kind, such as the California condor or the whooping crane, are particularly susceptible to extinction.

Case Histories

The Whooping Crane

A consideration of the factors that lead to success or failure is best approached through examples. The whooping crane is a good example to begin with because it almost became extinct, and its population is still low in spite of 60 years of conservation efforts. Why has this long effort not produced a dramatic result?

Although the whooping crane never completely disappeared, its population reached a low of 15, so that the attempt to save it from extinction is in essence equivalent to an attempt to reintroduce it. The endangered status of the whooping crane (*Grus americana*) is widely known, and its history has been well documented (Allen, 1952, 1957). The species was placed under protection of the Migratory Bird Treaty in 1916, but the population continued to decrease from 47 individuals in 1918 to a low of 15 reached in 1941. The whooping crane is North America's tallest bird and has survived in its present form through eons of environmental change. Fossilized bones from the Upper Pliocene period, indistinguishable from the bones of whooping cranes living today, have been dated at about 3.5 million years old (Fisher et al., 1969). During the Pleistocene period the whooping crane had a broad continental distribution through most of the North American continent, but whooping cranes are extremely shy and intolerant of human disturbance. They retreated before the steady advance of colonization in North America until they had been reduced to a migrant population of about 1400 birds in the mid-nineteenth century (Allen, 1952).

About 90% of this population was lost, mostly through hunting and habitat disruption between 1870 and 1900, and the total population was reduced to between 80 and 100 individuals by 1912. When its wintering ground was discovered on the Blackjack Peninsula of Texas, more than 47,000 acres were purchased by the federal government to establish the Aransas National Wildlife Refuge in 1937 (Howard, 1954). Intensive conservation efforts following World War II resulted in better protection during spring and fall migration (Novakowski, 1966). The breeding grounds were not discovered until 1955 in Wood Buffalo Park, Northwest Territories, Canada (Allen, 1957), which is a remote, isolated area.

In spite of unprecedented conservation efforts and almost total protection, the recovery of the wild population, reaching 51 in 1972 (Fig. 1), has been slow and somewhat erratic. That this recovery has been slow in spite of the bird's protection suggests that intrinsic characteristics of the whooping crane combined with current habitat conditions prevent the population from undergoing a rapid increase. What are these characteristics, and how might one best act to protect the species and increase the chance of its survival?

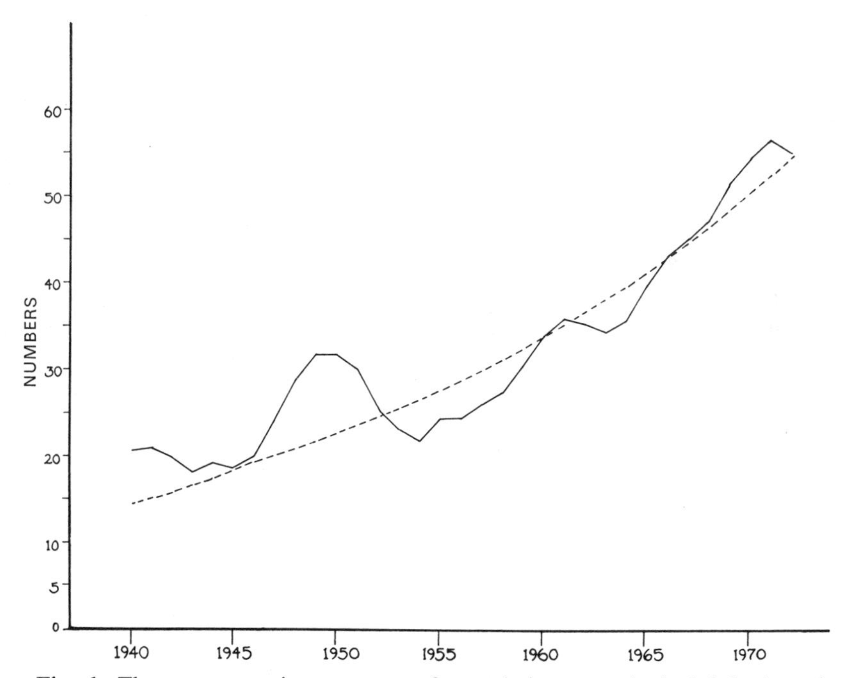

Fig. 1. Three-year moving average of population growth (*solid line*) and smooth curve (*dotted line*) of population growth for the whooping crane population. (Miller et al., 1974)

Although there has been considerable interest in preserving the whooping crane and promoting an increase in its population, little analysis of the basic species characteristics and requirements has been made. Efforts have concentrated on monitoring the population levels, locating and protecting summer and wintering grounds, and attempting to incubate eggs and raise chicks in artificial conditions.

Since 1955 Wood Buffalo Park has been kept under close surveillance during the breeding season, and annual counts have been made of the number of nesting pairs, the young that leave the park at the end of the breeding season, and the number that arrive at Aransas at the end of the fall migration (Table 1).

Intrinsic population characteristics are little known. Maximum longevity is unknown, but whooping cranes probably have a maximum lifetime greater than 30 years. They do not breed until 5 or 6 years old (Miller and Botkin, 1974). A breeding pair usually lays two eggs and seldom raises more than a single young. Nesting pairs are highly territorial and widely dispersed, so that nesting densities are low compared with most birds. Thus, the whooping crane illustrates the *K*-selected strategy. Under constant environmental conditions, the low mortality rates and low reproduction would result in a relatively constant population size. This strategy is clearly disadvantageous when the population has been reduced to abnormally low levels—as it has been through the interference of man.

In spite of the long interest in the whooping crane and the careful census of total numbers, annual production of young, and annual mortality, almost no analysis of the population dynamics of this species has been carried out. This is particularly surprising because conservationists and biologists have long been interested in the behavior of a species with an extremely low population, and it is unusual to have the kind of data available which exists for the whooping crane—data for the entire population, not for a sample.

Biologists have analyzed these data and used them to model the growth of the whooping crane population for the past 35 years. This population has been viewed as the realization of a linear stochastic birth-death process in which the expected value for this model is the same as that for a deterministic model of exponential growth:

$$N_t = N_0 \exp(rt)$$

where N_0 is the initial population, t is time, and r the net rate of increase (births-deaths) (Miller et al., 1974). Of course, the real population would not increase exponentially indefinitely, but the current

Table 1. Whooping cranes counted at Aransas National Wildlife Refuge from 1939 to 1972

Year	Adults	Young	Total
1938	10	4	14
1939	16	6	22
1940	21	5	26
1941	13	2	15
1942	15	4	19
1943	16	5	21
1944	15	3	18
1945	14	3	17
1946	22	3	25
1947	25	6	31
1948	27	3	30
1949	30	4	34
1950	26	5	31
1951	20	5	25
1952	19	2	21
1953	21	3	24
1954	21	0	21
1955	20	8	28
1956	22	2	24
1957	22	4	26
1958	23	9	32
1959	31	2	33
1960	30	6	36
1961	33	5	38
1962	32	0	32
1963	26	7	33
1964	32	10	42
1965	36	8	44
1966	38	5	43
1967	39	9	48
1968	44	6	50
1969	48	8	56
1970	51	6	57
1971	51	5	56
1972	46	5	51

Source: Unpublished data from R.C. Erickson, reprinted from Miller et al., 1974.

population size is so small that this model produces a reasonable fit for short-term changes in the population. An advantage of a stochastic, rather than a deterministic, model is that it can account for variations that occur in the observed population growth; one can examine the probability of extinction of the model population. If the variations are within one or two standard deviations, the model can be accepted as an accurate representation of observed events (Chiang, 1968). The modeling was done with the APL programming language on an IBM 360/67 computer. Three-year moving averages of the birthrate have been graphed from the 35-year period given in Table 1 (Fig. 2). The graph of the birthrate shows that it has tended to oscillate around a mean value but has decreased quite markedly in recent years. Except for slight increases around 1960 and 1965, the overall trend in the birthrate shows a slight decline (Fig. 2*a*). The death rate shows a different pattern (Fig. 2*b*). Oscillations are present in the early years but appear to be damping out, and the death rate seems to be stabilizing at about 0.1.

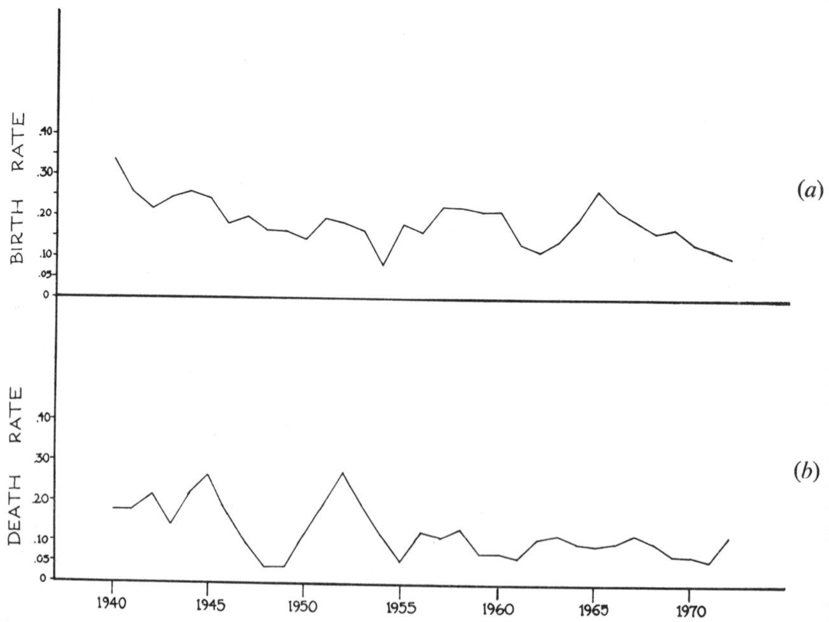

Fig. 2. Three-year moving averages of birthrates and death rates for the whooping crane population. (Miller et al., 1974)

For the 35-year period the mean values are $b=0.2$, $d=0.12$, giving the net rate of increase $r=(b\text{-}d)=0.08$. This value of r gives expected population levels (N) that are much too large. A better fit to the observed values is obtained if d is set equal to 0.16 so that $r=0.04$. This adjustment can be interpreted as a heavy weighting by the death rates in the early years, implying that early fluctuations in the death rates strongly affected the subsequent rate of growth, an effect which is still present. With this lower rate for r, the predicted values fall mostly with one standard deviation of the three-year moving averages of the observed values. A Pierson goodness-of-fit test gives a chi-square of 23.3 with 32 degrees of freedom, indicating that the model cannot be rejected with 90% confidence. These results suggest that at present the population is undergoing an exponential increase, representing an early stage in a logistic curve. The upper asymptote (the K-value) and the inflection point cannot be estimated from present data. One of the most interesting points of this simple analysis is that the exponential increase is the result of a decreasing birthrate and a stabilizing death rate, rather than the result of an increase in the birthrate accompanying an increase in total population.

Table 2 shows the number of breeding pairs counted at Wood Buffalo Park. Such counts are subject to error, particularly in the early years when the grounds were first discovered. For example, for the period 1954-66 an average of only 3.2 nesting pairs was counted, but an average of 5.1 young reached Aransas. Because twinning is unusual, it is assumed that a number of breeding pairs were not located during those years (Novakowski, 1966). All annual recruitment for 1967-72 can be accounted for by the observed number of nesting pairs. Although this period is too short for reliable statistical analysis, it appears that the total number of breeding birds has not changed significantly during this period, except possibly for 1972. The absolute number of young has remained relatively constant, averaging 5.6 for the 19-year period 1954 to 1972. In 1972, 16 breeding pairs (the maximum ever recorded) laid between 30 and 32 eggs, but only 5 young were produced. Even in this year of apparently high breeding success, annual production remained near the long-term average.

Our comparatively simple but careful analysis of the population dynamics of the whooping cranes raises some important questions. There are no obvious reasons why the number of nesting pairs cannot increase more rapidly, or why nesting success is so low. Those interested in the preservation of the whooping cranes should investigate these questions and try to determine what unidentified factors maintain

Table 2. Number of nesting pairs counted at the Sass River, Klewi River, and Nyarling areas of Wood Buffalo Park, 1954 to 1972

Year	Sass River	Klewi River	Nyarling	Total
1954	3			3
1955	4			4
1956	1			1
1957	5			5
1958	2			2
1959	2			2
1960	4			4
1961	3			3
1962	0			0
1963	2			2
1964	5			5
1965	5			5
1966	5			5
1967	6	3		9
1968	6	4		10
1969	5	7		12
1970	6	8	1	15
1971	7	5	1	13
1972	8	7	1	16

Source: Miller et al., 1974.

annual recruitment at a constant value, regardless of population size. If these trends continue, we can expect a continued decline in the birthrate and a corresponding slower rate of population increase.

What can be said about the future persistence of the whooping cranes? Inspection of the population growth curve (Fig. 1) suggests that marked decreases have occurred at approximately 10-year intervals. A time-series analysis of the data showed little correlation between events 10 years apart (Miller et al., 1974), so that we have no reason at present to regard these as anything other than random fluctuations. These fluctuations are important, however. It has frequently been said that when a species reaches a low value, near to or less than 50 individuals, it is likely to become extinct merely from random variations such as one observes for the whooping crane. Although the decreases do not have a significant periodicity, they might be explained by the effect of early oscillations in the death rate on the age structure of the population and the particularly high mortality experienced by certain year classes.

This would mean that the present population is still experiencing the effects of the reduction of the population to the recorded minimum level. Such effects of initial or early events in the introduction of a species must be considered carefully in evaluating the progress of the population at subsequent times. Another explanation has been proposed by Novakowski (1966), who suggests that large decreases tend to follow years of high production and that the observed losses are suffered mostly by inexperienced young birds that have recently left the family group. Although this seems logical, the data are yet insufficient to yield significant correlations between high production one year and high mortality the next. Continued study and time series analysis of the whooping crane population could help clarify the causes of these variations.

Use of a stochastic model of population growth allows an estimate of the probability of extinction. The model predicts a probability of extinction of only 0.018 for 1972, consistent with the observed continued survival of the species since then. Given a population of 51 birds in 1972, the model predicts a probability of extinction of only 5.12×10^{-9} in 1992, when the expected value of the population is 114. The doubling time for this modeling population is approximately 18 years (Miller et al., 1974).

Such projections must, of course, not be taken literally. They are merely linear extrapolations assuming that all conditions, both of the environment and of the intrinsic characteristics of the population, remain as they have been since observations began. The projections assume that the climatic conditions maintain the same averages and variability. The projections assume that there will be no disasters more severe than those in the past few decades. They assume no catastrophic climatic events at Buffalo Park or Aransas, no introduction of a new disease or of new human disturbances to the population at its winter and summer grounds.

The analyses do have uses. They point out studies that are important for evaluating the future success of the whooping cranes. They suggest that a *K*-strategist species with low reproductive rates can have a reasonable chance of survival in spite of a slow rate of increase and relatively high variations in population and regeneration from year to year. The analyses discussed here are comparatively simple and straightforward. They involve merely careful consideration of the available information. Such analyses could be an integral part of any attempt to preserve, introduce, or reintroduce a species. This is all the more true because there are cases where the attempts to reintroduce animals have failed.

The Indian Lion

An interesting example of such failures is the attempt to reintroduce the Indian lion into a part of its range where it had previously been eliminated (Negi, 1969). The Indian lion once occurred over west, north and central India, but it exists now only in the Gir forest preserve in Saurashtra, northwest of Bombay, where there are approximately 250 to 300 lions. The disappearance of the lion elsewhere in India is usually attributed to sports hunting and the widespread clearing of jungle for agriculture. It has also been argued that the lion has suffered from encroachment by the tiger into its preferred habitat. In 1957 one male and two female lions from the Gir forest were introduced into a 37-square-mile preserve in the Chakia forest called the Chandraprabha Sanctuary. The introduction was carried out carefully, with the lions kept for the first year in a 10-acre enclosure to allow for a transition from their previous temporary caging to a completely wild state. The lions were carefully monitored at first and the number increased to 4 in 1958, 5 in 1960, 7 in 1962, and 11 in 1965. They disappeared afterward and were never seen within or near the sanctuary again. The reasons for their disappearance have been variously attributed to (1) poisoning and shooting by villagers, (2) disturbance because the forest around the sanctuary was open to heavy grazing, (3) failure to restrict the movements of the lions, and (4) designation of only 37 square miles to maintain a viable population.

Whatever the influence of the first three factors, it seems clear that the sanctuary was indeed too small to maintain a lion population of sufficient size for long-term persistence. According to Berwick (pers. comm.), territorial male lions tend to live in pairs requiring approximately 50 square miles. Within this area females, juveniles and young, and occasional young nonterritorial males will also exist. A viable population therefore requires hundreds of square miles, and an adequate preserve should constitute at least 250 to 500 miles. The lessons of this case seem to be that a much larger area should have been set aside, with all grazing and farming eliminated within the area, and that the population should have been monitored for a longer period.

Bighorn Sheep

Small, isolated populations are subject to problems of inbreeding. This may be an important problem when one attempts to reintroduce a small population of a species in a small, isolated area which lies within what was once a large range for the species. The kind of problem that can arise is illustrated by attempts to preserve the Rocky Mountain bighorn sheep (*Ovis canadensis canadensis*). Berwick (unpublished manuscript) studied such problems for a bighorn sheep population in a 10-square-mile range in the Palouse grassland winter range bordering Rock Creek, Montana. This population is a remnant of previously widely distributed bands which lived in contiguous ranges throughout much of western Montana. In the nineteenth century, the population was large enough so that a party of 145 men lived on wild sheep during the spring of 1824 at Sula, Montana, about 230 miles southwest of Rock Creek. The Rock Creek herd has been isolated since the turn of the century and has undergone wide population fluctuations. According to Berwick (unpublished manuscript), "the population has experienced declines to about 10, 25, and 12 head in 1915-1918, 1936, and 1965-68, respectively." The last occurred after the population had reached approximately 150 animals.

The recovery of this population from each of its three minimum levels presents a problem similar to that of the introduction of a species into an isolated section of its former range. When a population reaches a size as small as ten individuals, it passes through a "genetic bottleneck," which can result in decreased variability in the population and fixation of deleterious genes. The genetic variability lost during these periods of minimum population is not quickly regained except by new mutation and subsequent recombination so that the inbreeding is carried forward. Although the small size of the Rock Creek population makes statistical tests difficult, an examination of skull characteristics suggests that the Rock Creek sheep population might already show differences from populations found elsewhere.

Moose and Wolves at Isle Royale National Park

That interactions among species may be important is illustrated by a natural introduction of moose and wolves which occurred during the twentieth century at Isle Royale National Park, Michigan. Moose were not recorded on this 550-km^2 island in Lake Superior before the early 1900s (Mech, 1966) at which time a few animals apparently swam over from the Ontario mainland. The island contained an abundance of their preferred food species and lacked their major predator, the timber wolf. Moose have a complex diet on Isle Royale, including more than 20 species of terrestrial woody plants and perhaps as many aquatic macrophytes. Their food habits change with age and season. Adult moose feed in the summer on aquatic plants and on the leaves of deciduous woody land plants. Calves do not use aquatic vegetation. Winter food consists of twigs of deciduous plants and the leaves and twigs of balsam fir, yew, and cedar. Moose show marked preferences within each of these categories.

After introduction, the moose population increased rapidly. Their impact on the vegetation was reported as early as 1913 by Cooper. He indicated that moose were in the process of eliminating water lilies, one of their preferred resources. Within 15 years their browsing impact was so striking that Murie (1934) predicted that the moose would run out of food and that heavy mortality was imminent. Die-offs occurred in the early 1930s. A major fire in 1936 led to the growth of many young trees and increased the carrying capacity of the island. The moose population recovered a second time, but die-offs, along with marked suppression of forest growth, occurred again in the mid-1940s (Krefting, 1951).

Wolves reached the island in 1947-48, probably traversing the ice from Canada. Their population increased within a few years to an average of 23 individuals (Mech, 1966), a level maintained until the last few years. Wolf predation appeared to have a significant effect on the moose population. From the arrival of the wolves until the early 1970s, the moose herd maintained a level higher than elsewhere but below that at which major mortality occurred. The total effect of the moose on the forest is now far less than it was before the appearance of the wolves. Their impact upon many tree species, however, is still sufficient to curb normal growth, and their impact on other species, such as yew, remains extreme.

The effect of the wolves is at least to dampen population oscillations of the moose. The moose's interactions with the community are complex; their population levels are affected by the abundance of a variey of plant species, by their own abundance and wolf hunting success, and by environmental catastrophes such as windstorms and fires.

Wolf predation upon moose appears to increase the herd vigor, since it is concentrated on nonreproductive members (the young, infirm, and old) and maintains the population below the carrying capacity (Jordan et al., 1971).

It is difficult to imagine a moose herd existing in an exact equilibrium with its predators and its food. Moose can feed on vegetation no higher than 2.8 m. Their preferred food species are characteristic of young forests. Under intense browsing pressure, young trees die; under too little pressure, they grow beyond reach of the moose. Major species typical of mature forests are utilized little or not at all by moose, and slowly increase while moose suppress other preferred species. Unless catastrophic storms or fires remove the large trees and unpalatable species, a forest inevitably becomes unsuitable for moose.

The complex mineral nutritional requirements of moose force them to use a mixture of different plants. For example, Botkin et al. (1973) have shown that the terrestrial plants could not provide sufficient sodium for the moose herd. The aquatic vegetation is capable of providing the major source of this ion but is only available for a brief period during the year, whereas land plants provide the bulk of the diet.

The moose introduction to Isle Royale in the absence of their predator led to wide population fluctuations; subsequent addition of the wolf appears to have dampened these oscillations. This example illustrates that (1) the complex interactions among species must be considered in any attempt to reintroduce a species into a part of its former habitat; (2) successful reintroduction of a prey may not be possible without the presence of its predator; (3) the mineral nutrition of available food must be considered for herbivores and a set of complex food resources may be necessary. A moose preserve including only upland boreal and eastern deciduous forest, no matter how large, appears to be an insufficient habitat for the moose in terms of mineral nutrition.

Initial population explosions need not always imply disaster. Riney (1964) stated that "introduced populations of large herbivores, if undisturbed, normally follow a pattern of adjustment to the new environment which consists of a single eruptive oscillation." In other words, an overshoot is a typical response of a population of herbivores in a new habitat. Thus, a temporary rise above the carrying capacity may not spell disaster; it must, however, be monitored carefully, so that die-off ending with extinction may be prevented.

Conclusion

Many introductions, reintroductions, and transplants of species have been attempted; most have failed. Success can never be guaranteed, but careful analysis of genetic, physiological, and ecological factors can facilitate the decision-making process as to where, when, and how attempts may be made. Ecological concepts and theory, including the concepts of *r* and *K* selection, niche theory, population interactions, and ecosystem dynamics, help clarify the problems to be faced. As with the whooping crane, even comparatively simple mathematical and computer models can be useful in attempts to preserve and reintroduce species by making clear the implications and the limitations of our knowledge and our assumptions. As this case also illustrates, too often in the past such simple analyses as the careful plotting and examination of population data have been overlooked.

It has been the purpose of this chapter to illustrate the utility of these considerations by case histories. The failures of an attempted reintroduction of the Indian lion illustrates the need to consider habitat size; the bighorn sheep illustrate problems of inbreeding in small populations; the whooping crane illustrates that analyses of population data can point out questions that require field research; the whooping cranes and the moose at Isle Royale illustrate the importance of analysis of population fluctuations; the history of Isle Royale also illustrates the need to consider interactions among species and the role of mineral cycling and availability of crucial nutrient ions in the success of a species.

This chapter's purpose has been to illustrate the usefulness of certain general approaches; it is not meant to be a definitive compendium of all such approaches. Many ecologists will no doubt recognize areas completely omitted from this discussion which have direct application to the problems of reintroduction. For example, the theories of island biogeography are useful in the consideration of habitat size and probabilites of extinction. The central theme of this chapter is that we must always take careful account of natural history, in the most quantitative form available to us. We must approach the reintroduction and the preservation of species with the warning of Paul Sears (1935) clearly in mind. "Nature," he wrote "is never conquered save on her own terms. Man is welcome to outnumber and dominate the other forms of life provided he can maintain order among the relentless forces whose balanced operation he has disturbed." Ecological principles and quantitative techniques can help us understand these forces and seek the order that is necessary for our management of natural ecosystems.

Acknowledgment

Some material in this paper has been reproduced from *Biological Conservation*, Volume 6, pages 106-111, published by Applied Science Publishers, Ltd.

Literature Cited

Allen, R. P. 1952. The whooping crane. Nat. Audubon Soc. Res. Rep., No. 3. pp. 1-246.

_____. 1957. A report on the whooping crane's northern breeding grounds. Nat. Audubon Soc. Suppl. Res. Rep., No. 3, pp. 1-60.

Berwick, S. Unpublished manuscript. Inbreeding in a Montana population of bighorn sheep–implications for endangered species. 9 pp.

Botkin, D. B. 1975. Functional groups of organisms in model ecosystems: analysis & predictions. Pages 98-102 *in* Simon A. Levin, ed. Ecosystems Analysis & Prediction. Soc. for Industrial & Applied Math., Philadelphia.

_____. In press. A functional approach to the niche concept in forest communities. *In* G. Innis, ed. Systems Analysis in Ecology: Where Do We Go from Here? Symposium Proceedings, Logan, Utah, 1975.

Botkin, D. B., P. A. Jordan, A. S. Dominski, and H. L. Lowendorf. 1973. Sodium dynamics in a northern ecosystem. Proc. Natl. Acad. Sci. 70(10):2745-2748.

Botkin, D. B., and R. S. Miller. 1974. Complex ecosystems: models and predictions. Am. Sci. 62:448-453.

Chiang, C. L. 1968. Introduction to Stochastic Processes in Biostatistics. John Wiley & Sons, New York. 313 pp.

Cooper, W. S. 1913. The climax forest of Isle Royale, Lake Superior, and its development. Bot. Gaz. 55:1-44, 115-140, 189-234.

Ehrenfeld, D. W. 1972. Conserving Life on Earth. Oxford University Press, New York. 360 pp.

Fisher, J., H. Simno, and V. Vincent. 1969. Wildlife in Danger. Viking Press, New York. 368 pp.

Howard, J. A. 1954. Aransas, a national wildlife refuge. U.S. Dep. Inter. Fish & Widl. Serv., No. 11, pp. 1-12.

Hutchinson, G. E. 1957. Concluding remarks. Cold Spring Harbor Symposium on Quantitative Biology XXII, pp. 415-427.

Jordan, P. A., D. B. Botkin, and M. L. Wolfe. 1971. Biomass dynamics in a moose population. Ecology 53:147-152.

Krefting, L. W. 1951. What is the future of the Isle Royale moose herd? Trans. N. Am. Wild. Nat. Resour. Conf. 16:361-470.

Mech, L. D. 1966. The Wolves of Isle Royale. U.S. Natl. Park Serv. Fauna, No. 7. 210 pp.

Miller, R. S. and D. B. Botkin. 1974. Endangered species: models and predictions. Am. Sci. 62:172-181.

Miller, R. S., D. B. Botkin, and R. Mendelssohn. 1974. The whooping crane (*Grus americana*) population of North America. Biol. Conserv. 6:106-111.

Murie, A. 1934. The Moose of Isle Royale. Univ. of Mich. Mus. Zool. Misc. Publ., No. 25. 44 pp.

Negi, S. S. 1969. Transplanting of Indian lion in Uttar Pradesh. Cheetal. 12:98-101.

Novakowski, N. S. 1966. Whooping crane population dynamics on the nesting grounds, Wood Buffalo Park, Northwest Territories, Canada. Can. Wildl. Serv. Rep. Ser. No. 1, pp. 1-20.

Riney, T. 1964. The impact of introductions of large herbivores on the tropical environment. UNCN Publications, new series, No. 4, pp. 261-273.

Sears, P. 1935. Deserts on the March. Univ. of Oklahoma Press, Norman. 231 pp.

Silcock, B. W., H. C. Hammit, and J. E. Crawford. 1973. Endangered wildlife on the natural resource lands. Mimeo. Bureau of Land Management, Washington, D.C.

Fire: A Destructive Menace or a Natural Process?

Richard J. Vogl

Abstract

Many natural systems have become adjusted to fire, containing organisms that have developed recovery mechanisms and fire adaptations proportional to fire frequencies. Some systems are fire dependent, possessing organisms that require fire to maintain health and complete life cycles, with fires functioning as essential decomposition and recycling agents, and with frequent light fires often preventing fuel buildups and thus severe fires. Fire intensities are usually inversely proportional to frequencies. Other systems contain relatively fire-free environments in which natural ignition and the conditions conducive to burning rarely occur. If fires do occur they act as catastrophic forces.

The biota, soils, and local environments of some systems were modified by primitive man's use of fire. Recent upset and damage have occurred because of indiscriminate and profuse burning and, conversely, because of attempts to exclude fire completely from systems in which fire was an important natural component.

Introduction

Nature strives to maintain balances. When the existing quasi-equilibrium is naturally disturbed or upset, phenomenal forces are amassed in the recovery and restoration of stability. Man's perturbations are often so frequent that instability prevails, or are so severe and unnatural that degraded and undesirable steady states become established.

Wildland fires and the control of wildfires have become problems for most nations. A number of terrestrial communities or ecosystems have evolved without the presence or influence of fire. When fire is introduced, it can be considered a perturbation, and damage assessed and recovery rates determined. But in a larger number of plant communities, which may constitute a majority of systems, fire is an inherent environmental component, ranging from a minor and infrequent to a major and limiting factor. Considerations of damage and recovery become inappropriate in systems in which fire is a natural process, and should rather be evaluated according to the effects of fires and the responses of organisms to them.

Comprehensive information on the impact of fire on fire-free ecosystems is not generally available because of a paucity of recorded observations, quantitative studies, and field experiments and because of the uncommon occurrence of fires and complications created by associated disturbances.

A continuum exists between fire-independent and fire-dependent ecosystems. Arbitrary, but definable, natural fire-related systems (fire dependent, fire initiated, fire maintenance) were selected to illustrate some of the variable fire responses among a myriad of responses that are complicated by man-altered fire frequencies.

The object of this paper is to evaluate the general effects and responses to fire in various systems, and to attempt to synthesize them in an ecological context. This treatise is not complete or definitive, but is intended to stimulate further thought and investigation.

The terminology traditionally used in discussing plant succession was purposefully avoided because of the proliferation of terms and differing definitions and confusing usage of them and because the classical concepts of succession have become stereotyped and are particularly limiting when considering fire as a natural process (Vogl, 1970*a*, 1974*b*; Niering and Goodwin, 1974).

Fire-independent Ecosystems

Fire-independent ecosystems are systems with organisms and environments that are usually free from fire. As a result, adaptations and mechanisms to withstand fire and to recover from it are absent. Natural ignition by lightning, volcanism, spontaneous combustion, and other sources and the conditions conducive to burning rarely occur. Even man-caused fires usually occur only when they are purposely and skillfully set.

Fire-free systems are common in the tropics and subtropics. Examples include different types of evergreen rainforests, lowlands dominated by various dipterocarps or other trees, Australian vine forests (Webb, 1968), Indonesian *Agathis* and ironwood forests, many kinds of montane forests, heath forests of Malaysia-Indonesia, and various cloud and mist forests, including dwarf or elfin woodlands. Permanent swamps, including peat and nonpeat forests, mangrove swamps, and herbaceous marshes, including reed "swamps," are generally fire independent. Some gallery forests and hammocks, including pockets of softwood rainforests or other forests amid combustible vegetation types, as well as certain thorn woodland and desert scrub, also exist without fire.

Temperate areas contain fire-free salt marshes, hardwood swamps, some floodplain and riparian forests, mature or mesic deciduous or coniferous forests or forest pockets, alpine fellfields or rock tundra, coastal strand and sand dunes, seaside bluffs, open desert, playas, desert washes, and some desert scrub types. Northern temperate and subarctic regions contain fire-independent peatlands and bogs, tundra barrens, and moss heaths.

Such properties as the species diversity of the vegetation, plant age variation, high plant densities, dense vegetational canopies, restricted air circulation, low evaporation, high transpiration, substrates with high water-retaining capacities, impaired drainage, high water tables, light litter fall and accumulations, discontinuous plant cover, undeveloped understories, bare ground, landform location, stage of vegetation development, presence of shade-tolerant species, high humidities, persistent cloud cover, perpetual wetness, and continual and/or high amounts of precipitation all discourage or exclude fires.

The physical and chemical nature of the vegetation or fuels and the mesic environments in which they exist promote rapid and complete decomposition, and not dry situations, fuel accumulations, and combustion (Mutch, 1970). Allelopathic substances and decay inhibitors are generally inoperative or absent. Flammable waxes, oils, and resins that might also resist decay are uncommon. Standing, as well as fallen, plant parts tend to decay quickly. Thus fuels that might be ignited by lightning and sustain fires during heavy rains usually do not exist. Finely divided and other highly combustible plant parts are rare. Available nutrients are rapidly assimilated by the living plants.

Effects of Fire

Although the prevailing biotic and environmental conditions of fire-free ecosystems effectively exclude fires, occasional fires can and do occur. The vast majority of these are man induced, and usually involve prefire preparation of the vegetation. This usually involves girdling, defoliation, flooding, or poisoning which results in the killing and subsequent drying of the standing vegetation, or involves cutting. Cutting may be used to partially open the canopy or to remove the entire canopy, thus adding large amounts of fuels on the ground where it is subject to the drying sun and winds. The killing and cutting is sometimes augmented by the browsing of domestic or feral herbivores. The entire process, including the burning, is often conducted during brief weather conditions most favorable for drying and combustion (Bartlett, 1956).

Occasionally, exceptional environmental conditions can result from atypical droughts, drainage, lowered water tables, flooding, erosion, high winds, volcanic activity, insect or herbivore damage, or loss of natural firebreaks, and natural fires occur. Ignition is usually caused by lightning, volcanism, or spontaneous combustion (Vogl, 1969*b*). In some areas, agricultural, urban, industrial, and recreational development have led to surface water drainage and lowered water tables. Erosional undermining of root systems, or water kill, caused by the inundation of the otherwise fire-independent plant communities also permits the occurrence of unintentional fires.

Fire-free systems in tropical, temperate, boreal, subarctic, and alpine regions have been intentionally and accidentally subject to fire as man has altered the existing organisms and environmental conditions and introduced fire. In many subtropical and tropical countries man has been burning fire-independent systems for many centuries under ambient or slash-and-burn agricultural systems (Watters, 1960; Conklin, 1969).

If and when fires do occur, they act as catastrophic forces, destroying the existing vegetation directly and the associated animals and soils directly or indirectly (Davis, 1959). The organisms are unadapted to these unusual fires and cannot survive or recover from them. As a result, fires function as retrogressive agents, setting back the vegetational development to some earlier stage (Vogl, 1964*a*, 1969*a*). The postfire vegetation usually consists of species not found in the unburned vegetation, even in the form of seeds.

Fires are often severe in that they consume organic soils and roots along with the surface fuels and crown in the standing vegetation, which results in a total plant kill. Sometimes the standing vegetation is toppled in the process, resulting in the fire consumption of the large-sized and living fuels. Even light surface fires can produce high mortality and have profound effects.

Recovery

Postfire erosion and subsequent changes in water quality and quantity are often dramatic because of the lack of recovery mechanisms and the long time that burned sites remain unvegetated and exposed to the elements.

Reinvasion of burned sites is slow because of the absence of pioneer species in the preburn sites and the surrounding unburned vegetation. The pioneer species that eventually invade tend to be less opportunistic and aggressive than those found in fire-dependent systems. Some sites are eventually occupied by monotypic stands of weeds or other undesirable plants, some of which persist for long periods. The loss of species diversity, in turn, encourages further disturbances which tend to perpetuate pioneer conditions. Some tropical forest sites may never return to the diverse preburn vegetational compositions, even after hundreds of years (Thomas, 1956). Other vegetation types change through a progression of various plant species back to the preburn compositions but seldom reach those preburn compositions and conditions in less than several hundred years (Sauer, 1958; Fosberg, 1960).

Fire-induced changes in soils and water can also substantially lengthen the recovery period, particularly when mesic sites are converted to xeric or hydric sites. In some places, mesic sites revert to hydric sites as fire destroys the standing vegetation and terminates its water-pumping and transpiration activities which previously prevented accumulating waters. Occasionally, the resulting swampy and open-water sites become semipermanent and even acid and boggy as in some subtropical and tropical areas (K. Kartawinata, pers. comm.), as well as in temperate and boreal forest areas (Wilde, 1958; Vogl, 1964*b*). Irreversible changes also occur in places where man has substantially lowered water tables and fires then sweep the weakened and vulnerable fire-free communities, often destroying the plants, substrates, and inherent animals. This is happening to the unique hardwood hammocks of the Florida Everglades (Egler, 1952; Hofstetter, 1974), as well as swamp bogs elsewhere.

Fire produces irreparable losses in many fire-independent communities with the death of unique and endemic organisms. These specialized plants and animals represent storehouses of diverse genetic materials that have evolved without interruption or termination for thousands of years, only to be lost in fires that are becoming more frequent with increasing human populations, developments, and perturbations.

Fire-independent communities are common in subtropical, tropical, desert, and tundra regions, places that have remained largely undeveloped until recently. Increased demands for space, food, water, energy, and other resources have now placed human pressures on these systems. These activities have resulted in the subjection of many previously fire-free communities to fires, and along with rampant and wholesale land-clearing efforts of an unprecedented nature, are irreparably destroying part of the earth's biological variability and evolutionary heritage.

Fire-dependent Ecosystems

Fire-dependent ecosystems are systems possessing organisms that require fire for their survival and continuance, with fire an essential part of the environment. The plant species that dominate are not only adapted to fire but possess fire-dependent structures, mechanisms, and functions. Fire often serves as the key or driving force of life cycles. Fires may be the cause as well as the effect or result of the existence of a species. Many animals are also fire dependent by relying on postfire plants and habitats. Fire often serves as the important decomposition or nutrient recycling agent. Natural ignition, climatic conditions, landforms, fuels, and other environmental characteristics are conducive to frequent fires.

Examples of this type occur in a variety of climates and vegetation types. Fire is common in tropical and subtropical grasslands, savannas, and dry forests. *Melaleuca leucadendra* lowland savannas in Malaysia-Indonesia, chir pine (*Pinus roxburghii*) in India, *Heteropogon contortus* grasslands throughout the South Pacific region, pine stands in southeast Asia and the Philippines, various *Eucalyptus* forests of Australia, open dipterocarp forests in Thailand, and *Acacia, Eucalyptus, Casuarina,* palm, and other species savannas or woodlands in various countries and continents are tropical and subtropical examples.

Fire-dependent ecosystems in the temperate zone include many grasslands and savannas (Vogl, 1974*b*), chaparral formations and scrublands of several continents, deciduous forests such as aspen (*Populus* spp.), some oaks (*Quercus* spp.) and others, and coniferous forests composed of such species as lodgepole pine (*Pinus contorta*), jack pine (*P. banksiana*), Aleppo pine (*P. halepensis*), pitch pine (*P. rigida*), pond pine (*P. serotina*), sand pine (*P. clausa*), and the closed-coned pines of California (Vogl, 1973*a*). Boreal areas contain black spruce (*Picea mariana*) forests and *Calluna* and ericad-dominated heathlands (Specht et al., 1958).

Fire Adaptations

Plants occurring in fire-reliant systems possess a variety of fire adaptations (Kozlowski and Ahlgren, 1974). Many species require mineral or exposed soils for seedling establishment or reestablishment, or pH substrate changes, fungal "blooms" (Wicklow, 1973), nitrogen increases, neutralization of allelopathic or growth-inhibiting chemicals in litter or soil, removal of the litter layer and damping-off fungi, heat-treated or semisterilized substrates, an ash layer (Vogl and Schorr, 1972), increased nutrient salts, altered water relationships, a heat-absorbing carbon-charcoal layer, and reduced species competition created by burning (Vogl, 1974*b*).

Some plants have fruiting bodies that are stimulated to release diaspores upon exposure to fire. The seeds or reproductive bodies of some species remain dormant until treated by fire, thereby making them obligate fire types. Many of these latter species have a high reproductive capacity or are prolific seeders. Some possess mechanisms to accumulate seeds such as closed cones, capsules, or pods, so that they are present in substantial members when the next fire occurs. Sometimes vertebrates and invertebrates assist in the dispersal and storage of seed until fire treatment. Physical forces may also perform these functions. The duration of seed viability generally exceeds that of comparable nonfire plants.

Some common adaptations possessed by fire-dependent plants are accelerated growth, precocity, rapid development of extensive and/or storage root systems, burls, basal plates, and ground-level or below-ground meristems, and the ability to resprout or produce adventitious stems from roots or root crowns after burning (Kozlowski and Ahlgren, 1974). Aboveground plant parts are commonly fire adapted with thick bark, the absence of surface cambium tissues, fire-protected meristems, ability to survive defoliation, ability to produce epicormic stems after fire, strong vegetative habits, and fire-stimulated growth, flowering, and fruiting. In a few species such as wire grass (*Aristida stricta*) and knobcone pine (*Pinus attenuata*), sexual reproduction is controlled exclusively by fire treatment (Vogl, 1972, 1973*a*).

Fire-type communities usually produce abundant standing and fallen fuels that accumulate faster than they decompose because of decay resistance or climatic conditions. The litter is generally very flammable or becomes flammable before burning. Some species are deciduous, and others such as grasses and some ferns die back to ground level each year, contributing to the combustibility of the standing vegetation as well as the litter on the ground. Some evergreen plants produce litter continuously, or large amounts, or retain dead leaves. The portions of dead standing and fallen plant materials usually increase with time.

The physical and chemical compositions of the living and dead fuels can also enhance combustion. Anatomical features such as finely divided or small leaves, numerous fine stems, abundant growth, and exfoliated bark promote burning (Countryman and Philpot, 1970; Philpot, 1973). Behavioral characteristics such as life form, growth form, and seasonal aspects facilitate fires.

Fire-dependent community components are often even-aged and short-lived. Classical examples are lodgepole pine (*Pinus contorta*), jack pine

(*P. banksiana*) forests (Vogl, 1964*c*), trembling aspen (*Populus tremuloides*); (Vogl, 1969*a*), and southwestern United States chaparral (Vogl and Schorr, 1972), communities in which the majority of individuals date back to the last fire. Fire-controlled environments are sometimes extreme environments that can result in simple systems. Species diversities are often low, and many community compositions approach monocultures. These simple systems generate ideal burning conditions since the majority of plant species are uniformly and simultaneously susceptible to droughts, diseases, physical damage, or death. This results in quick, massive, and contiguous fuel buildups that promote the unimpeded spread of fires.

In reality, all of the previously mentioned features, along with climate, act in an integrated fashion to create fire dependency in a species and fire-dependent environments (Habeck and Mutch, 1973).

Fire Climates

Fire-dependent ecosystems possess environments that promote ignition and combustion conditions infrequently, sporatically, periodically, or continuously. Vegetation cycles are often superimposed on the climatic cycles (Vogl, 1970*a*). When combustion conditions are not synchronized with ignition, individual fuel sources are usually present that can sustain fires until they can spread. Ignition sources are natural, and in some vegetation types natural fires have been purposely augmented by man-caused fires.

Lightning is the most important cause of fires (Taylor, 1973). Fire-dependent and fire-maintained communities subject to frequent fires usually occur in regions with a high lightning incidence. Lightning serves as an important ignition source even where it is nearly always accompanied by heavy rain or where it has a very low incidence. In some areas today, lightning is a very minor cause of ignition because of man's alterations and interventions, whereas in these same environments in pristine times a little lightning could go a long way and was an effective ignition source.

Climates are usually subject to droughts, either consistently or sporadically. Even regions with very high amounts of precipitation are interrupted by droughts which facilitate burning. In some places such as the southeastern United States and Hawaii, just a few rainless hours or days will allow the free spread of fires in otherwise wet environments (Vogl, 1969*b*).

Ignition from volcanism can often proceed without favorable climatic conditions because volcanic activities create ideal fuel conditions for burning, or generate their own climates. In some places, heavy fuel accumulations not only lend themselves to being readily burned but are the ignition source as well, by the process of spontaneous combustion (Viosca, 1931; Vogl, 1974*b*).

The geomorphology of fire-dependent landscapes also contributes to the unimpeded spread of fires. Uniform landscapes such as rolling hills or sweeping plains that are uninterrupted by landform features that might serve as firebreaks are common. Vegetation types and the fuels they produce also tend to be contiguous. The fuel types are often open, allowing wind circulation which dries the fuels and pushes the fires (Vogl, 1974*b*). In some types, such as dense chaparral or forest, fire storms commonly occur whereby the fire develops its own transport system. Fuels are also present in some types that assure the spread of fires by spotting ahead, that is, airborne embers that carry the fire ahead and across physical impasses.

Effects of Fire

The natural occurrence of fire in fire-dependent systems is a usual and necessary event that cannot be considered damaging (Vogl, 1973*c*). Response times tend to be directly proportional to the natural fire frequencies; that is, the more frequent the fires, the more rapidly vegetation recovers. Response in frequently burned communities, such as some grasslands, is rapid, with preburn conditions returning in one to several years. Regrowth is often phenomenal, becoming evident a few weeks after the fire, being initiated as much by fire as by other environmental factors (Vogl, 1974*b*). The postfire response is more a renewal or rejuvenation process than one of recovery, in that fire initiates new life, stimulates growth, and promotes vigor (Vogl, 1965).

Increased fire frequencies seldom occur in vegetation types already subject to frequent fires because of the low fuel or growth conditions that exist after fire. Communities that readily reburn tend to have varying fire frequencies. Increased fire frequencies occur more commonly in those types that have longer natural intervals between fires. If the fire occurs before the vegetation has a chance to develop reproductive structures, the existing plants can be eliminated, resulting in drastic vegetational changes. Otherwise, premature fires tend to selectively favor the more fire-tolerant over the less tolerant species. Substantial reductions in the time between fires can produce results that range from reduced species diversity to complete type conversions. Sometimes forest can be converted to brush, and brushland to grassland. This is often accomplished by a series of fires occurring more and more frequently, as one fire creates conditions for the next (Vogl, 1970*a*). With the return of fires to normal frequencies, most systems return to their original vegetational compositions.

Decreased fire frequencies and fire exclusion usually cause more drastic effects than increased frequencies. Postponing burning usually produces abnormal fuel buildups that result in atypical fires (Wilson and Dell, 1971). The severity of the fire and the resulting damage increase with the number of years beyond the normal burning time. A few persistent vegetation types make dramatic comebacks after burning, in spite of postponed fires (Vogl, 1964*c*). In most, fires that were formerly necessary and beneficial become destructive and even catastrophic. The vegetation is altered, and recovery is often slow. If fires are completely excluded, which is difficult and seldom accomplished because of decadence, fuel buildups, and fire environments, fire-dependent communities can be sometimes replaced by vegetation changes in favorable climates. The present goal of many fire protection agencies is fire exclusion, which is an unrealistic management objective for those communities that are adapted to and dependent upon recurring fires to maintain balance. Controlled burning is a more reasonable management goal as it approximates natural processes inherent to a given region. We must consider burned landscapes to be just as normal and acceptable as unburned ones in fire-related systems, each being equally important and related parts of the same ecosystem.

Fire-initiated Systems

Fire-initiated systems are ecosystems in which very infrequent catastrophic fires simultaneously terminate and initiate long-lived species. During the postfire period, which is usually free from severe fires for several hundred years, mesic species often become established as the original less-mesic pioneers are reduced by competition, lightning, snow, ice, wind, disease, and insects, sometimes to a few persistent individuals. These survivors usually create conditions that lead to an eventual fire which destroys them, as well as the associated species, but concurrently creates conditions necessary for their reestablishment and the continuance of the species and system.

Fire-initiated systems are common in temperate and boreal regions. North American examples include eastern (*Pinus strobus*) and western white pine (*P. monticola*), eastern (*Tsuga canadensis*) and western hemlock (*T. heterophylla*), eastern white cedar (*Thuja occidentalis*), western red cedar (*T. plicata*), eastern (*Larix laricina*); (Johnston, 1973) and western larch (*L. occidentalis*), coastal Douglas fir (*Pseudotsuga menziesii*), Jeffrey pine (*Pinus jeffreyi*), and red pine (*P. resinosa*) forests (Daubenmire, 1936; Curtis, 1959; Fowells, 1965; Vogl, 1967*b*; Frissell, 1973). Other gymnosperms function similarly elsewhere, particularly in Europe and Asia.

Angiosperm forest species in diverse climates, including tropical and subtropical associations, are also initiated by fire. These include *Fagus silvatica, Quercus ilex,* and *Q. pubescens* of European forests; tall *Eucalyptus* coastal forest and mixed temperate forests of Australia (Gilbert, 1959); and the mixed evergreen broadleaf, podocarp, and *Nothofagus* forests of New Zealand.

Numerous other species could be included as fire-initiated types, but were not because they are either maintained by additional fires once established or possess specific adaptations to fire; these latter types will be discussed separately.

Little or no successful reproduction of these initiating species takes place during the fire-free period; reproduction is limited or does not keep pace with mortality. Mortality of the established individuals is high in the early years, largely because of competiton, but declines sharply to very low levels. These pioneers can be classified as fire-dependent species in that fire is usually necessary for seedbed preparation and widespread seedling establishment, but it is not a strict necessity. Because of the long intervals between fires, obscure interrelationships between senescent individuals and infrequent fire, and limited amounts of largely unsuccessful non-fire-related reproduction, fire dependency is not always apparent (Maissurow, 1941; Curtis, 1959; Little, 1974).

Fire-initiated plants are typically shade intolerant and require mineral soils with minimum competition for establishment (Fowells, 1965). Some species are shade tolerant as adults, and even maintain themselves on a limited basis, usually as gaps are created in mesic stands by minor disturbances. Species such as beech (*Fagus* spp. and *Nothofagus* spp.), hemlock (*Tsuga* spp.), and cedar (*Thuja* spp.) are usually components of old-aged mesic forests, are able to invade mesic sites, and appear capable of maintaining themselves under certain circumstances, and are generally regarded as fire-independent species. Because of this, their fire origins and the importance of fire in their life cycles have been minimized or ignored (Blaisdell et al., 1973). Although these species may be considered as questionable examples of fire-initiated species, I believe that sufficient evidence exists in the form of field observations and personal communications (J. R. Bray, R. L. Dix, R. Godfrey, J. Habeck) to consider these as fire-initiated species. Further investigation, however, would be needed to substantiate this hypothesis. Some species, such as American basswood (*Tilia americana*), chestnut (*Castanea dentata*), and some oaks (*Quercus* spp.), maintain their positions in mesic forests by vegetative reproduction, which is initiated, among other ways, by fire (Curtis, 1959; Fowells, 1965). Apparently sexual reproduction sometimes also occurs in these species after fire.

Fire-initiated species tend to grow rapidly until reaching maturity, often attaining dominant positions in the canopy and thus assuring their survival. During this time, these pioneer plants modify the microenvironment and mesic species begin to appear. These species tend to be long-lived, with life spans that are commonly 500 or more years long. The mature individuals are particularly tenacious, surviving most biotic and environmental ravages. As they become decadent, structural and chemical changes occur that make them particularly susceptible to ignition, especially by lightning. Often the oldest individuals are also the tallest individuals, thereby being particularly vulnerable to lightning. When ignited, they tend to burn readily and continue burning despite wet conditions. Sometimes lightning acts as a predator (DuCharme, 1973) attacking the old and weak, as it does with Jeffrey and ponderosa pine, since it most commonly strikes old trees and thus eliminates overmature, insect- or disease-infested, or heavily damaged individuals (Vogl and Miller, 1968; Biswell et al., 1973). Sometimes because of a lack of understory or continuous surface fuels, burning only removes one individual or a small group, and the site is reinvaded by seedlings of the same species. But in most situations, the ignited individual leads to a severe fire that spreads over a large area where others and associated species are burned. Occasionally, a light surface fire may occur during the early postfire years, usually supported by fuels created when the previous fire killed but did not consume plants or plant parts. These fires usually reduce or thin the stands of new pioneers or promote further propagation of the fire-initiated species.

Fire Damage

Upset can occur in fire-initiated systems by shortening or interrupting the natural fire-free intervals and also by substantially lengthening them. These communities are normally relatively fireproof because of the physical and chemical nature of the plants, the stand microenvironment, and the general climate; and fire-vulnerable periods rarely coincide with natural ignition times (Curtis, 1959; Fowells, 1965).

Increased fire frequencies can perpetuate and even increase fire-initiated species, while the associated mesic species decrease and are even excluded. Increased fire frequencies reduce individual fire intensities, thereby minimizing damage. Many pioneers, and almost all the mesic species, are not adapted to survive fires, and even light fires take their toll. Some fire-initiated pioneer plants have commercial value, and fire can be used as a management tool, promoting the pioneers at the expense of those species that cannot survive such disturbance.

Total fire exclusion results in the eventual elimination of the fire-initiated species, since once the last individuals die, they are gone because they seldom survive in the form of seeds. In other words, to terminate life cycles a fire rather than old age or other factors, including harvesting by man, must occur. In reality, total fire exclusion is seldom accomplished, except on small tracts. A loss of pioneers in one area can be regained in time by a reintroduction of individuals from other areas, except where only isolated stands of vegetation remain. In these latter cases, a dilemma often exists because these remnant stands are sometimes the only remaining preserves of given vegetation types. Burning to maintain a preserve is difficult to accept, but by being given full protection, the fire-initiated components are finally protected out of existence. This is the case in some of the remaining stands of virgin white pine and hemlock in the eastern Lake states (Nelson, 1958).

Fire Recovery

Full recovery after the naturally infrequent fires is an extended process usually taking one or more centuries. Revegetation of the burned sites, however, is fairly rapid because of temporary fire-following plants that quickly invade, but which are usually locally absent before the fire. A number of animals will generally respond to these plants. The prefire presence of the surviving pioneer species often leads to the immediate re-seeding of the burned area even though they are killed in the fire. Surviving pioneers on the burned, but more commonly on the adjacent unburned, areas also contribute seeds during the postfire years. The net result is that within a few years after the fire, the burned area is completely revegetated, and sometime thereafter it will be dominated by the long-lived pioneers. If the community continues to be free from fire and other disturbances, shade-tolerant species will eventually become reestablished until a quasi-equilibrium of mesic species is reached. Recovery to preburn vegetational compositions is slow and generally takes as many years to recover as the time that elapsed since the previous fire. The recovery time can be extended considerably by a particularly severe fire created by unusual natural conditions or by logging and the like (Vogl and Ryder, 1969) and by excessive postfire erosion. Man-caused perturbations can so modify burning conditions and recovery rates that areas are permanently altered.

When fire is excluded beyond the life spans of the fire-favored pioneers and then occurs at some later time, abrupt vegetational changes occur. Existing plants are often decadent and fuel loads heavy, so that fires consume seeds, rootstocks, and soils. Sometimes site conversions occur, whereby forest is converted to semipermanent brush, grassland, or marsh (Vogl, 1964*a*, 1964*b*). Revegetation is often slow because of the absence of the adaptable pioneers and the destruction of the soil mantle, and unstable conditions prevail as weedy species become established. Extending the occurrence of fire beyond normal can result in vegetational transformations rather than the natural recovery of the previously existing type.

Fire Maintenance

Some plant communities are dependent upon frequent (annually to once every decade) light fires to prevent excessive fuel buildups and control plant invasion-succession which otherwise result in severe fires that destroy the vegetation (Biswell et al., 1973; Weaver, 1974). Fires in these types serve as the principal decomposing agents since the fuel composition and environmental conditions deter decomposition by the usual bacteria, fungi, and invertebrates. On pine barrens and other edaphically infertile sites, it appears that these light fires play a key role by recycling critical nutrients back into the overstory by conversion of the understory cover and meager litter to available ash (Vogl, 1970*b*).

Light fires reduce the occurrence of destructive fires by selectively thinning and pruning the standing crop and controlling the invasion of more mesic and fire-intolerant species while pertuating pioneer conditions. Many fire-maintained species are prolific seeders, and natality and recruitment are regulated by recurring fires. Among those plants that become established, additional fires eliminate the slow-growing individuals, those growing on poor sites, and the genetically inferior ones. Fires usually favor the faster-growing and larger trees that are more resistant tò surface fires by nature of their thicker bark that protects the cambium layer and because shoot meristems are farther above the heat and fire than those of small trees. Fire-intolerant species are selectively checked or eliminated. Annual surface fires in the southeastern United States longleaf pine (*Pinus palustris*) type, for example, once maintained parklike stands of stately pines while checking the invasion or succession of thin-barked hardwood species (Vogl, 1972).

Fire maintains teak forest; deciduous and dry evergreen monsoon forests and savannas; *Pinus merkusii* forests; palm savannas in many regions; the Guinea tree savannas of west Africa; grasslands, including *Imperata cylindrica* and other derived grasslands (Tothill, 1971); and *Acacia, Eucalyptus,* and other species savannas in tropical and subtropical regions. Various temperate zone grasslands and savannas are also maintained by frequent fires, ranging from moist pine savannas and tall-grass prairies in the southeastern United States (Vogl, 1972) to semiarid oak and pine savannas in the West and Southwest.

Maintenance fires usually occur as surface fires, quickly burning the surface litter accumulations along with the aboveground portions of existing herbaceous plants and shrubs. The fires are not usually intense or excessively hot and seldom crown or burn into the tree canopies. The fires remain on the ground because of insufficient surface fuels to generate the necessary heat to sustain crown fires and by nature of the usual low tree densities and clean tree trunks. The tall branchless boles of such conifers as redwood (*Sequoia sempervirens*), giant sequoia (*Sequoiadendron giganteum*), ponderosa pine (*Pinus ponderosa*), red pine (*Pinus resinosa*), longleaf pine (*Pinus palustris*), and Caribbean pine (*Pinus caribaea*) are produced by dominance growth, shade, and fire pruning. Species such as giant sequoia also have a relatively nonflammable, fibrous, and thick bark. Many pines possess a thick bark composed of laminated scales shaped like pieces of a jigsaw puzzle. Some palms and other monocots are without bark and peripheral cambium tissue, and other species have a smooth and tight bark. The various bark types provide effective barriers between fires burning in the surface fuels and the potentially flammable and usually fire-vulnerable tree canopies. Fires climbing the trunks of the puzzle-barked pines are often extinguished or drop back to the ground when ignited scales fall off the trunks as the resins that loosely cement the laminations melt from the heat of the fire (Vogl, 1967*b*).

Many landscapes capable of supporting forest are converted by frequent fires to savannas, that is, grasslands with an open overstory of widely spaced trees, or parklands containing discrete and sometimes dense groves or galleries of forest and/or brush surrounded and separated by open grasslands. Members of the Leguminosae contribute to the woody components of many Old and New World savannas. Monocotyledonous palms are common in African, Central and South American, Australian, and Asian savannas. Tree ferns and arborescent monocots are common to many overstories. Northern Hemisphere savannas support species of fire-resistant oaks (*Quercus* spp.), as well as pioneer species of the genus *Pinus* and other gymnosperms (Vogl, 1974*b*).

Grasslands cover the ground of savannas and parklands. These areas, along with open grasslands devoid of trees or other woody plants, usually occupy large and uninterrupted level plains or rolling hills that lend themselves well to the free spread of surface fires. Grassland fuels burn readily. Most grassland plants are surface deciduous hemicryptophytes, with the aboveground portions dying back at least once a year. Grassland plant debris often accumulates faster than it decomposes as entire plant tops are added to the litter layer at the end of each growing season. The fire consumption of the litter accumulations permits the development of vigorous growth. Litter otherwise becomes a stifling mantle that suppresses growth or physically impairs growth by depriving the grassland plants of space and light. Chemical substances leached from undecomposed plant remains may further inhibit growth. Frequent fires remove these detrimental accumulations while leaving the ground-level or below-ground meristems of the perennial grassland species unharmed. At the same time, fire injures or kills most living woody plant tops while generally leaving the living portions of grassland species undamaged. Thus competing woody plants are held in check directly by fires and indirectly by the vigorousgrassland growth that they promote. Other herbaceous species subject to recurring fires are also generally adapted to them and require periodic fires for their maintenance (Vogl, 1974*b*).

Unbalanced Systems

Changes occur in ecosystems that are maintained by recurring fires when the fire frequencies are substantially increased or decreased. The most dramatic alterations occur when the intervals between fires are extended, and particularly when attempts are made at complete fire exclusion. Under natural conditions, the frequent fires maintain the vegetation, recycle nutrients, and drive the vegetation cycles rather than set them back, and, therefore, recovery is not a consideration (Vogl, 1970*a*).

Conditions become unstable when fire is eliminated and vegetational adjustments occur. Excessive fuels begin to accumulate on the ground. The amounts of dead standing vegetation increase, often at the expense of the living portions. After a time overmature plants become senescent and decadent, succumbing to other catastrophes such as insects, diseases, or physical factors, or become weak without fire stimulation. Fire-intolerant species begin to invade, greatly increasing plant densities and providing convenient bridges or ladders for fires to carry from surface fuels into tree crowns. In grasslands, woody plants invade and shade out and physically compete with the grasses and other herbaceous plants. Prodigious numbers of fire-tolerant species often survive, that were formerly reduced and thinned by passing fires. This results in severe overcrowding, which usually leads to growth stagnation. Growth rates slow and even prematurely terminate as plants compete for limited resources. Stand stagnation is common in southwestern United States ponderosa pine forests because of fire exclusion (Biswell et al., 1973).

These changes result in excessive fuel loading and extreme burning conditions which lead to inevitable fires even with the best fire protection systems. The degree of damage relates to the length of time that a fire has been postponed beyond its natural occurrence. The relationship is usually not directly proportional but rather increases exponentially because of the compounding effects previously described. Recovery responses also correspond to the degree of damage or amount of time beyond the normal burning sequence; that is, the longer the time, the poorer the recovery.

In some vegetation types, once the overmature vegetation has reached a general state of decadence, there is little chance of normal recovery after fire because of the severe damage. The sexual or asexual reproductive potential, including seed stored by plants or seed remaining on or in the ground, is often destroyed. A fire under these circumstances can produce site conversions whereby the prefire vegetation type is replaced by a different type. Some brush fields in California's Sierra Nevada were formerly forests which were converted by unusually hot fires. Such intense fires certainly occurred in the past under natural conditions but appear to be becoming more widespread with present-day fire exclusion programs.

Fires occurring shortly after the natural burning sequence result in partial to full recovery. The major difference between fire maintenance and fire recovery in these cases is the time it takes to replace those individuals killed in the postponed fire. Species compositions usually do not change, however, because many fire-maintained species also reproduce vegetatively and/or sexually following fire.

Fire prevention that does not make provisions for the periodic occurrence of fire, under natural or controlled conditions, is probably more detrimental and produces more irreparable damage to fire-maintained systems than if all fire control efforts were terminated. In most of these communities the intervals between fires cannot be substantially shortened because the early postfire years do not contain enough fuels to sustain a fire. The usual time for the next fire is often the first year that there are sufficient fuels present to carry a fire. Increased fire frequencies, when possible, generally cause only subtle changes in vegetational compositions. In longleaf pine communities, for example, burning every year will favor longleaf pine reproduction and dominance, while burning every two to three years will favor slash pine (*Pinus rigida)* and loblolly pine (*P. taeda*) (Vogl, 1972).

Erosion and Fire

The relationships between fire and soil erosion have often been misrepresented. Although fires in nonfire systems and those out of sequence in fire types can cause extensive erosion, natural fires occurring under normal circumstances produce minimal amounts or no erosion (Viro, 1974), and some fire systems are even adjusted to and dependent upon the erosion process. Most accelerated or heavy erosion can be traced to man-caused fires in vegetation types in which they would not naturally occur or in those that are extra-severe because of abnormal fuel buildups and vegetation decadence as a result of fire prevention. Fires in many grasslands, for example, actually contributed to and built soil profiles until man-caused fires, livestock overgrazing, and the deterioration of the vegetation reversed the process (Vogl, 1974*b*).

In communities where erosion is associated with natural fires, fires usually do not cause the erosion. Rather it is caused by uplifting, steep slopes above the angle of repose, water cutting, friable substrates, and gravitational movements. In many parts of the West, for example, mountains are uplifting and valleys and plains are subsiding, and as long as this continues there will be erosion with or without fires (Vogl, 1973*c*). In these and other communities, fire and erosion cannot be viewed as detrimental, but rather as inherent and essential parts of those environments.

Fire and Water

Little attention has been given to the effects of fire on streams, rivers, marshes, and lakes, particularly in watersheds supporting fire-related systems. Watersheds of forests and shrublands reduced in woody components or converted to grasslands by fire produce increased water yields and surface flow (Adams et al., 1947; Rycroft, 1947; Rowe, 1963; Berndt, 1971), but less is known about changes in water quality (Johnson and Needham, 1966; Likens et al., 1969; Lotspeich et al., 1970; Longstreth and Patten, 1975).

The detrimental effects of fire on lakes and streams, including the losses of aquatic life, have been emphasized to the point where the beneficial aspects or natural roles of fires have been overlooked. Eutrophication has become a major concern of developed nations, and yet the role of fire in checking and even reversing this process is seldom considered.

Natural fires, whether mild or severe, must affect bodies of water in more than negative ways. The ash and soil surfaces of burned landscapes, for example, tend to have slightly higher pH values due to the release of alkaline earth metals. These usually enrich waters and contribute to eutrophication, but may as often decrease nutrients and retard eutrophication as the drift and fly ash of charcoal, carbon, and minerals precipitate out nutrients, organic matter, and organisms in suspension. Acid bog lakes, for example, with their coffee-colored waters were perhaps made hospitable to aquatic organisms by fire ash treatment in a way similar to the present-day treatment of lakes with lime.

The runoff and floods resulting from the fires temporarily ravage streams and rivers and destroy stream bank vegetation. But they also serve to renew by scouring streambeds, recutting channels, and rejuvenating the aquatic and terrestrial life. Riparian plant communities are characterized by prolific growth and soon become choked and stagnated, and depend upon disturbances such as fires and floods for renewal. If vegetation life cycles are adapted to such perturbations and require this type of stimulation, the aquatic fauna must also benefit and be adjusted to it. Future investigations should examine the relationships of fires and floods in aquatic communities in more objective and comprehensive ways.

Fire directly affects hydric systems by burning out wetlands and lake bottoms during droughts or periods of low water (Vogl, 1964*a*, 1969*a*; Heinselman, 1970; Cypert, 1972; Whitehead, 1972). Fire is one of the most important agents checking eutrophication and re-creating bodies of water as it reforms open and deep water holes grown over and silted in. Fire may be a critically important factor in preventing lake mortality. Fire exclusion policies have contributed to lake senescence and marsh deterioration in many regions by directly eliminating essential retrogressive agents or by indirectly contributing to the eventual outbreak of unusual fires.

Burned wetlands and lake bottoms often stimulate aquatic plants and animals upon reflooding. But sometimes fires appear to reverse eutrophication and reestablish oligotrophic conditions. Since lakes serve as nutrient traps gathering sediments from the surrounding watershed as well as those produced internally, there are few ways to reverse this process outside of the uncommon and major landform changes produced by glaciation, volcanism, uplift, and the like. Fire, however, is a common event that can account for the removal of nutrients from a lake basin. Smoke, containing a number of combustion products including nitrogen, and particulate matter generated by a fire are given off to the atmosphere and often carried long distances (Vogl, 1974*b*). The remaining ash, which can be deep, is subject to general wind removal, which is usually augmented by rising thermals or "dust devils" that form over the blackened surfaces. In these ways, quantities of accumulated nutrients are removed from their "traps" and redistributed or recycled to the same or other watersheds. A factor that may further contribute to eutrophication reversal is the process of postburn nitrification which is usually brought about by physical changes (Heilman, 1966; Christensen, 1973), bacterial and fungal "blooms," and nitrogen-fixing plants (Vogl, 1974*b*). When reflooding takes place before these organisms become active, nitrogen increases which normally promote eutrophication may be arrested.

Response of Animals to Fire

Despite a paucity of objective information on the responses of animals to fire, fire is commonly believed to be destructive to animals (Vogl, 1973*c*). Recent research indicates that this belief is more of an exaggeration than a generalization. The impression that fires are usually lethal to warm-blooded vertebrates appears to have been based on incomplete, subjective, and biased information or atypical fires, because recent studies report minimal mortality (Vogl, 1973*b*; Bendell, 1974). Birds and mammals usually do not panic or show fear in the presence of fire, and are even attracted to fires and smoking or burned landscapes. Rather, the presence of fire fighters and the occurrence of unusual fires is more apt to cause alarm in animals.

The absence of immediate animal mortality is often countered by predictions that even if animals survive, they are still doomed because of the destruction of their habitats. In some cases this is true, with predation the most common cause of mortality. Yet many animals survive on burned sites without the necessity of periodic retreats to unburned areas, and many even prefer postburn habitats. Some animals find temporary refuge in unburned coverts left by fires, since natural fires function as variables superimposed on varied landscapes and do not uniformly decimate habitats. Only when unusual or adverse fires occur, such as those resulting from unnatural fuel buildups, do the fires more consistently and uniformly destroy the vegetation. Other animals retreat to the unburned edges. In these situations, survival eliminates any needs for recovery (Bendell, 1974).

Even when deaths do occur, the resulting beneficial effects of fire on fire-dependent habitats usually more than compensate for any losses (Vogl, 1967*a*). Those claiming that fire is destructive to wildlife forget that almost all vegetation types that burn readily have been previously burned and fail to realize that animal populations thought to be in jeopardy are often a product of a previous fire. The prolonged or long-term benefits of a fire must be considered before deciding that it is detrimental (Vogl and Beck, 1970). Animal response or recovery following fire often corresponds directly to the vegetational responses (Beck and Vogl, 1972). In vegetation types stimulated by fire, animal numbers are also stimulated. In those plant communities adapted to fire, the majority of animals are also adapted to survive, or possess mechanisms to recover quickly. Even slowly recovering communities attract and support animal populations. Plant association completely altered by fire often contain animals that readily perish because of lost habitat or are slow to return. Fortunately, these latter communities usually contain low numbers of animal species. Even though fire-independent communities generally contain low animal numbers, the mortality inflicted by fires (usually man-caused), are nevertheless disruptive and sometimes significant because the animals lost are often uncommon and represented by low numbers of individuals per species.

Invertebrate and cold-blooded vertebrate populations are also affected by fire in a selective fashion (Vogl, 1973*b*). As a result, fire-independent communities contain vulnerable species. In fire types, burning can regulate and stimulate, just as it does with plants.

The greatest arrays of higher animal species, and the largest numbers per unit area, are associated with fire-dependent, fire-maintained, and fire-initiated ecosystems. Within these systems, the highest animal densities usually occur in the early postfire stages. In other words, the majority of animals, including domestic animals, have evolved with fire and fire systems. The great herds of herbivores, and their predators-scavengers, of every continent are associated with grasslands or savannas (Vogl, 1974*b*). Because of Smokey the Bear and Bambi propaganda, most Americans believe that animals require unburned forests for their well-being. They find it difficult to accept that most North American wildlife, as wildlife elsewhere in the world, prefers open grasslands, burned forest, forest edges, forest openings, or fire-controlled landscapes and is only residually present in protected systems because more favorable habitats are nonexistent (Vogl, 1973*c*).

Summary and Conclusions

Recovery from fire varies with the plant community or vegetation type and the time of burning. Burning time, that is, the environmental and fuel conditions at the time of the fire, the season, and the stage of vegetational development or time since the last fire, is directly related to fire intensity. A number of systems have evolved with fire and are inseparably related to it, so that under natural conditions damage and recovery are irrelevant. These fire systems are common in every continent, dominate temperate and subtropical areas, and occur in tropical and subarctic regions. Although the introduction of fire into fire-free systems disrupts the existing stability and causes damage, fires differ from many other man-caused perturbations, because fire is a natural process even when it is out of place and sequence. As a result, some form of response or recovery from fire always takes place, even in systems in which fire adaptations or recovery mechanisms are nonexistent, in that fire does not produce toxic persistent effects that man-made chemicals, synthetics, or unnatural or alien substances produce. The end products of fire cannot be considered contaminants any more than can the products of decay, fermentation, and digestion. The long-range effects of fire are rather that some systems take long periods to return to preburn conditions or may never return because of altered vegetational compositions and/or basic changes in the substrate or species compositions. The presence of exotic or nonnative plants is often an additional complicating factor, since fires out of sequence or fires in nonfire types create unstable conditions which promote the expansion of the undesirable, and usually troublesome, aliens at the expense of the natives.

Problems have arisen as a result of modern man's promiscuous use of fire in nonfire systems and in man's attempts to disrupt or terminate the natural fire process. Fire prevention education and protection systems have been developed in many countries to counter the indiscriminate and unnecessary uses of fire and have enjoyed varied degrees of success. Unfortunately, most fire control programs did not differentiate between fire-independent and fire-dependent systems or discriminate between man-caused or natural fires, which, in turn, created new problems that are sometimes worse than those caused by overburning. Fire exclusion attempts in fire communities ignore the principle that the use or uses and management of an area are dictated basically by the organisms and environments present, and not by monetary, social, resource, or political demands. Fortunately, current programs are being developed to take into account the natural roles of fire on an operational basis.

When efforts are made to eliminate fires from systems in which they are an integral part of the environment, direct, indirect, predictable, as well as unpredictable, effects occur, since everything, living and nonliving, is related or connected to everything else. The recent felling of monarch giant sequoia trees in California appears to be related to an unusual increase in trunk excavations made by abnormally high populations of carpenter ants. The ant population increases may be related to increases in their food supplies, which include aphids that feed on white fir (*Abies concolor*) meristems. White fir is a fire-intolerant conifer that has become a dominant plant since fire protection was established. Fire control was established in a well-meaning but naive attempt to protect the giant sequoias. In reality, the fire control efforts have now placed the giant sequoia in greater jeopardy, by allowing a more serious threat to arise. When this is coupled with the lack of sequoia reproduction and the massive fuel buildups that can now support fires that totally destroy the big trees, it is obvious that protection has become a form of destruction. Preservation, unfortunately, is usually equated with protection, when protection is only one restricted form of preservation management. Emphasis must be placed on a holistic ecological approach where all factors are considered, rather than deciding that one factor is universally good or bad and then blindly attempting to eliminate or promote it.

Because of fire protection-exclusion in many fire ecosystems, fuel accumulations and general decadence of the vegetation have led to a decline or loss of renewable resources because of inevitable and destructive fires. Even where fires have been prevented thus far, substantial reductions in productivity have occurred because of the absence of the stimulating and renewing effects of fire. Many parts of western North America, for example, have now deteriorated to less productive levels and contain superabundant concentrations of fuels and critical burning conditions as a result of fire protection. Because of this, accidental and incendiary fires, as well as potential sabotage burning by guerillas or terrorists or fires associated with war, pose local and regional threats to lives, property, resources, and national security.

In many regions, additional fire problems are being created as management practices, particularly those of foresters and ranchers, destroy nature's diversity to create monocultures of one plant species, usually an exotic, over large areas, forgetting that diversity provides stability (Vogl, 1971). Tree monocultures, for example, are now commonplace in the southeastern United States, Hawaii, New Zealand, Australia, Europe, Israel, and South Africa, and widespread pasture grass plantings occur in almost every country. As a result, these uniform plantings are more susceptible to natural catastrophes, including fire, and in some cases have created fire problems in areas where they were formerly nonexistent.

The attitude that effective natural resource management should be primarily or solely concerned with reducing or eliminating natural extremes or catastrophes must be abandoned, as it has been in some current programs. Attempts have been made to shield natural systems and the organisms present from such natural and often essential events as lightning-caused fires, floods, droughts, predation, starvation, erosion, and insect depredations. Often the organisms in a given region are there because of these extremes or limiting factors, and not in spite of them. The preoccupation with fire protection at any cost has often taken the form of trying to undo nature and fortunately is fading. Many environments are now in trouble or unbalanced because well-meant practices have resulted in the disruption of inherent natural processes. More and more land managers are realizing that proper management consists of working with nature, rather than disrupting or terminating natural processes. This occurs when we accept natural fires as an environmental process that cannot be replaced without severe and far-reaching consequences (Vogl, 1974*a*). The restoration of controlled fires or supervised wildfires to some systems unfortunately will result in losses or damage that relate to the number of years of fire protection. We must go beyond the present discussions and studies of whether fire is good or bad and attempt to determine the natural frequencies and burning conditions of various ecosystems. In doing this, research discrepancies will be resolved and the discussions will become irrelevant. Above all, we must learn from our mistakes, and not continue to proceed without considering the future.

Acknowledgments

I thank N. C. W. Beadle, J. Roger Bray, A. Malcolm Gill, Nola Hannon, R. Rose Innes, Kuswata Kartawinata, Gordon R. Miller, John Raison, Benjamin C. Stone, and Louis Trabaud for providing examples of the various systems and references that relate to them. I appreciate the helpful comments and suggestions made by Robert W. Mutch and William A. Neiring.

Literature Cited

Adams, F., P. A. Ewing, and M. R. Huberty. 1947. Hydrolic aspects of burning brush and woodland grass ranges in California. California Div. Forest., Sacramento.

Bartlett, H. H. 1956. Fire, primitive agriculture and grazing in the tropics. Pages 692-714 *in* W. L. Thomas, ed. Man's Role in Changing the Face of the Earth. Univ. of Chicago Press.

Beck, A. M., and R. J. Vogl. 1972. The effects of spring burning on rodent populations in a brush prairie savanna. J. Mammal. 53:336-346.

Bendell, J. F. 1974. Effects of fire on birds and mammals. Pages 73-138 *in* T. T. Kozlowski and C. E. Ahlgren, eds. Fire and Ecosystems. Academic Press, New York.

Berndt, H. W. 1971. Early effects of forest fire on streamflow characteristics. U.S. Dep. Agr. For. Serv. Res. Note PNW-148.

Biswell, H. H., H. R. Kallander, R. Komarek, R. J. Vogl, and H. Weaver. 1973. Ponderosa Fire Management. Tall Timbers Research Station Misc. Publ. No. 2.

Blaisdell, R. S., J. Wooten, and R. K. Godfrey. 1973. The role of magnolia and beech in forest processes in the Tallahassee, Florida, Thomasville, Georgia area. Pages 363-397 *in* Proc. 13th Annual Tall Timbers Fire Ecol. Conf. Tall Timbers Research Station, Tallahassee, Fla.

Christensen, N. L. 1973. Fire and the nitrogen cycle in California chaparral. Science 181:66-68.

Conklin, H. C. 1969. An ethnoecological approach to shifting agriculture. Pages 220-223 *in* A. P. Vayda, ed. Environment and Cultural Behavior. Natural History Press, New York.

Countryman, C. M., and C. W. Philpot. 1970. Physical characteristics of chamise as a wildland fuel. U.S. For. Serv. Pac. Southwest For. and Range Exp. Stn. Res. Paper PSW-66.

Curtis, J. T. 1959. The Vegetation of Wisconsin. Univ. of Wisconsin Press, Madison.

Cypert, E. 1972. The origin of houses in the Okefenokee prairies. Am. Midl. Nat. 87:448-458.

Daubenmire, R. F. 1936. The "Big Woods" of Minnesota: its structure and relation to climate, fire and soils. Ecol. Monogr. 6:233-268.

Davis, K. P. 1959. Forest Fire: Control and Use. McGraw-Hill, New York.

DuCharme, E. P. 1973. Lightning–a predator of citrus trees in Florida. Pages 483-496 *in* Proc. 13th Annual Tall Timbers Fire Ecol. Conf.

Egler, F. E. 1952. Southeast saline Everglades vegetation, Florida, and its management. Vegetatio: Acta Geobot. 3:213-265.

Fosberg, F. R. 1960. Nature and detection of plant communities resulting from activities of early man. Pages 251-262 *in* Symposium on the Impact of Man on Humid Tropics Vegetation. Goroka, Territory of Papua and New Guinea, Sept. 1960.

Fowells, H. A., ed. 1965. Silvics of Forest Trees of the United States. U.S. Dep. Agr. Handbook 271.

Frissell, S. S., Jr. 1973. The importance of fire as a natural ecological factor in Itasca State Park, Minnesota. Quat. Res. (N.Y.). 3:397-407.

Gilbert, J. M. 1959. Forest succession in the Florentine Valley, Tasmania. Proc. Roy. Soc. Tasmania 93:129-151.

Habeck, J. R., and R. W. Mutch. 1973. Fire-dependent forests in the northern Rocky Mountains. Quat. Res. (N.Y.). 3:408-424.

Heilman, P. E. 1966. Change in distribution and availability of nitrogen with forest succession on north slopes in interior Alaska. Ecology 47:825-831.

Heinselman, M. L. 1970. Landscape evolution, peatland types, and the environment in the Lake Agassiz Peatlands Natural Area, Minnesota. Ecol. Monogr. 40:235-261.

Hofstetter, R. H. 1974. The ecological role of fire in southern Florida. Fla. Nat. (April) 8 pp.

Johnson, C. M., and P. R. Needham. 1966. Ionic composition of Sagehen Creek, California, following an adjacent fire. Ecology 47:636-639.
Johnston, W. F. 1973. Tamarack seedlings prosper on broadcast burns in Minnesota peatland. North Central For. Serv. Res. Note NC-153.
Kozlowski, T. T., and C. E. Ahlgren, eds. 1974. Fire and Ecosystems. Academic Press, New York.
Likens, G. E., F. H. Bormann, and N. M. Johnson. 1969. Nitrification: importance to nutrient losses from a cutover forested ecosystem. Science 163:1205-1206.
Little, S. 1974. Effects of fire on temperate forests: northeastern United States. Pages 225-250 *in* T. T. Kozlowski and C. E. Ahlgren, eds. Fire and Ecosystems. Academic Press, New York.
Longstreth, D. J., and D. T. Patten. 1975. Conversion of chaparral to grass in central Arizona: effects on selected ions in watershed runoff. Am. Midl. Nat. 93:25-34.
Lotspeich, F. B., E. W. Mueller, and P. J. Frey. 1970. Effects of Large Scale Forest Fires on Water Quality in Interior Alaska. U.S. Dep. of Interior, Fed. Water Pollut. Control Adm., Alaska Water Lab.
Maissurow, D. K. 1941. The role of fire in the perpetuation of virgin forests of northern Wisconsin. J. For. 39:201-207.
Mutch, R. W. 1970. Wildland fires and ecosystems–a hypothesis. Ecology 51:1046-1051.
Nelson, J. C. 1958. Dying of mature eastern hemlock in Cook Forest, Pennsylvania. J. For. 56:344-347.
Niering, W. A., and R. H. Goodwin. 1974. Creation of relatively stable shrublands with herbicides: arresting "succession" on rights-of-way and pastureland. Ecology 55:784-795.
Philpot, C. W. 1973. The changing role of fire on chaparral lands. Pages 131-150 *in* M. Rosenthal, ed. Symposium on Living with the Chaparral Proceedings. Sierra Club, San Francisco, Calif.
Rowe, P. B. 1963. Streamflow increases after removing woodland-riparian vegetation from a southern California watershed. J. For. 61:365-371.
Rycroft, H. B. 1947. A note on the immediate effects of veld burning on storm flow in a Jonkershoek stream catchment. J. S. Afr. For. Assoc. 15:80-88.
Sauer, C. O. 1958. Man in the ecology of tropical America. Proc. Ninth Pacific Sci. Congr. 20:104-110.
Specht, R. L., P. Rayson, and M. E. Jackman. 1958. Dark Island heath (Ninety-mile Plain, South Australia). VI. Pyric succession: changes in composition, coverage, dry weight and mineral nutrient status. Aust. J. Bot. 6:59-89.
Taylor, A. R. 1973. Ecological aspects of lightning in forests. Pages 455-482 *in* Proc. 13th Annual Tall Timbers Fire Ecology Conf.
Thomas, W. L., ed. 1956. Man's Role in Changing the Face of the Earth. Univ. of Chicago Press.

Tothill, J. C. 1971. A review of fire in the management of native pasture with particular reference to northeastern Australia. Trop. Grass. 5(1):1-10.

Viosca, P., Jr. 1931. Spontaneous combustion in marshes in southern Louisiana. Ecology 12:439-442.

Viro, P. J. 1974. Effects of forest fire on soil. Pages 7-45 *in* T. T. Kozlowski and C. E. Ahlgren, eds. Fire and Ecosystems. Academic Press, New York.

Vogl, R. J. 1964*a*. The effects of fire on a muskeg in northern Wisconsin. J. Widl. Manage. 28:317-329.

———. 1964*b*. The effects of fire on the vegetational composition of bracken-grasslands. Trans. Wis. Acad. Sci. Arts Lett. 53:67-82.

———. 1964*c*. Vegetational history of Crex Meadows, a prairie savanna in northwestern Wisconsin. Am. Midl. Nat. 72:157-175.

———. 1965. Effects of spring burning on yields of brush prairie savanna. J. Range Manage. 18:202-205.

———. 1967*a*. Controlled burning for wildlife in Wisconsin. Pages 47-96 *in* Proc. 7th Annual Tall Timbers Fire Ecol. Conf.

———. 1967*b*. Fire adaptations of some southern California plants. Pages 79-109 *in* Proc. 7th Annual Tall Timbers Fire Ecol. Conf.

———. 1969*a*. One-hundred and thirty years of plant succession in a southeastern Wisconsin lowland. Ecology 50:248-255.

———. 1969*b*. The role of fire in the evolution of the Hawaiian flora and vegetation. Pages 5-60 *in* Proc. 9th Annual Tall Timbers Fire Ecol. Conf.

———. 1970*a*. Fire and plant succession. Pages 65-75 *in* Symposium Role of Fire Intermountain West. School of Forestry and Intermountain Fire Research Council, Missoula, Montana.

———. 1970*b*. Fire and the northern Wisconsin pin barrens. Pages 175-209 *in* Proc. 10th Annual Tall Timbers Fire Ecol. Conf.

———. 1971. Monotonous monocultures. Ecology Today 1(7):43-45.

———. 1972. Fire in the Southeastern grasslands. Pages 175-198 *in* Proc. 12th Annual Tall Timbers Fire Ecol. Conf.

———. 1973*a*. Ecology of knobcone pine in the Santa Ana Mountains, California. Ecol. Monogr. 43:125-143.

———. 1973*b*. Effects of fire on the plants and animals of a Florida wetland. Am. Midl. Nat. 89:334-347.

———. 1973*c*. Smokey's mid-career crisis. Saturday Rev. Sci. 1(2):23-29.

———. 1974*a*. Ecologically sound management: modern man's road to survival. West. Wildlands 1(3):6-10.

———. 1974*b*. Effects of fire on grasslands. Pages 139-194 *in* T. T. Kozlowski and C. E. Ahlgren, eds. Fire and Ecosystems. Academic Press, New York.

Vogl, R. J., and A. M. Beck. 1970. Response of white-tailed deer to a Wisconsin wildfire. Am. Midl. Nat. 84:270-273.

Vogl, R. J., and B. C. Miller. 1968. The vegetational composition of the south slope of Mt. Pinos, California. Madrono 19:225-234.

Vogl, R. J., and C. Ryder. 1969. Effects of slash burning on conifer reproduction in Montana's Mission Range. Northwest Sci. 43:135-147.

Vogl, R. J., and P. K. Schorr. 1972. Fire and manzanita chaparral in the Jacinto Mountains, California. Ecology 53:1179-1188.
Watters, R. F. 1960. The nature of shifting cultivation: a review of recent research. Pac. Viewpoint 1(1):59-99.
Weaver, H. 1974. Effects of fire on temperate forests: western United States. Pages 279-319 *in* T. T. Kozlowski and C. E. Ahlgren, eds. Fire and Ecosystems. Academic Press, New York.
Webb, L. J. 1968. Environmental relationships of the structural types of Australian rainforest vegetation. Ecology 49:296-311.
Whitehead, D. R. 1972. Developmental and environmental history of the Dismal Swamp. Ecol. Monogr. 42:301-315.
Wicklow, D. T. 1973. Microfungal populations in surface soils of manipulated prairie stands. Ecology 54:1302-1310.
Wilde, S. A. 1958. Forest Soils. Ronald Press, New York.
Wilson, C. C., and J. D. Dell. 1971. The fuels buildup in American forests: a plan of action and research. J. For. 69:471-475.

Environmental Factors in Surface Mine Recovery

Ronald D. Hill and Elmore C. Grim

Abstract

Major environmental damages can occur if coal is not surface mined in the proper manner. Major pollutants that occur are sediment and acid mine drainage. Factors involved in the formation of these pollutants are discussed. Methods for preventing their formation are detailed. The recovery of the disturbed area following mining and reclamation are reviewed.

The Destruction

The construction of a road to facilitate mining and prospecting is the beginning of the damage to the ecosystem by surface mining. The vegetation and topsoil rich in organisms are pushed aside or covered with subsoil and rock. All too often in hilly terrain the "cut and fills" are not balanced: cut material is pushed over the hillside, and more fill material must be borrowed, thus expanding the amount of disturbance. Runoff from adjacent land, road surfaces, and berms if not properly controlled causes accelerated erosion, producing the first off-site damage. Prospecting is often performed with dozers by ripping the soil surface to reveal outcrops of minerals or fuels. These scars are additional sources of sediment. In some states, control of sediment from the actual mining operations is required, but haul roads and prospectings which are not controlled may still contribute as much sediment as the mine itself.

Next in the mining sequence is the clearing and grubbing of trees, brush, etc., from the mine site. In some situations the trees are harvested. At other times the material is windrowed below the mine disturbance to serve as a filter for sediment. These large areas of bare soil are exposed and subject to accelerated erosion.

A massive earth-moving operation then begins. Some states require that the topsoil and/or subsoil be stockpiled for later spreading during final grading. The overburden rock is usually fractured by blasting before being ripped apart and moved by shovels, draglines, dozers, front-end loaders, or scrapers. The mining process loosens the overburden material, making it more susceptible to erosion. In addition, breaking up the consolidated material exposes new surfaces. This subjects ions on the overburden surfaces to weathering and leaching. An increase in ionic strength of the water discharging from the area can be noted. Ions that typically increase in concentration are sulfate, calcium, magnesium, manganese, aluminum, and iron. The fracturing process may expose pyrite to oxygen, and the resulting oxidation process releases sulfuric acid and iron. The acid accelerates leaching of heavy metals such as copper, zinc, aluminum, and manganese from the overburden rock. These metallic ions and the acid place an additional strain on the water ecosystem.

After exposing the ore body or fuel, appropriate equipment is used to remove it for market. Low-grade mineralized waste material is often associated with the seam being mined. This material can be a major source of chemical pollution and should be specially handled. Following removal of the commodity, the backfilling of the excavation begins. Similar environmental degrading can accompany overburden removal. The final process, termed *reclamation*, involves grading, replacing the topsoil where practical, and revegetation. The practices used during reclamation have a major bearing on the recovery rate of the surface mine.

Erosion and Sediment

Sediment in water reduces light penetration and alters the temperature, directly affecting aquatic flora and fauna. Fish production is hindered because food organisms are smothered, spawning ground destroyed, and pools filled. Sediment places an additional burden on treatment plants and cost as well as on aesthetic and economic values (McKee and Wolf, 1963). Sediment deposit in navigable streams must be removed at a high cost. Streams choked with sediment have reduced carrying capacity and are subject to flooding.

The mechanisms of soil erosion by water consist of soil detachment by raindrop, water scoring, and transported by surface flow. Sediment yield can best be described by the Universal Soil Loss Equation (USLE), which combines the principal factors that influence surface soil erosion by water (Roth et al., 1974). The equation takes the form:

$$A = RKLSCP$$

where A is the soil loss in tons/acre,
R is the rainfall factor,
K is the soil erodibility factor,
L is the length of slope factor,
S is the steepness of slope factor,
C is the cropping and management factor,
P is the erosion control practice factor.

The U.S. Department of Agriculture has prepared a handbook for utilizing this equation (Wischmeier and Smith, 1965).

The rainfall factor R reflects the potential of raindrop impact and runoff turbulence to dislodge and transport soil particles. This factor is dependent upon the rainfall pattern and intensity of a specific locality. The mine operator has no control over this factor.

The soil erodibility factor K reflects the most significant soil characteristics affecting soil erodibility. For surface soils, the soil texture, organic matter content, soil structure, and permeability are the key parameters. Recently Roth et al. (1974) have shown for subsoils low in organic matter and high in clay that the chemical properties of the material, such as its iron and aluminum content, have a major bearing on its erodibility. The K factors for the bedrock material fractured during mining and later weathered remains to be defined. The mining company has limited control of this material during the mining operation but can take advantage of the soil erodibility during reclamation by making every effort to place that material with the lowest erodibility factor on the surface. This material should have a high organic matter content and high permeability and should be granular. In many situations, the practice of topsoiling is a step toward obtaining a lower K factor if the soil is properly handled and placed.

The length of slope factor L reflects increased sediment detachment and transport as runoff velocities and volume increase with increased slope length. Spoil should be stored and final-graded to prevent long slopes. Where long slopes must be left, control structures should be installed to break up the slope (see P factor).

The degree of slope factor S also recognizes that runoff velocity increases as the slope of the land increases. For example, increasing the steepness of a slope from 10% to 40% doubles the flow velocity. Steep slopes should be discouraged in mining operations. Slopes of less than 33% are recommended but are sometimes difficult to obtain in mountain surface mining where the original slope equals or exceeds 33%.

The cover management factor *C* reflects the influence of the type of vegetative cover, seeding method, soil tillage, disposition of residues, and general management level. The *C* value ranges from near zero for excellent sod to 1.0 for bare soil. Mining firms can take advantage of this factor by establishing a quick cover of grasses and legumes and using proper tillage techniques. The *C* values decrease as the percentage of ground cover and the percentage of canopy cover increase and the height of the canopy decreases. The application of mulches such as straw, woodchips and chemicals also decreases the *C* factor.

The practice factor *P* accounts for erosion control structures such as terraces and diversions. These practices are used to influence the drainage patterns, runoff concentration, and runoff velocity. Numerous such practices are available, and the mining company should take advantage of those applicable to their situation.

A study by Collier et al. (1964) showed the average annual sediment production from a surface-mined watershed was 42 tons/acre, more than 1,000 times higher than an unmined watershed. Curtis (1974) studied three watersheds in Kentucky and found the sediment yield to range between 0.84 and 1.27 area-inches of area disturbed. He could find little correlation between sediment yield and the amount of land disturbed and concluded that methods of mining and handling the overburden are major factors controlling sediment yield. He also noted that the highest sediment yields were measured during the first six-month period after mining. This indicates the need for more attention to activities during and immediately after mining and the importance of an adequate cover of vegetation and of establishing control structures as quickly as possible after mining ceases.

Acid Mine Drainage (AMD)

The removal of overburden often exposes rock materials containing pyrite (iron disulfide). As shown in Equations 1 and 2, the oxidation of pyrite results in the production of ferrous iron and sulfuric acid. The reaction then proceeds to form ferric hydroxide and more acid, as shown in Equations 2 and 4.

$$2\,FeS_2 + 2H_2O + 7O_2 \rightarrow 2FeSo_4 + 2H_2SO_4 \quad (1)$$
$$\text{(Pyrite)} \rightarrow \text{(Ferrous Iron)} + \text{(Sulfuric Acid)}$$

$$FeS_2 + 14Fe^{+3} + 8H_2O \rightarrow 15Fe^{+2} + 2SO_4^{-2} + 16H^+ \quad (2)$$
$$\text{(Pyrite)} + \text{(Ferric Iron)} \rightarrow \text{(Ferrous Iron)} + \text{(Sulfate) (Acid)}$$

$$4FeSO_4 + 2O_2 + 2H_2SO_4 \rightarrow 2Fe_2(SO_4)_3 + 2H_2O \quad (3)$$

$$Fe_2(SO_4)_3 + 6H_2O \rightarrow 2Fe(OH)_3 + 3H_2SO_4 \quad (4)$$

The amount, and rate of acid formation, and the quality of water discharged are a function of the amount and type of pyrite in the overburden rock, ore, and coal, the time of exposure, the characteristics of the overburden, and the amount of available water (Moth et al., 1972). Crystalline forms of pyritic material are less subject to weathering and oxidation than amorphic forms. Since oxidation by oxygen is the primary reaction during early acid formation, the less time pyritic material is exposed to air, the less acid is formed. Thus, a positive preventive method is to cover pyritic materials as soon as possible with earth, which serves as an oxygen barrier. In terms of mining, this step is accomplished by current reclamation techniques and by operating with only a small pit.

If the overburden also contains alkaline material such as limestone, acid water may not be discharged even though it is formed, because of in-place neutralization by the alkaline material. Discharges from this situation are usually high in sulfate.

Enough water to satisfy Equations 1, 2, and 4 is usually available in the overburden and coal material. Water also serves as the transport medium that removes oxidation products from the mining site into streams.

All techniques for preventing acid formation are based on the control of oxygen. There are two mechanisms by which oxygen can be transported to pyrite–convective transport and molecular diffusion (Ohio State University, 1971).

The major convection transport source is wind currents that can easily supply the oxygen requirement for pyrite oxidation at the spoil surface. In addition, wind currents against a steep slope provide sufficient pressure to drive oxygen deeper into the spoil mass. A factor to consider is the degree of slope after regrading. This is especially important on slopes subject to prevailing winds, since the wind pressure on the spoil surface increases as the slope increases. Thus, the depth of oxygen movement into the spoil would increase as the slope increases.

Molecular diffusion occurs whenever there is an oxygen concentration gradient between two points, e.g., the spoil surface and some point within the spoil. Molecular diffusion is applicable to any fluid system, either gaseous or liquid. Thus, oxygen will move from the air near the surface of the spoil, where the concentration is higher, to the gas- or liquid-filled pores within the spoil, where it is lower. The rate of oxygen transfer is strongly dependent on the fluid phases and is generally much higher in gases than in liquids. For example, the diffusion of oxygen through air is approximately 10,000 times greater than through water. Therefore, even a thin layer of water (several millimeters) serves as a good oxygen barrier.

The most positive method of preventing acid generation is the installation of an oxygen barrier. Artificial barriers such as plastic films, bituminous layers, and concrete would be effective, but these have high original and maintenance costs and would be used only in special situations.

Surface sealants such as lime, gypsum, sodium silicate, and latex have been tried, but they too suffer from high cost, require repeated application, and have only marginal effectiveness. The two most effective barrier materials are soil, including nonacid spoil, and water. The minimum thickness of soil or nonacid spoil needed is a function of the soil's physical characteristics, soil compaction, moisture content, and vegetative cover. Deeper layers would be needed for a sandy, dry, granular material with large grain size and porosity than would be required for a tightly packed, saturated clay that is essentially impermeable. Soil thickness should be designed on the basis of the worst situation–such as a dry soil where oxygen can move more readily through cracks and pore spaces devoid of water. A "safety factor" should be included to account for soil losses from such causes as erosion.

Vegetation not only serves to control erosion but, after it dies, becomes an oxygen user through the decomposing process. This further aids the effectiveness of the barrier. The organic matter that is formed also aids in holding moisture in the soil.

Water is an extremely effective barrier when the pyritic material is permanently covered. Allowing the pyrite to pass through cycles where it is exposed to oxidation and then covered will worsen the AMD problem. Water barriers should be designed to account for water losses such as evaporation and should include at least 30 centimeters (1 foot) of additional depth as a safety factor.

Additional measures to control AMD are water control and in-place neutralization. Water not only serves as the transport media that carries the acid pollutants from the pyrite reaction sites, but also erodes soil and nonacid spoils to expose pyrite to oxidation. Facilities such as diversion ditches that prevent water from entering the mining area and/or carry the water quickly through the area can significantly reduce the amount of water available to transport the acid products. Sediment and erosion control are needed both during and after mining. Terraces, mulches, vegetation, etc., used to reduce the erosive forces of water are effective measures to prevent further pyrite exposure. These measures usually are performed during reclamation.

Alkaline overburden material and agricultural limestone can be blended with "hot" acidic material to cause in-place neutralization of the acid and assist in establishing vegetation. In some cases, grading directs acid seeps to drain through alkaline overburden. These techniques are more applicable to abandoned surface mines than to current mining, where proper overburden handling should prevent acid formation. The major exception may be those situations where an underground mine was breached and an acid discharge formed.

A problem that has been noted in certain locations is the acid seep. These may occur even though the surface mine has been properly graded and is supporting a good vegetative cover. Although the cause of acid seeps is not fully understood, it has been postulated that seeps occur when water percolating through the backfilled spoil reaches the impermeable underclay beneath the coal seam and is forced to move horizontally along the pit bottom and discharge at the toe of the backfill (Fig. 1). In many cases the highly pyritic material associated with the coal is buried in the pit bottom where it is subject to leaching by the horizontally flowing water. One solution to this problem is depicted in Figure 2, where the acid-forming material has been placed above the pit floor and covered with an impermeable material. A diversion ditch constructed along the top of the high wall keeps surface runoff from entering the fill area. Another solution would be to thoroughly mix the acid-forming material with an alkaline overburden material and place it in the fill above the pit floor.

Mineral Pollutants

Fracturing the overburden material exposes numerous new surfaces to the oxidation and leaching process. Those ions subject to leaching are transported by water percolating through the spoil.

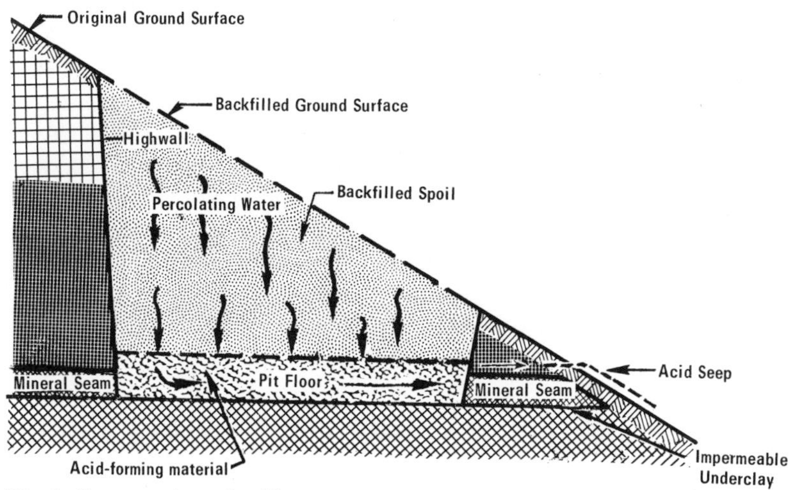

Fig. 1. Cross section of acid seep

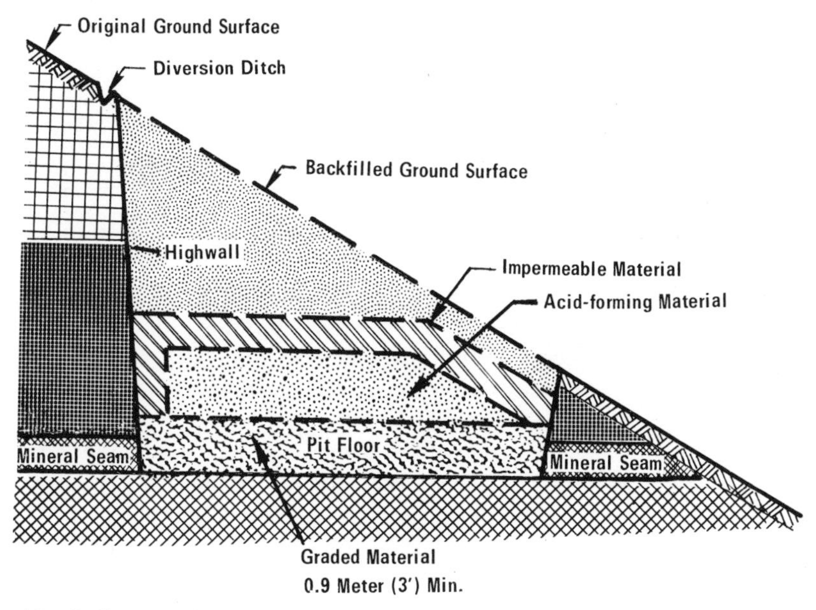

Fig. 2. Cross section of acid-forming material burial

Increases in concentration of calcium, magnesium, aluminum, sulfate, specific conductance, sodium, chloride, alkalinity, zinc, and pH have been reported (Curtis, 1972; McWhorter et al., 1974; Plass, unpublished manuscript). The leaching process can be expected to continue for years because under normal conditions the rate is very slow. In most mines the rate of release is low enough that the increase in dissolved solids is not detrimental. However, where extensive areas have been mined, the accumulation of the leachate may create problems. There is little that can be done to control this problem other than keeping the water that would cause leaching to a minimum and thus extending the process over a longer period.

Environmental Control of Surface Mining

We have prepared a comprehensive report (1974) on *Environmental Protection in Surface Mining of Coal*, which describes methods of preplanning, mining, and reclaiming surface mines.

In summary, environmental control starts in the mine-planning stage. Data collected at the proposed site on the topography, overburden characteristics, ground and surface water, and climatic conditions are used to plan the mining technique to insure (1) that proper erosion control is established, and (2) that selected overburden handling is accomplished to prevent acid mine drainage and insure revegetation. Major irreversible problems have resulted when environmental concerns have not been considered along with extraction procedures. Good premining data collection and mine planning are the key to final recovery of the disturbed area.

Major advances in surface mining methods have been realized in the last few years. The collection of topsoil and subsoil for later spreading over the reclaimed spoil has been a major step in accelerating the recovery of stripped areas. Selective handling of the toxic overburden and its proper burial have reduced both acid and erosion problems. For mined areas on rolling to flat land, the grading of the land to a topography similar to that before mining has not only resulted in land more pleasing to the eye but has made the land more usable and reduced the erosion threat because it is easier to revegetate. In the mountain areas, the advent of mining methods with the colorful names of "head-of-hollow fill," "over the shoulder," "put and take," "block out," and "mountain top removal" have led to ways of reducing the silt and landslides associated with steep-slope mining. Erosion control practices are being adopted but far more could be done in this area. Many states are now requiring sedimentation ponds below mining operations. They are a major step forward in reducing the off-site sediment load; however, design modifications will be needed to make these ponds more effective.

Reclamation and revegetation "know-how" have increased significantly. The days of the pine tree on six-foot centers are gone. Careful consideration of the nutrient deficiency of the spoil and needs of vegetation are being taken into consideration. Lime, fertilizers, and mulchers are being used. Better balanced mixtures of grasses and legumes are being planted. All-weather seeding, instead of only twice a year, is reducing erosion that occurred while awaiting the planting season. Nurse crops to hold the soil and assist permanent species are being utilized. Hydroseeding is more common in difficult areas. Good seedbed preparation is insuring better and faster establishment of vegetation. These practices lead to establishing a permanent vegetative cover more quickly, which is the first step toward recovery.

Recovery

Recovery of a surface-mined area has different meanings. Several states, by law, have specified recovery as a specific density of vegetation or cover. In other states, it means that the mining operator attempted to revegetate the area in a "workman-like manner." On the other end of the scale, there are those who believe that recovery has not been accomplished until the microflora and vegetation have returned to the premining condition. Recovery also can be measured as the return of the water quality to premining conditions.

Several investigators (Hill, 1969; Curtis, 1972; Plass, unpublished manuscript) have reported that the dissolved ion concentration of water discharging from a disturbed area does not return to premining levels as soon as vegetation is established. The ions that are leachable will continue to be discharged for years. The time required depends on the nature of the spoil material and the climatic conditions. Spoils with high infiltration and percolation rates, loosely bound ions, and which are easily weathered will leach quickly. Leach rates will be higher in high precipitation areas. At the Elkins, West Virginia, mine reclamation project, sulfate, aluminum, acidity, iron, and manganese concentrations were still above background levels some seven years after a good vegetative cover had been established.

Smith et al., (1974) have recommended that spoil be considered as a young soil and treated accordingly. They recommend that spoil be classified under the USDA soil classification system as *Spolents* under the order level of *Entisols,* which are defined as recent soils that have little or no evidence of development of pedogenic horizons. An intent here would be to bring to bear the knowledge of soils on the development of manageable soil units. Some investigations have studied the development of soil from spoil. Smith et al. (1971) studied 70- to 130-year old spoils and reported that natural soil proved superior to old spoils in bulk densities, porosity, infiltration, nitrogen, and organic matter, whereas the old spoils were superior in depth of plant rooting, total available water-holding capacity, and certain plant nutrients. Others (Krause, 1964) have reported increased nutrient levels in spoils because nutrients had been leached and utilized by plants from the original soil, while certain strata in the overburden contained large quantities of potassium and other nutrients. Selective placement of the overburden material is critical to recovery. Acid, toxic, and physically poor material must not be placed near the surface. Material that is capable of supporting plant growth is required. The growth-supporting material may require additions of soil amendments such as lime and fertilizer. The most recent trend has been the removing of "topsoil" and then its replacement after grading. *Topsoil* here may be defined as the horizons of natural soil that are capable of supporting plant growth. Several state laws now require topsoiling and the Federal Surface Mining Control and Reclamation Act may require it for coal mining throughout the United States. Although topsoil and subsoil may not always be ideal, they usually have certain advantages over other materials, including higher organic matter, better moisture-holding capacity, natural seed source, and soil microorganisms. Soil microorganisms are a key to the development of soil material, and fresh bedrock materials have insufficient organisms.

Recovery will be much faster and less fragile in the humid eastern United States than in the arid and semiarid western states. The low rainfall and saline soils of the West places additional stresses on recovery. The available plant species that will grow in these areas are limited, slow growing, and often hard to establish. A popular belief in these states is that the acceptable cover should be native vegetation; however, the know-how on culturing native vegetation is limited, seed sources are scarce, and germination may be as low as 10%.

There are striking examples of surface-mined areas that have recovered to productive use as forestland, high-yield cultivated land, pasture, recreational parks, and even home and industrial sites. On the other hand, there are even more examples of land that is still drastically disturbed. The key to recovery has been the attitude and reclamation techniques of the mine operator and the physical and climatic conditions at the mining site.

Literature Cited

Collier, C. R., R. J. Pickering, and J. J. Musser. 1964. Influence of Strip Mining on Hydrologic Environment of Beaver Creek Basin, Kentucky, 1955-1959. Dep. of Interior, USGS Professional Paper 427-B. Washington, D.C.

Curtis, W. R. 1972. Chemical changes in streamflow following surface mining in eastern Kentucky. *In* Fourth Symposium on Coal Mine Drainage Research, Monroeville, Pa.

_____. 1974. Sediment yield from strip-mine watersheds in eastern Kentucky. Pages 80-100 *in* Second Research and Applied Technology Symposium on Mined-Land Reclamation. National Coal Association, Washington, D.C.

Grim, E. C., and R. D. Hill. 1974. Environmental Protection in Surface Mining of Coal. National Environmental Research Center, Environmental Protection Technology Series EPA-670/2-74-093. Cincinnati.

Hill, R. D. 1969. Reclamation and revegetation of 640 acres of surface mines, Elkins, West Virginia. Pages 417-450 *in* Ecology and Reclamation of Devastated Land. Gordon and Breach Publishing Company, New York.

Krause, R. D. 1964. Spoil bank goes from waste to fodder. Coal Mining and Processing 1(5):29-31.

McKee, J. E., and H. W. Wolf. 1963. Water Quality Criteria. California State Water Quality Control Board, Sacramento, Publication No. 3-A.

McWhorter, D. B., G. V. Skogerboe, and R. K. Skogerboe. 1974. Water pollution potential of mine spoils in the Rocky Mountain region. Pages 25-38 *in* Fifth Symposium on Coal Mine Drainage Research. National Coal Association, Washington, D.C.

Moth, A. H., E. E. Smith, and K. S. Shumate. 1972. Pyritic Systems: A Mathematical Model. U.S. Environmental Protection Agency, Environmental Protection Technology Series, EPA-R2-72-002. Washington, D.C.

Ohio State University. 1971. Acid Mine Drainage Formation and Abatement. U.S. Environmental Protection Agency, Water Pollution Control Research Series, No. DAST-42, 14010 FPR 04/71. Washington, D.C.

Plass, William T. Unpublished manuscript. Changes in water chemistry resulting from surface mining coal on four West Virginia watersheds. U.S. Forest Service, Northeastern Forest Experiment Station, Princeton, W. Va.

Roth, C. B., D. W. Nelson, and M. J. M. Romkens. 1974. Prediction of Subsoil Erodibility Using Chemical, Mineralogical and Physical Parameters. U.S. Environmental Protection Agency Publication No. EPA-660/2-74-043. Washington, D.C.

Smith, R. M., E. H. Tryon, and E. H. Tyner. 1971. Soil Development on Mine Spoil. West Virginia University Agricultural Experiment Station Bull. 604T. Morgantown.

Smith, R. M., W. E. Grube, T. Arkle, and A. Sobek. 1974. Mine Spoil Potentials for Soil and Water Quality. U.S. Environmental Protection Agency, Environmental Protection Technology Series EPA-670/2-74-070. Cincinnati.

Wischmeier, W. H., and P. D. Smith. 1965. Predicting Rainfall-Erosion Losses from Cropland East of the Rocky Mountains. U.S. Department of Agriculture, Agricultural Handbook No. 282. Washington, D.C.

Ecosystem Development on Coal Surface-Mined Lands, 1918-75

Charles V. Riley

Abstract

Studies were conducted of ecosystem development on lands disturbed in 1918 by surface mining for coal in southeastern Ohio. The overburden, removed during mining, remained in the ridge-ravine topography, typical of that mining era. Both terrestrial and aquatic ecosystems were included in the study.

Ecological processes functioning in ecosystems development after landscape disturbances by surface mining for coal are dependent upon and influenced by several variables, including geochemistry of the overburden material, position of the various strata materials within the spoil column, topography of the mined overburden, physical characteristics of the spoil surface, microtopography, and degree of compaction. These variables influence rainfall retention, infiltration, leaching, and erosion, thus contributing to and altering the site characteristics.

The physical and chemical conditions present resulted in critical sites for the initial invasion and etablishment of vegetation. Natural processes, operating over a period of 57 years and during a period of 29 years (1946-75) in which data were collected, have resulted in the development of ecosystems of fairly diverse and numerous biotic components but of relatively low biomass. Small but very significant changes in the chemical conditions of the abiotic component, i.e., the spoil or rooting medium, also were revealed.

Species diversity of adjacent plant communities, distance of seed sources to the mined land site, and method of seed dispersal were significant to initial plant establishment and ecosystem composition and development.

Introduction

Surface or open-pit mining for minerals and energy resources has been practiced by man for the past several hundred years. As his population grew and his technologies advanced, the disturbance of the natural environment increased. During the present century, as the world's population continued its dramatic growth and as the industrial technologies of the western nations expanded and proliferated to other parts of the globe, especially after World War II, energy, the base of that expanding industrial growth, was required in ever-increasing abundance. In the United States electrical energy demand has been doubling every decade. Per capita use of electricity in the United States rose from about 2000 kwh in 1950 to 7800 kwh in 1971 and is projected to reach 32,000 kwh by the year 2000 (Dunham, et al., 1974).

While oil and natural gas have, in recent years, provided for an increasingly greater share of the world's energy requirements, coal has continued to be an important energy resource. Although total coal production increased very little in the United States through the 1950-60 period, coal production by surface mining increased greatly as compared to conventional underground mining. This has been especially true in the United States, the West German Federal Republic, and some of the countries of eastern Europe. Coal production by surface mining in the United States increased from about 9.4% in 1940 to 44% in 1971 (Krause, 1972) to about 52% in 1974 (Overton, 1974). Surface mining in the United States seldom occurs in overburden beyond 125 ft, while in West Germany open-pit mining for lignite reaches the 1000-ft level. Studies by the National Academy of Science (Hammond, 1974) predicted serious energy problems for the next several years and proposed alternatives such as coal gasification to strengthen the possibilities for increased surface mining and thus increased effects upon the associated ecosystems.

In the process of removing overburden or earth strata from above the coal seam, sedimentary materials having a developmental geo- and biochemical history of some 250 to 300 million years are removed. In the areas of lignite mining, the earth strata may have a considerably shorter history of development (e.g., about 20 million years in the Rhineland of West Germany). During the millenia, as the coal measures were being laid down, sedimentary materials of sands, silts, clays, and other mineral and chemical constituents were being deposited to form the associated earth strata. These periods of deposition of sands, silts, and clays were followed by intervening upheavals during which the exposed surface materials were subjected to the physical, chemical, and biological processes characteristic of the climate and environmental conditions existing at a specific time. Surface-mining processes involving strata removal, deposition, and grading produce a growth medium for vegetation of extreme and diverse geochemistry. Seldom will the ecologist, in his research of the ecosystem dynamics of natural landscapes, encounter the diversity and the extremes of geochemistry present in the surface medium or in the drainage that may follow the disturbance and exposure of the strata of the various coal measures.

This book, dealing with the restoration and recovery of damaged ecosystems, includes chapters on environmental disturbances resulting from a variety of human activities. Environmental disturbances, modification, and damage such as those resulting from surface mining, oil spills, chemical effluents, forest and grassland fires, etc., are quite obvious, but the ecological details of the recovery processes are very little understood.

An ecosystem is an environmental unit or community consisting of biotic and abiotic components which are related by the interchange of minerals and a flow of energy, and by interactions within each, as well as between the two components. While semantics may differ as to definition, I believe we are in reasonable agreement as to intent of the definition. This chapter, titled Ecosystem Development, will provide data on certain processes and changes resulting from both natural processes and human effort operating toward restoration and recovery of ecological systems developing on surface-mined land sites and in associated aquatic environments.

The Environment or Mining Site before 1918

The ecosystems studied and reported in this chapter are those that have developed on a disturbed landscape resulting from surface mining for coal in 1918. They are located in southeastern Ohio on the unglaciated Allegheny Plateau. The ecosystems include (1) the natural or volunteer ecosystem consisting of those plant and animal species that have developed on a mined landscape since 1918 (Fig. 1—N.Ec.V.) and (2) the black locust ecosystem, consisting of volunteer locust from an initial planting of this species in 1923, plus additional volunteer deciduous species (Fig. 1—B.L.V.Ec.). Adjacent to and including a part of the above natural ecosystem is the third ecosystem, modified by man with planting of trees, primarily conifers and shrubs, between 1947 and 1953 (Fig. 1—T.S.Pl.Ec.). The aquatic ecosystem (Fig. 1—Im 5.5) located within the mined landscape is a 5.5-acre mine water impoundment formed in 1918.

Natural ecosystem is defined, in this chapter, as the landscape mined in 1918 containing naturally established or volunteer species of plants indigenous to the area and the animal populations present within the existing habitat.

The original vegetation, before settlement, was described by Gordon (1969) as that of the mixed oak forest. A variety of subtypes or associations,

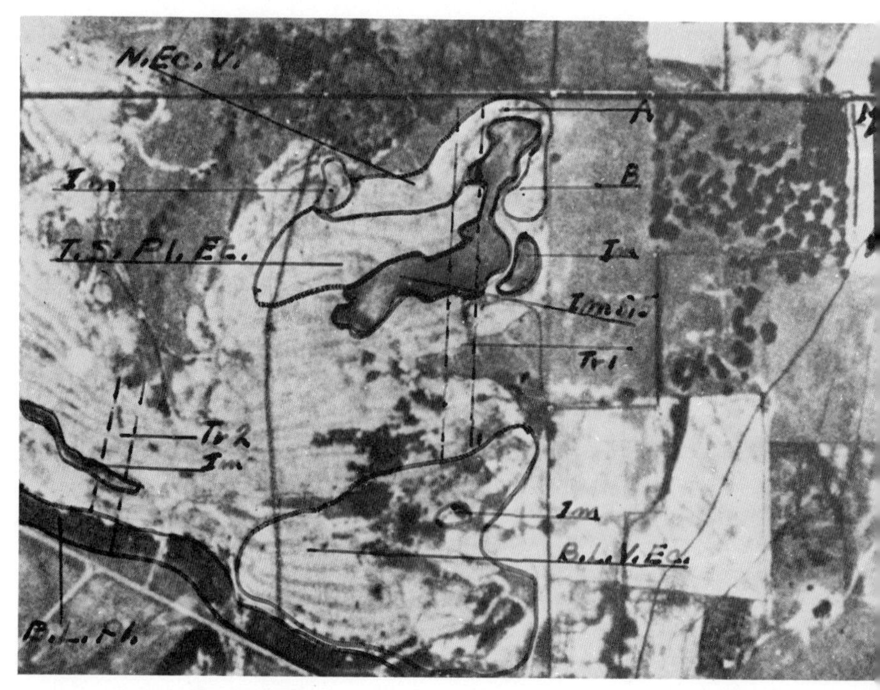

Fig. 1. Land surface mined for coal in 1918. Key: N.Ec.V. = natural ecosystem (1918); T.S.P1.Ec. = tree-shrub (conifer) planted ecosystem; B.L.V.Ec. = black locust volunteer ecosystem; Im 5.5 = aquatic ecosystem; Tr1, Tr2 = transects; B.L.P1. = black locust planting (1923); Im = mine water impoundments; *A* = location of Fig. 10 (1975); *B* = location of Fig. 11 (1975). (Photo U.S.S.C.S. 1935)

however, occupied the numerous sites presented by the naturally dissected portions of the plateau. Sites containing these subtypes varied from fertile river terraces, till plains, and swamp habitats, to steep-sided coves with exposed outcrops of acid-to-calcareous sandstones, limestones, shales, and clays, to hilltops and ridges, sometimes of clays but often sandy and well drained. Dependent upon the particular geological formation and amount of dissection present, the soil chemistry of those sites reflects the effects of the outcropping of the various strata.

The "witness trees" recorded in the original land survey of 1831, for 120 quarter sections surrounding the present study area, revealed a forest type consisting of 22 species of which the oaks and hickories totaled approximately 70%. Included in the 70% were white oak (48.57%), black oak (8.21%), northern red oak (2.5%) and hickory (10.71%). (Scientific names of plants are listed in Appendixes A, C, D.) In addition, red maple made up 7.14% and sugar maple 3.92%. Although covered with climax forest communities, this unglaciated Ohio region, containing the coal measures, presented an erosion landscape. The original forest vegetation had been removed many years before the mining, and conventional agriculture practices had been instituted. Agriculture on the western foothills of the Allegheny Plateau consisted of the 160-acre family farm unit and included generalized agriculture of dairying, grazing of beef cattle and sheep, grain farming, and woodlots. Agricultural activities began to decline in this region about the turn of the century, with increasing impetus during the depression years before World War II.

Natural Ecosystem Development, 1918-46

Natural Revegetation (Fig. 1—N.Ec.V.)

Approximately 600 acres of landscape described as very gently rolling to ridge topography were surface mined in 1918 and left in the typical giant furrows, or "spoil banks," characteristic of that mining era (Fig. 1*a*). Mining was done for the underlying Lower Kittanning coal seam of the Allegheny Formation. Overburden ranged up to 60 ft in depth and consisted of strata of acid clays, sandstones, and shales. Resulting spoil banks contained slopes of 15° to 45°; however, after 28 years of settling and erosion, most of the slopes ranged from 20° to 26°. Slope length at the greater overburden depth ranged from 80 to 100 ft. Several small mine water impoundments up to 5.5 acres and coal waste disposal sites were located within the mined land unit. Four small tree plantations were established in areas some distance from the present study area. The plantings included black locust (1923), red oak (1926), red pine (1926, 1932), and Scotch pine (1936). However, no activities such as grading, spoil amendments, tree planting, or seeding were conducted on the mined land discussed in this chapter until the late 1940s, when the present studies were initiated. The 1935 aerial photo (see Fig. 1*a*) reveals the topography and amount of naturally established vegetation present after a period of 17 years of development while the 1942 aerial photo (Fig. 2) reveals the growth and development of vegetation after a period of 24 years.

By 1946, when the study was initiated, volunteer black locust had invaded approximately 20 acres of the area (see Figs. 1*a*, 2, 6). A limited number of species including wild black cherry, large-toothed aspen, American elm, red maple, white ash, and black gum were present in the

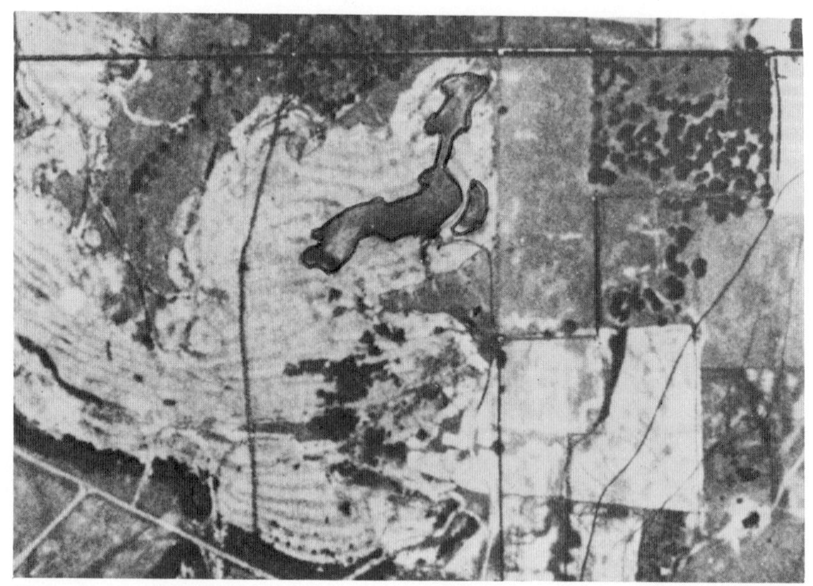

Fig. 1*a*. Land surface mined for coal in 1918. Note sparse volunteer vegetation on ungraded spoil banks 17 years after mining. (Photo U.S.S.C.S. 1935)

volunteer locust unit (see Fig. 1—B.L.V.Ec.). The canopy of the locust area reached a height of 18 to 30 ft, and the trees ranged in age from 12 to 20 years. A few large, isolated hardwoods were present on the mined landscape, probably the result of seedlings being uprooted and partially buried at the time of mining. Diameters recorded for these individuals included American elm, 16.7 in.; domestic cherry, 10.0 in.; black gum, 10.7 in.; and wild black cherry, 15.0 in. (Diameters were measured 1 ft from the soil line.) The remainder of the mined landscape contained very sparse communities of herbaceous, shrub, and tree species.

No detailed studies concerning the spoil chemistry were done on the mined lands in the early years following cessation of mining. During the mining of 1918, no effort was made regarding placement or burial of the toxic or the more acidic spoil in the lower portion of the spoil column. To determine what the chemistry of the original spoil material may have been, a one-acre spoil plot was established (1957) in which the spoil banks were graded 6 to 12 ft using a bulldozer, thus exposing raw earth material. Assuming 30 ft of the original over-burden above the coal seam and 2000 tons per acre ft, some 60,000 tons per acre of raw earth materials were removed. Before the soil plot was graded, the spoil banks supported

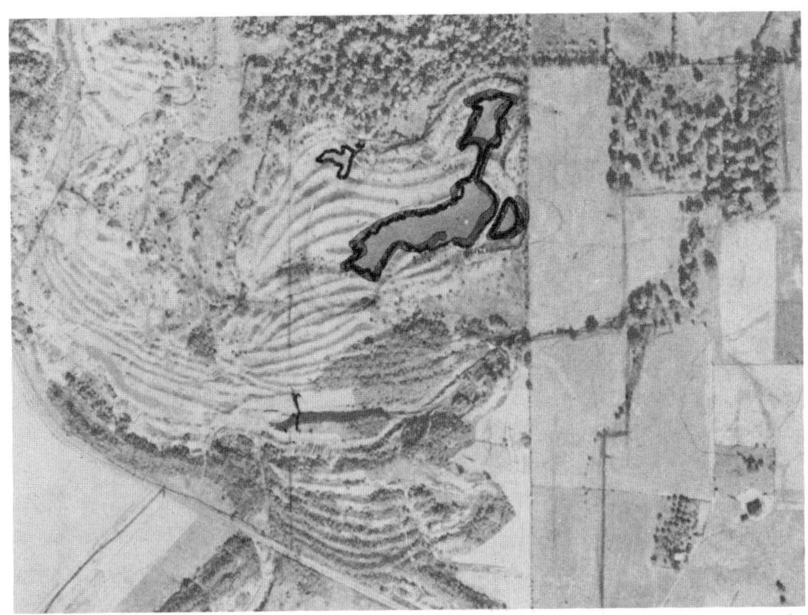

Fig. 2. Surface-mined lands of Figs. 1, 1a, 1918. Note density of vegetation in B.L.V.Ec., 24 years after mining. (Photo U.S.S.C.S. 1942)

black locust trees 12 to 20 years old. The spoil surface was covered by leaf litter one to two inches in depth. The spoil plot was graded to a plateau-type of topography.

Samples were collected (Jackson, 1958) from the weathered, ungraded spoil surface and from the newly graded area for comparison (Table 1). From this data one might speculate that the chemistry of the surface spoil in 1918 was similar to that reported in column 1957A. Data reported in column 1957B were for surface spoil collected from the plot after it had been modified by various physical and biochemical processes for 39 years. Indications are that the pH of the original surface spoil was probably 3.6 or lower and contained total soluble salts of about 1.5 tons per acre. Soil samples have been collected regularly from this plot since 1957, and the data reveal a pH of 4.0 in 1972 with total soluble salts and sulfates considerably lower than in 1957. Note also that aluminum, zinc, and manganese were extremely high in the spoil interior in 1957. Iron, zinc, and aluminum show a decline by 1972. For comparison, samples of undisturbed top soil were collected and analyzed from an old field area within 300 ft of the soil plot previously discussed. The old field soil had not been tilled or cropped since 1918. Vegetative cover consisted of broom sedge, poverty grass, and dewberry. Some of the macronutrients in the mined spoil (see Table 1) exceeded the amounts present in the old field topsoil (Table 2).

In September and October 1946 two belt transects were established on the mined landscape; plant species were identified and enumerated, and trees were measured for diameter (dbh) (see Fig. 1—Tr1, Tr2). The first transect (Tr1) measured 50 ft wide and 1050 ft long (52,500 ft^2); the second transect (Tr2) measured 30 ft wide and 1410 ft long (42,300 ft^2). The transects were laid out approximately north and south, perpendicular to the ravine, slope, and ridge sites. The lengths of slopes within the transects ranged up to 65 ft and from 15° to 26°. The transects culminated near a woodlot. The species composition of the woodlot was very similar to that of the original land survey of 1831 as defined by the "witness trees" recorded at that time.

The transect studies revealed sparse populations of several species of herbaceous plants. Plant communities having the greatest density were located in the ravines and usually consisted of a single species of redtop grass, timothy, broom sedge, yellow or white sweet clover, or alsike clover (Fig. 3).

Canada bluegrass and poverty grass were the dominant grass species on the slopes. One 50-ft-square quadrat contained a stand of Canada bluegrass (10-12 stems/ft^2) covering 30% of the plot; a few quadrats contained stands of the same species covering 10% of the surface of the plot. Poverty grass occurred in scattered patches of 10 to 15 ft in diameter. This species, because of its growth characteristics, provided 100% spoil surface cover wherever established. Patches of mosses, lichens, common club moss, and ground pine 2 to 6 ft in diameter also were growing on some slopes. Stands of smartweed and barnyard grass were common in the moist ravine sites.

Broad-leaved herbaceous species were extremely scattered and sparse over the various sites within the transects. More common species included wild carrot, sour dock, goldenrod, common milkweed, yarrow, dogbane, evening primrose, field bindweed and whorled loosestrife (Appendix A).

Poison ivy, wild grape, blackberry, dewberry, red raspberry, etc., were present but extremely sparse. Dewberry occurred in a few isolated patches 10-20 ft in diameter on the slopes and in the ravines (Figs. 4 and 5).

Tree species including wild black cherry, red maple, white ash, sassafras, large-toothed aspen, red oak, American elm, and dogwood were measured and mapped within each quadrat (see Fig. 5). Other plant species were located on the map as well. Due to the small size of most tree species,

Table 1. Chemistry of surface spoil in 1957 and 1972

	1957A*	1957B†	1972
pH	3.60	4.40	4.00
		(Available lbs/acre)	
NO_3	1.00	9.00	23.40
NH_3	32.00	35.00	48.00
P_2O_5	1.20	0.50	1.10
K_20	126.00	44.00	99.00
Ca	739.00	362.00	302.00
Mg	56.00	94.00	94.00
T. S. salts	2800.00	40.00	400.00
S as SO_4	219.00	25.00	91.00
Fe^{+3}	7.50	0.50	1.90
Mn^{+2}	11.00	6.50	16.00
Cu^{+2}	3.30	1.60	3.10
Zn^{+2}	32.00	13.00	2.50
Cl^-	2.40	0.80	8.70
Al^{+3}	43.00	32.00	2.10
B^{+3}	0.33	0.35	0.31
Mo^{+6}pp2B	20.00	20.00	125.00

Note: Soil plot is located in Tuscarawas County, Ohio. Sample collected for analyses consists of 15 subsamples per acre to 6-in. depth. Overburden type is acid sandstones, silty shales, and clays associated with Lower Kittanning coal seam.

* 1957A: Sampled interior of spoil bank immediately after grading off 6 to 10 ft of original ungraded banks. Chemistry may be similar to that of original spoil in 1918.

† 1957B: Sample collected from surface of ungraded spoil banks after weathering and being modified for 39 years.

Table 2. Chemical analyses of topsoil

				(Available lbs/acre)				
pH	NO_3	NH_3	P_2O_5	K_20	Ca	Mg	T.S. salts	S as SO_4
5.50	4.70	5.00	0.80	65.00	653.00	48.00	200.00	46.00
	Fe^{+3}	Mn^{+2}	Cu^{+2}	Zn^{+2}	Cl^-	Al^{+3}	B^{+3}	Mo^{+6}pp2B
	0.30	0.26	4.00	3.10	27.00	0.06	0.26	50.00

Note: Topsoil has been idle since 1918.

Fig. 3. Volunteer plant communities (Tr2) of redtop grass, timothy, and sweet clover occupied the ravine sites of the 1918 mined lands. N-S exposures. (Photo 1947)

measurements were made 1 ft from the soil line. Only trees of 0.50 in. in diameter and above were measured. Maximum diameters recorded for tree species in 1946 within 100 ft of the woodlot were red maple, 3.50 and 3.75 in.; wild black cherry, 2.00 and 4.25 in.; white ash, 0.75 and 1.50 in.; and black gum, 1.25 in. Between 100 and 200 ft from the woodlot measurements were red maple, 2.00, 4.50, and 5.00 in.; white ash, 1.25, 2.75, and 6.50 in.; black gum, 1.75 and 3.00 in.; wild black cherry, 1.00, 2.75, and 3.00 in.; and sassafras, 0.50 in. Between 250 and 350 ft from the woodlot measurements were red maple, 2.00 in.; wild cherry, 2.25 and 2.50 in.; large-toothed aspen, 1.00 and 2.00 in.; American elm, 1.00 and 3.25 in.; and white ash, 1.25 and 1.75 in.

Percentages of trees of each species recorded on the transect included red maple, 38.20%; wild cherry, 36.50%, white ash, 10.40%; large-toothed aspen, 4.35%; domestic cherry, 3.91%; black gum, 1.74% slippery elm, 1.74%; American elm, 1.30%; and domestic apple and sassafras, the remainder. Percentage of trees by diameter class were 1.00 in. and less, 46%; 1.10 to 2.00 in., 31.3%; 2.10 to 3.00 in., 12.6%; 3.10 to 4.00 in., 7.39%; 4.10 to 5.00 in., 2.6%; and 5.10 in. and above, 0.11%. Diversity and numbers of volunteer tree species on the transect appeared to be determined by

Fig. 4. Volunteer plant communities (Tr1) of broom sedge, dewberry, blackberry, poison ivy, wild grape, and seedlings of wild black cherry and white ash. N-S exposures. (Photo 1947)

distance from seed source and means of seed dispersal; in certain species another factor was use as food by birds and mammals and the resulting scattering of the seeds as waste. Extremely small amounts of organic matter were present on the spoil surface. Within the approximately 2.1 surface acres surveyed in the transects, an estimated 85% of the spoil surface was barren after 28 years.

Volunteer Black Locust Ecosystem Development, 1923-46
Fig. 1–B.L.V.Ec.

Volunteers from the original black locust plantation of 1923 invaded approximately 20 additional acres (see Fig. 1–B.L.V.Ec.) and have been described earlier as a 12- to 20-year-old stand (Fig. 6). Density was estimated at about 600 trees per acre ranging from 2 to 8 in. in dbh. Many of the trees were severely damaged by infestations of the locust borer. Other volunteer tree species similar to those listed earlier were present and totaled about 10% to 15% of the community. Dense patches of blackberry, raspberry, pokeweed, and burdock were present in some areas of the plantation along with rather dense mats of bedstraw over much of the

Fig. 5. Volunteer plant communities (Tr1) of timothy, Canada bluegrass, various species of forbs, and wild black cherry. N-S exposures. (Photo 1947)

Fig. 6. Volunteer black locust ecosystem (B.L.V.Ec.), containing trees 12 to 20 years of age. Stand contains about 600 stems per acre. (Photo 1947)

forest floor. A thick mat of locust leaf litter, 2 to 5 in. in depth on the slope and up to 6 in. in the ravines, covered the entire spoil surface (Fig. 7). Decomposed organic material was distributed abundantly into the surface 3 in. of the spoil. One area of the locust ecosystem having an open canopy contained an extremely dense stand of Canada bluegrass, approximately three-quarters of an acre in size.

Vertebrate Populations, 1946-47

Studies of the animal populations (Riley, 1954) revealed the cottontail rabbit, whitetail deer, red and grey fox, woodchuck, and ruffed grouse present as residents or utilizing the plant communities on that part of the mined land containing the vegetation designated as the natural ecosystem between 1918 and 1946. The cottontail rabbit and the woodchuck occupied or utilized the more dense herbaceous plant communities found in the ravines. Live trapping revealed greater cottontail rabbit populations on the mined land–natural ecosystem unit than on adjacent agricultural lands (Riley, 1954). The ruffed grouse were observed in the vicinity of wild grape and poison ivy vines. The song sparrow, field sparrow, vesper sparrow, eastern goldfinch, killdeer, and mourning dove were observed in the open herbaceous plant communities. The belted kingfisher, killdeer, woodcock, common snipe, lesser yellow-legs, American bittern, and green heron were observed along the shore of the mine water impoundments. The kingbird was also observed at the juncture of the forest and mine water impoundment. The muskrat inhabited the mine water impoundment, while the raccoon was a transient along the shore of the ponds. Observed on the impoundments were the common and hooded mergansers, canvas-back, mallard, pied-billed grebe, and horned grebe. The Cooper's hawk was observed over the same unit. In the black locust ecosystem, the above listed mammals were present plus the longtail weasel, common shrew, opossum, and the fox squirrel.

Mammal populations inhabiting the mined land environments as residents probably included only the cottontail rabbit, woodchuck, common shrew, and red fox. One red fox den was located in the black locust unit. Woodchuck dens were almost three times as numerous in the black locust communities as in the natural or volunteer plant communities.

The ruffed grouse were commonly observed in the black locust during the winter season; one covey of bobwhite quail also overwintered here, while the ring-necked pheasant was probably only a transient.

During the spring of 1947 a three-hour bird census in the black locust

Fig. 7. Leaf litter 2 to 5 inches in depth, covered the spoil surface in the black locust ecosystem. Organic matter was present in the surface 3 inches of spoil. (Photo 1947)

unit included 13 species, or a total of 41 individuals. Included were the red-eyed vireo, indigo bunting, robin, grey catbird, towhee, eastern brown thrasher, field sparrow, yellow-billed cuckoo, vesper sparrow, song sparrow, red-headed woodpecker, killdeer, and American bittern. The latter two species were near a small impoundment.

Natural Ecosystem, 1947-75, Planted plus Tree-Shrub Ecosystem, 1947-75

White pine, red pine, Norway spruce, multiflora rose, bicolor lespedeza, Japanese lespedeza, Sericea lespedeza, autumn olive, coral berry, and various species of deciduous trees were planted on several acres of the mined lands previously described as the natural ecosystem (see Fig. 1—T.S.Pl. Ec.; Fig. 8). Much of the black locust area was underplanted to black alder, red oak, white ash, multiflora rose, tartarian honeysuckle, and autumn olive. A total of about 70,000 tree and shrub seedlings were planted between 1947 and 1975 on the mined lands, with the exception of approximately four acres which were maintained as a natural or volunteer plant community (see Fig. 1—N.Ec.V.).

Survival rates for the seedlings planted were extremely high, 90% to 98%; by 1956, 90% of the spoil surface was covered by either vegetation or leaf litter and organic matter. At the present time (1975) approximately 95% of the spoil surface is covered by litter and vegetation. Very obvious changes in the density and growth of the vegetation are revealed in the photographs of 1948 (Fig. 8) and 1971 (Fig. 9). Depth of litter on the spoil surface beneath the pine measured 1.00 to 1.50 in. on the slopes and up to 3.00 in. in the ravines. The white pine (1975) range from 28 to 31 ft in height, with 9.00 in. dbh; white ash measure 18 ft in height, 4.00 and 5.00 in. dbh, and black locust 28 to 35 ft in height and 9.00 to 11.00 dbh. Bicolor lespedeza (1949-50), multiflora rose (1949), and autumn olive (1959) remain as vigorous species on those sites where the canopy is open. The shrubs produce an abundance of seed and fruit annually.

On the four-acre mined land unit described as the natural ecosystem very significant changes have occurred since the earlier vegetation survey (1946) was made (see Fig. 5). There are great numbers of seedlings (less than 1.00 in. dbh) of red oak, red maple, white ash, black gum, and flowering dogwood and lesser numbers of tulip poplar, American beech, large-toothed aspen, and white oak. Wild black cherry seedlings are absent from the seedling community. The older wild black cherry trees recorded in the 1946 survey are in a decadent condition. Ecological changes (Feb. 1975) in the vegetation of the natural plant community are revealed by dbh measurements of trees growing on or near the transect areas (Tr1) as compared to data recorded in 1946. Measurements of trees located within 50 ft of the center of the original (1946) transect (Tr1) and within 200 ft of the woodlot revealed the following: tulip poplar, 18.10/65 (dbh in in./estimated height in ft); American elm, 9.40/60; red oak, 4.30/18, 7.20/50; white oak, 6.40/22; large-toothed aspen, 7.20/45, 8.00/50; black gum, 7.60/32, 11.40/38, 13.00/45; white ash, 5.60/30, 6.40/32, 8.80/35, 12.00/30; red maple, 8.00/32, 8.40/45, 9.60/33, 12.20/45, 16.80/40; wild black cherry, 6.40/26, 12.00/36; and flowering dogwood, 6.00/15 (Fig. 10). The measured trees cited above were selected for maximum diameters. The trees were growing on north, northeast, and south exposures.

Approximately 100 ft east of the original transect the following species were present on banks having an east-west exposure. Measurements included tulip poplar, 8.80/35, 16.00/60; red oak, 16.40/65; red maple, 7.20/40; black gum, 17.00/37; sassafras, 9.40/38; wild black cherry, 11.10/45; and white ash, 11.60/65 (Fig. 11). The spoil surface beneath the volunteer hardwoods was covered by litter 1 to 4 in. deep or by patches of poverty grass, ground pine, and club moss. Scattered wild grape, bittersweet, poison ivy, and Virginia creeper were also present.

Fig. 8. Mine water impoundment (aquatic ecosystem Im 5.5) formed in 1918. Mined land in background is location of 1947 conifer-shrub plantings (T.S.P1.Ec.). The pH of the water ranged from 3.1 to 5.5. (Photo 1948)

Vertebrate Populations 1974-75

During the period of 28 years from 1947 to 1975, very obvious changes have occurred in the growth and development of the vegetation, in both the mined landscape containing the naturally established communities and those tree-shrub communities established by man. Growth and development of the vegetation are vividly illustrated in comparing the 1935 aerial photo (Fig. 1*a*) with another photo of the same area taken in 1973 (Fig. 12). As the plant communities evolved from the pre-1946-47 pioneer community stages into the early stages of a forest community, the resulting habitat or environment was more favorable for a greater diversity of species.

Mammalian species that have become established since 1947 include the beaver, red squirrel, eastern chipmunk, white-footed mouse, meadow jumping mouse, amd meadow vole. Those species inhabiting the area in 1947 were also present. As the tree species of the ecosystem continued to develop and reached the sapling pole stage, the cottontail rabbit population declined; however, the whitetail deer population increased. I am not implying that the mined land environment served as the total habitat for

Fig. 9. View of the same area shown in Fig. 8 in 1971; planted tree-shrub ecosystem (T.S.P1.Ec.) in background and aquatic ecosystem (Im 5.5) in foreground. See water chemistry data for 9/28/73 (Table 8). (Photo 1971)

deer, since the area is of limited acreage and the home range of the species is considerably greater than the area of the ecosystem. The deer population in Ohio has increased greatly in recent years. Land management has improved and the habitat resulting from changes in the use of lands adjacent to the study area have created a more favorable environment for the deer. Improved agricultural practices are in operation on adjacent croplands; formerly overgrazed lands have reverted to primary forest communities; and woodlots are protected from fire and grazing. A pair of beaver was first observed in the 5.5-acre impoundment in 1973. The muskrat was a common inhabitant of the impoundments and resulted in a harvest of 2 individuals per surface acre in the aquatic ecosystem in 1965. The ruffed grouse, a common inhabitant in the tree-shrub (conifer) and natural ecosystem, was first observed nesting in the latter area in 1960. Throughout the fall and winter season they have been observed feeding on multiflora rose and autumn olive fruits and apparently searching for food in the leaf litter of the black locust community. The red squirrel probably was a transient, having been observed in both the natural and the black locust ecosystems. The fox squirrel and chipmunk have been observed regularly in both of these areas. The meadow vole and white-footed deer

Fig. 10. Natural or volunteer forest ecosystem (Fig. 1, N.Ec.V.-A) containing tulip poplar, red maple, red oak, white ash, etc. Forest community represents 57 years of development. Compare to Fig. 4, 1947. (Photo 1975)

mouse are inhabitants of the locust area, the natural ecosystem, and the coniferous ecosystem. The meadow jumping mouse was observed in the natural ecosystem.

Since the mid-1950s waterfowl have become more abundant on the 5.5-acre mine water impoundment, especially during the spring and fall migrations. A total of 14 species of waterfowl have been observed on the impoundment during the migration periods. Species recorded include the mallard, woodduck, Canada goose, gadwall, American widgeon, canvas-back, common goldeneye, lesser scaup, buffle-head, hooded merganser, common merganser, coot, horned grebe, and pied-billed grebe. Fairly large rafts of ducks were observed during the migration periods in 1949, numbering an estimated 43 mallards, 100 woodducks, and 75 lesser scaup. The mallard nested in the leaf litter of the natural ecosystem in 1963 and 1973. In 1974 they also nested in the black locust plantation. Mallard nests were all located within 50 ft of the water. Broods of woodducks have been observed each year on the impoundments since 1965.

Within the black locust, tree-shrub (conifer), and natural forest habitats, the population diversity of the song and insectivorous species of birds

Fig. 11. Black gum (17 in. dbh) growing on ridge site of lands mined in 1918. For location, see Fig. 1, N.Ec.V.-B). (Photo 1975)

Fig. 12. View of land mined in 1918 and ecosystem development. Compare to Figs. 1, 1a (1935). (Photo 1973)

has increased. Species more commonly observed in recent years have been the dark-eyed junco, black-capped chickadee, brown creeper, tufted titmouse, white-breasted nuthatch, bluejay, golden-crowned kinglet, crested flycatcher, downy woodpecker, hairy woodpecker, common grackle, crow, starling, house wren, yellow-breasted chat, mourning dove, chipping sparrow, nighthawk, common flicker, eastern phoebe, red-wing blackbird, eastern goldfinch, American kestrel, and eastern bluebird. The most notable increase since 1960 has been in the bluebird population. Groups of 6 to 12 have been observed at the juncture of the locust-conifer plantations and a nearby tree-shrub fencerow containing decadent and dead wild cherry trees. The fencerow also contains a heavy growth of poison ivy. Observations indicate the species is inhabiting primarily the fencerow environment and utilizing the mined land habitat as a part of its home range.

A pair of red-tailed hawks has been observed nesting in the conifers each year since 1969. The osprey was observed in 1975. The mourning dove is frequently observed in the conifer plantation although no nesting has been observed. The great blue heron was first observed in 1973, along the shore of the larger lake.

The chimney swift and tree swallow were observed over the lake area for the past 20 years.

While not utilizing the mined land habitat, per se, the pileated woodpecker is often observed in the woodlot adjacent to the former unit.

Two amphibians, the tree toad and spring peeper, were first observed in 1970 in the black locust areas adjacent to the lake, while the black snake was found in all three ecosystem types.

Ecosystem Development, Microenvironments, and Changes in Spoil Chemistry

It is quite evident that succession and other ecological processes have been occurring during the 57-year period from 1918 to 1975 resulting in the development of more complex and dynamic communities or ecosystems (see Fig. 12). The development and changing aspects of the plant and animal communities are documented for (1) a natural ecosystem, consisting of native species of trees, shrubs, and herbaceous plants, (2) a volunteer black locust ecosystem, plus other volunteer species indigenous to the area, and (3) a coniferous ecosystem containing some deciduous trees and shrubs.

While the changes in the biotic components, flora and fauna, are quite obvious, the changes in the abiotic components of the system, especially the physical and chemical parameters of the spoil, as a plant growth medium are not as easily recognized or understood. Surface mining, in the removal process and deposition of the earth strata, not only creates a diverse mixture of spoil material but results in a diversity of microsites, or microenvironments. The resulting physical, chemical, and biological processes may function quite differently at each specific site. The effects of the diverse site parameters upon the biotic members (living systems) of any community and at any given time may be quite different. As these processes function and modify the growth medium, specific to the diverse micro-site conditions, interactions and interchanges occur between the biotic and abiotic aspects of the ecosystem which are not completely understood.

During the past 29 years of studies on the ecology of mined lands and the soil environment, a total of over 350 samples of various types of spoil have been collected and chemically analyzed. A total of 15 spoil plots approximately one acre in size were established and sampled annually or biannually. The earliest plots were established in 1957. The plots have been selected to include the various types of surface topography present on mined lands, i.e., plateaulike areas having a rough surface and also a smooth surface, furrow-graded surfaces, and smooth spoil surfaces having various degrees of slope and length. Sampling has also been done by selecting sites supporting different types of vegetation; areas having different amounts of litter and organic matter on the spoil surface; spoil areas devoid of vegetation, litter, and organic matter; and spoil that has been covered with treated sewage sludge.

The spoil, for the mined land area being discussed in this chapter, consists of heavy clays, shales, sandstones, and ironstones; possesses a moderate to extremely acid pH; and contains very high levels of soluble salts and sulfates and high to very high levels of certain metals (Jackson, 1958). The spoil provides a medium that can be defined as critical for seed germination, growth, and reproduction. Extremely diverse site conditions exist in mined lands and are not unique to the present study. Physical conditions existing at the spoil surface relative to temperature, texture, soil structure, presence or absence of vegetative cover, and chemical conditions such as the low pH, high levels of soluble salts and sulfates, and heavy metals undoubtedly create stress conditions in the physiological behavior and responses of the various plant species.

The diversity and extremes present in the chemical conditions of the spoil result in microsites, or microenvironments. These chemically diverse conditions may exist within a distance of a few inches between the sites or within the rooting zone of most plant species. Table 3 data indicate differences in the chemistry of surface spoil (0.5 in. depth), and approximately a distance of 12 in. between sampling sites. Samples were collected from an exposed site devoid of vegetation and litter and from an adjacent site covered by vegetation and litter. Variations in spoil chemistry exist between the surface spoil, the spoil immediately adjacent to the root at a depth of 6 in. (Table 4), and at greater rooting depths (Table 5). At the acidic pH levels recorded for these spoil materials, nutrient availability is limited; nutrients are also leached to lower levels and lost by erosion. The presence or absence of litter and/or vegetation, the degree of slope and topography, as well as the amount of surface compaction, all affect rainfall retention and loss and thus are important in release, retention, and loss of nutrients. Spoil moisture relations may also be affected by such conditions. Wilde (1954) points out that soil reaction (pH) exerts a definite influence on the life functions of organisms as well as the physical property of soils. He also states that profound adverse effects of H^+ and OH^- ions per se are revealed only at the pH extremes of 3.0 and 9.0.

Within the spoil column, the transport and movement of the various chemical elements may result in the complexing of those elements either similar in structure to the original materials or a quite different compound may result. Undoubtedly the complexing of elements to compounds will be affected differently at various levels in the spoil column by the presence or absence of certain elements, by spoil structure, and spoil moisture. Although the spoils data are reported in available pounds per acre, the analyses were conducted under a given set of laboratory conditions and thus the reported concentrations may not be available under

Table 3. Chemistry of surface spoil: microenvironments

	(Available lbs/acre)								
	pH	NO_3	NH_3	P_2O_5	K_2O	Ca	Mg	T.S. Salts	S as SO_4
A*	4.4	0.70	37.00	1.70	44.00	174.00	60.00	200.00	76.00
B†	5.0	2.10	65.00	4.80	153.00	560.00	114.00	300.00	46.00

	Fe^{+3}	Mn^{+2}	Cu^{+2}	Zn^{+2}	Cl^-	Al^{+3}	B^{+3}	Mo^{+6}pp2B
A	2.90	16.00	4.50	6.50	4.80	13.20	0.34	75.00
B	0.30	15.00	1.60	2.80	8.70	4.90	0.39	25.00

* Barren spoil surface since 1918; sample depth ½ in.; 1973.

† Spoil sample collected from beneath cover of *Lycopodium clavatum*. Distance between sample sites 12 in.

Table 4. Chemistry of surface spoil: shallow root environment

	(Available lbs./acre)								
	pH	NO_3	NH_3	P_2O_5	K_2O	Ca	Mg	T.S.Salts	S as SO_4
A*	3.7	0.70	60.00	1.80	26.00	90.00	42.00	320.00	112.00
B†	3.7	0.50	50.00	3.50	298.00	3920.00	58.00	3500.00	310.00
		Fe^{+3}	Mn^{+2}	Cu^{+2}	Zn^{+2}	Cl^-	Al^{+3}	B^{+3}	Mo^{+6}pp2B
A		7.80	5.70	3.80	4.20	10.90	18.00	0.40	50.00
B		7.50	11.30	7.40	1.90	4.80	14.00	0.33	75.00

* Sample: ½ in. of surface spoil above 6 ft long, large-toothed aspen root and shoots, 1973.

† Sample: spoil material collected adjacent to aspen root at 6 to 8 in. depth.

actual field conditions. Levels of soluble salts, sulfates, and metals have been established as being toxic or limiting for the survival and growth of various species of plants and for normal physiological responses, but the specific level at which the element or substance becomes toxic or limiting may be modified by soil moisture, texture, and species tolerance. Plant physiological responses are also affected by environmental factors such as temperature, light, CO_2, H^+ concentration, and ion interaction. Plant tolerance to the physical, biological, and chemical parameters of the site is also related to the structural characteristics of the species such as shoot-to-root ratio, depth of root systems, and growth rates (Sutcliffe, 1962).

Spoil chemistry may be significantly altered by the configuration of the surface and the topography, since both control and modify the amount and rate of runoff and the rate and amount of infiltration and leaching. A topography which reduces or eliminates runoff and erosion of the surface spoil and prevents the constant exposure of the raw unoxidized spoil enhances the site and the possibility for successful germination of seeds and plant growth. Even very minor surface depressions, such as the furrows resulting from disking, tracks of heavy equipment, or depressions made by hoofed animals, result in conditions that determine or influence microsite environments. Such physical characteristics of the spoil surface result in improved site conditions and less physiological stress upon the plant.

Furrow grading (Riley, 1969, 1973), which consists of grading the spoil into a series of ridges and ravines on the contour, is one method of water management and erosion control which also improves the chemistry of

Table 5. Chemistry of spoil: deep root environment

Site depth (in.)	pH	NO_3	NH_3	P_2O_5	K_2O	Ca	Mg	T.S. Salts	S as SO_4
		(Available lbs/acre)							
A 0- 2	3.7	1.30	40.00	1.20	35.00	90.00	75.00	400.00	250.00
B 8-10	3.5	1.50	43.00	0.50	159.00	3640.00	270.00	3400.00	310.00
C 24-26	3.5	2.80	52.00	2.30	172.00	3780.00	340.00	4400.00	302.00
		Fe^{+3}	Mn^{+2}	Cu^{+2}	Zn^{+2}	Cl^-	Al^{+3}	B^{+3}	Mo^{+6}pp2B
A 0- 2		2.90	5.40	1.00	1.20	8.70	2.50	0.43	125.00
B 8-10		2.50	41.20	3.50	5.50	9.80	2.60	0.42	175.00
C 24-26		0.50	43.40	5.00	4.80	10.90	2.10	0.39	150.00

Note: Spoil: blue shale, mined 1965, barren surface, slope flat to $<$ 1.0%. Sample collected 1972.

the site for the establishment and growth of vegetation (Table 6). Soil analyses data collected over a 10-year period from a plot having a furrow-graded surface reveal an increase in the pH of the ravine spoil over that of the ridge spoil and also a very noticeable decrease in amounts of soluble salts and sulfates. The distance between the ridge and ravine sites at the time of grading was approximately 3 ft. At the time of seeding and planting (March 1963), the distance was reduced to 18 in., while after 10 years, the distance between the ridge and ravine sites was approximately 2 to 6 in. Seedings on the plot included blackwell switch grass, orchard grass, and Japanese lespedeza, while plantings included autumn olive, bicolor-natob lespedeza, tall and medium purple willow, silver buffalo berry, western sandcherry, memorial rose, amd black alder (Appendix D II). High seedling survival and germination rates resulted for the plants on the lower one-half of the slope and in the ravine sites (Figs. 13, 14).

Additional evidence of the effects of a rough surface condition upon rainfall retention, leaching, and site improvement are indicated in Table 7 over a 14-year period. The area was mined in 1935 and remained in conventional spoil banks until July 1958, when 8 to 10 ft of the banks were graded to a plateau type of topography, with a rough spoil surface. The spoil was planted in March 1959 to a mixture of black locust, white oak, and red oak (Fig. 15). On the perimeter of the forest plantation, bicolor lespedeza, multiflora rose, white pine, and Scotch broom were planted. By 1966 the entire spoil surface was vegetated by Canada bluegrass, poverty grass, and broom sedge. Leaf litter measured 1.0 to 1.5 in. in depth. In 1974 the forest floor was covered by either grasses and/or leaf litter and measured 2 to 3 in. deep. Data of Table 7 reveal improvement

Table 6. Chemical changes of microsite spoil

	Yrs. of weathering and leaching							
	4		6		8		10	
	Ridge	Ravine	Ridge	Ravine	Ridge	Ravine	Ridge	Ravine
pH	3.40	3.60	3.50	3.60	3.50	3.80	3.70	3.80
	(Available lbs/acre)							
T.S. Salts	3700.00	500.00	4200.00	720.00	2200.00	560.00	400.00	200.00
S as SO_4	232.00	100.00	310.00	184.00	253.00	122.00	200.00	137.00
Ca	2800.00	142.00	3500.00	198.00	1330.00	142.00	222.00	110.00
Mg	110.00	27.00	114.00	60.00	69.00	50.00	60.00	52.00
NO_3	2.80	0.90	1.00	1.40	1.30	1.90	1.30	1.50
NH_3	38.00	31.00	48.00	40.00	53.00	48.00	46.00	36.00
P_2O_5	0.60	0.45	0.30	0.50	1.70	0.80	0.80	1.20
K_2O	166.00	48.00	330.00	40.00	99.00	31.00	35.00	26.00

Note: Spoil surface graded to 3-ft-deep furrows in winter 1962-63. Sites: series of ridges and ravines.

Fig. 13. Furrow-graded spoil surface (pH from 3.2 to 4.1) seeded to blackwell switchgrass spring in 1963. Growth occurred primarily in ravine sites. (Photo 1963)

Fig. 14. Blackwell switchgrass (see Fig. 13) after six growing seasons. (Photo 1968)

Fig. 15. A one-year-old stand of mixed hardwood species on graded spoil with a rough surface. (Photo 1959)

Table 7. Chemical analyses of surface spoil

	1958	1961	1972
pH	3.70	3.80	4.10
		(Available lbs/acre)	
NO_3	nil	2.10	1.70
NH_3	42.00	26.00	49.00
P_2O_5	0.90	1.70	1.20
K_2O	73.00	22.00	73.00
Ca	326.00	210.00	142.00
Mg	1.20	60.00	118.00
T.S. Salts	2100.00	1100.00	240.00
S as SO_4	128.00	264.00	76.00
Fe^{+3}	1.30	1.30	0.40
Mn^{+2}	4.50	20.80	7.90
Cu^{+2}	22.60	17.80	1.20
Zn^{+2}	15.00	33.40	0.90
Cl^-	9.00	6.60	12.00
Al^{+3}	2.00	4.20	2.40
B^{+3}	0.25	0.34	0.34
Mo^{+6}pp2B	200.00	175.00	50.00

Note: Soil plot located in Stark County, Ohio. Overburden type is acid sandstone, silty shales, and clays associated with Middle Kittanning coal seam. Mining completed 1935. Spoil graded to flat-rough surface July 1958.

of the spoil relative to pH, soluble salts, and sulfates as well as reduced amounts of metals, especially copper and zinc.

In 1972 a young forest ecosystem (Fig. 16) was present on the spoil, which in 1958 contained only scattered herbaceous species and a few wild black cherry seedlings.

From the time of removal and placement of the raw overburden materials (earth strata) into the spoil mass, and upon exposure of the surface materials to the atmosphere, the spoil components are subjected to various physical and chemical processes which create a site possessing specific characteristics. As the surface spoil matures, as the essential factors for living systems develop, as organisms invade and occupy the site, as seedlings are planted or seeds are placed on the site, germination occurs, root systems develop, litter is deposited, and ecological processes begin to operate and result in the establishment of a specific stage of the ecosystem, to be followed by further changes and still another more advanced ecosystem. Such systems continue to develop, to modify and alter the sites, while the various species are replaced by others better adapted to existing conditions of the site or habitat. For occupancy by a

Fig. 16. Mixed hardwood plantation of 1959 (see Fig. 15) as a young forest ecosystem after 13 years of growth and development. (Photo 1972)

species the site must contain a soil medium of adequate pH, levels of soluble salts not in excess of the species tolerance, available nutrients adequate in amount and kind, base exchange capacity, soil moisture, and proper soil structure all being essential to insure germination, survival, growth, and reproduction of the species.

After 57 years of vegetative development on the mined lands, a climax ecosystem is not present, although many of the biotic species indigenous to the region and those recorded in the early surveys of 1831 are present in the natural ecosystem. While the abiotic aspect of the system is quite different from that of the premining period (before 1918), it appears to be quite adequate for the invasion and establishment of many native species as well as many species introduced by man.

Aquatic Ecosystem Development, 1918–46

Upon completion of mining in 1918, a 5.5-acre mine water impoundment formed within the interior of the land unit (see Fig. 1 - Im 5.5). The area of the watershed is approximately 10 acres. The chemistry of the soil material of the watershed in 1918 was probably similar to that indicated by the analyses discussed earlier in the chapter (see Table 1, 1957A). Approximately 300 ft of the coal seam remained exposed at the water level; 1/10 acre of the coal seam remained exposed on the watershed. Waste coal covered approximately 1.5 acres of the lake basin. A clay seam, lying beneath the coal, formed the lake basin.

During the period August 1946 through July 1947 a total of 39 water samples were collected, and tests indicated a pH range of 3.1 to 5.5. A Beckman pH meter was used in the tests. Of the 39 samples tested for pH, 22 samples ranged 3.1 to 4.0, 11 ranged 4.1 to 4.5, 4 ranged 4.6 to 5.0, and 2 ranged 5.1 to 5.5.

Submerged vegetation included needle spike rush, bur-reed, and an aquatic moss (Appendix A VII). Approximately 30% of the lake basin was covered by the two species of submerged aquatics. Emergent aquatics included narrow-leaf cattail, soft rush, sharp fruited rush, arrowhead, fox sedge, and bulrush (Appendix A VI, VII).

Studies of plankton in 1946, in a pond immediately adjacent to the 5.5-acre impoundment and with the pH of the water very similar at the time the sample was collected and organisms were identified, indicated 4 classes of algae including 11 species and 1 class of diatoms present (Appendix E I). Bottom sampling resulted in the collection and identification of 14 species (2 species not defined) of insects including the orders Coleoptera, Hemiptera and Odonata (Appendix E II). Plankton identified in May 1948 from the same impoundment included 16 species of algae, 2 Protozoa, 1 Arthropoda, and 1 Rotifera. Sampling in the larger 5.5-acre impoundment (see Fig. 1- Im 5.5) revealed 5 species of algae (Appendix E IV). Vertebrates inhabiting the lake or utilizing the shore areas included the American toad, green frog, leopard frog, snapping turtle, water snake, and the muskrat. The bird population and other mammal species present in 1946–47 were discussed earlier. Eighteen red-wing blackbird nests were located in the cattails along the shore of the lake in the summer of 1949.

During the period 1918–46 only sparse populations of herbaceous and woody species had developed on the watershed (see Figs. 1, 4, 8). With the limited amounts of vegetation and lack of litter and humus, the surface spoil material was constantly exposed to chemical reactions, especially those oxidative in nature, resulting in the formation of acids, sulfates, salts, etc. Subsequent runoff containing those materials along with solids and other spoil constituents entered the lake after each period of precipitation. With each precipitation-runoff-erosion cycle, the result was newly exposed surface spoil material to be affected by the same processes.

Although only a single parameter (pH) was used to determine the chemistry of the impounded waters presented earlier in the chapter, it does reveal the harshness of this environment for aquatic organisms during the early years of ecological development.

Plantings were commenced on the watershed in 1949 and completed in 1953. By 1956 the chemistry of the lake water was suitable to support limited populations of the bluegill sunfish and the largemouth black bass. Both species spawned sucessfully in 1956, although growth was limited, probably due to the low fertility of the water.

During the period June 1958 to July 1959 bimonthly sampling and analyses of the water revealed a pH range of 4.8 to 8.4, total acidity of 5.0 to 26.0 mg/1, total alkalinity of 0.0 mg/1 to 8.0 mg/1, and total solids of 504.0 mg/1 to 1128.0 mg/1. Extremes in sulfates ranged from 270.0 mg/1 to 625.0 mg/1 (Riley, 1965). Analyses of the lake bottom materials indicated a pH of 5.7, low phosphorus (P_20_5), and high levels of manganese and copper. Calcium and magnesium levels were higher than in the topsoil of adjacent undisturbed soils (see Table 2). The improvement in the chemistry of the water (Table 8) and the ecological changes in the lake both appear to reflect the increasing growth and development of the vegetation, humus, and litter in the spoil surface of the watershed. The role of vegetation and organic matter in decreasing the exposure of the surface spoil material to oxidative reactions, reducing runoff and erosion, retaining precipitation, increasing infiltration and leaching, increasing soil moisture, and modifying spoil temperature improves the spoil environment for microbial and other biological processes to function at increasing rates.

The role of vegetation and organic litter in altering spoil chemistry is revealed by comparing the chemical conditions of barren spoil with those of adjacent spoil covered by vegetation and organic matter (see Table 3). A slightly higher pH and higher levels of nitrates, ammonia nitrogen, and phosphorus were present. The increased nitrogen components reflect increased biological activity, while the higher pH may reflect the effect of moisture retention and leaching. The higher pH level (5.0) probably is also a factor in the release of phosphorus. There were also lower levels of sulfates and heavy metals in the surface one-half inch of spoil beneath the plants and litter.

A similar spoil site study was conducted on a nearby (300-m distance) spoil, mined in 1965, where the spoil surface was covered by 2 inches of black locust litter. The data indicated an almost indentical trend, except in the test for nitrates; the amount of nitrate tested slightly higher for the site devoid of cover. The physical, chemical, and biological processes and changes occurring as the biomass increased on the watershed undoubtedly enhanced the chemistry of the runoff entering the lake.

Table 8. Chemical quality of mine water impounded in 1918

	Sample dates			
	9/11/58	8/30/68	9/28/73	12/7/74
pH*	4.90	6.90	8.60	7.64
Sp. cond. (μmhos)	1080.00	1200.00	1020.00	1120.00
C.O.D. (mg/l)	–	0	47.71	–
T. Sol. "	–	1450.00	1001.00	924.00
T. Ac. "	14.00	0	0	3.40
T. Alk. "	2.00	40.00	80.00	72.00
T. SO_4 "	560.00	740.70	579.60	500.00
Fe^{+3} "	–	17.00	0.75	0.22
T. Fe "	< 1.00	17.00	0.81	0.32
Mn "	–	2.10	0.02	1.26
Al "	–	0.01	0.01	0.22
SiO_2 "	2.80	4.50	6.80	–
Mg "	–	111.80	38.90	88.00
Ca "	–	154.00	127.00	104.00

*pH of water 1946-47; 39 samples (3.1-5.5 range). Lake 5.5 acres; watershed area 5.5 acres.

The runoff from mined lands covered by vegetation and litter contains lower levels of certain harmful or limiting chemical substances and places less stress upon the biotic inhabitants of the aquatic ecosystem. An improved aquatic environment permits increased species diversity as well as increased biotic productivity.

By 1960 very rapid ecological change had occurred, probably the result of an increased nutrient pool, increased species diversity, and larger numbers of organisms. Aquatic vegetation, vascular and nonvascular, became so abundant that chemical control was required to permit use of the lake for recreation. Approximately one-half of the lake basin was occupied by coontail and water milfoil along with lesser amounts of pondweed (Appendix D I). The latter species has now become the dominant vegetation in the impoundment, and chemical control is required annually.

Two additional species of fish were successfully stocked in 1960, the eastern chain pickerel and the redear sunfish. Both species have successfully reproduced.

The chemical and biological conditions of the impoundment during the early years, 1918–27, may have been comparable to the conditions present in an adjacent impoundment formed in 1966. The 1966 impoundment, approximately the same size, is located about 800 yards from the older lake. Mining was completed in December 1965. Overburden materials were similar to those described for the older lake. The watershed spoil was planted to mixed hardwoods in 1966. Survival of seedlings was 80%. About 10% to 15% of the spoil surface is presently covered by vegetation and leaf letter. Sampling and analyses of the water during the period 1968–74 indicated an extremely acid pH (3.25), specific conductance (5600.0 μmhos), total solids (6400.0 mg/1), total acidity (390.0 mg/1), sulfates (4400.0 mg/1), total iron (45.0 mg/1), and manganese (168.0 mg/1), as shown in Table 9. Analyses in 1975 reveal similar conditions in water quality.

A survey in 1969 resulted in 7 species of aquatic insects being collected and identified. Included were the giant water bug (adult), water boatmen (adult), damsel fly (naiad), green darners (naiads), predaceous diving beetle (adult), whirly-gig beetle (adult), and blood worm midge. The water boatmen, diving beetles, and blood worms were listed as common (Wolfe, 1969). Richards (1974) identified 9 species of algae, including 2 classes and 7 genera (Appendix F I). Richards also surveyed the algae of an adjacent mine water impoundment in which the chemical conditions of the water were much less severe: pH (4.0–6.68), specific conductance (600.0 μmhos), total solids (535.0 mg/1), total sulfates (288.0 mg/1), total iron (0.51 mg/1), and manganese (1.5 mg/1). A total of 12 species of algae were collected from this impoundment and identified, including 4 classes and 9 genera (Appendix F I).

In the very acid impoundment Greenshields (1973) studied the diatom populations and identified 1 class, including 3 genera and 3 species. From diatom populations of the second impoundment, a less harsh environment, he identified 4 classes, 5 genera, and 11 species (Appendix F II).

Summary

Ecological processes functioning in the development of natural ecosystems after the disturbances of coal surface mining are dependent upon and influenced by a number of variables. Such variables include type of overburden materials, i. e., acid versus alkaline strata, and amounts of the mineral components. The placement of strata components in the spoil column, topography of the spoiled overburden, type of spoil surface, and the degree of surface spoil compaction are important factors in rainfall retention, infiltration, leaching, and erosion. The latter conditions are important in altering and modifying the chemistry of the surface spoil.

Table 9. Chemical quality of mine water impounded in 1966

	Sample dates			
	5/14/68	5/11/71	9/28/73	12/7/74
pH	3.25	2.98	3.10	3.12
Sp. cond. (μmhos)	5600.00	3700.00	4800.00	4500.00
C.O.D. (mg/l)	27.70	27.10	20.40	–
T. Sol. "	6409.00	5377.00	6977.00	5922.00
T. Ac. "	390.00	265.00	310.00	171.00
T. Alk. "	0	0	0	0
T. SO_4 "	4424.80	3446.00	4330.60	3620.00
Fe^{+3} "	44.00	23.20	31.50	19.90
T. Fe "	45.00	23.40	31.90	21.00
Mn "	168.80	235.00	139.00	133.00
Al "	0.85	3.80	4.18	0.14
SiO_2 "	8.75	12.20	11.50	–
Mg "	619.80	568.10	770.70	670.00
Ca "	550.80	401.60	451.60	260.00

Note: Lake 5.0 acres; watershed area 7.5 acres. Vegetation-trees 80%, planted 1966; mixed hardwoods. Mining completed 1965.

Species diversity of adjacent plant communities and distance of the seed source from the mined land site influence and determine the rate of invasion, succession, ecosystem development, species diversity and numbers, and thus community dynamics. The rate of ecological change and the developmental processes of the ecosystem and the various interim communities that may develop will be greatly modified by human activities, such as the reclamation techniques applied to the mined lands. Reclamation activities, including overburden placement, replacement of topsoil, addition of soil amendments, grading of the land to varying types of topography, spoil surface conditions, and planting and seeding, influence the rate of biotic development and the associated physical, chemical, and biological processes.

This study and the supporting data reveal the biotic components and the ecological changes that have occurred in specific terrestrial and aquatic ecosystems on mined lands, under a given set of abiotic parameters, primarily the chemistry of the spoil or growth medium and the impounded waters occurring within the affected land area.

Data concerning the rooting medium for plants reveal a very acid pH, high levels of soluble salts, high levels of sulfates, and low nutrient pool, the last influenced both by the low pH as well as by the low levels of some major nutrients.

The physical and chemical conditions present resulted in critical sites for the initial invasion and establishment of vegetation. Natural processes, operating over a period of 57 years and during a period of 29 years (1946–75) in which data were collected, have resulted in the development of ecosystems of fairly diverse and numerous biotic components, but of relatively low biomass, yet with small but very significant changes in the chemical conditions of the abiotic component, i.e., spoil or root medium.

The very rapid rate of ecological change following the establishment of vegetation by man, the increased abundance of vegetation, organic matter, and litter on the spoil surface, and the very obvious increase of biomass in the natural ecosystem are evidences of ecosystem development.

The chemistry of the mine water impounded reflects to a degree the conditions of the surface spoil of the watershed resulting from the ecological changes occurring over the 57-year period as well as from the influence of man through revegetation over a period of 29 years.

Improvements in the chemical quality of the water of the aquatic ecosystem and the increased abundance of vegetation undoubtedly reflect the changes or improvements occurring in the adjacent terrestrial environments.

Further evidence of ecological change and improved environmental conditions on the mined landscape and in the aquatic environment is given by the increasing numbers and diversity of species of the faunal community.

Acknowledgments

I wish to express sincere appreciation to Kent State University and the Ohio Mining and Reclamation Association for financial support, without which certain aspects of the long-term study and research could not have been accomplished. Special thanks are due Dr. G. Dennis Cooke, Associate Professor of Biology, Kent State University, for his critical review and comments of the manuscript. Grateful acknowledgment is due Dr. Russell Rhodes, Associate Professor of Biology, Kent State University, and his graduate students who contributed much to the algal and diatom studies. Many other individuals and colleagues have provided assistance in plant taxonomy and other phases of the study.

Literature Cited

Bailey, L. H. 1949. Manual of Cultivated Plants, Rev. ed. Macmillan Co., New York, 1116 pp.

Burt, W. H. 1957. Mammals of the Great Lakes Region. Univ. Michigan Press, Ann Arbor. 246 pp.

Conant, R. 1951. The Reptiles of Ohio. 2d ed. Univ. of Notre Dame Press. 284 pp.

Dunham, J. T., C. Rampacek, and T. A. Henrie. 1974. High sulfur coal for generating electricity. Science 184 (4134): 346–351.

Fernald, M. L. 1950. Gray's Manual of Botany. 8th ed. American Book Co., New York, 1630 pp.

Gordon, R. B. 1969. The Natural Vegetation of Ohio in Pioneer Days. Ohio Biol. Sur. Ser. 3(2). Ohio State University, Columbus. 109 pp.

Greenshields, J. 1973. The algal flora of acid mine water impoundments in Tuscarawas County, Ohio with special reference to the Bacillariophyta. M. S. thesis, Kent State University, Kent, Ohio. 120 pp.

Hammond, A. L. 1974. Academy says energy self-sufficiency unlikely. Science 184(4140): 964.

Jackson, M. L. 1958. Soil Chemical Analyses. Prentice-Hall, Englewood Cliffs, N.J. 498 pp.

Krause, R. R. 1972. Recovery of mined land. Coal Mining and Processing. 9(1): 51–56.

Overton, J. A. 1974. From the earth: abundant energy. Coal Mining and Processing. 11(4):43.

Peterson, R. T. 1947. A Field Guide to the Birds. 25th imp. Houghton Mifflin Co., Boston. 290 pp.

Richards, J. N. 1974. Studies of the phytoplankton and soil algae of two acid strip-mine impoundments in Tuscarawas County, Ohio. M. A. thesis, Kent State University, Kent, Ohio, 56 pp.

Riley, C. V. 1954. The utilization of reclaimed coal striplands for the production of wildlife. Trans. 19th N. Am. Wildl. Conf. 19:324–337.

———. 1960. The ecology of water areas associated with coal strip-mined lands in Ohio. Ohio J. Sci. 60(2):106–121.

———. 1965. Limnology of acid mine water impoundments. Pages 175–187 *in* Symposium on Acid Mine Drainage Research. Sponsored by Coal Industry Advisory Commission to Ohio River Valley Water Sanitation Comm. Mellon Inst., Pittsburgh, Pa.

———. 1969. Chemical alterations of strip-mine spoil by furrow grading–revegetation success. Pages 315–331 *in* Ecology and Reclamation of Devastated Land. Vol 2. Gordon and Breach for Sci. Publ. Ltd., New York and London.

———. 1973. Furrow grading–key to successful reclamation. Pages 159–177 *in* Research and Applied Technology Symposium on Mined-Land Reclamation. Nat. Coal Assoc. and Bituminous Coal Res., Pittsburgh.

Sutcliffe, J. F. 1962. Mineral Salts Absorption in Plants. Vol. 1. Int. Ser. Monogr. Pure Appl. Biol. Vol. 1. Pergamon Press, New York. 194 pp.
Trautman, M. B. 1957. The Fishes of Ohio. Ohio State Univ. Press, Columbus. 683 pp.
Walker, C. F. 1946. The amphibians of Ohio: the frogs and toads. Ohio State Mus. Sci. Bull. 1(3). Ohio State Archiol. and Historical Soc. Columbus. 109 pp.
Wilde, S. A. 1954. Reaction of soils: facts and fallacies. Ecology 35:89–92.
Wolfe, G. R. 1969. A survey of aquatic insects in one acid coal strip-mine pond in Tuscarawas County, Ohio. Research Rep. Kent State Univ., Kent, Ohio. 9 pp.

Appendix A

Plant Species (Fernald, 1950) Recorded on Transects and Adjacent Mined Land in 1946

	Common name	Scientific name
I.	Trees	
	Red maple	*Acer rubrum*
	Wild black cherry	*Prunus serotina*
	White ash	*Fraxinus americana*
	Black gum	*Nyssa sylvatica*
	American elm	*Ulmus americana*
	Black locust	*Robinia pseudo-acacia*
	Slippery elm	*Ulmus fulva*
	Sassafras	*Sassafras albidum*
	Tulip poplar	*Liriodendron tulipifera*
	Red oak	*Quercus rubra*
	Large-toothed aspen	*Populus grandidentata*
	Flowering dogwood	*Cornus florida*
	Domestic apple	*Malus* sp.
	Domestic cherry	*Prunus* sp.
II.	Shrubs and Vines	
	Smooth sumac	*Rhus glabra*
	Hazel nut	*Corylus americana*
	Elderberry	*Sambucus canadensis*
	Blackberry	*Rubus allegheniensis*
	Red raspberry	*Rubus strigosus*
	Black raspberry	*Rubus occidentalis*
	Dewberry	*Rubus flagellaris*
	Poison ivy	*Rhus radicans*
	Wild grape	*Vitis* sp.

	Common name	Scientific name
	Service-berry	*Amelanchier canadensis*
	Virginia creeper	*Parthenocissus quinquefolia*
III.	Forbs	
	Daisy-fleabane	*Euerigeron annuus*
	Wild carrot	*Daucus carota*
	Sour dock	*Rumex acetosella*
	Field-bindweed	*Convolvulus arvensis*
	Common cinquefoil	*Potentila canadensis*
	Yarrow	*Achillea millifoleum*
	Golden rod	*Solidago canadensis*
	Ironweed	*Veronia altissima*
	Whorled loosestrife	*Lysimachia quadrifolia*
	Evening-primrose	*Oenothera biennis*
	Spanish needles	*Bidens bipinnata*
	Dogbane	*Apocynum cannibinum*
	Common milkweed	*Asclepias syriaca*
	Pokeweed	*Phytolacca decandra*
	Joe pye weed	*Eupatorium purpureum*
	Burdock	*Arctium minus*
	Bedstraw	*Galium* sp.
	Smartweed	*Persicaria pensylvanicum*
IV.	Grasses	
	Red top	*Agrostis alba*
	Timothy	*Phleum pratense*
	Canada bluegrass	*Poa compressa*
	Broom-sedge	*Andropogon virginicus*
	Poverty grass	*Danthonia spicata*
	Barnyard-grass	*Echinochloa crusgalli*
	Fox tail	*Setaria viridis*
	Hairy panic grass	*Panicum dichotomiflorum*
V.	Legumes	
	Alsike clover	*Trifolium hybridum*
	White sweet clover	*Melilotus alba*
	Yellow sweet clover	*Melilotus officinalis*
VI.	Emergent Aquatics	
	Fox sedge	*Carex vulpinoidea*
	Sharp fruited rush	*Juncus acuminatus*
	Soft rush	*Juncus effusus*
	Arrowhead	*Sagittaria latifolia*
	Bulrush	*Scirpus cyperinus*

	Narrow-leaf cattail	*Typha angustifolia*
	Common cattail	*Typha latifolia*
VII.	Submerged Aquatics	
	Aquatic moss	*Drepanocladus fluitans* (Dill.) Warnst
	Needle spike rush	*Eleocharis acicularis*
	Bur-reed	*Sparganium* sp.
VIII.	Others	
	Sensitive Fern	*Onoclea sensibilis*
	Common club moss	*Lycopodium clavatum*
	Ground pine	*Lycopodium flabelliforme*
	Scouring rushes	*Equisetum arvense*
	Moss	*Ceratodon purpureum*
	Scarlet-crested cladonia	*Cladonia cristatella Tuck*

Appendix B

Vertebrate Species

I.	Mammals (Burt, 1957)	
	Cottontail rabbit	*Sylvilagus floridanus*
	Whitetail deer	*Odocoileus virginianus*
	Red fox	*Vulpes fulva*
	Grey fox	*Urocyon cinereoargenteus*
	Woodchuck	*Marmota monax*
	Longtail weasel	*Mustela frenata*
	Opossum	*Didelphis marsupialis*
	Fox squirrel	*Sciurus niger*
	Red squirrel	*Tamiasciurus hudsonicus*
	Eastern chipmunk	*Tamias striatus*
	Muskrat	*Ondatra zibethica*
	Beaver	*Castor candensis*
	Raccoon	*Procyon lotor*
	Meadow vole	*Microtus pennsylvanicus*
	Meadow jumping mouse	*Zapus hudsonius*
	White-footed mouse	*Peromyscus leucopus*
	Shorttail shrew	*Blarina brevicauda*
II.	Birds (Peterson, 1947)	
	Horned grebe	*Colymbus auritus*
	Pied-Billed grebe	*Podilymbus p. podiceps*

Great blue heron	*Ardea herodias*
Green heron	*Butorides v. virescens*
American bittern	*Botaurus lentiginosus*
Canada goose	*Branta canadensis*
Mallard	*Anas p. platyrhynchos*
Gadwall	*Anas strepera*
American widgeon	*Mareca americana*
Wood duck	*Aix sponsa*
Canvas-back	*Aythya valisineria*
Lesser-scaup	*Aythya affinis*
Common goldeneye	*Glaucionetta clangula americana*
Buffle-head	*Glaucionetta albeola*
Hooded merganser	*Lophodytes cucullatus*
Common merganser	*Mergus merganser americanus*
Cooper's hawk	*Accipiter cooperii*
Red-tailed hawk	*Buteo jamaicensis*
Osprey	*Pandion haliaetus carolinensis*
American kestrel	*Falco sparverius*
Ruffed grouse	*Bonasa umbellus*
Bob-white	*Colinus virginianus*
Ring-necked pheasant	*Phasianus colchicus torquatus*
Coot	*Fulica americana*
Killdeer	*Charodrius v. vociferus*
Woodcock	*Philohela minor*
Common snipe	*Capella gallinago delicata*
Spotted sandpiper	*Actitus macularia*
Lesser yellow-legs	*Totanus flavipes*
Mourning dove	*Zenaidura macroura*
Yellow-billed cuckoo	*Coccyzus a. americanus*
Nighthawk	*Chordeiles minor*
Chimney swift	*Chaetura pelagica*
Belted kingfisher	*Megaceryle a. alcyon*
Common flicker	*Coloptes auratus*
Pileated woodpecker	*Hylatomus pileatus*
Red-headed woodpecker	*Melanerpes e. erythrocephalus*
Hairy woodpecker	*Dendrocopus villosus*
Downy woodpecker	*Dendrocopus pubescens*
Eastern kingbird	*Tyrannus tryannus*
Crested flycatcher	*Myiarchus crinitus*
Eastern phoebe	*Sayornis phoebe*
Tree swallow	*Iridoprocne bicolor*

Blue jay	*Cyanocitta cristata*
Crow	*Corvus brachyrhynchos*
Black-capped chickadee	*Parus atricapillus*
Tufted titmouse	*Parus bicolor*
White-breasted nuthatch	*Sitta carolinensis*
Brown creeper	*Certhia familiaris*
House wren	*Troglodytes aëdon*
Grey catbird	*Dumetello carolinensis*
Eastern brown thrasher	*Toxostoma r. rufum*
Robin	*Turdus migratorius*
Eastern bluebird	*Sialia sialis*
Eastern golden-crowned kinglet	*Regulus s. satrapa*
Starling	*Sturnus v. vulgaris*
Red-eyed vireo	*Vireo olivaceus*
Yellow-breasted chat	*Icteria v. virens*
Red-wing	*Agelaius phoeniceus*
Common grackle	*Quiscalus versicolor*
Cardinal	*Richmondena cardinalis*
Indigo bunting	*Passerina cyanea*
Eastern goldfinch	*Spinus t. tristis*
Towhee (chewink)	*Pipilo erythrophthalmus*
Vesper sparrow	*Poaecetes g. gramineus*
Dark-eyed junco	*Junco hyemalis*
Chipping sparrow	*Spizella p. passerina*
Song sparrow	*Melospiza melodia*
Field sparrow	*Spizella p. pusilla*

III. Reptiles and Amphibians (Conant, 1951; Walker, 1946)

Water snake	*Natrix s. sipedon*
Black snake	*Coluber c. constrictor*
Snapping turtle	*Chelydra serpentina*
Painted turtle	*Chrysemys belliimarginata*
American toad	*Bufo a. americanus*
Green frog	*Rana clamitans*
Pickerel frog	*Rana palustris*
Spring peeper	*Hyla c. crucifer*
Tree toad	*Hyla v. versicolor*

IV. Fish (Trautman, 1957)

Largemouth blackbass	*Micropterus s. salmoides*
Bluegill sunfish	*Lepomis m. machrochirus*
Redear sunfish	*Lepomis microlophus*
Chain pickerel	*Esox niger*

V. Spring Bird Census 1947 in Black Locust Ecosystem

Red-eyed vireo	*Vireo olivaceus*
Indigo bunting	*Passerina cyanea*
Grey catbird	*Dumetello carolinensis*
Towhee (chewink)	*Pipilo erythrophthalmus*
Eastern brown thrasher	*Toxostoma r. rufum*
Yellow-billed cuckoo	*Coccyzus a. americanus*
American bittern	*Botaurus lentiginosus*
Killdeer	*Charodrius v. vociferus*
Red-headed woodpecker	*Melanerpes e. erythrocephalus*
Song sparrow	*Melospiza melodia*
Vesper sparrow	*Poaecetes g. gramineus*
Field sparrow	*Spizella p. pusilla*
Robin	*Turdus migratorius*

Appendix C

Species Planted on Study Area, 1947-75

I. Trees (Fernald, 1950)

Black alder	*Alnus glutinosa*
Red oak	*Quercus rubra*
White ash	*Fraxinus americana*
Black locust	*Robinia pseudo-acacia*
White oak	*Quercus alba*
Tulip poplar	*Liriodendron tulipitera*
White pine	*Pinus strobus*
Red pine	*Pinus resinosa*
Scotch pine	*Pinus sylvestris*
Norway spruce	*Picea abies*
Bald cypress	*Taxodium distichum*

II. Shrubs (Bailey, 1949)

Multiflora rose	*Rosa multiflora*
Autumn olive	*Elaeagnus umbellata*
Tartarian honeysuckle	*Lonicera tartarica*
Bicolor lespedeza	*Lespedeza bicolor*
Japanese lespedeza	*Lespedeza japonica*
Thunbergii lespedeza	*Lespedeza thunbergii*
False indigo	*Amorpha fruticosa*
Scotch broom	*Cytisus scoparius*
Coral-berry	*Symphoricarpos orbiculatus*

Appendix D

Other Plant Species

I. Volunteer Species (Fernald, 1950) Recorded on Study Area since 1946 Survey

American beech	*Fagus grandifolia*
Climbing bittersweet	*Celastrus scandens*
Pondweed	*Potamageton foliosus*
Coontail	*Ceratophyllum demersum*
Water-milfoil	*Myriophyllum* sp.

II. Species (Bailey, 1949) Planted on Adjacent Study Areas

Tall purple willow	*Salix purpurea gracilis*
Medium purple willow	*Salix purpurea lambertiana*
Western sand cherry	*Prunus besseyi*
Silver buffalo berry	*Shepherdia argentea*
Memorial rose	*Rosa wichuraiana*
Blackwell switchgrass	*Panicum virgatum*
Orchard grass	*Dactylis glomerata*
Natob lespedeza	*Lespedeza bicolor natob*

III. Witness Trees, Land Survey, 1831

Sugar maple	*Acer saccharum*
Black oak	*Quercus velutina*
Hickory	*Carya* sp.

Appendix E

Algae and Invertebrates in Mine Water Impoundments of the Study Area, 1946 and 1948 (Riley, 1960)

I. Plankton, 1946
- Algae
  - Chlorophyceae
    - *Chlamydomonas* sp.
    - *Closterium* sp.
    - *Cosmarium* sp.
    - *Oedogonium* sp.
    - *Pleurotaenium* sp.
    - *Spirogyra* sp.
  - Englenophyceae
    - *Euglena* sp.
    - *Phacus* sp.
    - *Peranema* sp.
  - Dinophyceae
    - *Peridinum* sp.
  - Myxophyceae
    - *Gomphosphaeria* sp.
  - Bacillariophyceae
    - Diatoms

II. Arthropods, 1946

- Insecta
  - Coleoptera
    - *Berosus striatus*
    - *Enemidotus edentulus*
    - *Enemidotus muticus*
    - *Coptotomus interrogatus*
    - *Dineutes americanus*
    - *Dytiscus fasciuentris*
    - *Tropisternus* sp.
  - Hemiptera
    - *Aretocorixa parshley*
    - *Belostoma fluminea*
    - *Pelacoris femoratus*
    - *Sigara atopodonta*
  - Odonata
    - *Aeshna sitchensis*
    - *Aeshna* sp.
    - *Arglatibialis* sp.

III. Plankton, 1948

- Algae
  - Chlorophyceae
    - *Ankistrodesmus* sp.
    - *Chaetosphaeridium* sp.
    - *Closterium* sp.
    - *Cosmarium* sp.
    - *Gonatozygon* sp.
    - *Gonium* sp.
    - *Microthamnion* sp.
    - *Mougeotia* sp.
    - *Oedogonium* sp.
    - *Pleurotaenium* sp.
    - *Spirogyra* sp.
  - Bacillariophyceae
    - *Diatoma* sp.
    - *Gomphonema* sp.
    - *Navicula* sp.
    - *Tabellaria* sp.
  - Dinophyceae
    - *Peridinium* sp.

IV. Others

- Protozoa
  - *Difflugia* sp.
  - *Dinoflagellate* sp.
- Rotifera
  - *Cathypna* sp.
- Arthropoda
  - *Cyclops* sp.

Appendix F

Algae of Mine Water Impoundments Formed in 1966 Adjacent to the Study Area

I. Algae

- Extremely Acid Pond (Richards, 1973)
  - *Chlorella saccharophila*
  - *Chlorella* sp. (strain 1-a)
  - *Ochromonas* sp.
  - *Chlorella vulgaris* var. *autrophica*

Chlamydomonas acidophila
Ulothrix variabilis
Lobosphaera tirolensis
Chlorococcum sp.
Stichococcus bacillaris

Moderately Acid Pond (Richards, 1973)

Chlorella protothecoides
Ochromonas sp.
Peridinium sp.
Chlorella saccharophila
Ulothrix variabilis
Ulothrix subtilissima
Microthamnion strictissimum
Stichococcus bacillaris
Chlorella saccharophila var. *ellipsoidea*
Radiosphaera minuta
Neospongiococcum perforatum
Botryokoryne simplex

II. Diatoms

Extremely Acid Pond (Greenshields, 1973)

Eunotia exigua
Navicula sp.
Nitzschia confinis

Moderately Acid Pond (Greenshields, 1973)

Eunotia exigua
Navicula gregaria
Navicula mutica f. intermedia
Navicula sp.
Nitzschia confinis
Nitzschia parvuloides
Pinnularia biceps f. petersonii
Pinnularia microstauron
Pinnularia obscura
Pinnularia viridis
Surirella ovata

Nutrient Accumulation by Vegetation during the First Year of Recovery of a Tropical Forest Ecosystem

P. A. Harcombe

Abstract

Regrowth of successional vegetation retards nutrient loss after clear-felling of a tropical forest. However, this retardation of nutrient loss is not strongly related to vegetation species composition, to biomass, or to vegetation nutrient content. This suggests either that the nutrient capital in the soil in this particular ecosystem is high enough to compensate for high losses after disturbances or that the ecosystem is acquiring sufficient nutrients from external sources.

Introduction

In mature terrestrial ecosystems, the nutrient capital is probably relatively constant. The biota regulates nutrient loss by controlling the "total volume of water passing out of the system, and the pattern of discharge rates, and the erodibility of the system" (Bormann et al., 1969). Such control is generally thought to be great, especially in the tropics: "in the rainforest, the vegetation itself sets up processes tending to counteract soil impoverishment, and under undisturbed conditions there is a closed cycle of plant nutrients" (Richards, 1952).

When vegetation is damaged or destroyed, loss of the accumulated nutrients occurs through decomposition of organic matter and leaching of inorganic ions. Likens et al. (1970) have shown that complete destruction of a temperate forest and suppression of higher plant regeneration result in rapid flushing of nutrients from the ecosystem. This nutrient loss may be followed by destruction of soil nutrient-holding capacity (Popenoe, 1960; Nye and Greenland, 1964). Furthermore, in some tropical areas this flushing may be of great consequence to ecosystem recovery because of reputedly small reserves of nutrients in rock minerals (Richards, 1952; Van Baren, 1961; Stark, 1971).

Retention of site nutrients by incorporation into rapidly growing vegetation after disturbance could be important in reestablishment of the mature ecosystem (Bormann et al., 1969). Marks (1974) suggests that the degree of nutrient retention is determined by the "rapidity of successional regeneration," to which the rates of both site colonization and vegetation growth contribute. Bartholomew et al. (1953) stress "prevention of nutrient losses by virtue of plant immobilization and storage of nutrients in plant tissue." Alternatively, Pomeroy (1970) states that "succession is seen as a process through which populations accumulate enough nutrients to make possible the rise of succeeding populations," rather than as a process involving nutrient retention.

Thus, at least two factors may be involved in ecosystem recovery—retention of available nutrients and accumulation of nutrients. Retention, as used here, refers to holding the nutrients that are present at the time of disturbance. It is prevention of nutrient loss. Accumulation involves both capture of nutrient inputs and utilization of nutrients made available through weathering.

The effect of successional vegetation on the stock of nutrients has never been studied by direct plot experimentation. This study represents a preliminary analysis of nutrient loss after disturbance and of nutrient retention or accumulation by successional vegetation in a humid tropical forest region.

Disturbance and Recovery

Loss of Nutrients after Disturbance

The loss of nutrients following forest removal is a result of increased leaching and erosion, plus rapid deterioration of soil nutrient-holding capacity (Nye and Greenland, 1960). Soil nutrient-holding capacity depends primarily on cation exchange capacity (Nye and Greenland, 1960; Buckman and Brady, 1969), which may be supplied in large measure by soil organic matter. Thus, since forest removal results in loss of soil organic matter due to cessation of input and to continued or accelerated decomposition, cation exchange capacity and nutrient-holding capacity should also decrease. However, the presumed relationship between loss of nutrient-holding capacity and loss of nutrients has not been tested directly in the field.

Likewise, there is little information on the hydrologic and erosional aspects of nutrient loss after destruction of tropical forests. Sioli et al.

(1969) state that volcanic soils of El Salvador are highly erodible and have been completely destroyed in some areas of intensive use. However, latosols usually have good physical structure and low erodibility (Kellogg, 1963); so removal of particulate matter should be low. In any case, leaching of dissolved substances should be high, since removal of the forest cover would decrease evapotranspiration, thus increasing water percolation and runoff.

Though mechanisms of loss are only poorly understood, it is clear that significant nutrient loss does occur. Reported values for the magnitude of this loss (Table 1) vary greatly, due to differences in such factors as soils, climates, land use history, and experimental technique.

Nutrient Retention by Successional Vegetation

Successional vegetation may retain nutrients in the system by (1) rapid incorporation of dissolved nutrients into biomass, (2) regeneration of soil nutrient-holding capacity by adding organic matter to the soil, (3) diminution of leaching by increasing evapotranspiration, and (4) protection of soil structure from breakdown by heavy rains. Each of these processes has

Table 1. Soil nutrient loss following forest removal (percentage of decrease per year, averages of 1-3 years)

Place	Years of observation	C	N	K	Ca	Mg	Reference
Trinidad	1	9					Duthie et al., (1937)
Guatemala	1	37	27	—*	6	9	Popenoe (1960)
Guatemala	3	7	6	19	15	4	Cowgill (1961)
Ghana	3	13	12				Cunningham (1963)
Ghana	2	25	27		33	27	Nye & Greenland (1960)
Solomon Islands	1	54	46	52	73	78	Van Baren (1961)
Costa Rica	1	4	9	3	39	44	This study (bare plots, unfertilized)

* Net increase due to nutrient input from burned vegetation.

been shown to contribute to nutrient retention under different circumstances.

Storage of nutrients in biomass has been emphasized as an important process in the humid tropics where soils are highly weathered, the nutrient supply is mobile, plant growth is rapid, and organic matter decomposition rate is high (Richards; 1952; Bartholomew et al., 1953; Nye and Greenland, 1960). Nutrient storage in vegetation is clearly a function of amount of biomass and of concentration of nutrients in living tissue. The high rate of biomass production by successional vegetation is well documented (Snedaker, 1970; Ewel, 1971*a*, 1971*b*; Tergas and Popenoe, 1971); the concentration of nutrients in successional species appears to be no higher than in other species (Nye, 1958), despite some suggestions to the contrary (Budowski, 1961; Stark, 1970). Species composition and growth form undoubtedly also influence rate of nutrient uptake (Bartholomew et al., 1953).

Increases in soil aggregation and nutrient-holding capacity through addition of organic matter to soil do contribute to regeneration of soil fertility (Greenland and Nye, 1959). However, during the first years of natural succession, vegetation uptake of nutrients is more likely to deplete the soil of available nutrients rather than to increase the nutrient supply through soil improvement (Bartholomew et al., 1953; Popenoe, 1960; Nye and Greenland, 1964).

High evapotranspiration is reported to have resulted in a two-thirds reduction of soil calcium loss, despite soil loss due to plant uptake (Nye and Greenland, 1964).

Physical protection of the site from high insolation and high raindrop impaction has been studied by Cunningham (1963). He covered bare soil with perforated tin roofing which permitted passage of water but provided shade and broke the impact of raindrops. This decreased average loss of C and N by 50%, and the author indicated that cation losses were probably similarly reduced.

The conspicuous position of fast-growing trees in succession should be mentioned in the context of nutrient retention. Marks and Bormann (1972) postulate that in northern hardwood forests of the temperate zone, the wide dispersal, rapid germination, and rapid growth of a successional tree, pin cherry (*Prunus pensylvanica*), make it especially important in conservation of nutrients after forest removal. Since the tropical successional trees exhibit all these characteristics, it is reasonable to suppose that they, too, could be important in minimizing nutrient losses from a disturbed ecosystem (Harcombe, 1973; Farnworth and Golley, 1974). Budowski (1961) implied that *Cecropia obtusifolia* conserved nutrients by storage in nutrient-rich stem wood. Van Steenis (1958) presented the idea most dramatically: "the nomads [successional trees] fulfill a function in the rainforest similar to the role of lymph coagulating to form a scab for the skin which falls off after having completed its function of closing a wound." It was this possibility that led to the inclusion of tree plantations in my experiments.

The Experiment

Experimental Design

A factorial experiment was set up to test the effects of vegetation cover types and of fertilization on site nutrient loss after clear-felling. Three covers were used:

1. Bare (B)–soil was maintained free of vegetation by hand removal of seedlings and sprouts.
2. Natural vegetation (NV)–succession was allowed to proceed naturally except that stump and root sprouts were continuously cut back.
3. Tree plantations (TP)–dense plantations of a single tree species, *Cecropia obtusifolia,* were established (other sprouts and seedlings were continuously cut back).

Two fertility levels were used:

1. Unfertilized (0)–natural soil fertility.
2. Fertilized (F)–commercial fertilizer added.

A randomized block design with four blocks was used. The natural vegetation treatment was replicated within blocks, but the others were not.

It was assumed that two processes, adsorption of inorganic nutrients in the soil and uptake of nutrients by plants, were the primary determinants of ecosystem nutrient content. Thus measurements of soil and plant C, N, K, Ca, and Mg were made at 0, 4, 8, and 12 months, and losses of nutrients over time were calculated by subtraction. Nutrients in litter were arbitrarily considered to be a part of the plant contents.

It is assumed that the fractions which were measured constitute the nutrients participating in plant uptake, recycling, and storage. This ignores internal system changes whereby nutrients become available for uptake through soil organic matter decomposition, root decomposition, and mineral weathering.

After the successional experiment was begun, a reference study was set up to determine seasonal patterns of soil nutrients in the forest, since it was thought that this might be helpful in interpreting trends in the experimental plots.

Four 10 by 10 meter plots were staked out in areas of the remaining forest which were most similar in appearance to the experimental area. Soil samples were taken from each plot at 0, 4, 8, and 12 months, by the same methods used on the experimental plots (described below).

The Study Area

The experiment was performed at the Instituto Interamericano de Ciencias Agrícolas, Centro Tropical de Enseñanza y Investigación, Turrialba, Costa Rica (83°40′ W, 9°53′ N).

The Turrialba region is a short and steep valley at the eastern end of the central plateau of Costa Rica, on the Atlantic side of the continental divide, in the drainage basin of the Reventazon River (Morrison and Leon, 1951). The experimental site is approximately 100 m above the valley floor on a gently rolling shoulder of the ridge which forms the southwest boundary of the Turrialba Valley. The site faces northeast at an altitude of 650 m.

The ridge is a late Pliocene lava flow originating in Irazu and Turrialba volcanoes (Hardy, 1961). The parent rock is an augite-hypersthene andesite, which corresponds to the lower part of the aguacate volcanic rock series (Dengo, 1962). Volcanic ash showers in historic times may have contributed some ash to soils forming on the older lava flows (Hardy, 1961).

The soil of the experimental site is an inceptisol (Aguirre, 1971) of the Colorado sandy clay series (Dondoli and Torres, 1954). High allophane content (Besoain, 1970) and high, stable organic carbon content (Harcombe, 1973) suggest andic tendencies.

The coarse soil fraction is primarily iron and aluminum oxides (Hardy, 1961); the clay fraction is mostly kaolinite and gibbsite (Bornemisza et al., 1967).

Total nutrient element content of the soil is low, but exchangeable cations are present in moderate quantities (Table 2; Hardy, 1961; Martini, 1969*a*). Organic matter, cation exchange capacity, and nitrogen are high and stable (Harcombe, 1973), presumably due to organic matter–allophane complexation (see Martini, 1969*b*).

The soil is deep and free of rocks or concretions. Soil drainage is rapid, and water retention moderate, despite the high clay content (Gavande, 1968).

Table 2. Chemical and physical soil characteristics at Florencia Norte experimental site, taken at time of initiation of experiment, after forest clearing and site preparation. Data from this study (average of 32 plots), except as noted

	0-15 cm	15-30 cm
Carbon (%)	7.18	3.59
Total nitrogen (%)	0.605	0.300
Available cations (μg/g)		
Potassium	209	54
Calcium	586	152
Magnesium	120	29
Total cations (μg/g)*		
Potassium	1069	784
Calcium	1147	439
Magnesium	701	697
Cation exchange capacity (me/100g)	30.14	27.01
pH	4.5	4.6
Bulk density (g/cm^3)	0.682	0.746

	0-20 cm
Texture (%) †	
Sand	7.5
Silt	5.5
Clay	87.0
Water-holding capacity (% moisture) ‡	
0.1 bars	58.33
0.5 bars	53.09
15.0 bars	42.85

* This study, averages of 5 plots.
† Hardy (1961).
‡ Gavande (1968).

The climate of the Turrialba region is warm, wet, and humid. Mean monthly temperature is 22.3°C, annual precipitation is 2683 mm, average humidity is 87.7%, evaporation is 1393 mm per year, and the average daily sunshine is 4.52 hours (Aguirre, 1971). The least rainy season (Jan.-Mar.) is characterized by an excess of 9.6 mm/mo of evaporation over precipitation, and for the rest of the year precipitation exceeds evaporation by 146.5 mm/mo (Aguirre, 1971).

The Turrialba region is within the tropical rainforest zone of Richards (1952) and Wagner (1964). In the Holdridge Life Zone system, it is transitional between Tropical Moist Forest and Premontane Wet Forest (Holdridge et al., 1971). Forest types in these zones have been described by Sawyer and Lindsey (1971) and by Budowski (1961).

The Bosque de Florencia, where the experimental site is located, is a 60-to-70-year-old secondary forest (30-40 m tall) which probably regenerated after use as a pasture (Holdridge, pers. comm.) or as a coffee plantation (Budowski, 1961).

Methods

An area of homogeneous forest about 2 hectares in extent was cleared in February-April 1970. Heavy timber was skidded from the site, and the slash was piled in windrows around the edge of the site. Then the site was divided into 4 blocks, each containing eight 20 x 20 meter squares. Centered in each square was a 10 x 10 meter plot, from which all litter was removed. In all cases sampling of vegetation and soil was done within the 10 x 10 meter plots. In each block, half the plots were fertilized (7.0 g N, 8.7 g P, 16.6 g K/m^2 in commercial inorganic fertilizer) in May 1970, and half were not. For each fertility level in each block, one plot was maintained bare, two plots were allowed to regenerate naturally from seed (sprouts removed), and one plot was planted to *Cecropia obtusifolia*. Sprouts and seedlings were removed from bare and plantation plots at 15-day intervals.

Twenty soil cores (30 cm deep by 2.5 cm diameter) were taken randomly from each plot in May 1970. They were divided into 0-15 cm and 15-30 cm layers, then composited for each layer. These soil samples were analyzed for total N (Kjeldahl method) and exchangeable (ammonium acetate extractable) K, Ca, and Mg. Laboratory analysis was performed using semimicro versions (Harcombe, 1973; Müller, 1961) of standard techniques (Black et al., 1965).

The vegetation on the experimental plots was sampled at 4, 8, and 12 months. Species density (number of stems/m^2) and cover (m^2 ground surface covered by canopy projection/100 m^2 ground surface) were recorded from ten quadrats (1 m^2) on each plot. Four of the quadrats on each plot were selected randomly for harvesting. Plant material in the vertical projection of the area was collected, sorted by species or life form and by tissue (woody vs. herbaceous), and analyzed in the laboratory for dry weight and nutrient content (N, K, Ca, and Mg). Root biomass was estimated from root-to-shoot ratios of six plants excavated outside the plots for 4- and 8-month samples and was determined directly by excavation at 12 months. Tree dry weight was estimated by dimension analysis

(Whittaker and Woodwell, 1968) on 12-month natural vegetation plots and on tree plantation plots of all ages (4, 8, and 12 months).

At each sampling time the number, kind, and weight of seedlings and sprouts which had germinated on bare plots in the 14-day interval between cleanings were recorded.

Species were grouped by life forms for analysis of succession data because of the large number of species and low number of individuals per species. Observed habit on the experimental plots was the basis for assigning the species into the following groups: ephemeral herbs, ephemeral shrubs, persistent herbs, tall shrubs, grasses, vines, trees, and scitaminae. Ephemerals were herbs with short life cycles (1-4 months) or low shrubs maturing within 6 months. Persistent herbs grew to more than one meter tall, generally did not reproduce before 6 months, and lived up to 18 months. Tall shrubs grew to more than one meter tall and began to reproduce at 8-12 months. Scitaminae were bananas and related families. Details of the successional sequence are reported elsewhere (Harcombe, 1973).

Results and Discussion

Nutrient Loss on Bare Plots

The bare plots represent the extreme of disturbance; so they should provide an indication of the maximum rate of nutrient loss after disturbance. The loss, expressed as the difference between soil nutrient content at 0 months and at 12 months, was statistically significant for N, Ca, and Mg. Relative to initial nutrient content, loss was roughly 4% C, 10% N, 11% K, 44% Ca, and 39% Mg (Table 3). Fertilization appears to have enhanced element losses slightly, but the differences were not significant. Cation exchange capacity and pH did not change significantly.

In comparison with data from the literature (see Table 1), K loss was low, losses of Ca and Mg were somewhat high, and losses of C and N were slightly low. Low loss of K on bare plots, despite its high solubility and presumed mobility in soils (Fried and Broeshart, 1967; Buckman and Brady, 1969), may have been related to low base saturation or to the existence of part of the exchangeable K in a form resistant to rainwater leaching.

The high loss of Ca and Mg from soil may have been due to formation of mobile complexes with organic acids (Martini, 1969*b*), since these elements have a fairly low ionic mobility in the soil solution (Fried and Broeshart,

Table 3. System nutrient contents: amount present initially, amount of change in 12 months (mean ± SE). Plus denotes increase, minus indicates loss. Italicized values are significantly different from zero (two-tailed t-test, $p < 0.05$); i.e., change was significant. Asterisks indicate significant difference from corresponding bare plot average (two-tailed t-test, $p < 0.05$); i.e., retention by vegetation was significant. *O*=unfertilized; *F*=fertilized.

	Carbon			Nitrogen			Potassium		
	initial g/m²	change g/m²	change %	initial g/m²	change g/m²	change %	initial g/m²	change g/m²	change %
Bare									
O	11170 ± 504	*-387 ± 112*	*-3.4 ± 1.0*	917 ± 23	*81 ± 12*	*-8.8 ± 1.1*	21 ± 1	-0.7 ± 0.5	-3.1 ± 2.3
F	11560 ± 690	-576 ± 289	-5.0 ± 2.4	977 ± 44	*110 ± 13*	*-11.4 ± 1.6*	25 ± 3	-5.6 ± 2.9	-19.1 ± 9.0
Avg	11370 ± 402	-481 ± 209	*-4.2 ± 0.9*	946 ± 28	*96 ± 10*	*-10.1 ± 1.0*	23 ± 2	-3.1 ± 1.6	-11.1 ± 5.2
Natural vegetation									
O	11010 ± 342	+*786 ± 211*	+*6.9 ± 1.7**	972 ± 25	-7 ± 22	-0.9 ± 2.2*	28 ± 3	-0.7 ± 2.9	+6.2 ± 13.4
F	11610 ± 366	+*1057 ± 149*	+*8.9 ± 1.1**	997 ± 29	+16 ± 15	+1.4 ± 1.4*	34 ± 3	-2.4 ± 1.9	-7.7 ± 6.4
Avg	10110 ± 928	+*921 ± 129*	+*7.9 ± 1.0**	984 ± 19	+14 ± 13	+0.3 ± 1.3*	31 ± 2	-1.5 ± 1.7	-0.7 ± 7.4
Tree plantations									
O	10850 ± 656	1731 ± 696	+15.9 ± 6.1*	923 ± 28	+33 ± 31	+3.5 ± 3.2*	23 ± 2*	-0.7 ± 2.2	-0.1 ± 9.5
F	12080 ± 966	*1091 ± 180*	+*9.5 ± 2.3**	1010 ± 62	+6 ± 21	+0.9 ± 2.1*	27 ± 2	-2.0 ± 0.8	-7.2 ± 2.7
Avg	11470 ± 588	*1411 ± 354*	+*12.7 ± 3.2**	966 ± 35	+19 ± 18	+2.2 ± 1.8*	25 ± 2	-1.3 ± 1.1	-3.7 ± 4.8

Table 3 (*cont.*)

	Calcium			Magnesium		
	initial g/m^2	change g/m^2	change %	initial g/m^2	change g/m^2	change %
Bare						
O	50 ± 11	*−18 ± 5*	*−39 ± 9*	11.6 ± 3.2	*−5.7 ± 2.1*	*−44.3 ± 9.0*
F	57 ± 10	*−27 ± 4*	*−48 ± 5*	8.4 ± 0.7	*−3.0 ± 0.9*	*−34.0 ± 8.6*
Avg	53 ± 7	*−23 ± 4*	*−44 ± 5*	10.0 ± 1.6	*−4.4 ± 1.2*	*−39.1 ± 6.1*
Natural vegetation						
O	88 ± 17	*−20 ± 7*	*−20 ± 6*	19.8 ± 7.0	−2.3 ± 2.3	−12.3 ± 12.3
F	92 ± 11	*−21 ± 7*	*−20 ± 10*	18.2 ± 2.9	−0.7 ± 1.6	−1.6 ± 6.2*
Avg	90 ± 10	*−20 ± 5*	*−20 ± 6**	19.0 ± 2.5	−1.5 ± 1.3	−7.0 ± 6.8*
Tree plantation						
O	76 ± 22	8 ± 4	−8.2 ± 5.3*	13.2 ± 2.1	+ 1.1 ± 0.8	+ 10.4 ± 5.7*
F	75 ± 18	−22 ± 14	−23.6 ± 9.4	15.5 ± 5.2	−1.8 ± 3.7	−4.8 ± 19.7
Avg	75 ± 13	−15 ± 8	*−15.9 ± 5.8**	14.4 ± 2.6	−0.3 ± 1.8	−7.6 ± 9.5*

1967; Buckman and Brady, 1969). Most of the short-chain humic acids are highly water soluble (Kononova, 1961), and the complexing capacity of Ca and Mg is high, whereas that of K is low (Basolo and Pearson, 1958).

Formation of stable complexes of clay and organic matter, a process common to volcanic ash soils (Martini, 1969*b*), may account for the slow breakdown of organic matter observed here. Presence of such complexes may also be responsible for the high, stable cation exchange capacity and the low rate of N loss.

Soil nutrients did not decrease in forest reference plots; so losses on bare plots are probably attributable to forest removal rather than to any climatic fluctuation.

The high cation losses from bare plots indicate greater instability of available nutrients in tropical soils than anticipated. Even though soil nutrient-holding capacity (cation exchange capacity) was high and stable, large cation losses occurred simply as a result of leaching. This happened without significant change in soil pH, probably because of low base saturation (sum of K, Ca, and Mg is roughly 3.2 me/100 g, about 10% saturation).

Disappearance of nutrients from the upper 30 cm of soil was significant, but the ultimate fate of these nutrients is not known. There is some indication (Fig. 1) of slightly increased concentration of at least Ca at a depth of 90 and 100 cm. If cations or cation complexes accumulate at this depth, the losses recorded here may be temporary losses to the ecosystem, which will be recovered as roots penetrate to these depths. The possible nutrient accumulation in the subsoil needs further study.

Net loss of nutrients from vegetated plots, computed by subtracting nutrient content in vegetation-plus-soil at 12 months from nutrient content in soil at 0 months, was significant only for Ca on only one set of treatment plots (see Table 3). This was loss from the ecosystem, not loss from the soil alone. There was an increase in C and N, probably due to photosynthesis for C and to microbial fixation for N. Likewise, loss percentages were significantly less on vegetated plots than on bare plots for C, N, Ca, and Mg (see Table 3). Apparently, vegetation presence resulted in retention of site nutrients.

Nutrient Retention on Vegetated Plots

Nutrient retention on vegetated plots, defined as the difference between actual nutrient content and "expected" nutrient content, was significant for C, N, Ca, and Mg (Table 4; Figs. 2–5). The expected nutrient content

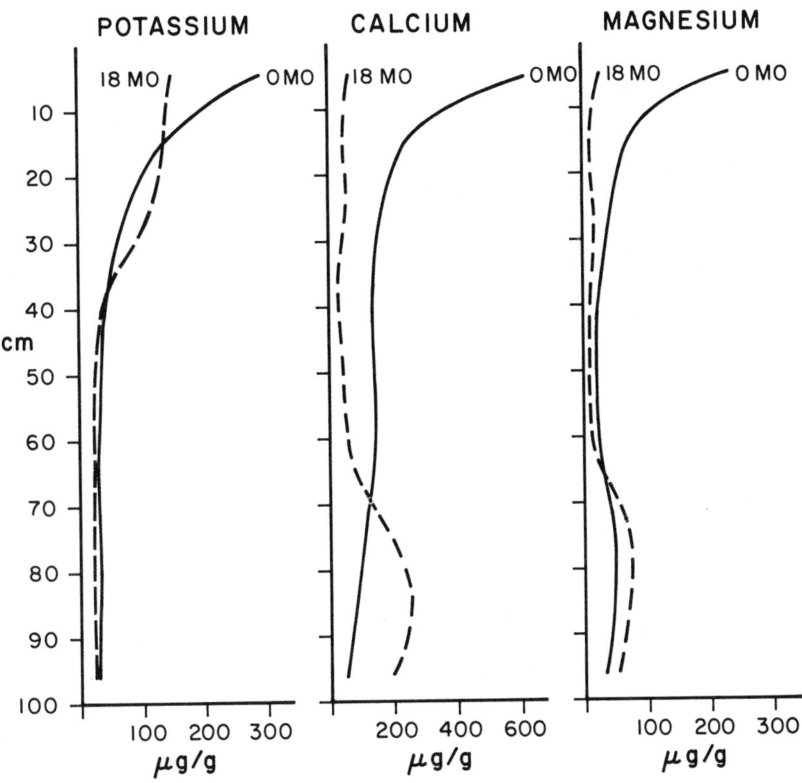

Fig. 1. Variation in available soil nutrients with depth at the beginning of the experiment and after 18 months on a bare plot (no. 33, unfertilized). Curves are based on a single sample from each 10-cm depth interval.

is the amount that would have been measured had the plot remained bare. It was calculated by multiplying the initial nutrient content of each vegetated plot by the loss rate on the corresponding bare plot (same fertilizer treatment, same block). This measure of retention was used to minimize interpretation problems due to variability in initial soil nutrients across the experimental site.

That some retention would occur as a result of vegetation presence is not surprising. The interesting question is whether the nutrients retained are important to ecosystem recovery. This question can be examined by attempting to establish a relationship between nutrient retention and some measure of system recovery. Three related measures of ecosystem recovery are used here: rate of biomass increase, rate of increase in vegetation nutrient content, and rate of attainment of specific successional stages.

To test the idea that ecosystem nutrient retention may be related to vegetation nutrient content, a direct comparison of ecosystem nutrient retention with vegetation nutrient content at 12 months is appropriate. The rank orders of vegetation nutrient content for the different elements among the treatments (Fig. 6), do not, however, correspond to the rank order retention of any element (see Table 4).

Even within treatments, where differences such as altered woody: herbaceous tissue ratios are minimized, ecosystem retention is not significantly positively related to vegetation nutrient content. Thus ecosystem nutrient retention does not appear to be related to the amount of nutrients bound in biomass.

Another possibility is that retention could be directly related to biomass alone, since biomass is in this case a measure of net primary production. However, there are no apparent relationships between nutrient retention (see Table 4) and biomass (Fig. 7) either in rank order comparison of treatments or in correlation tests within treatments.

The final possibility is that retention might be related to species composition, since one could define ecosystem recovery in terms of rate of species replacement in the normal successional sequence. Because of the experimental treatments and some unanticipated competitive interactions (Harcombe, unpublished manuscript), the rate of species replacement differed among the experimental plots–fertilized natural vegetation plots remained in the herb stage, unfertilized natural vegetation plots to the shrub stage, and tree plantation plots reached the tree stage.

The ordering of plots on the basis of stage in succession is likewise unrelated to the order of plots based on ecosystem nutrient retention.

Thus nutrient retention does not appear to be related to any of the three measures of ecosystem recovery–biomass, vegetation nutrient content, or rate of successional replacement.

This suggests two questions. First, can any of the variation in nutrient retention be explained without reference to ecosystem recovery; and second, are there other aspects of nutrient dynamics which might affect ecosystem recovery?

It is possible that nutrient loss and its modification by vegetation can be explained in terms of strong effects of soil chemical and physical processes. It may be, for example, that these soils are high in nutrients; so high nutrient leaching masks the relatively small effects of differences in vegetation. Central American soils are generally considered to be richer

Table 4. Nutrients retained* in soil-plus-vegetation at 12 months on vegetated plots ($\bar{x} \pm$ SE; g/m^2). *O* = unfertilized; *F* = fertilized.

	C	N	K	Ca	Mg
Natural vegetation plots					
0	*1158 ± 242.5*†	*78.27 ± 23.08*	0.25 ± 2.92	6.51 ± 7.05	*7.62 ± 2.09*
F	*1613 ± 196.0*	*129.65 ± 16.29*	3.76 ± 2.13	*22.05 ± 6.88*	*5.15 ± 1.31*
Avg	*1385 ± 161.6*	*103.95 ± 15.17*	2.01 ± 1.80	*14.28 ± 5.17*	*6.38 ± 1.23*
Tree plantation plots					
0	2103 ± 715.0	113.49 ± 36.46	0.10 ± 2.30	23.39 ± 12.50	*7.26 ± 1.89*
F	*1727 ± 259.6*	*120.64 ± 25.32*	3.37 ± 2.02	13.09 ± 6.06	*3.99 ± 1.21*
Avg	*1915 ± 359.3*	*117.07 ± 20.59*	1.73 ± 1.55	*18.24 ± 6.72*	*5.62 ± 1.21*

* Retention is amount of nutrients that would have been lost had the plots remained bare. See text for further explanation.

† Italicized values are significantly different from zero (two-tailed t-test, $p < 0.05$); i.e., retention was significant.

than the average tropical soil, especially those receiving periodic ash input. However, it is unlikely that soil fertility is so high as to make nutrient loss completely irrelevant to ecosystem recovery (see below).

Another possibility is that nutrient retention is due primarily to the physical effects of vegetation cover. That is, any plot covered with vegetation should show reduced leaching due to increased evapotranspiration and reduced mineralization of organic matter. Since all plots were completely covered by vegetation, one might not expect differences in retention.

Finally, nutrient dynamics seem, with one possible exception, to be consistent with our understanding of plant nutrient requirements and nutrient mobility to soils. Two examples illustrate this point. First, for those elements added as a fertilizer (N, K), retention appears to be higher on fertilized plots. It may be that the more readily available the nutrients are, the more likely they are to be retained, regardless of biomass or species composition or vegetation nutrient content.

Second, relative distribution of elements between vegetation and soil can be explained as easily on the basis of mobility in the soil as on the basis of vegetation adaptations for storage in plant tissue. Loss percentages of elements on bare plots were much lower for N and C than for Ca and Mg (see Table 3) due to difference in solubility, as discussed previously. This solubility difference probably also explains why the proportion of ecosystem (soil-plus-vegetation) nutrients existing in the vegetation was greater for Ca and Mg than C and N (Table 5); i.e., the higher the solubility, the greater the relative loss from the soil portion of

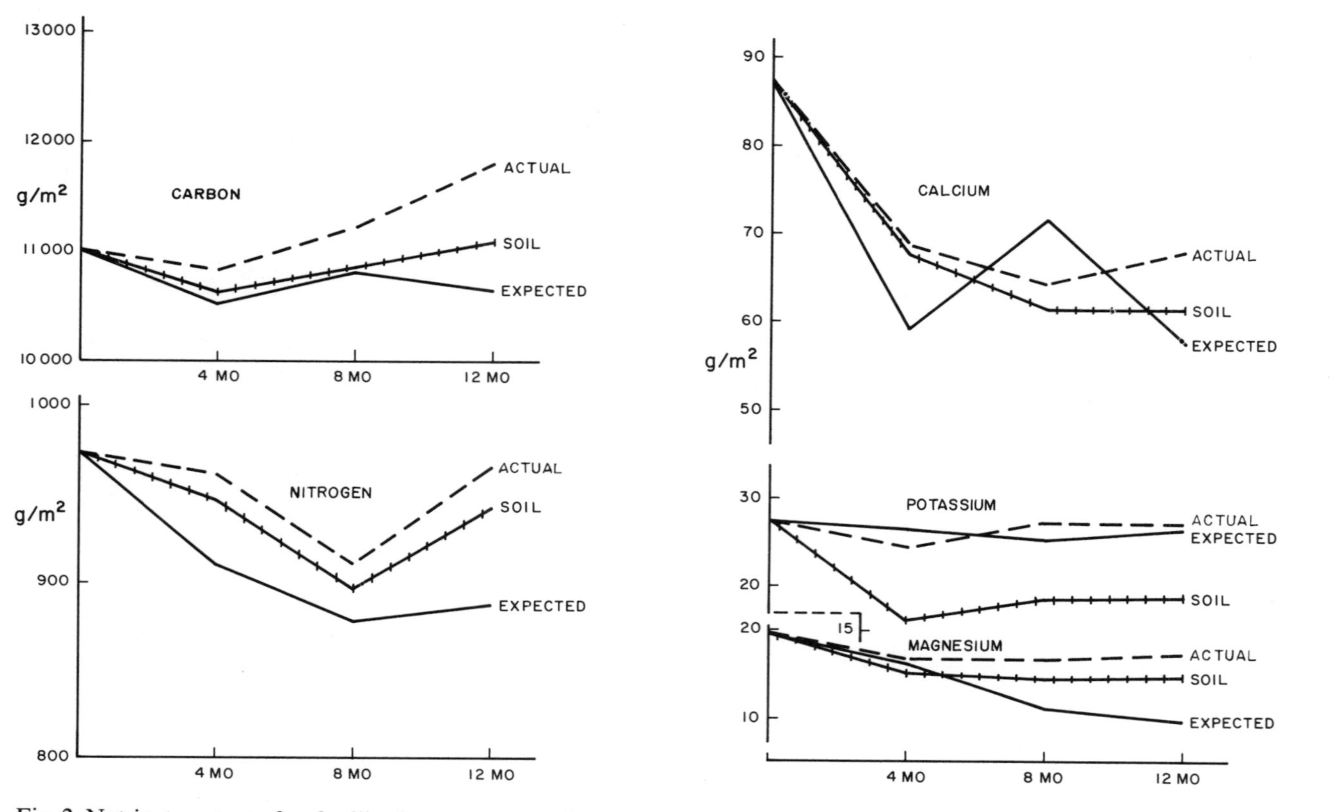

Fig. 2. Nutrient content of unfertilized natural vegetation plots (g/m^2; mean of 8 plots). Actual = content of soil-plus-vegetation. Soil = content of soil alone. Expected = predicted content of soil if plots kept bare. Retention = difference between Actual and Expected.

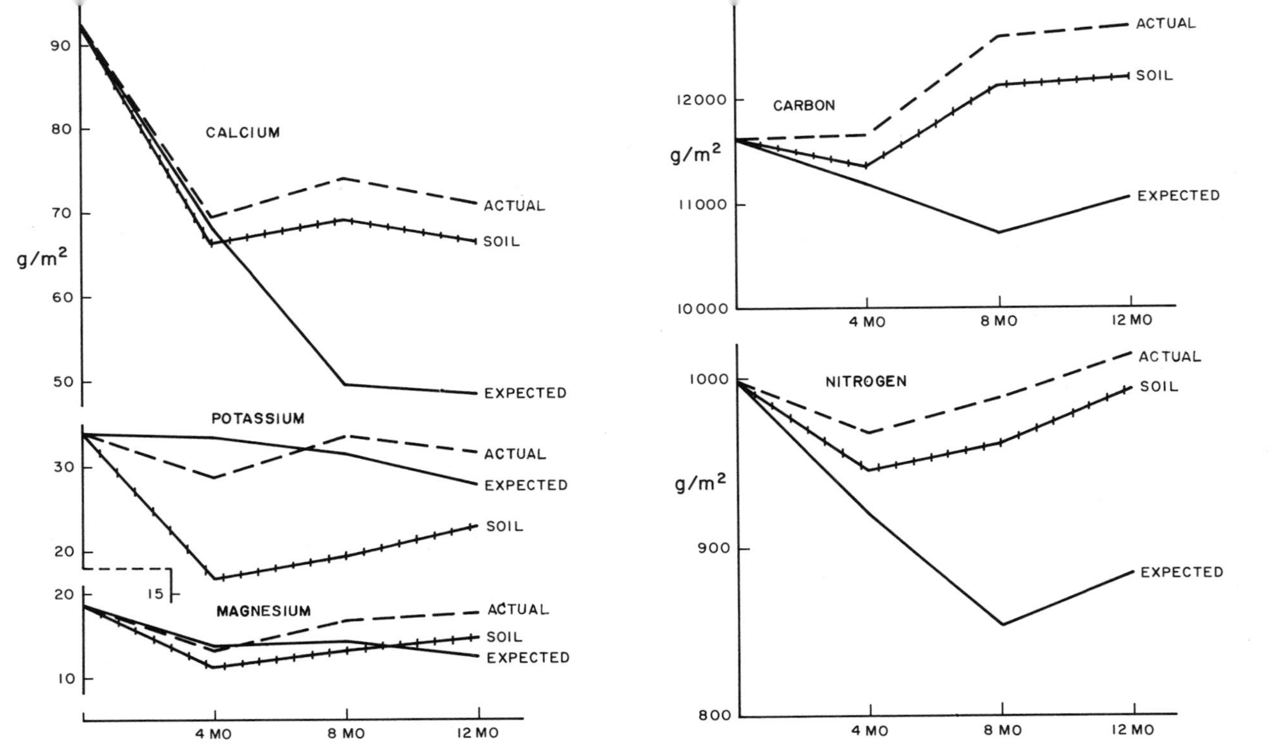

Fig. 3. Nutrient content of fertilized natural vegetation plots (g/m^2; mean of 8 plots). Actual = content of soil-plus-vegetation. Soil = content of soil alone. Expected = predicted content of soil if plots kept bare. Retention = difference between Actual and Expected.

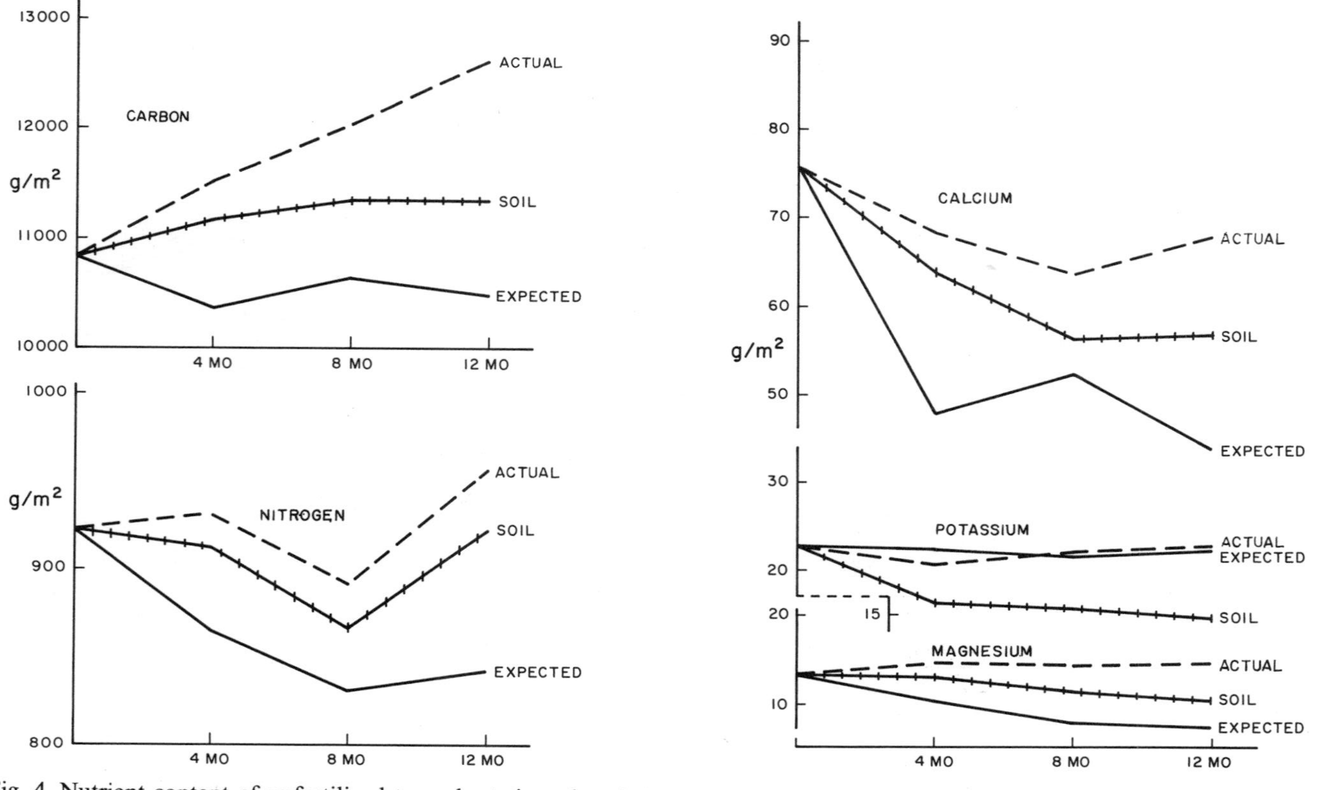

Fig. 4. Nutrient content of unfertilized tree plantation plots (g/m^2; mean of 4 plots). Actual = content of soil-plus-vegetation. Soil = content of soil alone. Expected = predicted content of soil if plots kept bare. Retention = difference between Actual and Expected.

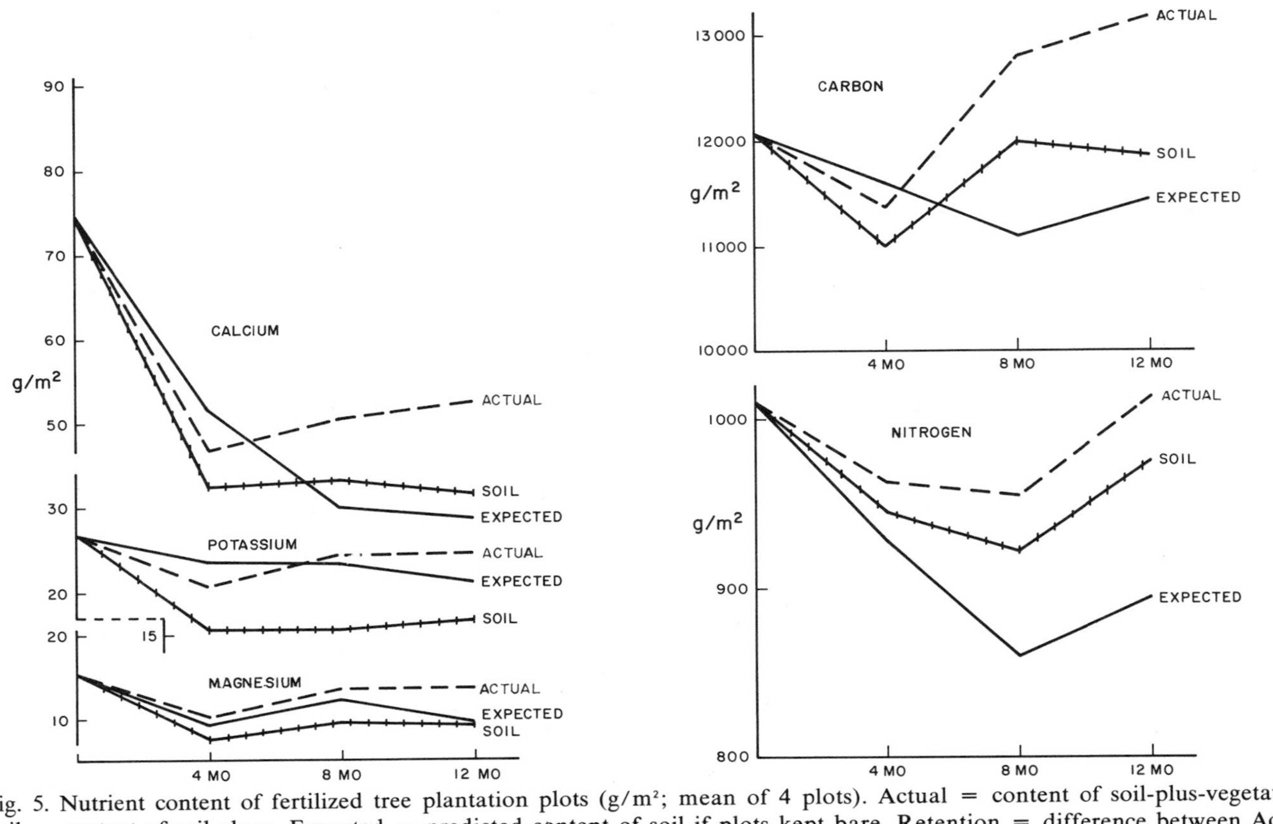

Fig. 5. Nutrient content of fertilized tree plantation plots (g/m²; mean of 4 plots). Actual = content of soil-plus-vegetation. Soil = content of soil alone. Expected = predicted content of soil if plots kept bare. Retention = difference between Actual and Expected.

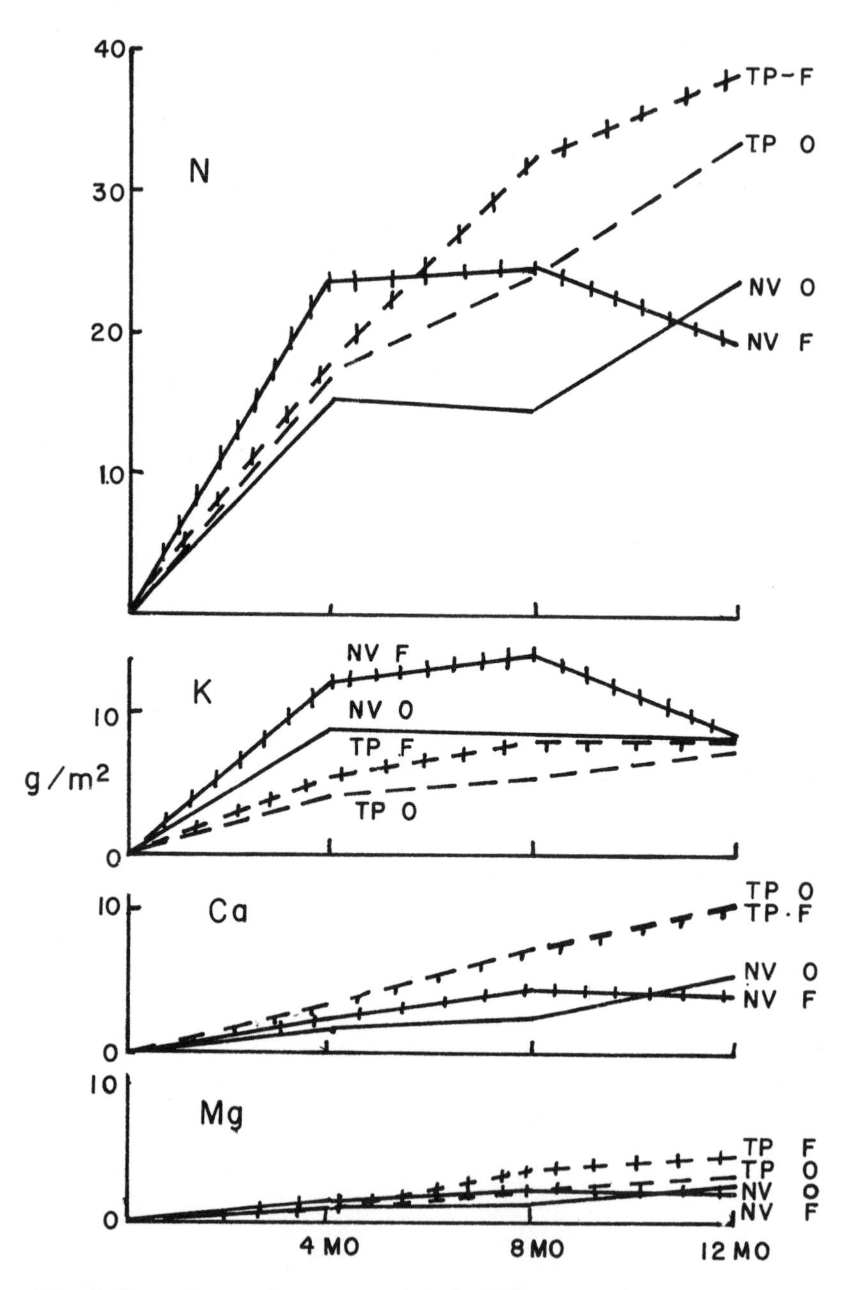

Fig. 6. Vegetation nutrient content (g/m²). *NV* = natural vegetation; *TP* = tree plantations; *0* = unfertilized; *F* = fertilized.

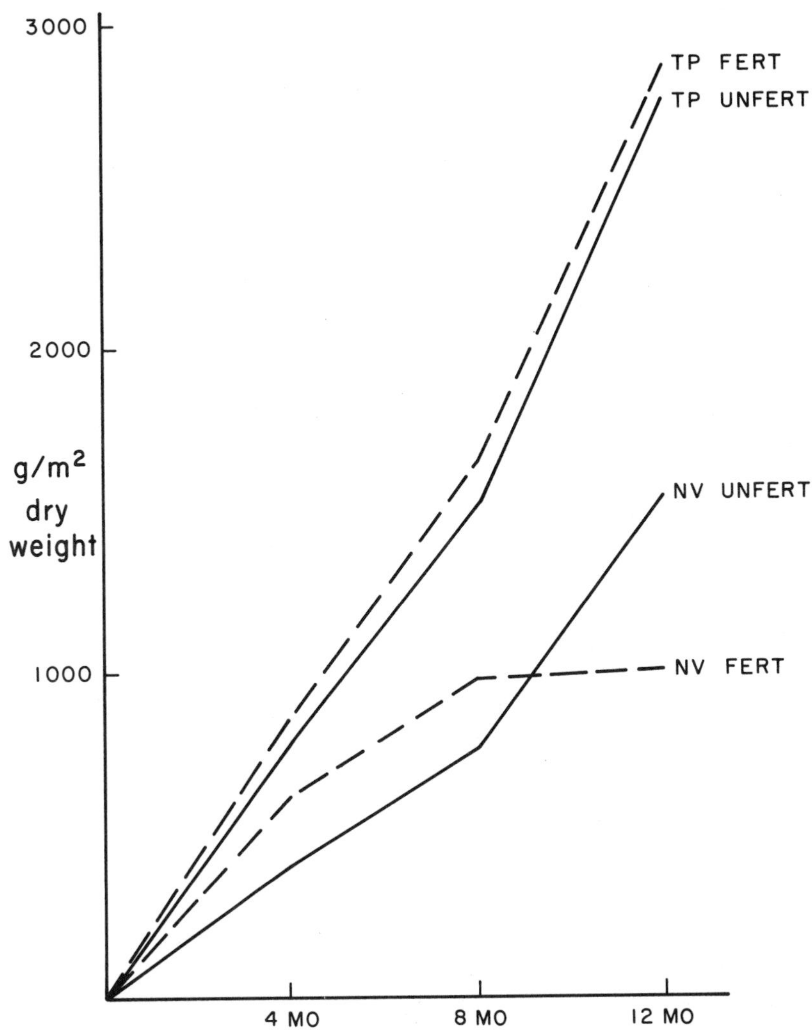

Fig. 7. Vegetation biomass (oven dry weight of plant tissue, including litter and roots). *NV* = natural vegetation plots ($n = 8$); *TP* = tree plantation plots ($n = 4$).

Table 5. Nutrients in vegetation alone at 12 months, expressed as percentage of nutrients in vegetation-plus-soil. *O*=unfertilized, *F*=fertilized.

	C	N	K	Ca	Mg
Natural vegetation					
0	5	3	31	9	15
F	3	2	28	6	15
Tree plantations					
0	9	4	35	16	27
F	9	4	34	21	32

the ecosystem. The alternative, that Ca and Mg were selectively concentrated in plant tissue, seems much less reasonable.

The anomalous behavior of potassium on vegetated plots requires mention, though it is difficult to interpret. Despite low mobility on bare plots (low loss from soil), potassium was highly mobile on vegetated plots (high loss from soil; see Figs. 2–5). Only for K was the soil content on vegetated plots less than "expected" ecosystem content at 12 months (see Figs. 2–5). Apparently, plants took up K which was not susceptible to leaching (i.e., not lost by leaching on bare plots), whereas they took up only Ca and Mg that otherwise would have leached out of the ecosystem. Greater understanding of the chemistry of "exchangeable" and other forms of nutrients is obviously necessary to our understanding of nutrient dynamics during succession.

Nutrient Capture

If retention of available nutrients is not important to ecosystem recovery, it is reasonable to consider other possible sources of nutrients and to ask whether the inputs from these sources are sufficient to sustain a reasonable rate of ecosystem recovery, or whether the initial capital is great enough that sufficient nutrients remain despite high nutrient loss.

Mature tropical forest vegetation may contain roughly 100 g/m^2 K, 200 g/m^2 Ca, and 40 g/m^2 Mg (Greenland and Kowal, 1960), which is two to four times as much as was present in available form in the soil (0-30 cm) at the outset of the experiment.

Thus, although it may be that on the short ter.n fertility is high enough to mask vegetation effects on nutrient dynamics, on the long term fertility does not appear to be high enough for complete regeneration of forest biomass. In this experiment, inputs are the only source for rebuilding the nutrient supply, since the forest vegetation was removed from the site. So it is worthwhile to consider whether rate of supply might limit forest regeneration.

Input of nutrients from the subsoil by root uptake and from the atmosphere by rain and dry fallout are probably the major processes whereby nutrients are acquired by the developing ecosystem. In addition, nutrients might become available through internal system changes such as decomposition of rock (weathering), mineralization of humus, or decomposition of dead roots. I have made rough estimates of these as follows.

Rain and dry fallout inputs were measured from nutrient content of rainwater and total precipitation volume (Table 6). Nutrients recovered from the subsoil (below 30 cm) were estimated by multiplying vegetation nutrient content by the proportion of in roots in the subsoil (Nye and Greenland, 1960). It was assumed that 10% of the total absorptive root surface was deeper than 30 cm.

Nutrients released by decomposition of dead roots were estimated using measurements of root nutrient content (Table 7) and an assumed decomposition rate of 10% per year. Other net internal changes were estimated by subtracting change in exchangeable nutrients from change in total nutrients (which includes exchangeable nutrients, nutrients bound in rock minerals; Table 8).

As a check on the above source-by-source estimate of nutrient inputs, I made a direct estimate of nutrients actually added to the experimental ecosystems during the year. This was done by subtracting ecosystem nutrient content at 4 months from ecosystem nutrient content at 12 months on unfertilized natural vegetation plots. The difference was always positive, indicating input (see Figs. 2–5). The two estimates agreed relatively well except for nitrogen, for which the great excess of addition over input must represent fixation (see Table 8).

If rain input were the only source, and all the input were retained in the ecosystem, regeneration of nutrient standing crop would require 250 years. This is a reasonable maximum estimate of the time necessary for a tropical forest to reach climax, though it is somewhat long for attainment of maximum biomass (Richards, 1952; Nye and Greenland, 1960; Schultz, 1960). If all inputs listed in Table 9 were captured, regeneration of nutrient content in climax vegetation could occur in 50–75 years (even though root decay would provide only 10 years of input). This is a reasonable minimum estimate for time to reach climax (Nye and Greenland, 1960).

Table 6. Precipitation amount (ml/m^2) and chemistry (μg/ml) for the experimental site. Input is computed from weighted average chemical composition: $\bar{x} = \Sigma$ (weekly vol. x conc.)/ Σ weekly vol.; input = ($\bar{x}$) (total volume).

Amount (mn; ml/m^2 x 10^3; 1970-71)

May	258	Sept.	243	Jan.	264
June	199	Oct.	205	Feb.	50
July	162	Nov.	342	Mar.	133
Aug.	246	Dec.	1096	Apr.	108

Total amount 3307 x 10^3 ml/m^2

Chemical composition for weekly samples (1971)

	Volume	K	Ca	Mg
13 Apr.	98	.06	.762	.035
25 May	44	.16	.912	.039
8 June	65	.10	.549	.059
23 July	98	.22	.549	.086
9 July	26	.06	.422	.035
	331			

Nutrient input

	K	Ca	Mg
Weighted average chemical composition (μg/ml)	.128	.651	.055
Nutrient input (g/m^2/yr)	.423	2.153	.181

The ranges of time-to-climax probably bracket real times for regeneration of ecosystem nutrient capital. Hence my data indicate that inputs alone may be sufficient to allow normal forest regeneration. Thus, high retention of available nutrients present at the time of disturbance may not be critical in insuring ecosystem recovery.

In this experiment, 25% to 100% of the nutrients incorporated into plant tissue in the first year (vegetation nutrient content; see Fig. 7) could have come from inputs; the rest must have been supplied from the initial stock of available soil nutrients.

General Discussion

Available nutrient loss was significant on bare plots, and natural vegetation did retard this loss. However, the significance of ecosystem nutrient retention in ecosystem recovery remains unclear, since retention seems to have been unrelated to vegetation nutrient accumulation, biomass accumulation, or rate of progression through successional stages. The implication is that within a modest range, loss of available soil nutrients does not strongly affect ecosystem recovery. Instead, nutrient input may play an important role in recovery.

Several observations support the importance of nutrient inputs. First, there was no correlation of nutrient accumulation in vegetation with initial soil nutrient levels; second, there was no stimulation of overall growth by fertilizer. These facts might indicate that growth of

Table 7. Calculation of nutrients released by decomposition of dead roots during the first year after forest removal

	Biomass	N	K	Ca	Mg
Root biomass (g/m^2), nutrient concentrations in root tissue (%), and average root nutrient content (g/m^2) in forest soil (0-30 cm) at time of forest removal					
Pit 1	1560	0.97	0.19	0.14	0.31
Pit 2	1606	0.79	0.14	0.23	0.18
Pit 3	1353	0.75	0.11	0.21	0.08
Pit 4	1719	0.81	0.16	0.19	0.19
Mean	1560	0.83	0.15	0.19	0.19
Weighted average root nutrient content at time zero (g/m^2)*		12.97	2.31	2.98	2.98
Estimated nutrient release by root decomposition within twelve months (g/m^2)†					
		1.30	0.23	0.30	0.30

* Weighted average nutrient content = Σ (individual pit biomass x nutrient concentration)/4.
† Nutrient release is assumed to be 10% of weighted average nutrient content.

Table 8. Estimated loss of soil nutrients (g/m²; 0-30-cm horizon) by weathering and mineralization

Treatment*	Plot no.	Total nutrients 0 mo.	Total nutrients 12 mo.	Total nutrients Change	Available nutrients 0 mo.	Available nutrients 12 mo.	Available nutrients change	Net change
Potassium								
B-F	22	166.65	154.37	-12.28	20.30	18.62	- 1.68	-10.60
TP-F	14	181.53	169.76	-11.77	23.58	15.88	- 7.70	- 4.07
NV-O	7	204.81	187.62	-17.19	31.06	19.78	-11.28	- 5.91
NV-O	8	175.62	181.98	+ 6.36	29.41	22.67	- 6.74	+13.10
NV-F	16	242.43	233.36	- 9.07	54.53	40.94	-13.59	+ 4.52
Avg				- 8.79			- 8.20	- 0.59
Calcium								
B-F	22	104.01	86.99	-17.02	46.26	35.97	-10.29	- 6.73
TP-F	14	138.88	134.65	- 4.23	29.91	20.77	- 9.14	+4.91
NV-O	7	170.39	140.79	-29.60	33.05	10.80	-22.25	-7.35
NV-O	8	163.05	133.65	-29.40	75.07	52.23	-22.84	-6.56
NV-F	16	300.31	321.34	+21.03	130.60	133.84	+3.24	+17.79
Avg				-11.84			-12.25	+ 0.41
Magnesium								
B-F	22	115.59	106.56	-9.03	5.29	4.07	-1.22	-7.81
TP-F	14	173.28	165.28	-8.00	8.70	7.05	-1.65	-6.35
NV-O	7	112.61	118.57	+5.96	16.51	10.86	-5.65	+11.61
NV-O	8	155.04	147.45	-7.59	11.74	12.62	+0.88	-8.47
NV-F	16	187.98	185.82	-2.16	24.93	26.80	+1.87	-4.03
Avg				-4.16			-1.15	-3.01

* *B* = bare plots, *NV* = natural vegetation plots, *TP* = tree plantations, *O* = unfertilized, *F* = fertilized. Because total nutrients were not part of the original sampling scheme, only a few representative samples were analyzed.

successional species was relatively constant, a strategy appropriate to utilization of inputs, rather than highly variable, a strategy appropriate to exploitation of the mobile supply of nutrients available in the soil after forest removal. In this regard, a point apparent from the data on biomass and vegetation nutrient content, that successional trees produce more biomass per unit of nutrients than herbaceous, might be significant. This could represent an adapation for maximum growth on a low nutrient supply.

Further support for the importance of input comes from the calculations that indicate inputs (including rainfall, weathering, recovery from

Table 9. Ecosystem nutrient budget for unfertilized natural vegetation plots ($g/m^2/yr$)

	N	K	Ca	Mg
Nutrients made available during the year (estimated as described in the text and in Tables 6-8)				
Inputs from outside				
Rain	1.40	0.42	2.15	0.18
Recovery from below 30 cm	2.37	0.84	0.58	0.27
Nutrients generated by internal changes				
Rock weathering		0.59		3.01
Root decay	1.30	0.23	0.30	0.30
Total	5.07	2.08	3.03	3.76
Nutrients added to the ecosystem (average of differences in ecosystem nutrient content between 4 and 12 months)	55.27	2.16	2.81	1.12

below 30 cm, or fixation from the atmosphere) might be sufficient to sustain vegetation nutrient accumulation. This suggests that retention of the initial supply is not critical in satisfying the nutrient needs of the developing vegetation.

The suggestion that the initial supply of available nutrients may not control ecosystem recovery appears at first glance to be at variance with the large body of evidence concerning fragility of tropical soil resources and the rapid loss of soil fertility.

For agricultural cash crops, which require amounts of nutrients in readily available forms, high levels of exchangeable nutrients are probably necessary for continued good growth and production. However, if native successional species are adapted to lower nutrient levels and to utilization of inputs, they may not depend so much on the exchangeable nutrients remaining after disturbance. Instead, these plants may depend more upon recovery of nutrients from lower horizons or capture of nutrients becoming available through weathering or entering from the atmosphere.

Thus, in agriculture the decrease of exchangeable nutrients brings effective loss of fertility; and given high plant nutrient needs and conditions conducive to rapid leaching, the loss of nutrients probably strongly affects vegetation growth. In natural succession, an initial decrease of exchangeable nutrients may not strongly alter ecosystem recovery. Prolonged soil exposure could, of course, eventually deplete the soil of the available nutrients stored below 30 cm and the bound nutrients released by weathering, and might also result in deterioration of soil structure and in erosion, which could severely retard natural ecosystem regeneration.

This explains the conclusion of Nye and Greenland (1960): "Long continued cultivation or the incidence of severe erosion might also reduce soil fertility sufficiently to limit the vigor of the growth of vegetation. Such an occurrence must, however, be regarded as exceptional; in all normal circumstances regrowth is, as a matter of experience, extremely rapid and vigorous."

This possible dichotomy in the behavior of natural successional vegetation and agricultural vegetation might also apply to temperate ecosystems versus tropical ecosystems. For example, in temperate ecosystems, increased solubilization of nutrients after disturbance (nutrient dumping) stimulates revegetation (Marks and Bormann, 1972; Marks, 1974). This accelerates reestablishment of control by vegetation over nutrient losses and erosion. In the humid tropical ecosystem, revegetation is rapid, but the process of nutrient accumulation seems to be little affected by a pulse of nutrients. Here, where erosion is less of a problem, but leaching of available nutrients is very rapid, adaptation of successional species for nutrient accumulation by capture of inputs may be a better strategy than adaptation for exploitation of a highly ephemeral pulse of nutrients.

The evidence I present is mostly circumstantial. There are, for example, other possible explanations for the absence of a strong direct relationship between ecosystem recovery and ecosystem nutrient retention in this study.

Nevertheless, I suggest that growth rates of successional species may be regulated for efficient acquisition of nutrients, rather than for retention of nutrients after disturbance. This regulation could determine the course of succession and the rate of ecosystem recovery.

Summary

1. One year after forest clearing, plots with successional vegetation retained 13% to 100% of the nutrients that would have disappeared had the plots remained bare. This was always significant for C, N, and Mg; it was significant for some treatments for Ca; and it was not significant for K.

2. Nutrient retention was not significantly affected by experimental

manipulation of vegetation which resulted in difference in biomass, vegetation nutrient content, and species composition.

3. Nutrient input from external sources (rain and dust) and internal sources (solubilization of bound nutrients in surface and subsoil) can account for the known rate of ecosystem development, and may provide more than half the nutrients incorporated into vegetation during the first year.

4. Therefore, it is concluded that although retention of available nutrients is significant, ecosystem recovery may also depend upon dynamics of semiavailable nutrients and slow incorporation of bound reserves into the available nutrient pool. This is in contrast to the original hypothesis that the ability of the vegetation to retain the initial stock of available nutrients was an important determinant of the rate of ecosystem recovery.

Acknowledgments

This study was supported by the Organization for Tropical Studies (Pilot Study Grant No. 69–20), by the National Science Foundation (Training Grant GB 12895 to G. E. Hutchinson, Research Grant to F. H. Bormann, and Graduate Fellowship to me), by the Society of the Sigma Xi (Grant-in-Aid of Research to me), and by the Department of Biology and the School of Forestry, Yale University. The manuscript was prepared during my tenure at Rice University.

I wish to thank the personnel of the Instituto Interamericano de Ciencias Agrícolas, Centro Tropical de Enseñanza y Investigación, Costa Rica, for use of laboratory facilities and land and for laboratory assistance. F. H. Bormann and P. L. Marks were instrumental in the development of the ideas. P. L. Marks, J. Ewel, and E. Harcombe read the manuscript and made many helpful comments.

Literature Cited

Aguirre, V. 1971 Classificación de suelos del I.I.C.A.-C.T.E.I., Turrialba, Costa Rica. M.S. thesis, Instituto Interamericano de Ciencias Argícolas, Turrialba, Costa Rica.

Bartholomew, W. V., J. Meyer, and H. Laudelot. 1953. Mineral Nutrient Immobilization under Forest and Grass Fallow in the Yangambe (Belgian Congo) Region. I.N.E.A.C. Publ. No. 57. Serie Scientifique.

Basolo, F., and R. G. Pearson. 1958. Mechanisms of Inorganic Reactions. John Wiley & Sons, New York.

Besoain, E. 1970. Clay mineralogy of some andosols of Costa Rica. Mimeo. I.I.C.A., Turrialba, Costa Rica.

Black, C. A., D. D. Evans, J. L. White, L. E. Ensminger, F. E. Clark, eds. 1965. Methods of Soil Analysis. Part I. Physical and Mineralogical Properties and Statistics of Measurement and Sampling. Part 2. Chemical and Microbiological Properties. Am. Soc. Agron., Madison, Wis.

Bormann, R. H., G. E. Likens, and R. S. Pierce. 1969. Biotic regulation of particulate and solution losses from a forest ecosystem. Bioscience 19:600–610.

Bornemisza, E., F. A. Laroche, and H. W. Fassbender. 1967. Effects of liming on some chemical characteristics of a Costa Rican latosol. Proc. Soil and Crop Sci. Soc. of Fla. 27:219–226.

Buckman, H. O., and N. C. Brady. 1969. The Nature and Properties of Soils. Macmillan & Co., London.

Budowski, G. 1961. Studies on forest succession in Costa Rica and Panama. Ph.D. diss. Yale University School of Forestry.

Cowgill, U. M. 1961. Soil fertility and the ancient Maya. Trans. Conn. Acad. Arts Sci. 42:1-56.

Cunningham, R. K. 1963. The effect of clearing a tropical forest soil. J. Soil Sci. 14:334-345.

Dengo, G. 1962. Tectonic-igneous sequence in Costa Rica. Pages 133-161 *in* Petrologic Studies: A Volume to Honor A. F. Buddington. Geological Society of America Memoir.

Dondoli, C., and J. A. Torres. 1954. Estúdio geoagronómico de la región oriental de la Meseta Central. Ministério de Agricultura y Industrias, San José, Costa Rica.

Duthie, D. W., F. Hardy, and G. Rodriguez. 1937. Soil investigations in the Arena Forest Reserve, Trinidad. A summarized account by R. L. Brooks. Imperial Forestry Institute, Univ. of Oxford, Paper 6. Mimeo. 16 pp.

Ewel, J. J. 1971*a*. Biomass changes in early tropical succession. Turrialba 21:110-112.

———. 1971*b*. Experiments in arresting succession with cutting and herbicides in five tropical environments. Ph.D. diss. Univ. of North Carolina, Chapel Hill.

Farnworth, E., and F. B. Golley, eds. 1974. Fragile Ecosystems. Springer-Verlag, New York. 258 pp.

Fried, M., and F. Broeshart. 1967. The Soil-Plant System. Academic Press, New York.

Gavande, S. A. 1968. Water retention characteristics of some Costa Rican soils. Turrialba 18:34-38.

Greenland, D. J., and J. M. L. Kowal. 1960. Nutrient content of the moist tropical forest of Ghana. Plant Soil 12:154-174.

Greenland, D. J., and P. H. Nye. 1959. Increases in the carbon and nitrogen contents of tropical soils under natural fallows. J. Soil Sci. 9:284-299.

Harcombe, P. A. 1973. Nutrient cycling in secondary plant succession in a humid tropical forest region (Turrialba, Costa Rica). Ph.D. diss., Yale University.

_____. Unpublished manuscript. Plant succession and ecosystem recovery.

Hardy, F. 1961. The soils of the I.A.I.A.S. area. Mimeo. I.I.C.A., Turrialba, Costa Rica. 76 pp.

Holdridge, L. R., W. H. Hatheway, T. Liang, and J. A. Tosi, Jr. 1971. Forest Environments in Tropical Life Zones: A Pilot Study. Pergamon Press, Oxford. 747 pp.

Kellogg, C. K. 1963. Shifting cultivation. Soil Sci. 93:221-230.

Kononova, M. M. 1961. Soil Organic Matter. Pergamon Press, New York.

Likens, G. E., F. H. Bormann, N. M. Johnson, D. W. Fisher, and R. S. Pierce. 1970. Effects of forest cutting and herbicide treatment on nutrient budgets in the Hubbard Brook watershed-ecosystem. Ecol. Monogr. 40:23-47.

Marks, P. L. 1974. The role of pin cherry (*Prunus pensylvanica* L.) in the maintenance of stability in Northern Hardwood Ecosystems. Ecol. Monogr. 44:73-88.

Marks, P. L., and F. H. Bormann. 1972. Revegetation following forest cutting: mechanisms for return to steady-state nutrient cycling. Science 176:914-915.

Martini, J. A. 1969*a*. Caracterización del estado nutricional de los principales "latosoles" de Costa Rica mediante la técnica del elemento faltante en el envernadero. Turrialba 19:394-408.

_____. 1969*b*. Geographic distribution and characteristics of volcanic ash soils in Central America. Pages A.5.1-17 *in* Panel on Volcanic Ash Soils in Latin America, I.I.C.A., Turrialba, Costa Rica.

Morrison, P. C., and J. Leon. 1951. Sequent occupance, Turrialba Central District, Costa Rica. Turrialba 1:185-198.

Muller, L. 1961. Un aparato micro-Kjeldahl simple para análisis rutinários rápidos de materiales vegetales. Turrialba 11:17-25.

Nye, P. H. 1958. The mineral composition of some shrubs and trees in Ghana. J. West Afr. Sci. Assoc. 4:91-98.

Nye, P. H., and D. J. Greenland. 1960. The Soil under Shifting Cultivation. Commonwealth Bureau of Soils, Harpenden, England Tech. Comm. No. 51.

_____. 1964. Changes in the soil after clearing tropical forest. Plant Soil 21:101-112.

Pomeroy, L. R. 1970. The strategy of mineral cycling. Ann. Rev. Ecol. Syst. 1:171-190.

Popenoe, H. 1960. Effects of shifting cultivation on natural soil constituents in Central America. Ph.D. diss., Univ. of Fla., Gainesville.

Richards, P. W. 1952. The Tropical Rain Forest: An Ecological Study. Cambridge Univ. Press, New York. 450 pp.

Sawyer, J. O., and A. A. Lindsey. 1971. Vegetation of the Life Zones of Costa Rica. Indiana Acad. Sci. Monogr. No. 2.

Schultz, J. P. 1960. Ecological studies on rain forest in Northern Surinam. Verhandlungen der kommklijke Nederlandse Akademic von wetenschnappen Afd Natuurkunde Tweede Reeks, Deel. 53(1):1-367.

Sioli, G., H. Schwabe, and H. Klinge. 1969. Limnological outlooks on landscape ecology in Latin America. Trop. Ecol. 10:72-82.

Snedaker, S. C. 1970. Ecological studies on tropical moist forest succession in eastern lowland Guatemala. Ph.D. diss., Univ. of Fla., Gainesville.

Stark, N. 1970. The nutrient content of plants and soils from Brazil and Surinam. Biotropica 2:51-60.

Tergas, L. E., and H. Popenoe. 1971. Young secondary vegetation and soil interactions in Izabal, Guatemala. Plant Soil 34:675-690.

Van Baren, F. A. 1961. The pedological aspects of the reclamation of tropical, and particularly volcanic, soils in humid regions. *In* Tropical Soils and Vegetation, Proc. of the Abidjan Symposium. UNESCO.

Van Steenis, C. G. G. J. 1958. Rejuvenation as a factor for judging the status of vegetation types: the biological nomad theory. Pages 212-218 *in* Study of Tropical Vegetation, Proc. Kandy Symp. UNESCO.

Wagner, P. L. 1964. Natural vegetation of middle America. Pages 216-263 *in* Wauchope and West, eds. Handbook of Middle American Indians, Vol. 1.

Whittaker, R. H., and G. M. Woodwell. 1968. Dimension and production relations of trees and shrubs in the Brookhaven Forest, N.Y. J. Ecol. 56:1-25.

Recovery of Tropical Lowland Forest Ecosystems

Paul A. Opler, Herbert G. Baker, and *Gordon W. Frankie*

Abstract

Data from successive studies of recovery by perturbed tropical lowland wet and dry forests in Costa Rica are presented. Species richness and mean seed weight of reproductive plants increase through time in both ecosystems. The proportion of plants possessing physically dispersed fruits is initially high but decreases asymptotically at both sites, while a concomitant asymptotic increase of plants with fleshy animal-dispersed fruits takes place contemporaneously. The intercession of a six-month rainless period in the dry lowlands causes a seasonal pulse to these long-term changes and may slow the overall rate of succession. Projecting from known rates of change in mean seed weight to those typical of nearby mature forests gives figures of 1000 and 150 years for full recovery of wet and dry lowland forest ecosystems, respectively. The increased rate of tropical forest destruction may mean that full recovery of these forests will only take place in evolutionary time.

Introduction

This century, particularly the last 25 years, has seen profound changes in the tropical world. With population growth rates triggered by human death control and the challenges of tropical living solved by technological advances, natural tropical ecosystems have been and are being transformed, and perhaps irreversibly damaged, at an astronomical rate (e.g., Croat, 1972; Gomez-Pompa et al., 1972; Allen, 1975). When some of us were younger we read, and sometimes dreamed, of the seemingly endless green expanses of lowland tropical forest as described in the writings of Beebe, Hudson, and Bates. It is now estimated by a number of tropical biologists that within 10 years, and certainly in no more than 25 years, relatively undisturbed tropical lowland forest will exist only in a few parks and preserves (e.g., MacBryde, 1972; Terborgh, 1974; Allen, 1975). Accompanying this transformation will be the extinction of thousands of species of animals and plants, many unrecorded (Gomez-Pompa et al., 1972). Recovery (here defined as a return to a mature forest community in relative equilibrium with regard to both biotic and abiotic factors; as for example, soil profile development and stable age-class distribution of plants, to name a few) of these tropical forests must now be viewed only as an event that might occur over future evolutionary time, after the present perturbations have ceased. Naturally, the unique combinations of species in the forests being destroyed could not be replicated, and will have been lost forever. Accompanying the present habitat changes are unwise agricultural practices which will have a further severe impact upon ecosystem recovery in the tropics (Janzen, 1973).

Given the optimistic view that such ecosystems might recover to a degree, or might be more intelligently managed in the terms of successional processes, there are several sets of conceptual frameworks we should grasp. The study of plant succession had its early beginnings in the late nineteenth century (Cowles, 1899), and its first concepts were formalized in 1916 by Clements. Since then, many studies have been published. The best recent summarization of ecosystem recovery is that by Odum (1969), who stressed productivity and mineral cycling, as well as a lot of other purported changes in life histories of species occupying various seral stages. Drury and Nisbet (1973) questioned the reality of Odum's idealized abstract of ecosystem processes, while Horn (1974) presents an alternative viewpoint, stressing random plant-by-plant replacement as a guiding principle. Odum also emphasized the relevance of "island equilibrium theory" (MacArthur and Wilson, 1963, 1967), pointing out, for example, that species richness and equitability increase with successional age. Recent treatments and discussion are becoming increasingly quantitative, and it appears that use of such guideposts as the works of Margalef (1968), MacArthur and Wilson (1967), and Odum (1969) will be focal points for such efforts. Each theory will allow us to view ecosystem recovery in a different, yet not necessarily conflicting, way; each enabling us to better grasp the various aspects of this complex process. Yet, we must remember that we are dealing with biological communities, which are exemplified by the interplay of organisms and the environment; validation of our concepts must of necessity always be based on the natural world and thoughtfully conceived experiments which measure actual environmental parameters.

Understanding of successional processes has, until now, been advanced first by temperate zone studies, often corroborated later in the tropics. This has probably been the case because temperate succession is relatively simple and because most biologists reside in those areas. To date, most tropical work on this subject has been anecdotal or descriptive, usually lacking in theoretical backing. Some studies have been based on agriculture in attempts to understand and control the "weeds" of pioneer stages. Budowski (1961) was one of the first to generalize on tropical succession from the standpoint of all successional stages. He provided a good review of the earliest literature on the subject. Lieth (1974) presents a partial, yet extensive, bibliography which includes many pertinent references more recent than those reviewed by Budowski. The recent work of Harcombe (in this volume) and Williams et al. (1969*a*) are examples of the as yet few attempts at quantification of tropical successional processes on a theoretical basis.

Many workers have described and quantified the various vegetative aspects of mature tropical forests (e.g. Richards, 1952; Ashton, 1969; Williams et al., 1969*b*; Odum, 1970; Holdridge et al., 1971); works such as theirs serve as the standard with which successional information must be compared and contrasted.

We now present the results of studies on the recovery of two lowland forest sites in Costa Rica (see Frankie et al., 1974, for site and climate descriptions). These studies were only a portion of a larger program intended to describe and contrast the reproductive ecology of plants occurring in these two ecosystems.

Procedure

At La Selva, 3 km south of Puerto Viejo, Heredia Province, the wet forest site, three contiguous, approximately 0.5 hectare blocks of 8- to 15-year-old forest were cut during early March 1971, 1972, and 1973. These plots were examined at least eight times each year, and all reproductive phanerogam species present were recorded. Nonreproductive species were not recorded. For most species diaspore dispersal type (Dansereau and Lems, 1957), dry seed weight, seeds/fruit, and life form were determined. On occasion relative abundance and dominance of each species were subjectively estimated and photographs were taken. It should be noted that regeneration from extant root systems by suckering was not prevented, and, as a result, successional processes may have been somewhat accelerated.

At La Pacifica, 5 km northwest of Cañas, Guanacaste Province, the dry forest site, a 0.17-hectare plot was established in June 1972, where a small piece of 40- to 50-year-old forest had been cut and burned a year previously and was grazed in the intervening year. The plot was divided into seventeen 10 m by 10 m subplots. A wide buffer area was left to eliminate possible edge effect. All reproductive phanerogam species present were recorded by subplot each month, and again diaspore dispersal type, dry seed weight, seeds/fruit, and life form were recorded.

An attempt was made to identify all plants to species through the collection of pressed voucher specimens and subsequent determination by taxonomic specialists (see Acknowledgments).

Results

Taxonomic identity and temporal occurrence of the reproductive species encountered are presented in Tables 1-4 at the end of this chapter. Species accumulation, diaspore dispersal type, and seed weight changes through time are presented graphically by Figures 1-7.

Turnover of Dominants

Although there were no sharply defined stages, we may distinguish several phases that each successional area underwent with regard to gross physical structure and dominant species. Budowski (1961) defines stages for Costa Rican wet forest that will allow the reader to extrapolate to the changes that might be expected in future years at La Selva, while our observations of older secondary forest in Guanacaste will allow us another blurred glimpse into the future there. It should also be noted that these phases were often overlapping at the La Pacifica site.

La Selva

Phase 1 at La Selva, the "low herb" stage (Fig. 1), extends from the onset of the secondary succession and persists for only three months. The plants of this phase probably all result from the previously extant dormant seed pool (cf. Guevera and Gomez-Pompa, 1972; Vasquez-Yanes, 1974) and are notably represented by *Erechtites*

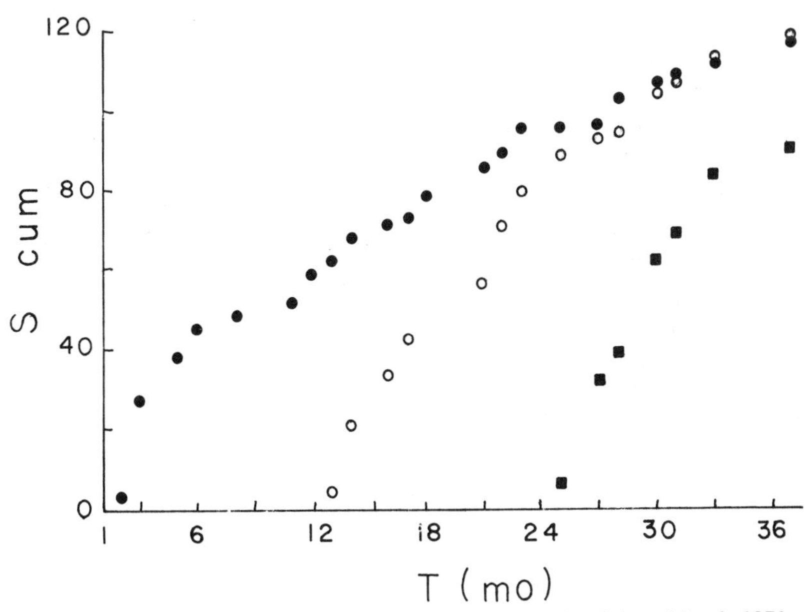

Fig. 1. Species accumulation for 0.5-hectare plots at La Selva, March 1971 to April 1974. *Solid circles:* Plot 1, established March 1971. *Open circles:* Plot 2, established March 1972. *Solid squares:* Plot 3, established March 1973

hieracifolia, Borreria (2 species), *Achyranthes sessilis, Lindernia* (2 species), and *Drymaria cordata*.

Phase 2, the "grass and sedge" stage, persists for four months and is dominated by several species of *Cyperus*, grasses (including *Panicum maximum, P. pilosum,* and *Axonopus compressus*), *Cyathula prostrata*, and *Blechum brownei*.

Phase 3, the "subshrub" stage, which is evident for seven months, is almost completely dominated by *Phytolacca rivinoides*. Important codominants include *Sida rhombifolia, Cyathula achyranthoides,* and *Hyptis capitata.*

Phase 4, the "shrub and herbaceous vine" stage, persists for approximately ten months and is dominated by a number of shrubs, including *Hamelia patens, Piper* (several species), Melastomataceae (several species), *Cassia fruticosa,* and *Solanum rugosum*. Herbaceous vines include *Cissampelos* spp., *Passiflora* spp., and *Ipomoea* spp. Through this stage almost all plants which came into reproduction were characterized by seed dormancy and photoblastic germination (cf. Vasquez-Yanes, 1974).

Phase 5, the "small tree" stage, begins approximately 2 years from the inception of succession. At its onset the vegetation begins to show a weak appearance of layers which become more pronounced through time. The first plants which may be presumed to lack dormancy become reproductive, although they are a minority, and the first few plants which may also be found in climax forests, usually in medium to large light gaps created by tree falls, make their appearance. According to Budowski (1961) this phase persists for up to 15 years, although he gauges its onset at 3 to 5 years after the initial disturbance. Perhaps the faster rate of the La Selva succession can be accounted for by the fact that burning did not take place and that suckering was not prevented. It should also be noted that our first four phases are accounted for by Budowski as a single "Pioneer Seral Stage."

La Pacifica

The succession of phases at La Pacifica (Fig. 2) in Guanacaste is very similar to that described for La Selva but is confused by the intercession of a six-month dry season.

Phase 1, the "herb stage," begins with the perturbation and persists for the first three months, during the wet season each of the first two years. Its primary dominants (nine or more subplots) are *Amaranthus spinosus, Baltimora recta, Tridax procumbens, Cyperus tenerrimus, Chamaesyce* (2 spp.), *Calliandra portoricensis, Oxalis neaei, Priva lappulacea, Hybanthus attenuatus,* and *Kallstroemia pubescens.*

Phase 2, the "grass and sedge" stage, was blurred in our study due to the one-year lapse between initial disturbance and the first observations. It overlaps with the herb stage during the first three months of the second wet season. It includes *Cenchrus pilosus, Dactyloctenium aegyptium, Digitaria bicornis, Echinochloa colonum,* and *Leptochloa filiformis* as codominants.

Phase 3, the "subshrub and herbaceous vine" stage, dominates the second half of the wet season, overlapping into the mid-dry season for each of the first three years. During the first year *Sida acuta, S. rhombifolia,* and *Malvastrum americanum* dominate, while no vines are present. During the second year these species

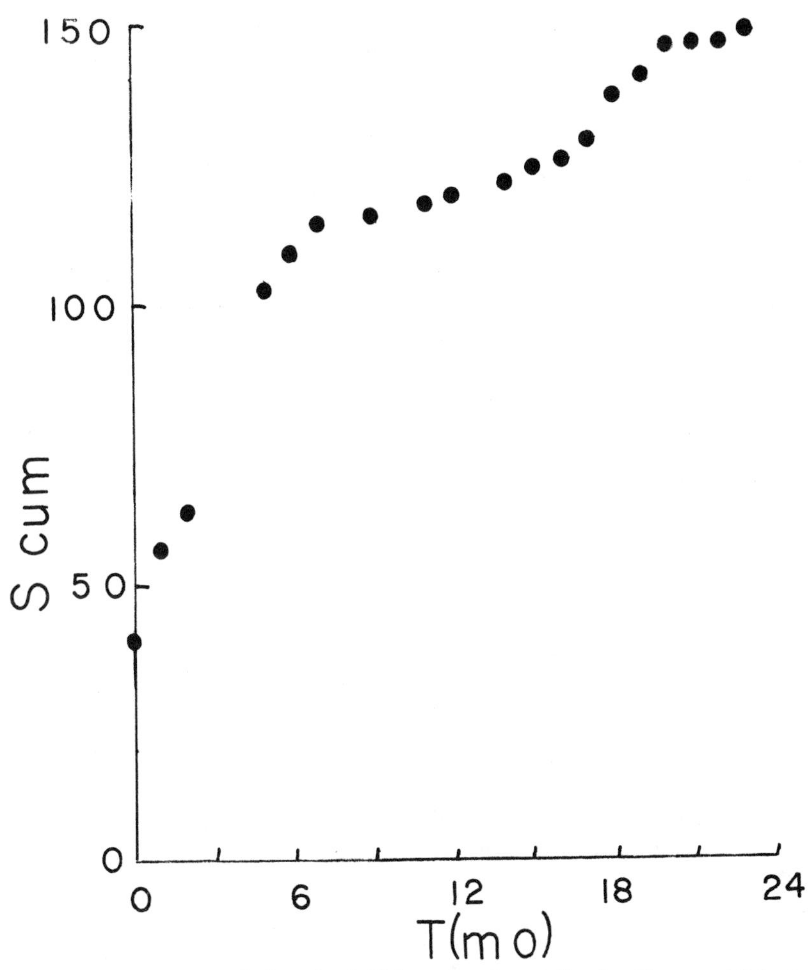

Fig. 2. Species accumulation for 0.17-hectare plot at La Pacifica established June 1972

retain importance, while *Melanthera nivea, Hyptis suaveolens, Melochia nodiflora,* and *Waltheria indica* become codominant subshrubs. Many vines, including several Convolvulaceae and many Papilionoideae, also become important codominants during the latter half of the second wet season. All these subshrubs and vines remain important during the third wet season as well.

Phase 4, the "shrub stage," is less important than in the wet forest succession, and was most notably represented in our study area by *Solanum hazeni.* This phase is entered during the second dry season and persists through the third year.

Phase 5, the "small tree" stage, became important during the third dry season and was dominated by *Guazuma ulmifolia* and to a lesser degree by *Trema micrantha.* It is expected that this phase will persist for several more years, with several species of *Cordia* and *Cochlospermum vitifolium* entering as codominants or subdominants.

Species Accumulation

The species accumulation curves for both La Selva and La Pacifica appear to be asymptotic, although they are not (see Figs. 1 and 2). The rate of accumulation at La Pacifica is more rapid and the inflection point more marked than at La Selva. Taking into account the disparate sizes of the two areas, the rapid accumulation of species at La Pacifica is even more remarkable. This is no doubt accounted for by the long history of human activity in Guanacaste Province (more than 200 years) and the resultant rich species pool of pioneer plants. Yet the total floristic richness of the two regions is clearly dominated by La Selva and the species/area accumulation curves are expected to cross over at some later stage, as the accumulation rate is expected to decrease more rapidly at La Pacifica.

Species Richness

At La Selva, species richness seems to be asymptotic through time (Fig. 3); however, at La Pacifica, species richness is seasonally cyclic with high peaks of flowering at the wet/dry season interfaces (Fig. 4). At La Pacifica, with increasing importance of woody species through time, it is expected that the seasonal oscillations will be damped, but that species richness will rise gradually. An upper asymptote

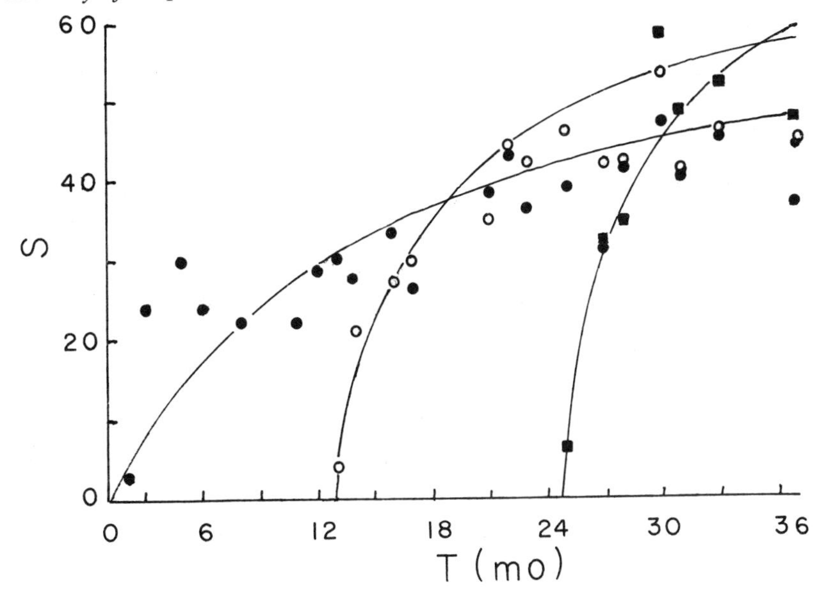

Fig. 3. Species richness for La Selva plots. See Figure 1 legend for symbols.

for species richness, including all phanerogam life forms, is estimated at 90 species/0.17 hectare or 100 species/0.5 hectare at La Pacifica. At La Selva, although complete species counts are not available, phanerogam species richness is minimally estimated at 200 species/0.5 hectare of mature forest.

Diaspore Dispersal Type

During the period of our study, only two diaspore dispersal types were of primary importance in both study areas. Sclerochores are those plants whose diaspores (the unit dispersed) are small-sized, morphologically undistinguished seeds which fall to the ground by gravity or can be blown by sharp wind (see Dansereau and Lems, 1957, for definition), although they may also be eaten and dispersed farther by graminivorous birds and mammals, e.g., doves, sparrows, and small rodents. Sarcochores are those plants whose diaspores are surrounded by or connected to a soft fleshy material, and are usually selected for ingestion and dispersal by some frugivorous birds, mammals, and reptiles.

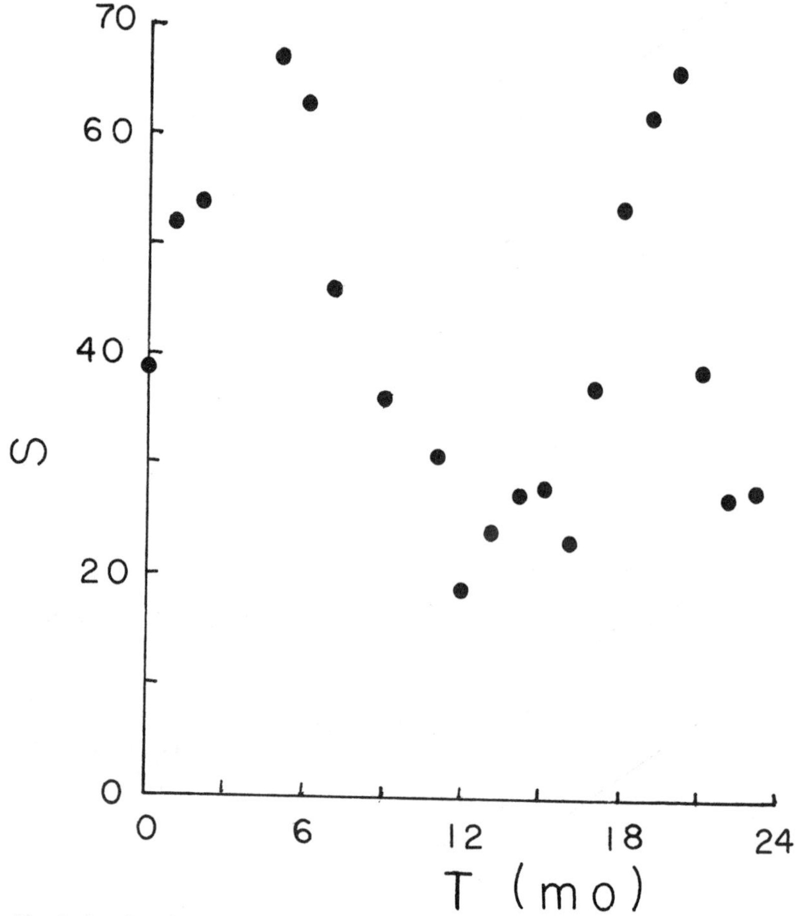

Fig. 4. Species richness for La Pacifica plot

At La Selva, all the first plants (Fig. 5) to become reproductive are sclerochores. Through time the proportion of sclerochores among the reproductive plants declines asymptotically, reaching a level close to zero in climax forest. Concomitant with this decline is the asymptotic rise in representation by sarcochores. At the three-year point, representation of sarcochores at La Selva had risen to about 80%, close to its representation in mature forest (Frankie et al., 1974), while sclerochore representation had fallen to about 5%. Reference to a more complete discussion of seed biology in these two forests (Baker, Frankie, and Opler, manuscript in prep.) reveals that sarcochores

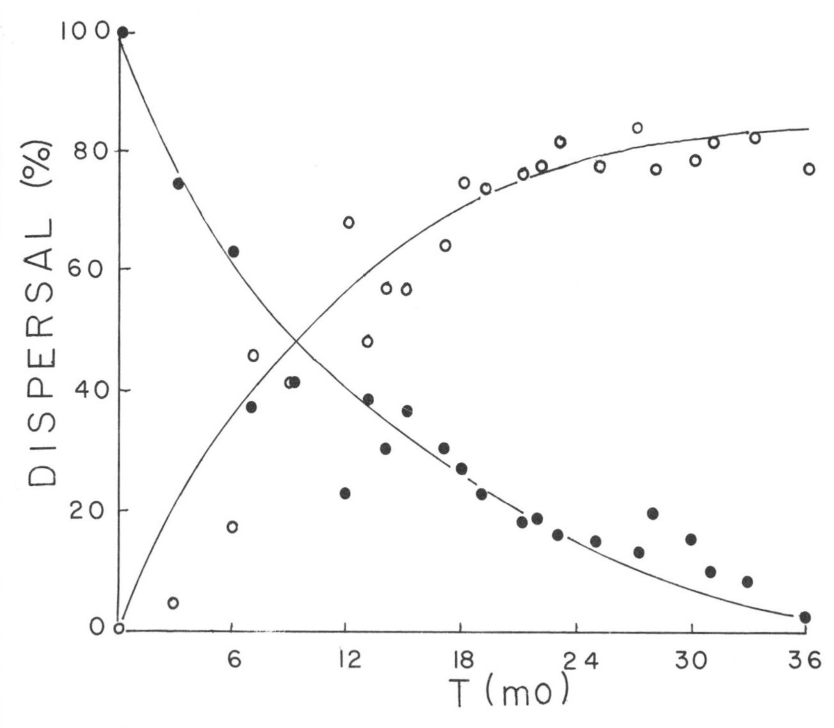

Fig. 5. Change in relative diaspore dispersal type representation for Plot 1 at La Selva, March 1971 to April 1974. *Solid Circles* indicate sclerochores; *open circles* indicate sarcochores.

constitute approximately 85% of mature forest plants, while the remaining 15% is accounted for by other dispersal types.

At La Pacifica, the same trends as at La Selva are revealed (Fig. 6), except that oscillations are introduced by the inherent seasonal cyclicity which, in turn, affects reproduction. This seasonal fluctuation is most marked for sarcochores; while ballochores (not shown), those plants whose diaspores are explosively propelled, display a similar cyclicity with peaks displaced by six months. This latter group is composed of Papilionoideae, Acanthaceae, and Euphorbiaceae. Solanaceae were the dominant reproductive sarcochores during the early La Pacifica succession. Climax dry forest consists of no more than 5% sclerochores while approximately 40% are sarcochores. Through time the oscillation of diaspore representation will become damped as woody plants become more important in the succession, just

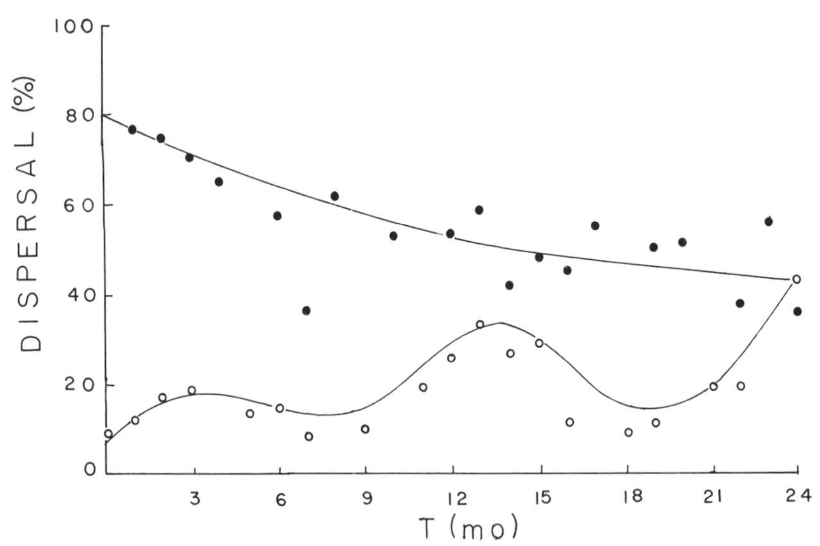

Fig. 6. Change in relative diaspore dispersal types representation at La Pacifica, June 1972 to May 1974

as was projected for species richness. Wind dispersal is increasingly important, and approximately 30% of reproductive plants have winged diaspores in climax forest.

Seed Weight

At both La Selva and La Pacifica mean seed weight (dry) for reproductive plants increases with successional age (Fig. 7). The abscissal value, seed weight class, is a logarithmic series of groupings devised by Baker (1972) to represent all possible seed weights within 14 groupings. Without the semilogarithmic transformation the two functions appear linear for both sites, although it is expected that with time the functions will appear more asymptotic. After three-years' succession at La Selva, mean seed weight (dry) was slightly more than 1 mg, while at La Pacifica a value of almost 2 mg was attained. Just as for the species accumulation and species richness curves, the final upper asymptote for La Selva is about twice that of La Pacifica, 60 mg versus 32 mg mean seed weight (dry), and the two curves are expected to cross over after some years. Salisbury

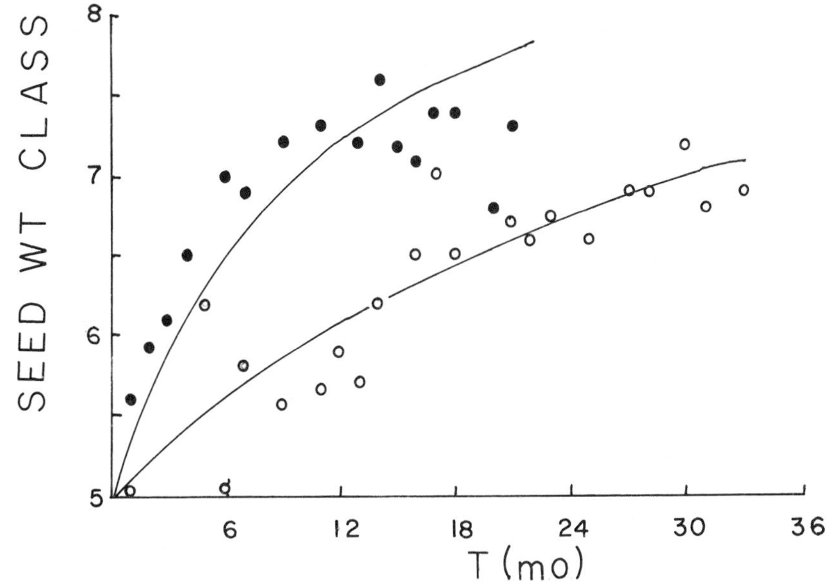

Fig. 7. Change in mean seed weight of reproductive plants through time at La Selva (Plot 1: *open circles*) and at La Pacifica (*closed circles*). Abscissa is seed weight class devised by Baker (1972).

(1942) found greater seed sizes for plants of increasingly advanced seral stages in Great Britain.

Discussion

In relating the significance of our results to the general process of tropical forest recovery, we find it most illuminating to point out the analogies with the theory of island biogeography (MacArthur and Wilson, 1963, 1967). Their theory, briefly stated, is that the richness of any island's biota is determined by an equilibrium between immigration and extinction rates. For increasingly large islands, biotic richness increases linearly with the logarithm of area. The biotic richness and nearness of continental source areas for colonists are also important determinants for the richness of an island's biota. For a depopulated island, initially the immigration rate of new colonists is high and extinction rate is low, but through time immigration rate declines and extinction rate rises, both exponentially, until an equilibrium between the two is reached. A colonization curve may be obtained by integration of the above two rates as they change through time. The first successful immigrants to arrive have high reproductive rates, large population fluctuations, and, of course, are good colonists. Gradually, these species are replaced by species that have relatively low reproductive rates, more stable population levels, and are poor colonists but good competitors. These contrasting syndromes are referred to as *r*- and *K*- strategies, respectively. It is worth considering the several aspects of island theory and how they may apply to ecosystem recovery, particularly in its vegetational aspects.

Diversity

Diversity is a somewhat ambiguous combined function of species richness and equitability (Peet, 1974). Species richness, the absolute number of species, is well known to increase with area, either as a function of island size (MacArthur and Wilson, 1967), or habitat size on continents (Vestal, 1949; Monk, 1971). Our results, which show increasing species richness with successional time, are in agreement with many studies for temperate successions (e.g., Reiners et al., 1971; Monk, 1974). Studies of postfire succession, however, show initial increases in species richness followed by declines (Shafi and Yarranton, 1973). Vogl (in this volume) points out basic differences between "fire-independent" and "fire-dependent" plant associations, and it will probably be found that increasing species richness is a general trait of the former, but not the latter.

Colonization

The seeds of many pioneer plants in successional sequences are characterized by innate dormancy (Barton, 1961; Numata et al., 1964; Wulff, 1973; Vasquez-Yanes, 1974), may be found in forest soils (Guevera and Gomez-Pompa, 1972), and germinate quickly in response to the increased light and thermal regimes created when a forest is cut, burned, or otherwise disturbed (Vasquez-Yanes, 1974). Thus, in a sense, a significant portion of potential colonists have already immigrated when a disruption occurs. These short-lived plants, often annuals, are gradually replaced over time by other plants whose seeds must be dispersed from some other area after the disturbance. The immigration curves for successional processes should not differ significantly in other ways from those of initially empty islands.

In our study, immigration includes only those plants which established at least one individual that grew to reproductive age. This differs from the definition of MacArthur and Wilson (1963, 1967), who include among immigrants all arriving immigrants or propagules not represented by an extant population. Needless to say, our restriction is a necessary pragmatic one, although, as we shall soon see, it has important consequences with regard to the shape of the extinction curve. Our species accumulation curves (see Figs. 1 and 2) most closely approximate immigration curves, the rate of accumulation declining through time as the pool of potential immigrant species is depleted.

Extinction

The classical extinction curve described by MacArthur and Wilson (1963, 1967) does not seem to fit our knowledge of turnover rates related to vegetation recovery. In an attempt to explain the seemingly nonasymptotic recolonization of Krakatau by plants as reported by Docters van Leeuwen (1936), MacArthur and Wilson (1967) postulated an initially high extinction rate followed by a decline and ultimately an increasing rate. However, our reasoning is not in agreement. At the onset of a successional sequence, extinction rate must be zero, since there are no species present to become extinct. It may be noted, however, that this conflict is only a semantic one, being a direct consequence of our definition of immigration. In plant succession, mean species survival time is initially brief, then gradually rises through time. We may thus propose an extinction curve for recovering ecosystems which begins at zero, rapidly rises exponentially to a very high level, and then decreases asymptotically at a slightly less rapid rate than does colonization until the two rates equilibrate (Fig. 8). Note that this extinction curve generally has the same shape as the gross and net production curves illustrated by Odum (1969). If colonists are accepted as those species which establish a breeding population, however small, then the colonization and extinction rates of long-established climax communities may both closely approach zero. That this may be the case is documented by the difficulty of demonstrating natural turnover of bird faunas on continental islands (Lynch and Johnson, 1974).

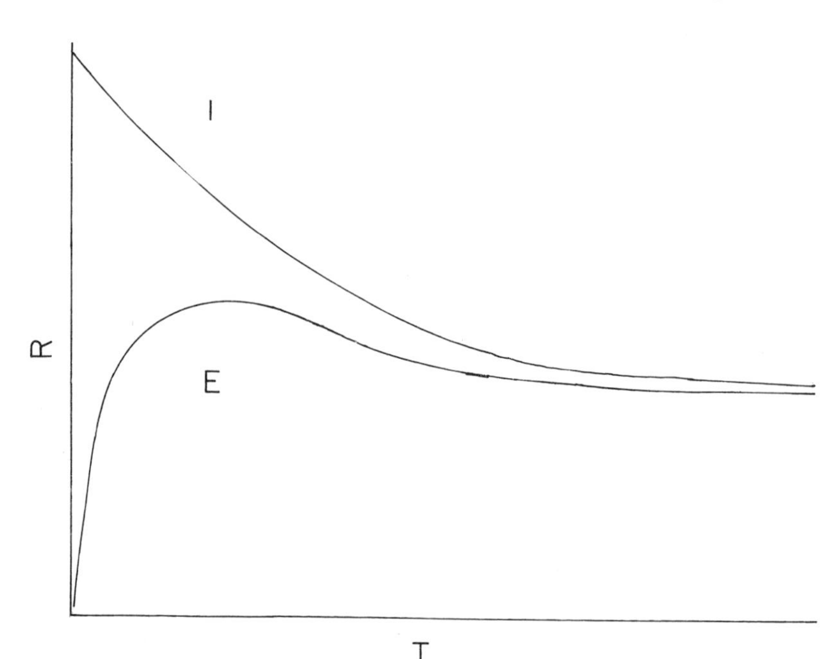

Fig. 8. Putative immigration and extinction curves for secondary succession.

Area

The effect of area on the richness of equivalent plant communities was clearly demonstrated by Vestal (1949). Increasing area has the effect of both altering the colonization rate upwards due to an increased sampling of immigrants and lowering the extinction rate by providing an increased arena for survival.

For ecosystem recovery processes, time acts in the same manner as does area for oceanic islands. Species richness, equitability (see Odum, 1969), and certain biological properties of the community (such as mean seed weight) are some examples of important factors which increase asymptotically with time (or linearly with logarithm of time).

Distance to "Mainland"

That distance of a disturbed area from colonization sources is important to reforestation has been anecdotally observed by Budowski

(1961) for tropical areas. This distance factor will be of increasingly critical importance in the tropics. As the ratio of disturbed areas to climax has increased, instead of small cultivated islands surrounded by forest, the historical tropical land use technique (Croat, 1972; Harris, 1972), small forest islands are being surrounded by vast expanses of perturbed areas, cultivated, grazed, or abandoned. In these circumstances, the potential rate of ultimate natural recovery is being constantly lessened, and the process ultimately leads to the formation of "derived savannas" (Richards, 1952; Budowski, 1956).

It can be seen from Figures 1 and 3 that the colonization rate curves for the La Selva plots 1 to 3 were increasingly steep. This was probably due to the increasing proximity of a diverse source for potential colonists. Thus, the immigration rate will vary for a recovering tropical forest depending upon its distance from, and the size of, a mainland source area (species pool). For a plot surrounded by forest, the immigration rate for species of early successional phases would be relatively low, but would be relatively higher in the late phases; while, concomitantly, for a plot surrounded by disturbed sites, the immigration rate will be high at first, relatively slow later.

Richness of Source Area

Concomitant with the increasing distance to source areas for recovery of disturbed sites is the diminution of source area species richness through the establishment of new lowered equilibria by increased extinction and lowered colonization rates. The best tropical documentation of this process is for the forest birds on Barro Colorado Island, Canal Zone, by Willis (see Terborgh, 1974).

r- and *K*-Strategies

From MacArthur and Wilson's (1967) first elaboration of the characteristics of early and later island colonists has developed a somewhat amorphous assortment of writings adding to, modifying, and criticizing prior statements with regard to *r*- and *K*-selection. As first stated, MacArthur and Wilson postulated that certain species had very high potential reproductive rates (*r*), but poor competitive ability (*r*-selected species); while other species had relatively low reproductive rates, but good competitive survivorship, thus maintaining population levels near environmental carrying capacity (*K*-selected species). On islands, early colonists are *r*-selected, but as community structure develops they are gradually replaced by increasingly *K*-selected species. Pianka (1970) proposed that all species are either *r*- or *K*-strategists, and listed additional characteristics of each grouping. Although some writers have decried the biological applicability of *r*- and *K*-selection (Hairston et al., 1970), most authors have accepted its basic formulation by regarding it as producing a continuum, rather than two discontinuous groupings (Gadgil and Solbrig, 1972). Most agree that it is necessary to determine potential reproductive rates before assigning selective strategy (Gadgil and Solbrig, 1972).

We feel that the application of *r*- and *K*-selection theory to vegetation recovery processes not only is appropriate but has predictive value. If we contrast the characteristics of the group of plants which first colonize a recovering area with those of the constituents of climax vegetation in a nearby area, then from data collected over a relatively brief period we may predict either the rate at which change will occur or the time required to reach a steady state at a given rate. This, of course, assumes that on the average pioneer species are *r*-selected, and that climax species are *K*-selected. In addition, we assume constant rates of change and that changing community characteristics will reflect the strategies of the included species. We do not wish to enter into lengthy discussion of each of these traits but refer the reader to the indicated literature. However, as an example, we have projected expected rates of change through time for mean seed weight at La Selva and La Pacifica using the respective climax values for an upper asymptote (Fig. 9). By this technique we obtain values of 1000 and 150 years as the time required to attain climax at La Selva and La Pacifica, respectively. Although the La Selva estimate might seem high, the fact that Sauer (1958) finds sites of former fields recognizable 400 years after abandonment in recovering wet forest lends credibility to the 1000-year determination. Equally, we could have projected other values such as species richness, equitability, etc. By doing so, the confidence of our recovery time estimates may be refined. For example, the above estimates for La Selva and La Pacifica might seem too large and too small, respectively. This technique requires knowledge of the initial and final values, together with sufficient data to discover initial rates.

As Golley (1975) has cautioned, we must be quite careful about applying any absolute values obtained during one study to the expected outcome of any other study. The same qualitative results may be

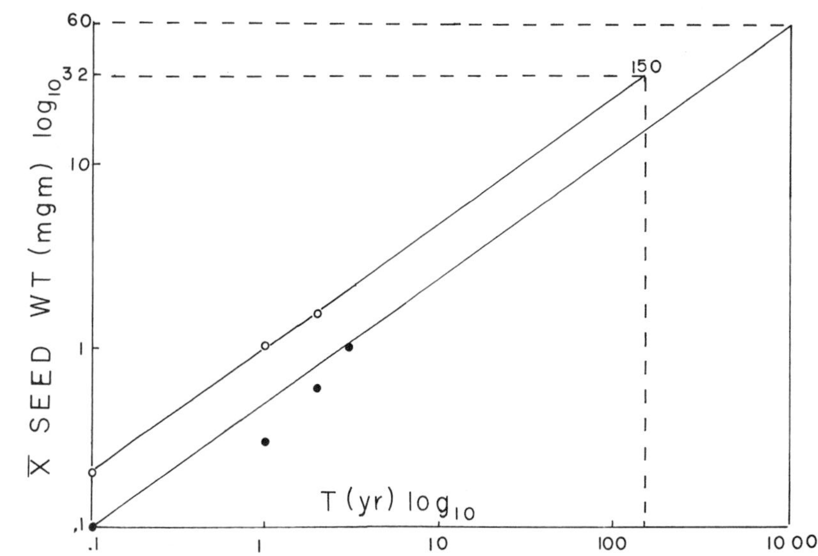

Fig. 9. Projection of observed rate of mean seed weight change at La Selva (*solid circles*) and at La Pacifica (*open circles*) through time to mean seed weight value for climax vegetation

predicted for comparable areas, but not their quantitative beginnings, rates, or end points. Rate of succession may vary widely due to difference in physical environment, length and nature of disturbance, size of area cleared, distance and richness of source area, etc. In addition, species with allelopathic qualities may unexpectedly enter a sere and create an arrested succession, at least temporarily (Wilson and Rice, 1968; Parenti and Rice, 1969). References to some values given for the time required to reach climax (Fig. 10) bring home the above point that there are no apparent correlations between altitude or longitude and the time required to attain climax.

Outlook for Tropics

At the present rate, the rapid diminution of climax forest in the tropics will result in (1) the near absence of climax tropical forests by the end of this century and (2) mass extinctions of biota associated with such forests. There will continue to occur expansions of range and speciation among pioneer-adapted taxa (see Baker, 1974). In the extremely unlikely event that tropical ecosystems are allowed to recover beginning with the year 2000, it will soon be discovered that such full recovery will become possible only over considerable evolutionary time.

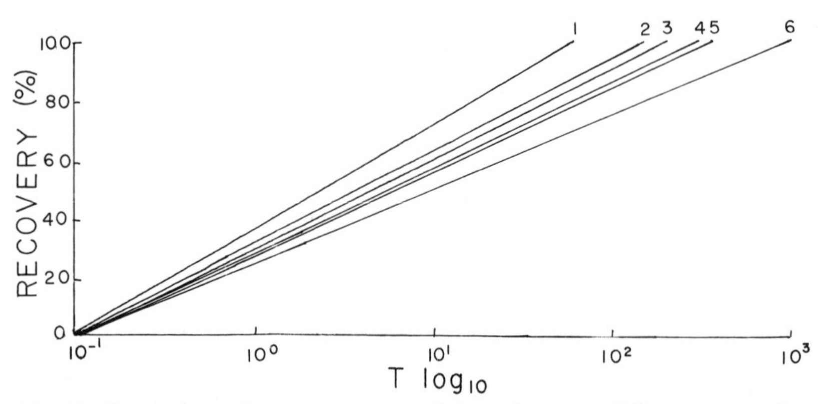

Fig. 10. Comparison of recovery rates and times for some different vegetation types. 1: *Adenostoma* chaparral, California, 60 years (Hanes, 1971). 2: Costa Rica dry forest, 150 years (present study). 3: Deciduous forest, Georgia Piedmont, 200 years (Monk, 1974). 4: Mixed deciduous forest, New Hampshire, 300+ years (Henry and Swan, 1974). 5: Conifer forest, Alberta, Canada, 355 years (Day, 1972). 6: Costa Rica, wet forest, 1000 years (present study)

Table 1. Systematic list of reproductive plants in La Selva secondary plots (alphabetical by family)

Acanthaceae
Bacopa sp.
Blechum brownei Juss.
Dicliptera sp.

Amaranthaceae
Alternanthera sessilis (L.) R. Br.
Borreria latifolia (Aubl.) Schum.
Borreria ocymoides (Burm.) DC.
Cyathula achyranthoides (HBK) Moq.
Cyathula prostrata (L.) Blume
Iresine celosia L.
Genus sp.

Araceae
Dieffenbachia sp. 1
Dieffenbachia sp. 2
Genus sp.

Aristolochiaceae
Aristolochia sp.

Begoniaceae
Begonia semiovata Liebm.

Bignoniaceae
Amphilophium paniculatum (L.) HBK
Anemopaegma orbiculatum (Jacq.) DC.

Bombacaceae
Ochroma pyramidale (Cav. ex Lam.) Urban

Boraginaceae
Cordia bifurcata R. & S.
Cordia lucidula I. M. Johnst.
Tournefortia sp.

Caricaceae
Carica papaya L.
Jacaratia spinosa

Cesalpinaceae (Leguminosae)
Cassia fruticosa Mill. vel. sp. aff.

Commelinaceae
Campelia zanonia (L.) HBK
Genus sp.

Compositae
Clibadium polygonum Blake
Erechtites hieracifolia (L.) Raf.
Erigeron bonariensis L.
Eupatorium iresinoides HBK
Eupatorium macrophyllum L.
Mikania guaco H. & B.
Pseudelephantopus spicatus (Juss.) Rohr
Genus sp. 1
Genus sp. 2

Convolvulaceae
Ipomoea sp. 1
Ipomoea sp. 2
Maripa panamensis

Costaceae
Costus sp. 1
Costus sp. 2

Cucurbitaceae
Anguria warscewiczii Hook.
Genus sp.

Cyperaceae
Cyperus diffusus Vahl.
Cyperus luzulae (L.) Retz
Cyperus odoratus L.
Cyperus tenuis Sw. vel. sp. aff.
Dichromena radicans Schl. & Cham.
Kyllinga sp.
Scleria pterota Presl.

Euphorbiaceae
Acalypha macrostachya Jacq.
Jatropha sp.
Manihot esculenta Crantz
Phyllanthus niruri L. vel. sp. aff.
Dalechampia sp.

Table 1 (*cont.*)

Heliconiaceae
Heliconia imbricata
Heliconia latispatha
Heliconia rostrata
Heliconia subulata

Labiatae
Hemidiodia ocimifolia (Willd.) Schum.
Hyptis capitata Jacq.
Hyptis vilis Kunth. & Bouché
Hyptis sp. 1
Hyptis sp. 2

Loganiaceae
Spigelia anthelmia L.

Malpighiaceae
Bunchosia macrophylla Rose ex. Donn. Sm.
Stigmaphyllon lindenianum Juss.

Malvaceae
Sida rhombifolia L.

Marantaceae
Calathea insignis
Calathea sp.
Genus sp.

Melastomataceae
Actiotis levyana Cogn.
Clidemia dentata D. Don
Conostegia subcrustulata (Beurl.) Triana
Leandra dichotoma (D. Don) Cogn.
Leandra mexicana (Naud.) Cogn.
Miconia barbinervis (Benth.) Triana
Miconia microcarpa DC.
Genus sp. 1
Genus sp. 2

Menispermaceae
Cissampelos andromorpha DC.
Cissampelos tropaeolifolia DC.

Monimiaceae
Siparuna nicaraguensis Heml.
Siparuna pauciflora (Beurl.) A. I

Moraceae
Cecropia obtusifolia Bertol.

Myrsinaceae
Ardisia nigropunctata Oerst.

Nyctaginaceae
Neea amplifolia Donn. Sm.
Neea sp.

Papilionaceae (Leguminosae)
Desmodium adscendens (Sw.) DC
Desmodium axillare (Sw.) DC.
Phaseolus peduncularis HBK
Phaseolus sp. 1
Phaseolus sp. 2

Passifloraceae
Passiflora auriculata HBK
Passiflora sp. 1
Passiflora sp. 2
Passiflora sp. 3

Phytolaccaceae
Phytolacca rivinoides Kunth. & Bouche

Piperaceae
Peperomia pellucida (L.) HBK
Piper arieanum C. DC.
Piper auritum HBK
Piper biolleyi C. DC.
Piper biseriatum C. DC.
Piper colonense C. DC.
Piper hispidum Sw.
Piper sancti-felicis Trel.
Pothomorphe peltata (L.) Miq.

Table 1 (*cont.*)

Poaceae
Axonopus compressus (Sw.) Beauv.
Digitaria adscendens (HBK) Henr.
Gynerium sagittatum (Aubl.) Beauv.
Ichnanthus pallens (Sw.) Munro
Ichnanthus sp.
Lasiacis scabrior Hitchc.
Oplismenus burmannii (Retz) Beauv.
Panicum maximum Jacq.
Panicum pilosum Sw.
Panicum polygonatum Schrad. ex. Schult.
Panicum trichoides Sw.
Paspalum conjugatum Berg.
Paspalum decumbens Sw.

Rhamnaceae
Colubrina spinosa Donn. Sm.
Gouania lupuloides (L.) Urban

Rubiaceae
Diodia sarmentosa Sw.
Drymaria cordata (L.) Willd.
Hamelia patens (L.)
Mannettia reclinata L.
Pentagonia donnell-smithii
Psychotria brachiata Sw.
Psychotria involucrata Sw.
Psychotria dispersa Standl.
Psychotria sp.
Sabicea villosa R. & S.

Sapindaceae
Cupania cooperi Standl.

Scrophulariaceae
Lindernia crustacea (L.) F. W. Muell
Lindernia diffusa (L.) Wettst.
Scoparia dulcis L.
Genus sp.

Solanaceae
Cyphomandra costaricensis Donn. Sm.
Solanum americanum Mill.
Solanum lanceifolium Jacq.
Solanum ochraceo-ferrugineum (Dunal.) Fern.
Solanum rugosum Dunal
Solanum sanctaeclarae Greenm.

Umbelliferae
Hydrocotyle leucocephala Sham. & Schl.

Urticaceae
Urera baccifera (L.) Gaud.

Verbenaceae
Aegiphila martinizensis Jacq.
Lantana trifolium L.

Violaceae
Gleospermum diversipetalum Standl. & L. Wars.

Vitaceae
Cissus sp. 1
Cissus sp. 2

Zingiberaceae
Renealmia cernua (Sw.) Macbride

Family undetermined
Genus sp. 1
Genus sp. 2
Genus sp. 3
Genus sp. 4
Genus sp. 5

Table 2. Systematic list of reproductive plants in the La Pacifica secondary plot (alphabetical by family)

Acanthaceae
- *Blechum brownei* Juss.
- *Elytraria imbricata* (Vahl.) Pers.
- *Ruellia* sp.
- *Tetramerium nervosum* Nees.
- Genus sp.

Aizoaceae
- *Mollugo verticiliata* L.

Amaranthaceae
- *Achryanthes aspera* L.
- *Amaranthus spinosa* L.
- *Gomphrena decumbens* Jacq.

Anacardiaceae
- *Spondias purpurea* L.

Apocynaceae
- *Rauwolfia tetraphylla* L.
- *Stemmadenia obovata* (H. & A.) Schum.

Asclepiadaceae
- *Asclepias curassavica* L.
- Genus sp.

Boraginaceae
- *Cordia dentata* Poir.
- *Heliotropium* sp.

Caesalpiniaceae (Leguminosae)
- *Cassia biflora* L.
- *Cassia stenocarpa* Vog.
- *Cassia tora* L.
- *Cassia leptocarpa* Benth.

Caricaceae
- *Carica papaya* L.

Cochlospermaceae
- *Cochlospermum vitifolium* (Willd.) Spreng.

Commelinaceae
- *Commelina erecta* L.

Compositae
- *Baltimora recta* L.
- *Bidens riparia* HBK
- *Erechtites hieracifolia* (L.) Raf.
- *Eupatorium odoratum* L.
- *Melanthera nivea* (L.) Small
- *Sclerocarpus divaricatus* (Benth.) Benth. & Hook.
- *Tridax procumbens* L.

Convolvulaceae
- *Ipomoea hederifolia* L.
- *Ipomoea meyeri* (Spreng.) G. Don.
- *Ipomoea nil* (L.) Roth.
- *Ipomoea trifida* (HBK) G. Don.
- *Ipomoea umbraticola* House
- *Merremia quinquefolia* (L.) Hallf.
- *Merremia umbellata* (L.) Hallf.

Cucurbitaceae
- *Cucumis anguria* L.

Cyperaceae
- *Cyperus hermaphroditus* (Jacq.) Stan
- *Cyperus mutisii* (HBK) Griseb.
- *Cyperus tenerrimus* Presl.

Euphorbiaceae
- *Acalypha arvensis* Poepp & Endl.
- *Chamaesyce hirta* (L.) Millsp.
- *Chamaesyce hyssopifolia* (L.) Small.
- *Cnidista urens*
- *Croton hirtus* L'Hèr.
- *Ditaxis guatemalensis* (Muell. Arg.) Pax. & Hoffm.
- *Euphorbia heterophylla* L.
- *Phyllanthus* sp.
- Genus sp.

Table 2 (*cont.*)

Flacourtiaceae
Casearia nitida Jacq.

Labiatae
Hyptis suaveolens (L.) Poit.
Salvia occidentalis Sw.
Salvia sp.

Loganiaceae
Spigelia anthelmia L.

Malvaceae
Abutilon crispum (L.) Medic.
Malvastrum americanum (L.) Torr.
Pavonia guanacastensis Standl.
Pseudabutilon spicatum (HBK) Fries.
Sida acuta Burm. "Orange flowered"
Sida acuta Burm. "White flowered"
Sida glutinosa Commers.
Sida rhombifolia L.
Sida savannarum K. Schum.
Sida urens L.
Wissadula aff. *hirsuta* Presl.

Marantaceae
Maranta arundinacea L.

Mimosaceae (Leguminosae)
Acacia collinsii Saffl.
Calliandra portoricensis (Jacq.) Benth.
Mimosa pudica L.
Mimosa quadrivalvis L.

Nyctaginaceae
Boerhaavia erecta L.

Oxalidaceae
Oxalis neaei DC.

Papilionaceae (Leguminosae)
Aeschynomene americana L.
Calopogonium mucunoides Desv.
Centrosema pubescens Benth.
Centrosema sagittatum (H. & B.) Brandg.
Cracca ochroleuca
Crotalaria pumila Orteg.
Desmodium glabrum (Mill.) DC.
Desmodium triflorum (L.) DC.
Indigofera suffruticosa Mill.
Indigofera sp.
Mucuna pruriens (L.) DC.
Phaseolus atropurpureus DC.
Rhynchosia minima (L.) DC.
Stylosanthes sp.
Teramnus uncinatus (L.) DC.
Genus sp. 4
Genus sp. 5
Genus sp. 6
Genus sp. 7
Genus sp. 8

Passifloraceae
Passiflora foetida (L.)

Poaceae
Anthephora hermaphrodita (L.) Kuntze
Bouteloua disticha (HBK) Benth.
Brachiaria reptans (L.) Gard. & Hubb.
Cenchrus pilosus HBK
Cynodon dactylon (L.) Pers.
Dactyloctenium aegyptium (L.) Richt.
Dichanthium annulatum (Forsk.) Staff.
Digitaria bicornis (Lam.) R. & S.
Echinochloa colonum (L.) Link.
Eleusine indica (L.) Gaertn.
Eragrostis ciliaris (L.) R. Br.
Hyparrhenia rufa (Nees.) Stapf.
Leptochloa filiformis (Lam.) Beauv.
Panicum fasciculatum Sw.
Panicum molle Sw.
Paspalum corypheum Trin.
Paspalum notatum Flugge
Genus sp. 1
Genus sp. 2
Genus sp. 3
Genus sp. 4
Genus sp. 5
Genus sp. 6

Table 2 (*cont.*)

Polemoniaceae
Loeselia ciliata L.

Portulacaceae
Portulaca oleracea L.

Rubiaceae
Richardia scabra L.

Scrophulariaceae
Capraria biflora L.

Solanaceae
Capsicum annuum L. var. *minimum* (Miller) Heiser
Cestrum sp.
Physalis ignota Britt.
Solanum americanum Mill.
Solanum ferrugineum Jacq.
Solanum hazeni Britt.
Solanum hirtum Vahl.
Solanum nudum Dunal

Sterculiaceae
Ayenia dentata Brand
Guazuma ulmifolia Lam.
Melochia nodiflora Sw.
Waltheria indica L.

Tiliaceae
Triumfetta lappula L.

Ulmaceae
Trema micrantha (L.) Blume

Verbenaceae
Lantana camara L.
Priva lappulacea (L.) Pers.

Violaceae
Hybanthus attenuatus (H. & B.) G. K. Schultze

Vitaceae
Cissus rhombifolia Vahl.

Zygophyllaceae
Kallstroemia pubescens (G. Don) Davi
Genus sp.

Family unknown
Genus sp. 1
Genus sp. 2

Table 3. Occurrence of reproductive plants in La Selva plots. Numbers refer to plots established March 1971, 1972, and 1973, respectively.

	1971					1972							1973								1974
	Mar.	June	Aug.	Sept.	Nov.	Feb.	Mar.	Apr.	May	July	Aug.	Dec.	Jan.	Feb.	Apr.	June	July	Sept.	Oct.	Dec.	Apr.
Bacopa sp.			1	1	1	1	1	1	1,2	1,2	2	1,2	1,2	1,2	1	2,3	2	1,2,3	1	1,2	1,2
B. brownei		1	1	1	1	1	1	1,2	1,2	2	2	2	2	2	2,3	2,3	2,3	2,3	3	2,3	2,3
Dicliptera sp.															2						
A. sessilis	1								2	2	2										
B. latifolia		1							2								3				
B. ocymoides		1							2							3	3	3	3		
C. achyranthoides		1	1	1	1	1	1	1	1	1	1	1,2	1,2	1,2	2	2	1,2,3	1,2,3	2,3	1,2,3	2,3
C. prostrata			1		1			2	2	2	2				3	3	3	3			
I. celosia							1	1					1,2	1	1	2					
Amaranth.	1																				
Dieffenbachia 1																				1	
Dieffenbachia 2																					1,2
Arac.												1	1	2		1	1,2	1		1	
Aristolochia							1	1	1									1,2		2	
B. semiovata			1											2							
A. paniculatum																	2				
A. orbiculatum																	1	2			
O. pyramidale																				1	1
C. lucidula																					1
C. bifurcata			1	1	1	1	1		1	1,2	1,2	1,2	1,2	1,2	1,2	1,2,3	1,2,3	1,2,3	2,3	2,3	1,2,3
Tournefortia sp.									1	1	1		1	1	1	1			1	1	
C. papaya												1,2	2	2	2	1,2	1,2	1,2,3	1,2,3	1,2,3	2,3
J. spinosa											1										

Table 3. Occurrence of reproductive plants in La Selva plots. Numbers refer to plots established March 1971, 1972, and 1973, respectively.

	1971					1972							1973								1974
	Mar.	June	Aug.	Sept.	Nov.	Feb.	Mar.	Apr.	May	July	Aug.	Dec.	Jan.	Feb.	Apr.	June	July	Sept.	Oct.	Dec.	Apr.
C. fruticosa										1	1	1,2				1,2	1,2	1,2,3	1,2,3	1,2,3	
C. zanonia			1																		
Commelinac.																	1				
C. polygonum													2							3	3
E. hieracifolia		1	1	1	1				2	2	2					3	3	3	3		3
E. bonariensis			1																		
E. iresinoides		1	1															3			
E. macrophyllum			1	1	1						2	2				3	3	3	3	3	
M. guaco												1,2	1,2	2					1,2	2,3	3
P. spicatus													2	2						3	
Composite 1											2										
Composite 2												2	2	2	2				2,3	2,3	
Ipomoea 1							1					1,2	2	1,2	1,2				2,3	2,3	2,3
Ipomoea 2											1	1	2		2					1	1,2,3
M. panamensis																	1	1,2	1	1	
Costus 1			1									2	2	2			1,2,3	1,2,3	1,2	2,3	2,3
Costus 2										1							2				3
A. warscewiczii							1					2									
Cucurbitac.																		3	3		
C. diffusus		1																			
C. luzulae		1	1	1					2	1,2	2					3	2,3	3	3	3	3
C. odoratus									2		2							3			
C. tenuis		1	1						2	2	2					3	2,3	2,3	3	3	3

D. radicans		1							2		2					3	3	3			
Kyllinga sp.			1													3					
S. pterota		1	1	1	1	1	1	1	1,2	1,2	2	2	2	2	1,2	2,3	2,3	3	3	3	3
A. macrostachya							1	1	1	1				2	1,2	1,2				1,2	1,2
? *Jatropha* sp.		1																			
M. esculenta				1	1				1	1	1	1	1		1	1	1	1	1		1
P. niruri		1	1	1			1	1		2						3	3	3	3	3	3
Dalechampia sp.																		2			
H. imbricata																		2	2		
H. latispatha												2				2	1,2,3	1,2	2	2	
H. rostrata												1	1	1	1	1	1	1	1,2	1,2	
H. subulata												1	1		2						
H. ocimifolia													2	2		2		2			
H. capitata		1	1	1	1	1	1	1	1	1,2	1,2	1,2	1,2	1,2	1,2	1,2,3	1,2,3	1,2,3	1,2,3	1,2,3	
H. vilis	2,3		1								2		2	2	2	2	2	2,3	3	2,3	
Hyptis 1														2	2						
Hyptis 2																			2,3		
S. anthelmia		1															3				
B. macrophylla																	1	1,2	2	1,2	
S. lindenianum										2					2	2,3	2	2,3	3	3	
S. rhombifolia		1	1	1			1	1	2	1,2	2	2	2	2	2,3	2	2,3	2,3	3	3	
C. insignis															2	1,2	1	1,2	1,2	1	
Calathea sp.																			1	1,2	
Marantac.													2	2	2	2				2	
A. levyana									1	1,2			1	1	1,2			3		1,2	
C. dentata						1	1	1	1	1	1	2	2	2	1,2	1,2	1,2	1,2	1,2	1,2,3	
C. subcrustulata														1		3	3	3	3	3	3
L. dichotoma												1	1	1,2	1	1,2			1,2	1	
L. mexicana								1						1	1,2		1,2	1,2	1,2	1,2,3	2

Table 3. Occurrence of reproductive plants in La Selva plots. Numbers refer to plots established March 1971, 1972, and 1973, respectively.

	1971					1972							1973								1974
	Mar.	June	Aug.	Sept.	Nov.	Feb.	Mar.	Apr.	May	July	Aug.	Dec.	Jan.	Feb.	Apr.	June	July	Sept.	Oct.	Dec.	Apr.
M. barbinervis																					2
M. microcarpa						1						1			1	1	1			1	1,2
Melastome 1							1														
Melastome 2														1	1			1			
S. nicaraguensis														2		2					2
S. pauciflora													1	1	1		1	1		1	1
C. andromorpha											2		1,2	1,2	1,2	1,2		2		1	
C. tropaeolifolia										1	1	1,2	1,2	1,2	1,2	1,2	1,2	1,2	2	1,2,3	1,2,3
C. obtusifolia																			1	1	
A. nigropunctata														1							
N. amplifolia				1		1	1	1	1	1	2	1,2	1,2	1,2	1,2	1,2,3	1,2,3	1,2,3	1,2,3	1,2,3	1,2,3
Neea sp.										1	1	1,2	1,2	1	1	2	1	1		1	
D. adscendens		1						1,2	1,2	1,2						3	3	2,3	3		
D. axillare										2											
P. peduncularis									1,2	2	2				1	3	1,2,3	2,3	1,3		1
Phaseolus 1												1									
Phaseolus 2																					
P. auriculata															2				3		
Passiflora 1													1	1	1	1	1	1,2,3	1,2	1	1,2
Passiflora 2																				3	
Passiflora 3																				3	
																				2	
P. rivinoides		1	1	1	1	1	1	1	1	2,3	2,3	2,3	2,3	2,3	2,3	1,2,3	1,2,3	1,2,3	2,3	1,2,3	1,2,3
P. pellucida		1																3			

P. arieanum															1			1,2		
P. auritum			1	1	1	1	1	1	1	1	1,2	1,2	1,2	1,2	1,2	1,2	1,2,3	1,2,3	1,2,3	1,2,3
P. biolleyi																				1
P. biseriatum											1	1	1	1			1	1	1	1
P. colonense			1	1	1		1	1	1	1	2	1,2	1,2	1,2	1,2	1,2	1,2,3	1,2	1,2,3	2,3
P. hispidum																	1		1,2	
P. sancti-felicis			1	1	1	1	1			1,2	1,2	1,2	1,2	1,2	1,2	1,2	1,2,3	1,2,3	1,2,3	1,2,3
P. peltata				1	1	1	1	1	1	2	1,2	2	2	2	3	3	2,3	3	3	3
A. compressus	1	1	1	1	1	1	1	1	1	1		2	1,2			3			3	
D. adscendens																				
G. sagittatum																	2	2		
I. pallens																	1,3			
Ichnanthus																				
L. scabrior												1	1,2	1,2						1,2,3
O. burmannii												2	2	2					3	
P. maximum	1	1	1	1		1	1	1	1,2	1,2	1,2	1,2	1,2	2	1,3	1,2,3	1,2,3	1,2,3	2,3	2,3
P. pilosum	1	1	1	1	1	1	1,2	1,2	1,2	1,2	1,2	1,2	1,2	2,3	3	1,2,3	1,2,3	1,2,3	2,3	2,3
P. polygonatum																				
P. trichoides	1	1							2											
P. conjugatum		1									2	2			3	1,3	3	3	3	3
P. decumbens													2	2						
C. spinosa											1	1,2	1,2	1,2				1	1,2	1,2
G. lupuloides																	2		1,2,3	1,2,3
D. sarmentosa		1																		
D. cordata	1			1				2	2	2	2	2		2,3	2,3	2,3	2,3	3	3	2
H. patens			1				1		1	1,2	1,2	1,2	1,2	1,2	1,2,3	1,2,3	1,2,3	1,2,3	1,2,3	1,2,3
M. reclinata												2								
P. donnell-smithii						1	1	1	1	1	1	1	1	1	1	1	1	1	1	1,2

Table 3. Occurrence of reproductive plants in La Selva plots. Numbers refer to plots established March 1971, 1972, and 1973, respectively.

	1971					1972							1973								1974
	Mar.	June	Aug.	Sept.	Nov.	Feb.	Mar.	Apr.	May	July	Aug.	Dec.	Jan.	Feb.	Apr.	June	July	Sept.	Oct.	Dec.	Apr.
P. brachiata									1	1	1	1	1	1	1	1,2	1,2	1,2	1	1	1,2,3
P. involucrata																1		1	1	1	
P. dispersa																					1
Psychotria sp.																	1	1	1		
S. villosa			1	1		1		1	1	1	1,2	1,2	1,2	1,2	1,2	2,3	1,2,3	1,2,3	1,2,3	1,2,3	1,2,3
C. cooperi																					1,2
L. crustacea									2												
L. diffusa									2												
S. dulcis		1	1																		
Scroph.																		3	3	3	
C. costaricensis																2	2				
S. americanum									2	2		2	2								
S. lanceifolium					1	1	1	1	1	1	1	2	1,2	1,2	1,2		3	1,2,3	2,3	2,3	2,3
S. echraceo- *ferrugineum*					1	1	1	1	1	1	1	1	1	1	1	2	2	2,3	3	2,3	2,3
S. rugosum						1	1	1	1	1	1	1	1,2	1,2	1,2	1,2	1,2	1,2,3	1,2,3	1,2,3	1,2,3
S. santaeclarae			1	1		1	1					1	1,2	1	1,2	1,2,3	1,2	1,2,3	1,2,3	1,2,3	1,2,3
H. leucocephala									2												
U. baccifera				1	1	1		1				2	1,2	1,2	2	2	1,2	2,3	1,2,3	1,2,3	
A. martinizensis														1				1,2,3	1,2,3	1,2,3	
L. trifolium													2	2	2	2		2,3	2,3		3
G. diversipetalum								1										1			
Cissus 1									1	1,2	1,2	2			2,3	2,3	2,3	2,3	2,3	3	2,3

Cissus 2									2	2,3	1,2,3	1,2,3	2	
R. cernua	1		1	1				1,2	1,2	1,2	1,2	1,2	1,2	1,2
Family 1		2												
Family 2					1	1		1			1		1,3	
Family 3								2						
Family 4											3			
Family 5											2			

Table 4. Occurrence of reproductive plants at La Pacifica. Numbers refer to number of subplots occupied by reproductive individuals of a species during a given month.

	1972						1973										1974				
	June	July	Aug.	Sept.	Nov.	Dec.	Jan.	Mar.	May	June	July	Aug.	Sept.	Oct.	Nov.	Dec.	Jan.	Feb.	Mar.	Apr.	May
B. brownei					1	3	5	1	1							1	1	4	2		
E. imbricata						1	1		1							2	1	2			
Ruellia sp.						3	2	2	1								5	8	4	2	
T. nervosum						1	1								1	2	1	2	2		
Acanth.																		1			
M. verticiliata	1	1	1																		
A. aspera	6	1	7	5	8	12	8	9	7				5	9	12	15	8	11	5	5	3
A. spinosa	8	10	11	4	1	1			1									2	2	2	
G. decumbens	3	5	5	2		2															
S. purpurea																		3	4		
R. tetraphylla																		1			
S. obovata			4	1	1	2			2		1		1			1	1		2		2
A. curassavica					1										1	1					
Asclepidac.																1	1	1			
C. dentata																					1
Heliotropium sp.	1																				
C. papaya														1							
C. biflora			1	4	6	3	3	1									2	4	4	3	2
C. stenocarpa					1	2		1							1	4					
C. tora					4																
C. leptocarpa																2	1			1	
C. vitifolium																					
C. erecta	9	4	1								1	1									2

B. recta		10	9								1	10	4								
B. riparia				3	9	6										5	2	1			
E. hieracifolia																		1			
E. odoratum																	1	1	1		
M. nivea					12	14	13	13	11	4	3				12	15	16	16	15	2	14
S. divaricatus		1	1	1		1											1				
T. procumbens	10	9	11	3	1	2		3	1			1		1							
I. hederifolia				1	2	2	1	1													
I. meyeri							5										9	5			
I. nil					4									1	3	3	3	4			
I. trifida				4	6	7	3								1		4	3			
I. umbraticola					4	8	2	4								4		1			
M. quinquefolia																1	1	1	1		1
M. umbellata					1	4	4	5	5	5							6	10	7	7	7
C. anguria		7	7	2	1	1	1	2				2	3								
C. hermaphroditus		8	8																		
C. mutissi		1	2																		
C. tenerrimus	10	1	1																		
Cyperus sp.		1	1																		
A. arvensis	3	5	6										1								
C. hirta	9	13	12																		
C. hyssopifolia	9	11	10	5	2	4		1													
C. urens		2	2	2	4	3	2	2	3	5	4	5	5	4	4	4	2	1	1	1	
C. hirtus		3	4	3							3	3	5								
D. guatemalensis							1	1	1	1	1	1	1		1	1	2				
E. heterophylla		1																			
Phyllanthus sp.	6	10	11	3	2			1													
Euphorb.														1							

Table 4. Occurrence of reproductive plants at La Pacifica. Numbers refer to number of subplots occupied by reproductive individuals of a species during a given month.

	1972						1973										1974				
	June	July	Aug.	Sept.	Nov.	Dec.	Jan.	Mar.	May	June	July	Aug.	Sept.	Oct.	Nov.	Dec.	Jan.	Feb.	Mar.	Apr.	May
C. nitida																					2
H. suaveolens				5	17	17	17	16	1	1					17	17	15	16	15	15	15
S. occidentalis						2	1										1				
Salvia sp.																		1	1		
S. anthelmia	1	1	1	1								1									
A. crispum			7	9	7	9	7	5		10	1	1	3		1	8	5	3	1	1	
M. americanum					8	11	11	9	9						7	9	12	13	6	6	
P. guanacastensis					3	5	1		1						2	4	4	4	2	1	
P. spicatum					1																
S. acuta Orange	15	17	13	11	9	8	2	2	2	2		1	1		1	3		1			
S. acuta White					2	1	1	1								2	1	1			
S. glutinosa					1																
S. rhombifolia	9	11	9	15	17	16	17	15	15	15			4	3	13	15	16	14			
S. savannarum					1																
S. urens					1																
W. hirsuta					2	2	2										1	2			
M. arundinacea		4	4									2									
A. collinsii									1										1	1	1
C. portoricensis	11	12	13	5	3	7	1	1	1		1								1		
M. pudica				1	3	2										1					
M. quadrivalis						2							1	5	4	5	2	1			
B. erecta	5	1																			
O. neaei	9	3	4	1	1	1	1		1												3

A. americana					1	1								3	1	1	1		
C. mucunoides					12	17	1						1	9	13	13	11		
C. pubescens															1	1	1	1	1
C. saggittatum															1		1		
C. ochroleuca				1	1										6	6	5	2	1
C. pumila					6	8	5	1						1	5	3	2		
D. glabrum					8	11	8	1					10	5	5	14	7		
D. triflorum							1							1	1				
I. suffruticosa							1						1	1	1	1	1	1	1	1
Indigofera															1	1	1	1	1
M. pruriens					3	2	1	1	2					4	4	4	4	4	4	4
P. atropurpureus					1	1									1	1			
R. minima					5	12	7	7	2				3	1	10	12	9	3	1	1
Stylosanthes sp.					7	9	1									1			
T. uncinatus					10	14	12	3						7	10	11	10		
Papilionoid. 4														3		4	8	3		1
Papilionoid. 5															1	2	3	1	1
Papilionoid. 6																1			
Papilionoid. 7																	1	1	
Papilionoid. 8																			
P. foetida	1	1	1	1	1	1				1	1								
A. hermaphrodita					1	1													
B. disticha					1	1													
B. reptans	5	4	4	1															
C. pilosus	13	16	16	9	16	11	11	2	3	7	10	6	8	9	10	8	8	3	3
C. dactylon	12	8	8	2															
D. aegyptium	1	15	15	9	1	1	1												
D. annulatum	5	2	2							4									
D. bicornis	17	15	15	7															

Table 4. Occurrence of reproductive plants at La Pacifica. Numbers refer to number of subplots occupied by reproductive individuals of a species during a given month.

	1972						1973										1974				
	June	July	Aug.	Sept.	Nov.	Dec.	Jan.	Mar.	May	June	July	Aug.	Sept.	Oct.	Nov.	Dec.	Jan.	Feb.	Mar.	Apr.	May
E. colonum	17	17	17	2																	
E. indica	1	8	8	1																	
E. ciliaris	1	3	3		1	1															
H. rufa				1	16	16	13	1													
L. filiformis	12	16	16	6	3	1					4	4	3								
P. fasciculatum		8	11	5																	
P. molle		1	1																		
P. corypheum		1	1																		
P. notatum	4	8	8	1							1										
Poac. 1						1															
Poac. 2												1	1	2							
Poac. 3													1								
Poac. 4													4	4	1		1	2			
Poac. 5													1								
Poac. 6																1		1			
L. ciliata							1									1	2	2			
P. oleracea	7	2	2																		
R. scabra	1	1																			
C. biflora			1			2	1		1	1	1	1	1	1				1			1
C. annuum	5	7	8	5	2	3	1	1			1	3	3	1		1	1	1			1
Cestrum sp.								1										1	1		
P. ignota			1	1	1	1															
S. americanum	1			2																	

S. ferrugineum										3	1										1
S. hazeni		4	6	6	4	5	1	1	1	7	9	10	10	8	9	9	6	7	8	7	12
S. hirtum	3	7	5	5	3	3			1	2	1	1	1	3	1	3	1	1	1	1	2
S. nudum		1	1									1	1	1			1				1
A. dentata															1	1	1	1			
G. ulmifolia									2											12	13
M. nodiflora					16	14	16	14	16	16						6	13	15	14	14	1
W. indica					12	14	10	7	11	11					1	1	3	5	5	5	
T. lappula					6	9	7	4	8	3	4	2	1			8	8	7	7	7	5
T. micrantha										3	3	4	4	4	1						3
L. camara			1	4	1	2	1		1	2	2	2	1	2	2	2	1	2	2	2	3
P. lappulacea	17	17	14	5	3	4				1	8	4	4	5	2	5	3	2			
H. attenuatus	10	10																			
C. rhombifolia					1	1						1					1	1	1		
K. pubescens	14	7																			
Zygophyll.					2																
Family 1	1		1																		
Family 2												1									

Acknowledgments

This study was funded by National Science Foundation grants GB-7805 and GB-25592 to the Organization for Tropical Studies' "Ecosystem Comparison Study."

Werner Hagnauer and family of Hacienda La Pacifica provided land and labor, not to speak of every encouragement, for the dry forest study; while Gary N. Hartshorn and the staff of the Organization for Tropical Studies cut the wet forest study areas. Jorge Campabadal and his staff, particularly Liliana Echevarria, provided constant logistical support during our Costa Rican stays.

Discussions during a seedling biology symposium at Barro Colorado Island, Canal Zone, sponsored by the Smithsonian Institution, with Sylvia Del Amo, Robert Dressler, Robin Foster, Gary Hartshorn, Manuel Rico, and Carlos Vasquez-Yanes were particularly appropriate to the ideas expressed herein.

The scientific names of plants included in the study were provided by the determination of specimens by countless botanists, most notably W. D'Arcy, D. Austin, R. Baker, W. Burger, T. Croat, G. Davidse, J. Dwyer, L. Fournier, P. Fryxell, A. Gentry, J. Gentry, L. Gomez, R. Leisner, R. Pohl, L. Poveda, D. Spellman, C. de Trejos, and J. Wurdack. Vouchers of these plants have been deposited at the Field Museum of Natural History, Missouri Botanical Garden, Museo Nacional de Costa Rica, Universidad de Costa Rica, and University of California, Berkeley. We thank Steve Hubbell, who reviewed an early version of this chapter.

Literature Cited

Allen, R. 1975. The year of the rain forest. New Sci. 66(946):178-181.

Ashton, P. S. 1969. Speciation among tropical forest trees; some deductions in the light of recent evidence. Biol. J. Linnean Soc. 1:155-196.

Baker, H. G. 1972. Seed weight in relation to environmental conditions in California. Ecology. 53:997-1010.

_____. 1974. The evolution of weeds. Ann. Rev. Ecol. Syst. 5:1-24.

Barton, L. V. 1961. Seed Preservation and Longevity. Leonard Hill, London. 216 pp.

Budowski, G. 1956. Tropical savannas, a sequence of forest felling and repeated burnings. Turrialba 6:23-33.

_____. 1961. Studies on forest succession in Costa Rica and Panama. Ph.D. thesis, Yale University. 189 pp.

Clements, F. E. 1916. Plant Succession: An Analysis of the Development of Vegetation. Carnegie Inst. Wash. Publ. 242. 512 pp.

Cowles, H. C. 1899. The ecological relations of the vegetation on the sand dunes of Lake Michigan. Bot. Gaz. 27:95-117, 167-208, 361-391.

Croat, T. 1972. The role of overpopulation and agricultural methods in the destruction of tropical ecosystems. Bioscience 22:465-467.

Dansereau, P., and K. Lems. 1957. The Grading of Dispersal Types in Plant Communities and Their Ecological Significance. Inst. Bot. Univ. Montreal, Contrib. 71. 52 pp.

Day, R. J. 1972. Stand structure, succession, and use of southern Alberta's Rocky Mountain forest. Ecology 53:472-478.

Docters van Leeuwen, W. M. 1936. Krakatau, 1833 to 1933. Ann. Jard. Botan. Buitenzorg 56-57:1-506.

Drury, W. H., and I. C. T. Nisbet. 1973. Succession. J. Arnold Arboretum 54:331-368.

Frankie, G. W., H. G. Baker, and P. A. Opler. 1974. Comparative phenological studies of trees in tropical wet and dry forests in the lowlands of Costa Rica. J. Ecol. 62:881-919.

Gadgil, M., and O. T. Solbrig. 1972. The concept of *r*- and *K*-selection: evidence from wild flowers and some theoretical considerations. Am. Natur. 106:14-31.

Golley, F. B. 1975. Structural and functional properties as they influence ecosystem stability. Pages 97-102 *in* Proc. 1st International Congress of Ecology. The Hague, Netherlands.

Gomez-Pompa, A., C. Vasquez-Yanes, and S. Guevara, S. 1972. The tropical rain forest: a nonrenewable resource. Science 177:762-769.

Guevara, S., S., and A. Gomez-Pompa. 1972. Seeds from surface soils in a tropical region of Veracruz, Mexico. J. Arnold Arboretum 53:312-335.

Hairston, N. G., D. W. Tinkle, and H. M. Wilbur. 1970. Natural selection and the parameters of population growth. J. Wildl. Manage. 34:681-690.

Hanes, T. L. 1971. Succession after fire in the chaparral of southern California. Ecol. Monogr. 41:27-52.

Harcombe, P. 1975. Nutrient accumulation by vegetation during the first year of recovery of a tropical forest ecosystem. *In* this volume.

Harris, D. R. 1972. The origins of agriculture in the tropics. Am. Sci. 60:180-193.

Henry, J. D., and J. M. A. Swan. 1974. Reconstructing forest history from live and dead plant material—an approach to the study of forest succession in southwest New Hampshire. Ecology 55:772-783.

Holdridge, L. R., W. C. Grenke, W. H. Hatheway, T. Liang, and J. A. Tosi, Jr. 1971. Forest Environments in Tropical Life Zones: A Pilot Study. Pergamon Press, Oxford.

Horn, H. S. 1974. The ecology of secondary succession. Ann. Rev. Ecol. Syst. 5:25-37.

Janzen, D. H. 1973. Tropical agroecosystems. Science 182:1212-1219.

Lieth, H. 1974. Primary productivity of successional stages. Pages 187-193, 293-356 *in* R. Knapp, ed. Handbook of Vegetation Science. Part VIII. Vegetation Dynamics. Junk, The Hague.

Lynch, J. F., and N. K. Johnson. 1974. Turnover and equilibrium in insular avifaunas, with special reference to the California channel islands. Condor 76:370-384.

MacArthur, R. H., and E. O. Wilson. 1963. An equilibrium theory of insular zoogeography. Evolution 17:373-387.

———. 1967. The Theory of Island Biogeography. Princeton Univ. Press. 203 pp.

MacBryde, B. 1972. Set-backs to conservation in Ecuador. Biol. Conserv. 4:387-388.

Margalef, R. 1968. Perspectives in Ecological Theory. Univ. of Chicago Press. 111 pp.

Monk, C. D. 1971. Species and area relationship in the eastern deciduous forest. J. Elisha Mitchell Sci. Soc. 87:227-230.

———. 1974. Plant species diversity in old-field succession on the Georgia piedmont. Ecology 55:1075-1085.

Numata, M., I. Hayashi, T. Komura, and K. Oki. 1964. Ecological studies on the buried-seed population in the soil as related to plant succession. Jap. J. Ecol. 14:207-215.

Odum, E. P. 1969. The strategy of ecosystem development. Science 164: 262-279.

Odum, H. T., ed. 1970. A Tropical Rain Forest. Div. Tech. Inf., U.S.A., A.E.C., Oak Ridge, Tenn.

Parenti, R. L., and E. L. Rice. 1969. Inhibitional effects of *Digitaria sanguinalis* and possible role in old-field succession. Bull. Torrey Bot. Club. 96:70-78.

Peet, R. K. 1974. The measurement of species diversity. Ann. Rev. Ecol. Syst. 5:8-15.

Pianka, E. R. 1970. On *r*- and *K*-selection. Am. Nat. 104:592-597.

Reiners, W. A., I. A. Worley, and D. B. Lawrence. 1971. Plant diversity in a chronosequence at Glacier Bay, Alaska. Ecology 52:55-69.

Richards, P. W. 1952. The Tropical Rain Forest: An Ecological Study. Cambridge Univ. Press, London. 450 pp.

Salisbury, E. J. 1942. The Reproductive Capacity of Plants: Studies in Quantitative Biology. G. Bell and Sons, London. 244 pp.

Sauer, C. O. 1958. Man in the ecology of tropical America. Proc. 9th Pacific Sci. Congr. 1957. Bangkok. 20:104-110.

Shafi, M. I., and G. A. Yarranton. 1973. Diversity, floristic richness, and species evenness during a secondary (post-fire) succession. Ecology 54:897-902.

Terborgh, J. 1974. The preservation of natural diversity: the protection of extinction-prone species. BioScience 24:715-722.

Vasquez-Yanes, C. 1974. Estudios sobre ecofisiologia de la germinacion en una zona calido-humeda de Mexico. Thesis, Facultad de Ciencias, Universidad Nacional Autonoma de Mexico, Mexico, D.F. 139 pp.

Vestal, A. G. 1949. Minimum Areas for Different Vegetations: Their Determinations from Species-Area Curves. Univ. of Illinois Press, Urbana. 129 pp.

Vogl, R. J. 1975. Fire: a destructive menace or a natural process? *In* this volume.

Williams, W. T., G. N. Lance, L. J. Webb, J. G. Tracey, and M. B. Dale. 1969*a*. Studies in the numerical analysis of complex rain-forest communities. III. The analysis of successional data. J. Ecol. 57:515-535.

Williams, W. T., G. N. Lance, L. J. Webb, J. G. Tracey, and J. H. Connell. 1969*b*. Studies in the numerical analysis of complex rain-forest communities. IV. A method for the elucidation of small-scale forest pattern. J. Ecol. 57:635-654.

Wilson, R. E., and E. L. Rice. 1968. Allelopathy as expressed by *Helianthus annuus* and its role in old field succession. Bull. Torrey Bot. Club. 95:432-458.

Wulff, R. 1973. Interpopulational variation in the germination of seeds of *Hyptis suaveolens*. Ecology 54:646-649.

Recovery of Disturbed Tundra and Taiga Surfaces in Alaska

Keith Van Cleve

Abstract

Tundra and taiga ecosystems differ radically from temperate latitude ecosystems because of several characteristics unique to northern environments: (1) the presence of permafrost and its potential for degradation following surface disturbance; (2) the presence of thick, slowly decaying organic layers covering the mineral soil, which are important nutrient reservoirs and are a source of insulation for frozen ground; (3) the cold-dominated nature of the environments and associated short growing season which presents unique problems for the use of introduced species in surface restoration.

Physical disturbance encompasses the entire range of conditions from minimal surface disruption to removal of large quantities of fill for road and work surface construction. Removal of plant cover, alteration of soil thermal regimes, loss of nutrient reserves, and possible severe erosion have been the result of past injudicious use of all-terrain vehicles on the tundra.

Promoting recovery of disturbed surfaces in arctic Alaska through use of introduced plant species may only be a temporary measure (three to five years). In most instances revegetation of disturbed surface in the taiga of Alaska has shown considerable success and does not appear to pose a long-term problem because of greater longevity of introduced plant species and relatively rapid reinvasion by native plant species.

Introduction

During the past five years human disturbance of the Alaskan arctic and subarctic has increased as a result of the development of oil and gas reserves. At the present time little quantitative information is available with regard to effects of surface disturbance on the soil-plant component of these northern ecosystems. It is obvious from qualitative field observations that certain types of disturbance have substantial short- and long-term effects (Hok, 1971).

Impact of surface disturbance on tundra and taiga ecosystems would be expected to be radically different from that on temperate latitude ecosystems because of the following features unique to the northern environment: (1) the presence of permafrost and its potential for degradation after surface disturbance: (2) the presence of thick, slowly decaying organic layers covering the mineral soil, which are important nutrient reservoirs and are a source of insulation for frozen ground; (3) the cold-dominated nature of the environment and associated short growing season which presents problems for the use of introduced plants species in surface restoration.

Plant growth provides a protective cover for the soil. Removal of plant cover alters surface albedo, and if the ground surface experiences sufficient disturbance, permafrost degradation may occur (Heginbottom, 1973; Hernandez, 1973*b*). For example, four summers after removal of only overstory black spruce and tall shrubs from permafrost-dominated sites in the Alaskan taiga, ground subsidence of 30 to 50 cm had occurred. Depth of thaw increased from 1 to 1.5 m (Viereck, 1973).

The soil is a nutrient storehouse, source of moisture, and location of microbial activities associated with recycling of nutrients incorporated in plant matter. Removal of surface organic layers may drastically deplete nutrient reserves and expose mineral substrates which are deficient in chemical elements critical for plant growth. Rates of restoration will be slowed on nutrient-poor surfaces unless a portion of the lost nutrient capital is replaced.

In contrast to the soil-plant component of these ecosystems, mobile components such as some large mammal and bird populations may disperse from the point of disturbance, thus avoiding immediate impact, but may experience subsequent long-term adverse effects if food supplies and migratory and breeding habits are altered or destroyed. For example, occurrence and patterns of movement of reindeer in Norway, Sweden, and Finland have been dramatically altered by human presence and construction of transportation corridors (Klein, 1971). The impact of long-distance pipelines on migratory behavior of caribou in the Alaskan arctic has yet to be determined. Simulated pipeline studies conducted at Prudhoe Bay have shown marked impact on movement of caribou (Klein, 1972; Child, 1974).

On the other hand, a change in vegetation composition caused by either natural or man-made disturbances may result in increases in some populations. Rare species may invade these sites and flourish in seral stages of plant succession.

The objectives of this chapter are to examine briefly some of the sources of disturbance and effects of perturbations on arctic and taiga soil-plant systems, to discuss factors important in revegetation of disturbed surfaces, and to discuss recovery of disturbed surfaces using introduced plant species and recovery through secondary plant succession.

Sources of Damage to Soil-Plant Component of Arctic and Subarctic Ecosystems

Probably the oldest and most widespread cause of substantial disturbance of arctic ground surfaces has been the movement of heavy track vehicles to remote sites in the course of geophysical exploration for oil and gas reserves. Summer bladed tractor trails are easily observed in many places on Petroleum Reserve No. 4, 20 to 30 years after the disturbance (Hok, 1971). Lightweight personnel carriers (weasels) have also been used in oil exploration and scientific endeavors in past times and with repeated passage over the same areas during summer months produced considerable disruption to plant communities and surface soil layers, especially in moist locations. In addition to the type of vehicles, time of year and type of terrain traversed had considerable influence on the extent of surface disturbance in arctic and taiga regions (Hok, 1971; Heginbottom, 1973; Radforth, 1973; Rickard and Slaughter, 1973; Rickard and Brown, 1974). On the North Slope of Alaska, present regulations appear to be quite effective in controlling movement of all terrain vehicles outside of areas specifically designated for their operation. Development of low surface pressure vehicles has also shown considerable promise with regard to reduced surface disturbance (Burt, 1970).

Generally, construction activities associated with roads, various types of work surfaces such as construction camps and drilling pads, and rock or gravel material sites have resulted in the covering over of the natural ground surface or the removal of the plant cover and, sometimes, the organic mat, the most nutrient rich portion of the ecosystem. Current Bureau of Land Management stipulations require revegetation of abandoned work surfaces and other sites of surface disturbance. In subarctic locations revegetation of logging sites may occur rapidly because of favorable temperature and moisture regimes and an abundant, dependable supply of inoculum in the form of seed and rootstock. However, attaining desired species composition may require special site manipulation (Zasada and Gregory, 1969). In addition, logged surfaces are generally not buried beneath gravel overburden or completely stripped to mineral subsoil (except in the case of access roads). Spillage of oil and other petroleum products is a constant although reasonably well-controlled potential source of surface damage. Evidence from both field and laboratory tests indicates varying tolerance of plant species to crude oil and varying longevity of effect with regard to impact of contaminated soil on plant growth (McCown et al., 1973).

Fire is an additional source of surface disturbance, generally of less importance in the arctic but highly important with regard to vegetation distribution and successional relationships in the subarctic of Alaska (Barney and Comiskey, 1973; Wein, 1974; Lutz, 1956). While the immediate impression is one of adverse effects to the aesthetic quality of ecosystems, fire has produced the mosaic of vegetation types which exist across the taiga. Periodic fire may be necessary to sustain vegetation types including habitats favorable for certain big game species such as moose. However, devastation of some habitats by fire has been suggested as a possible cause of decline of some big game populations (Scotter, 1971). Intense fire may result in site nutrient depletion by volatilization of nitrogen (Van Cleve, 1973*b*). Fire also acts as a rapid decomposer, releasing nutrients contained in relatively unavailable forms in organic matter for reuse by plants. Severe burning with loss of plant cover and surface-insulating organic layers can change surface albedo and result in thawing of frozen ground (Brown et al., 1969; Viereck, 1973).

Effects of Surface Disturbance on Soil-Plant Systems

Physical disturbance encompasses the entire range of conditions from minimal surface disruptions to removal of large quantities of fill for road and work surface construction. As in other regions, the extent of disturbance will have a major impact on rate of natural surface recovery.

Minimal physical effects associated with winter off-road travel may result in breaking of senescent and some green plant parts from the previous year's growth with little or no physical disruption of the surface organic layers along the line of travel. After several years little or no adverse effect is visible with regard to surface disturbance (Hok, 1971). At the other extreme are the bladed winter and summer trails where tussocks and other surface organic matter were removed to provide a level road for movement of heavy equipment. At the minimum several decades may be necessary for stabilization by natural revegetation of the sites if road-ways were constructed in wet terrain or intersected stream courses or were placed perpendicular to landscape contours. Removal of plant cover, alteration of soil thermal regimes, loss of nutrient reserves, and possible severe erosion have been the result of past injudicious use of all-terrain vehicles on the tundra.

Studies of the direct physical effects of light personnel carrier traffic on physical properties of wet coastal tundra soils at Barrow, Alaska, have been documented by Gersper and Challinor (1975). Higher soil bulk density and reduced soil moisture percentage and soil aeration were encountered in soil associated with vehicle tracks six years after disturbance. Increased soil temperatures and depth of thaw in track sample sites were attributed to exposed, dark-colored organic matter surfaces and the higher soil bulk density (greater thermal conductivity) and lower soil moisture content (reduced heat capacity).

Impact of this disturbance on soil chemical environment was documented by Challinor and Gersper (1975). Increase in soil temperature resulted in higher concentrations of exchangeable base elements and reduced acidity in disturbed soils compared with undisturbed soil. The concentration of nutrients in the soil solution was highest in track soils, and plant productivity and plant nutrient status was highest in the disturbed soil. Larger plants were encountered in the track soils.

Impact of disturbance on nutrient reserves depends on the extent of removal of vegetation and surface soil organic matter layers. In selected black spruce forests in the interior of Alaska, removal of all organic matter above the mineral soil including tree roots would result in loss of 28% to 30% of the nitrogen and 93% to 97% of the phosphorus contained in the soil-plant system. In these ecosystems plant roots are largely concentrated in the forest floor. The forest floor contained up to 21% of the nitrogen and up to 83% of the phosphorus in these ecosystems (unpublished data on file at the Forest Soils Laboratory, University of Alaska, Fairbanks, Alaska). Since system organic matter constitutes potentially the most readily available reserve of nutrients for plant growth, loss of this component may constitute impoverishment of the ecosystem with respect to selected nutrients. The amount of organic matter remaining after disturbance and the related impact on nutrient reserves determine to an important degree the formulation of fertilizers which may have to be applied to replenish lost nutrient capital, the nature of the seedbed, and the potential for regrowth from seed or rootstock of native plant species contained in the organic matter.

Relatively light disturbances such as that encountered in the Barrow track study may result in improved microenvironmental conditions for microbial activity and growth of higher plants. Severe disturbances, such as those associated with stripping of surface organic matter, may result in major shift in microbial populations. Scarborough and Flanagan (1973) report a change from fungal to bacterial domination of microbial populations at the University of Alaska heated-pipe test facility. Populations of fungi responsible for degradation and recycling of carbon from polysaccharides were obliterated in many cases. With regard to success of revegetation, increased soil temperatures over the heated pipe resulted in greater productivity for native and introduced plant species during two seasons of study (McCown, 1973).

Treatment of soil with crude oil resulted in reduced germination and production by native arctic and subarctic species after treatment (McCown and Deneke, 1973; McCown et al., 1973). Heavy saturation of soils, or application to both foliage and root systems, resulted in severe damage or death of plants, especially in the case of mosses and lichens. Tests with introduced species in the interior of Alaska indicated plants in these soils may be more sensitive to oil spillage. Reduction in growth and presence of mineral deficiency symptoms followed by 100% winterkill were shown for these treatments. Observations in interior Alaska indicate that refined fuel spills associated with established pipelines may be more destructive than crude oil. Depending on the site and degree of contamination, refined fuel spills remain in subsurface layers of the soil for extended periods (up to 15 years).

Studies of the impact of oil application on soil microflora of a meadow tundra soil at Barrow, Alaska, showed a stimulation in growth of hydrocarbon-decomposing flora, especially yeasts and several *Pseudomonas* species, whereas other soil microorganisms such as diatoms and higher fungi were inhibited (Campbell et al., 1973). Total microbial biomass was reduced slightly by heavy oil applications.

Factors Important in Recovery of Disturbed Surfaces

Abiotic Factors

In arctic and subarctic locations climatic factors, especially temperature, dominate rate of plant growth and determine success of introduced compared with native plant species as agents in surface recovery. Marked year-to-year variations in temperature and precipitation during the growing season, wind, and snow depth may contribute in varying degrees to the growth and reproduction for all plants. On the arctic coastal plain at Barrow, frost may occur any month during the period in which plant growth occurs. This interval (growing season) averages 87 days. Record summer temperatures of 26°C have occurred, but the average temperature of June through August is 3°C. In the subarctic near Fairbanks, frost may occur through the first week of June and after the first of August. The growing season is 60 to 90 days.

Land form characteristics also play critical roles in surface restoration. Because of low sun angles at arctic and subarctic latitudes, slope and aspect play a critical role in amount of solar radiation received at the ground surface. Air and soil temperature regimes undoubtedly reflect the impact of radiation, with north slopes being coolest and south slopes warmest. Steepness of slope is critically related to soil drainage. In the discontinuous permafrost zone of interior Alaska, gentle slopes or flat areas, except in the immediate vicinity of river courses, are generally poorly drained, underlain by permafrost, and could be subject to thermokarst and general ground subsidence with surface disturbance.

Soil nutrient status and texture are important factors in determining success of surface restoration. The cold-dominated character of arctic and subarctic soils reduces rates of mineral weathering, organic matter decomposition, and release of important nutrient elements such as nitrogen. These factors in addition to the general concentration of plant roots in surface soil organic layers emphasizes the importance of these layers as a nutrient reservoir.

Textural relations are important from the standpoint of soil moisture regime and also the ability of the soil to retain important nutrient elements. Coarse-textured mineral soil would tend to be droughty, have low retentive capacity for nutrient elements (especially cations), and be well aerated. Fine-textured mineral soil would tend to retain more moisture, have a higher nutrient-retaining capacity, but be less well aerated, especially on flat areas or gentle slopes. Poorly drained soils would be associated with greater incidence of high-ice-content permafrost. Compacted work pads, even though constructed of sand and gravel, may be difficult surfaces to revegetate because plant root systems may not be able to readily penetrate surface layers to more moist subsurface layers. Organic soil would display high nutrient-retaining capacity and be able to retain abundant moisture. However, surface soil layers would tend to dry rapidly, providing an unfavorable medium for seed germination and seedling establishment (Zasada and Gregory, 1969; Dabbs et al., 1974).

Plant Ecological Characteristics

Some of the general ecological characteristics of plant species important in surface restoration include adaptations to winter and summer survival, growth rate, reproductive habit, and competitive ability.

Winter survival in arctic locations generally refers to a plant's ability to withstand snow abrasion, desiccation, and the impact of fungal activity

(Saville, 1972). Because of generally calm conditions and deeper snowpack, snow abrasion and desiccation may not be as important in the taiga except in alpine locations. Many native arctic species have developed growth forms which tend to protect meristematic tissue from the effects of snow abrasion (Saville, 1972). Introduced grass species, some having a higher growth form, would be subjected to snow abrasion in exposed sites.

Summer survival includes the ability of plants to metabolize normally in the hardy state and avoid delay in resuming growth after freezing. Adaptations to low temperatures and the short growing season are important considerations in this regard.

Physical adaptations to low temperature, listed by Saville (1972), include growth forms (generally low, prostrate habit, mats, rosettes, cushions) which reduce wind movement and result in increased surface temperatures, pigmentation of plant parts (increase in surface temperature), phenotypic and genotypic dwarfing (related to energy consumption with regard to biomass and seed production), and pubescence (conservation of heat, conservation of water by reducing air movement near stomata).

Physiological adaptations to low temperature involve the demonstrated ability for higher photosynthetic rates at lower temperatures, higher respiration rates at all temperatures, and attainment of photosynthetic light saturation at lower light intensities for arctic plants compared with alpine plants (Mooney and Billings, 1961). Chapin (1974) concludes with regard to phosphate adsorption that species and ecotypes which have endured in colder climates differ from their warm-adapted counterparts in having (1) lower temperature optima for root initiation, elongation, and production; (2) larger surface-to-volume ratios of roots; (3) proportionately more nutrient-adsorbing tissue; (4) higher phosphate adsorption capacities at given measurement and acclimation temperatures; (5) lower potential of the phosphate adsorption system to acclimate in response to temperature change; and (6) less temperature sensitivity of the phosphate adsorption system.

Adaptations to a short growing season include rapid initiation of spring growth (presence in some species of semievergreen leaves), vegetative reproduction (increases in importance for certain species exposed to adverse or marginal growing conditions), perennial habit, and periodic growth (plant development is halted at a particular stage even where an abundant opportunity remains for further growth) and aperiodic growth (plants develop until halted by deteriorating weather).

Rates of cover production and production of soil-binding root systems are highly important in surface restoration. For erosion control, rapid production of cover and extensive production of root systems are necessary, with growth sustained over a number of years. In contrast to more rapidly growing annual or perennial grasses, slower-growing woody plant species may not be suitable for cover production within the first year or two after disturbance but could provide long-term assurance of stabilization.

Introduced species growing under arctic conditions probably will reproduce primarily by vegetative means. The dispersal of seed from native plant species over damaged sites from adjacent undisturbed areas or from plants growing on piles of bermed organic matter may be the principal means of initial cover establishment under certain conditions. Species also capable of vegetative reproduction would insure more rapid cover production and surface stabilization over an extended period. Proximity of disturbance to native plant seed sources which could assist in inoculation and size of disturbed area over which seed must disperse may also be important factors in rate of surface revegetation (Zasada and Gregory, 1969).

Management Objectives

Management requirements for surface restoration will be dictated by the particular situations of concern. Aesthetic objectives may require different measures than surface stabilization against erosion. Although use of native grasses or slower-growing herbaceous and woody plant species would undoubtedly be more aesthetically pleasing and assure long-term surface recovery, rapid-growing hardy annual and perennial grasses may be necessary for rapid control of erosion. Management of undisturbed plant communities adjacent to disturbed sites to increase native species seed production may be one strategy to promote surface recovery.

Recovery of Disturbed Surfaces

Recovery Using Introduced Species

Rate of recovery of disturbed tundra and taiga surfaces depends on the varying combinations of environmental and plant ecological factors considered above. Field observations of primary production in native plant communities indicate these conditions are generally favorable for considerably more rapid surface recovery at low elevations in the taiga. For example, net annual primary production in *Dupontia-Carex-Eriophorum* communities on the arctic coastal plain at Barrow, Alaska, may attain 102 g/m^2 (Tieszen, 1972). Yearly net primary production was estimated to be between 35 and 96 g/m^2 for aboveground vascular plants in cottongrass tundra communities in Alaska and the Yukon Territory (Wein and Bliss, 1974). However, net annual primary production of selected native grasses in subarctic Alaska may reach approximately 320 g/m^2 (native stand of *Calamagrostis canadensis* near Eklutna, Alaska; Klebesadel and Laughlin, 1964).

Results of phytometer tests conducted in the arctic and taiga of Alaska (lowland and alpine sites) showed that greatest production for all species utilized occurred at the lowland taiga site (Bonde et al., 1973). Better growth at this site was tentatively attributed to greater relative humidity and more favorable temperature regime associated with the site.

Annual net primary production in river bottom alder stands (*Alnus incana*) near Fairbanks was estimated to be approximately 473 g/m^2 averaged over the 20-year period of alder development (Van Cleve et al., 1971).

Work with over 20 species of introduced grasses has been conducted at nine revegetation study sites from Prudhoe Bay south along the pipeline corridor (Fig. 1; Table 1). Results from arctic sites generally indicate that lower growth form grasses such as arctared fescue and nugget bluegrass have been more successful than taller growth form grasses such as timothy and wheat grasses. Hernandez (1973*a*) found arctared and boreal creeping red fescues and nugget bluegrass to be consistently the most successful species in selected Canadian arctic tundra sites. For the most successful species, maximum cover production over periods of up to five years following seed application was obtained at sites south of Prudhoe Bay (Figs. 2 and 3; Table 2). Longevity of introduced species at southerly sites appears to be greater than at the most northerly site. However, species such as arctared fescue and nugget poa which have shown considerable success at northerly sites have not displayed equal success under supposedly less severe southerly environments. For the best species tested, the greatest number of failures in survival and growth appear at Fish Creek, Galbraith Lake, and Prudhoe Bay. A possible explanation is the fact that a limited number of experimental sites were established within any one physiographic province to assess the range of environmental conditions important with regard to success of introduced species (see Table 1). For example, more cover was produced in fertilized creeping red fescue and hard fescue plots at Happy Valley, in the arctic foothills province, than at Dall Summit, in the Kokrine-Hodzana province north of the Yukon River (see Table 2). Differences in chemical properties of the substrate between the two areas may have resulted in greater productivity in the supposedly more extreme arctic environment. The pH, total percentage of N and P, and organic matter content are markedly lower for the soil at Dall Summit than for the soil at Happy Valley (Table 3). The Happy Valley disturbance did not result in removal of all of the surface organic layer, exposing a nutrient-deficient substrate, as was the case with the Dall Summit disturbance. A better comparison of the potential of introduced species for restoration could be obtained between regions if field tests could be conducted on surfaces which experienced the same degree of disturbance. In addition, between-year differences in climatic conditions undoubtedly exert considerable influence on success of introduced species.

Physical analysis of study area soils shows that extent of surface disturbance and subsequent treatment of the surface probably have a significant effect on moisture-retaining properties of coarse-textured soils (Van Cleve, 1973*a*) The greater water-retaining capacity of fine-textured soils compared with coarse-textured soils has previously been mentioned. If organic matter removed during construction activities is mulched with coarse-textured mineral subsoil, the water-retaining capacity as well as the chemical properties may be improved. Coarse-textured soils were encountered at the Fish Creek, Wiseman, Galbraith Lake, and Happy Valley sites. These soils contained at least 65% sand, 11% silt, and 5% clay (see Table 2). The Fish Creek and Happy Valley soils containing the highest percentage of organic matter (17.2% and 17.5% respectively) also had the largest moisture retention ranges (21.6% and 19.8% respectively, for the ⅓ atm. to 15 atm. range) for the coarse-textured soils (see Table 3).

Fertilization has a marked favorable impact on the rate, amount, and longevity of cover production. In most cases growth of introduced species without fertilization has shown limited success or no success two to three years after seedling establishment. Extent of depletion of nutrient capital during disturbance coupled with generally low available contents of important nutrients in these cold-dominated soils necessitates the use of fertilizer. For the most vigorous species tested (arctared fescue), cover production after five years was up to

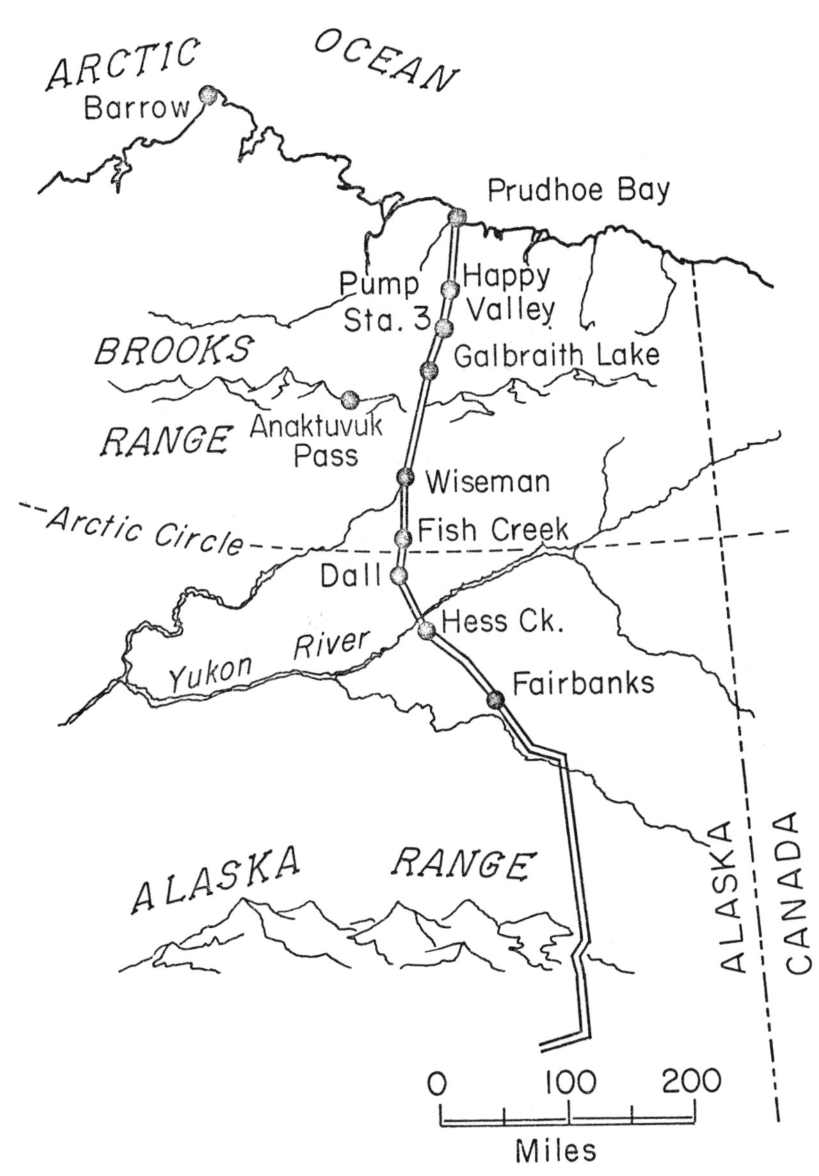

Fig. 1. Location of revegetation study plots along route of trans-Alaska pipeline between Prudhoe Bay and Fairbanks, Alaska

Table 1. Location of study sites along route of trans-Alaska pipeline

Study site	Elevation (meters)	Latitude	Longitude	Physiographic province
Prudhoe Bay (P.B.)	15	70°17′N	148°35′W	Arctic coastal plain
Happy Valley (H.V.)	330	69°10′N	148°50′W	Arctic foothills
Pump Station 3	335	68°50′N	148°50′W	Arctic foothills
Galbraith Lake (Gal. Dike)	802	68°27′N	149°30′W	Arctic mountain
Anaktuvuk Pass (Ak.P.)	640	68°8′N	151°45′W	Arctic mountain
Wiseman (Wise.)	488	67°25′N	150°5′W	Arctic mountain
Fish Creek (Fish)	445	66°28′N	150°33′W	Kokrine-Hodzana highlands
Dall Summit (Dall.)	320	66°20′N	149°56′W	Kokrine-Hodzana highlands
Hess Creek (Hess Min.)	244	65°40′N	139°12′W	Yukon Tanana uplands

23-fold greater for fertilized treatments (see Table 2 and Fig. 2). With the exception of arctared fescue and hair grass, fertilization has not offset the severe environment at Prudhoe Bay to result in extended cover production (see Table 2).

Detailed summaries of impact of fertilization on selected growth parameters for the best species after two growing seasons emphasize the importance of soil nutrient regime in cold-dominated, disturbed soils. At all study sites where greater than trace amounts of production occurred, fertilization produced marked increases in plant height, numbers of flowers, percentage of cover, biomass, and vegetative reproduction over controls (Table 4). The marked increases in vegetative reproduction (an index of sod formation) with fertilization are highly important from the standpoint of ground surface stabilization and reflect the ability of certain species to expand protective surface cover on at least a short-term basis. With the exception of arctared fescue, measurable flowering did not occur north of the Pump Station 3 site, and showed the least success at Galbraith Lake and Fish Creek. Success of seed production and seed viability were not determined; so the significance of flowering in the introduced species from the standpoint of cover maintainance remains unknown. Areas where vegetative reproduction showed the least success were Prudhoe Bay, Galbraith Lake, and Fish Creek. At Prudhoe Bay the two-year

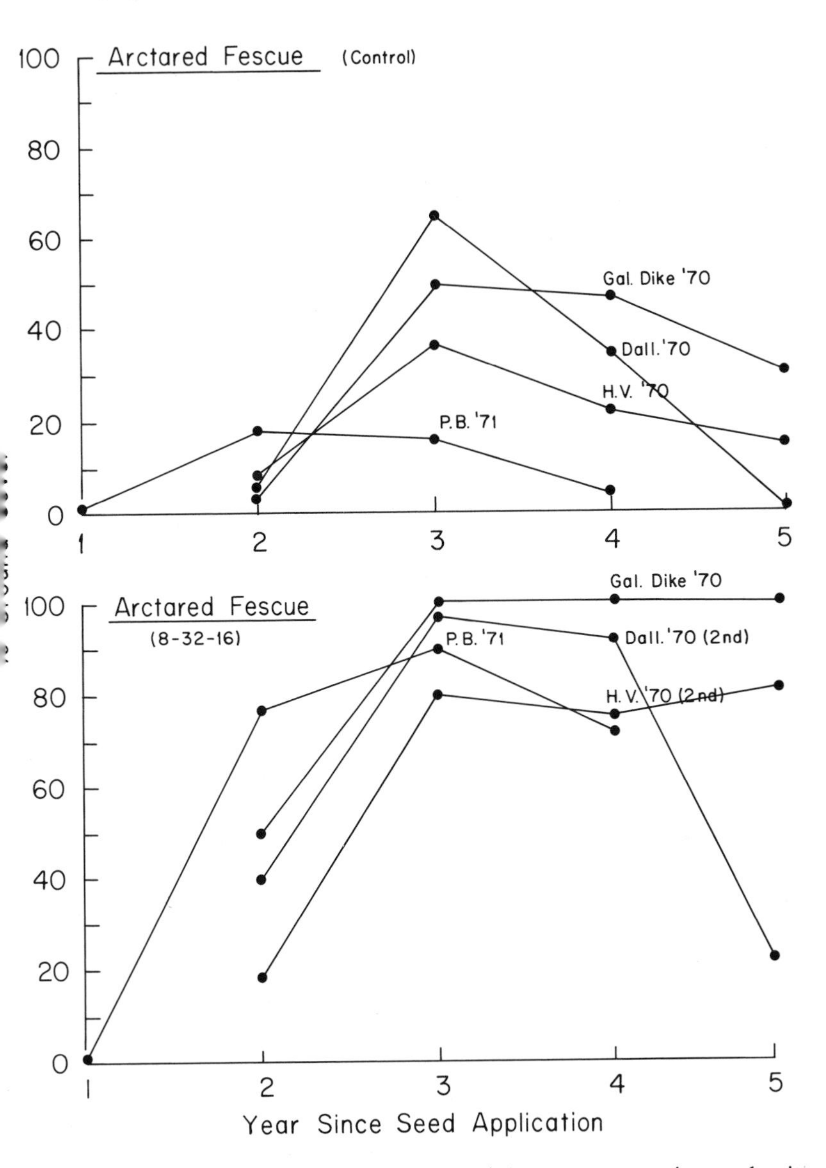

Fig. 2. Average cover production for arctared fescue at respective study sites with no fertilizer application and application of high phosphate fertilizer (8-32-16) at rate of 444 kg/ha at time of seeding. Seed applied at rate of 44 kg/ha. Second fertilizer application at same rate made at Happy Valley and Dall sites in 1972. Standard error of mean ranges from 10% to 20%. Site abbreviations defined in Table 1. Year of seed application is indicated beside site abbreviation.

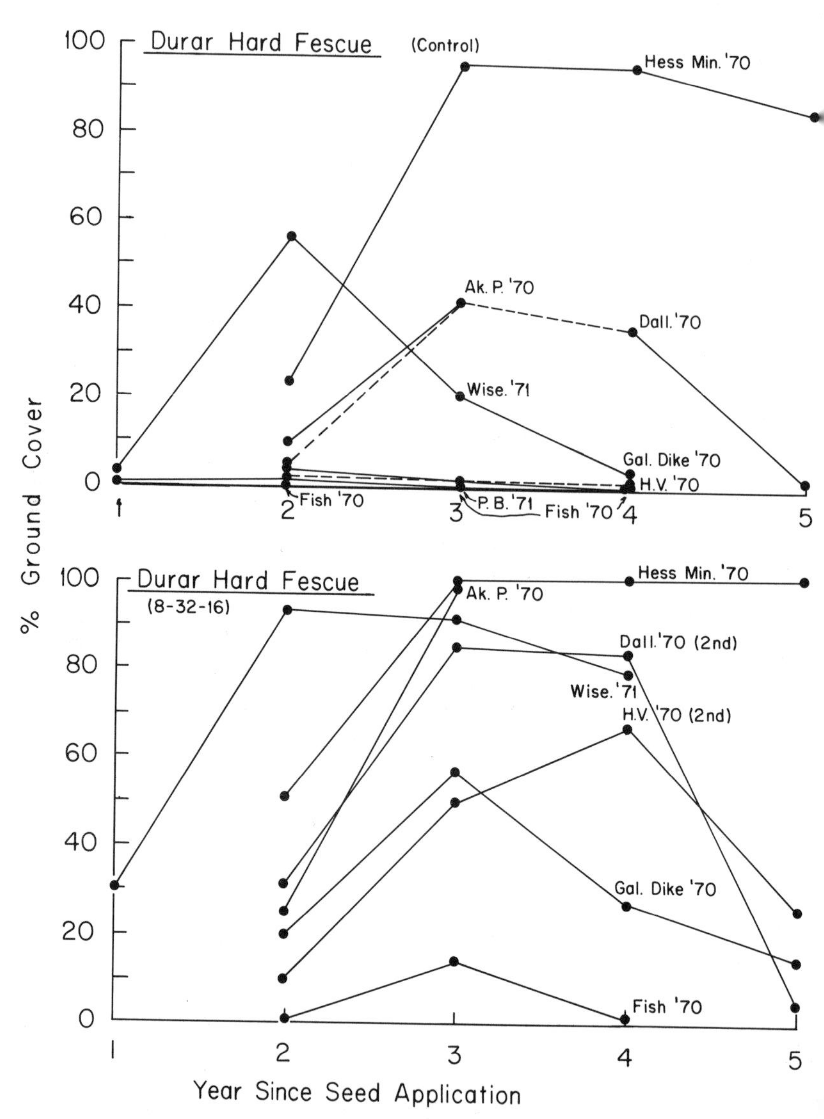

Fig. 3. Average cover production for durar hard fescue at respective study sites with no fertilizer application and application of high phosphate fertilizer (8-32-16) at rate of 444 kg/ha at time of seeding. Seed applied at rate of 28 kg/ha. Second fertilizer application at same rate made at Happy Valley and Dall sites in 1972. Standard error of mean ranges from 10% to 20%. Site abbreviations defined in Table 1. Year of seed application is indicated beside site abbreviation.

Table 2. Cover production (%) after at least 4 years for best species tested at revegetation study sites

Study area	Arctared* fescue	Boreal creeping red fescue	Durar hard fescue	Nugget poa	Meadow foxtail	Slender wheat grass	Hair grass	Climax timothy
Prudhoe Bay	72 ± 3† (7 ± 2)‡	0	0	—	0	—	62 ± 10 (trace)	trace (trace)
Happy Valley	82 ± 7 (14 ± 8)	24 ± 11 (trace)	27 ± 12 (0)	—	4 ± 2 (trace)	—	—	0
Pump Station 3	—§	—	—	44 ± 15 (2 ± 1)	—	62 ± 11 (6 ± 4)	—	—
Galbraith Lake	100 ± 0 (31 ± 17)	0	0	0	0	32 ± 16 (10)	95 ± 0 (13 ± 9)	0
Anaktuvuk Pass	—	—	—	100 ± 0 (60 ± 19)	—	76 ± 5 (10 ± 10)	—	—
Wiseman	—	13 ± 3 (7 ± 2)	78 ± 2 (4 ± 3)	0	53 ± 23 (6 ± 5)	—	0	2 ± 2 (trace)
Fish Creek	—	13 ± 4 (1 ± 1)	0	—	0	—	—	0
Dall Summit	23 ± 3 (1 ± 1)	3 ± 1 (2 ± 1)	6 ± 2 (2 ± 1)	1 ± 1 (trace)	0	trace (trace)	22 ± 12 (trace)	trace (trace)
Hess Creek	—	58 ± 14 (22 ± 4)	100 ± 0 (88 ± 5)	trace (trace)	23 ± 18 (12 ± 3)	0	—	10 ± 6 (trace)

* Seeding rates in kg/ha: arctared fescue, 44; boreal creeping red fescue, 42; durar hard fescue, 28; nugget poa, 22; meadow foxtail, 48; slender wheat grass, 32; hair grass, 14; climax timothy, 13.
† Mean ± standard error for fertilized treatment (8-32-16).
‡ Mean ± standard error for control.
§ Species not tested at this site.

Table 3. Chemical and physical properties for revegetation study site soils

	Hess Creek	Dall Summit	Fish Creek	Wiseman
% organic matter	33.1	9.2	17.2	5.0
% total nitrogen	0.46	0.09	0.21	0.09
% total phosphorous	0.08	0.02	0.07	0.09
Available phosphorous (ppm)	8.9	0.7	2.1	0.5
Fe (me/100g)	0.13	0.00	0.18	0.00
Mn ''	0.26	0.01	0.00	0.31
K ''	0.18	0.10	0.08	0.00
Zn ''	0.00	0.00	0.00	0.00
Mg ''	3.77	1.89	0.45	0.63
Ca ''	22.92	2.31	2.28	6.92
Cation exchange capacity (me/100g)	46.0	16.0	25.0	4.0
Exchangeable hydrogen (me/100g)	18.74	11.87	22.01	–
pH	6.4	5.7	5.7	7.6
% moisture content (at ⅓ atm)	58.1	28.7	31.7	12.5
% moisture content (at 15 atm)	24.4	7.4	10.1	3.2
Particle size (%)	Organic			
sand	soil	44	76	65
silt		44	15	30
clay		12	9	5
Bulk density (g/cm^3)	1.153	0.682	1.169	0.979

Table 3 (*cont.*)

	Anaktuvuk Pass	Galbraith Lake	Happy Valley	Prudhoe Bay
% organic matter	21.6	6.2	17.5	19.3
% total nitrogen	0.64	0.26	0.56	0.62
% total phosphorous	0.11	0.07	0.07	0.07
Available phosphorus (ppm)	0.7	0.7	0.7	0.5
Fe (me/100g)	0.02	0.00	0.01	0.02
Mn ''	0.19	0.16	0.12	0.14
K ''	0.10	0.08	0.10	0.06
Zn ''	0.03	0.00	0.01	0.02
Mg ''	1.67	0.73	0.84	0.92
Ca ''	83.73	6.01	52.37	67.05
Cation exchange capacity (me/100g)	14.0	7.5	14.5	13.5
Exchangeable hydrogen (me/100g)	—	0.52	—	—
pH	7.7	7.2	7.6	7.7
% moisture content (at ⅓ atm)	23.9	9.0	27.9	27.2
% moisture content (at 15 atm)	8.9	4.7	8.1	7.8
Particle size (%)				
sand	47	80	72	63
silt	39	11	20	31
clay	14	9	8	6
Bulk density (g/cm^3)	0.704	0.917	0.333	0.717

results generally reflect growth obtained after four years with regard to poor growth of introduced species. However, other investigators have shown greater success with varieties of bluegrass at this location (Mitchell and McKendrick, 1974). More species showed improved growth at southerly sites than at northerly sites (see Table 4). Differences in cover percentage and biomass at various sites probably reflect the impact of differing environments on species growth.

Field tests with various formulations have shown high phosphorus fertilizers generally produce the best growth response over a wide range of sites (Prudhoe Bay to Hess Creek). However, definitive studies have not been conducted which relate soil nutrient contents and plant response to recommended fertilizers. Observation of plant foliage coloration at test sites apparently indicates nutrient deficiencies for nitrogen and phosphorus the second and third year after seeding. A clearer picture of formulations, levels of application, and necessity of refertilization in relation to plant growth response is necessary in order to obtain maximum growth and longevity of introduced species on disturbed sites.

At the present time it appears that promoting recovery of disturbed surfaces in the arctic through the use of introduced plant species may only be a temporary measure. With the exception of a few grass species, surface cover production will probably not be maintained for longer than three to four years. The fate of the most vigorous species has yet to be determined since a maximum of only five to six years has passed since initial seeding. In most cases even these species are showing a marked decline in cover production at the majority of sites after five years (see Figs. 2 and 3). Most species tested are undoubtedly beyond the limits of climatic adaptation, although nutritional relationships may be a factor over which management agencies can exert more control in order to improve plant growth. Reseeding of some areas may be necessary to provide surface stabilization before native plants can effectively colonize disturbed sites.

In most cases revegetation of disturbed sites in the taiga of Alaska has shown considerable success and does not appear to pose a long-term problem because of greater longevity of introduced species and relatively rapid reinvasion by native species.

Table 4. Selected growth parameters for best species after 2 seasons of growth, selected at respective revegetation study sites (lower figure is control)

	Arcta red fescue	creeping red fescue	Durar hard fescue	Nugget poa	Meadow foxtail	Slender wheat grass	Hair grass	Engmo timothy
Prudhoe Bay								
Height (cm)	4.7*(0.3)†	5.1;0		0; 2.0	7.6; 7.6	5.1; 5.1	16.0*(0.8)	0; 0
	2.5 (0.0)‡						6.9 (0.5)	
Flowers	4	0; 0		0; 0	0; 0	0; 0	0; 0	0; 0
(no./m^2)	0							
% cover	90.0*(5.0)	T§		T; T	T; T	T; T	88.3 (4.4)	0; 0
	16.7 (7.3)						7.3 (6.3)	
Biomass	— ; —	—; 0		T; T	T; T	T; T	—; —	0; 0
(g/m^2)	0							
Vegetative	4#	0; 0		T; T	T; T	0; 0	4	0; 0
repro.	0						0	
Happy Valley								
Height	5.1**(.5)	5.1; 2.5	3.6**(0.5)		5.5 (0.7)			5.6 (1.1)
	2.5 (.3)		2.4 (.4)		2.8 (0.3)			
Flowers	0; 0	0; 0	0; 0		0; 0			T; 0
% cover	8.3 (3.6)	T; T	10.0 (3.9)		13.4 (4.4)			T; T
	18.3 (5.0)		3.3 (2.3)		6.7 (3.2)			
Biomass	11.0 (4.5)	T; T	4.5**(0.1)		7.1 (3.2)			T; T
	3.9 (4.5)		0		0.7 (0.0)			
Vegetative	3.0**(.6)	T; T	1.7 (.7)		0.7 (0.3)			0; 0
repro.	0.3 (.3)		0		0			

Table 4 (*cont.*)

	Arcta red fescue	Boreal creeping red fescue	Durar hard fescue	Nugget poa	Meadow foxtail	Slender wheat grass	Hair grass	Engmo timothy
Fish Creek								
Height		19.5**(0.9)	12.1*(1.2)		0; 0			0; 0
		9.32 (2.2)	0					
Flowers		3	0; 0		0; 0			0; 0
		0						
% cover		76.7 (8.3)	1.2*(4.1)		0; 0			0; 0
		37.0 (18.9)	0					
Biomass		—; —	34.9**(16.8)		0; 0			0; 0
			0					
Vegetative		3	1.0 (1.0)		0; 0			0; 0
repro.		1	0					
Dall Summit								
Height	10.6*(0.9)	7.4*(0.5)	10.4*(1.4)	8.4 (1.1)*	15.9*(1.6)	T; T	22.0**(3.4)	13.8*(2.5)
	5.1 (0.4)	4.5 (0.2)	3.4 (0.7)	3.8 (0.5)	6.1 (1.0)		9.3 (3.1)	4.5 (0.4)
Flowers	81.4 (54.3)	T; 0	27.8**(12.3)	64.6 (33.0)	27.8**(12.3)	T; T	3	22.0 (12.3)
	0		0	0	0		0	0
% cover	40.0*(0.0)	46.7*(6.4)	31.7*(6.0)	23.3*(5.5)	11.7*(4.1)	T; T	45.0**(14.4)	25.0 (5.6)
	5.8 (2.8)	18.3 (5.0)	5.0 (2.8)	8.3 (3.6)	T		2.3 (1.3)	11.7 (4.1)

Biomass	147.9*(33.0)	107.9*(16.2)	101.4*(23.9)	71.1**(18.1)	60.7*(13.6)	T; T	—; —	67.2**(18.7)
	12.9 (2.6)	39.4 (7.8)	10.3 (4.5)	16.0 (5.2)	16.5 (0.0)			16.8 (4.5)
Vegetative	4.0**(0.0)	4.0*(0.0)	3.0 (1.0)	2.3**(0.3)	2.7*(0.3)	T; T	2	2.3**(0.3)
repro.	1.7 (0.7)	1.7 (0.3)	1.0 (0.6)	0.7 (0.3)	0		0	0.7 (0.3)
Anaktuvuk Pass								
Height			16.3*(1.1)	6.4*(0.5)	9.0**(1.3)	23.2**(2.4)		7.0**(0.9)
			4.4 (0.6)	3.1 (.5)	5.2 (0.8)	16.3 (1.2)		3.3 (0.4)
Flowers			286.2*(96.3)	416.7 (183.5)	0; 0	177.7 (48.5)		5.8 (5.2)
			0	0		4.3 (0.9)		0
% cover			25.0**(5.6)	23.8**(4.8)	13.3*(4.4)	20.0*(4.5)		6.7**(3.2)
			10.0 (4.0)	7.5 (4.2)	T	15.0 (4.0)		0
Biomass			189.3*(38.1)	155.0**(47.8)	31.0*(9.0)	131.8**(31.0)		32.4*(7.1)
			22.0 (9.0)	17.4 (7.8)	1.9 (1.3)	63.3 (11.0)		4.5 (2.6)
Vegetative			4.0*(0.0)	3.0**(0.4)	1.3*(.3)	3.5*(.3)		1.3 (.3)
repro.			0.7 (0.3)	1.0 (0.0)	0	1.3 (.3)		0.3 (.3)
Wiseman								
Height		12.3**(1.8)	30.5 (3.1)	15.8**(3.3)	43.8*(7.0)		25.4*(2.9)	41.9*(9.0)
		7.1 (1.0)	10.9 (4.8)	4.8 (0.3)	7.3 (1.7)		8.5 (1.7)	6.8 (1.5)
Flowers		0; 0	4	0; 0	16.8 (11.6)		4	269.4*(89.8)
			0		0		3	27.1 (22.0)
% cover		63.3*(6.2)	91.7 (8.3)	—; —	33.3*(6.1)		18.3**(1.7)	23.3*(5.5)
		41.7 (6.4)	20.3 (17.4)		10.0 (3.9)		5.3 (2.6)	8.3 (3.6)
Biomass		253.9** (54.9)	—; —	—; —	419.3*(113.1)		—; —	823.7* (129.9)
		105.3 (29.1)			39.4 (17.4)			40.7 (14.9)
Vegetative		4.0 (0.0)	4	—; —	3.7*(0.3)		0; 0	3.3 (0.7)
repro.		27.7 (0.9)	2		1.0 (0.6)			1.3 (0.9)

Table 4 (*cont.*)

	Arcta red fescue	Boreal creeping red fescue	Durar hard fescue	Nugget poa	Meadow foxtail	Slender wheat grass	Hair grass	Engmo timothy
Pump Station 3								
Height				11.5*(1.5)		32.8*(2.5)		
				3.5 (0.3)††		7.6 (1.0)††		
Flowers				2.3*(0.6)		299.7*(58.1)		
				0		2.6 (1.9)		
% cover				28.8*(3.6)		10.6**(2.4)		
				3.1 (1.4)		5.0 (1.7)		
Biomass				68.5*(11.6)		129.9*(20.7)		
				5.2 (1.3)		7.8 (1.9)		
Vegetative repro.				3.3*(0.4)		3.3*(.4)		
				0.8 (0.3)		0.3 (0.2)		
Galbraith Lake								
Height	13.8*(2.0)	<1; <1	8.8*(0.5)		0; 0	21.4 (5.6)	57.6*(3.4)	0; 0
	3.2 (0.6)		2.2 (0.4)			8.3 (0.6)	13.6 (1.7)	
Flowers	T; 0	T; 0	T; 0		0; 0	146.0 (89.8)	4	0; 0
						0	3	
% cover	50.0*(6.5)	T; 0	20.0*(5.2)		0; 0	5.0 (2.81)	100.0*(0.0)	0; 0
	3.3 (2.3)		1.7 (1.7)			0	12.3 (11.3)	
Biomass	170.5*(32.3)	T; 0	37.5*(9.0)		0; 0	48.5 (26.5)	—; —	0; 0
	7.8 (3.2)		1.3 (0.7)			3.2 (1.9)		
Vegetative repro.	3.7*(0.3)	0; 0	2.0 (.6)		0; 0	1.7 (1.2)	4	0; 0
	0.3 (0.3)		0.3 (0.3)			0	1	

Hess Creek				
Height	22.1*(2.3)	24.7*(1.9)	0; 0	36.0 (6.7)
	8.4 (0.8)	10.9 (1.8)		15.1 (3.8)
Flowers	4	124.0**(34.2)	0; 0	274.6 (97.6)
	3	21.3 (16.8)		107.9 (43.9)
% cover	93.3**(4.4)	52.0**(6.0)	0; 0	18.3 (5.0)
	50.0 (12.6)	23.0 (5.0)		11.7 (4.1)
Biomass	–; –	492.9**(93.0)	0; 0	588.5 (181.5)
		228.0 (69.1)		158.9 (71.7)
Vegetative repro.	4	4.0 (0.0)	0; 0	2.7 (0.3)
	3	3.3 (0.3)		2.3 (0.3)

Note: Height, flowers, % cover, and biomass reflect only second season of growth; vegetative reproduction reflects accumulation of 2 seasons' growth.

* Difference between means significant at 1% level of probability.

† Mean ± standard error of mean for fertilized treatment (8-32-16).

‡ Mean ± standard error of mean for control.

§T = trace

‖ Parameter not tested, or species not tested at site.

Visual estimation of sod formation on scale 0-4 (none, poor, fair, good, excellent). Where flower counts could not be made, same scale was used.

** Difference between means significant at 5% level of probability.

†† Three-year data.

Recovery by Natural Sucession

Four to ten years after disturbance, the course of natural revegetation at the various study sites along the pipeline corridor reflects to a large extent the degree to which rootstock remains in the soil after disturbance and the influence of species which are spread by seed dispersal from areas adjacent to the disturbed sites (Table 5). Up to 100% surface cover has been produced at several sites. Woody plant genera important in this regard are *Salix, Populus,* and possibly *Vaccinium* and *Arctostaphylos* (Table 6). In addition mosses, lichens, and liverworts appear to be assuming an important role in revegetation at some sites (especially on fertilized subplots). Importance of moss with respect to ground cover is most evident at most northerly sites (Prudhoe Bay, Happy Valley, and Pump Station 3). Soil moisture content is probably an important factor with regard to occurrence of nonvascular plants on disturbed sites. At arctic sites and selected taiga locations shallow permafrost tables may impede soil drainage, resulting in higher soil moisture content and conditions favorable for growth of mosses, lichens, and liverworts. (Importance of moss in natural recolonization of other tundra and taiga sites has been reported by Hernandez [1973] and Zasada* [pers. comm.].) Willow, various grass and sedge species, and herbaceous plants assume importance in secondary succession south of Prudhoe Bay.

According to measurements obtained during the 1971 growing season, vascular plant families which appeared in greatest abundance on disturbed sites, according to biomass measurements, are the grasses, willows, rose, and legumes (see Table 6). Grasses (probably primarily through seed dispersal), willows and rosaceous plants through seed dispersal, and regrowth from rootstock appear to be effective species in recolonizing disturbed surfaces. In most cases ericaceous shrubs and sedges suffered the greatest impact from disturbance and had not, to any extent, become reestablished on disturbed sites at the time of sampling.

Disturbances where the substrate is mineral soil generally appear to require time intervals longer than the five to ten years noted for surfaces retaining some organic matter including viable rootstock. Soil moisture and nutrient regimes are undoubtedly important in this regard. Proximity of all sites to undisturbed areas from which seed may disperse

* Dr. J. C. Zasada, Institute of Northern Forestry, U.S. Forest Service, Fairbanks, Alaska.

over exposed surfaces and inability of some components of the native flora to produce seed are also important factors in succession. If disturbances are in exposed locations, native plant establishment may require longer time periods because of snow abrasion and desiccation.

Work in other areas of the Alaskan and Canadian Arctic also indicates the possibility for relatively rapid surface recovery depending on the degree of disturbance and the particular community affected. Sections of bladed trails at the Gubik Gas Field had achieved complete cover after 11 years (Hok, 1971). Windrow tops and flooded areas remained free of vascular plants. Similar conditions were encountered at Cape Simpson, Umiat, and Knifeblade, all of which were disturbed about 20 years before observation. However, as Hok points out, recovery of vegetation to an unrecognizable, predisturbance condition requires a great deal more time than the initial establishment of plant cover. He found that bladed summer tractor trails, while frequently revegetated, were clearly evident due to relatively permanent changes in microtopography, drainage, and the plant communities they supported. Nonbladed summer tractor trails were set apart from

Table 5. Estimates of percentage of ground cover produced by native plant reinvasion after disturbance, summer 1974

Study site	Nature of disturbance	Years since disturbance	% cover
Prudhoe Bay	Mineral soil, abandoned road	4	54
Happy Valley	Combination of mineral soil, organic layers remaining, temporary construction campsite	5	100
Pump Station 3	Surface organic layers left, temporary construction campsite	6	100
Galbraith Lake	Site one: winter haul road scar, mineral subsoil exposed. Site two: mineral soil berm	6–10	90 15
Anaktuvuk Pass	Haul road construction campsite, organic mat left intact	6–10	100
Wiseman	Survey campsites and gravel borrow areas, mineral soil	4–6	67
Fish Creek	Winter haul road scar, mineral soil	5	94
Dall Summit	Abandoned airstrip, mineral soil	5–7	80

Table 6. Biomass (g/m²) for selected plant species in natural and disturbed communities at study sites. Time since disturbance is indicated in Table 1.

	Hess Creek*		Dall Summit		Fish Creek	
Family and species	N	D	N	D	N	D
Gramineae						
Gramineae spp.			0.33	0		
Calamagrostis sp.	10.23	35.90	0	4.97		
Hordeum jubatum						
Elymus innovatus						
Litter					0	2.5
Salicaceae						
Salix spp.						
S. reticulata						
Ericaceae						
Vaccinium vitis-idea	43.28	7.96	34.82	0.03	3.77	0
V. uliginosum			24.64	0	41.80	0
Arctostaphylos ruba	1.03	0				
A. alpina						
Ledum spp.	10.47	0	9.40	0	23.30	4.67
Rosaceae						
Potentilla fruticosa						
Rubus chamaemorus					0	1.64
Rosa acicularis	0.43	7.08				
Leguminosae						
Oxytropis campestris						
Lupinus arcticus						
Hedysarum alpinum						
H. mackenzii						
Cyperaceae						
Eriophorum spp.						
Carex spp.					7.40	0

Table 6 (*cont.*)

Family and Species	Wiseman N	Wiseman D	Happy Valley N	Happy Valley D
Gramineae				
Gramineae spp.				
Calamagrostis sp.				
Hordeum jubatum	0	12.77		
Elymus innovatus	0	13.8		
Litter	0	46.2		
Salicaceae				
Salix spp.	0	63.84		
S. reticulata			11.62	28.58
Ericaceae				
Vaccinium vitis-idea	2.47	0		
V. uliginosum	9.60	0	19.69	11.23
Arctostaphylos rubra	10.57	0	20.48	0
A. alpina			0	12.58
Ledum spp.	12.61	0		
Rosaceae				
Potentilla fruticosa	3.82	4.17		
Rubus chamaemorus				
Rosa acicularis			9.79	11.86
Leguminosae				
Oxytropis campestris	1.91	16.81	0	2.30
Lupinus arcticus			0.41	2.38
Hedysarum alpinum	1.97	0		
H. mackenzii			0	2.75
Cyperaceae				
Eriophorum spp.			13.69	0.40
Carex spp.	7.18	0.51	14.93	4.53

Note: N = natural; D = disturbed.

*At Hess Creek, disturbance was forest fire, instead of heavy equipment operation.

undisturbed areas by surface depression, minor vegetation changes, and occasionally by complete absence of vascular plants where vehicle tracks had passed (Hok, 1971).

Work conducted on the Tuktoyaktuk Peninsula (McKenzie River delta region) of the Northwest Territories showed that secondary succession resulted in 30% to 50% plant cover six years after summer seismic lines were bladed to permafrost, exposing mineral soil (Hernandez, 1973*a*). *Arctagrostis latifolia, Calamagrostis canadensis, Poa arctica,* and *Luzula confusa* were the predominant pioneer species on upland mesic sites. Wet sites were colonized predominantly by *Arctophila fulva* and *Carex aquatitis*. The importance of *Arctagrostis* and *Calamagrostis* in recolonizing disturbed surfaces in the McKenzie Delta region has also been emphasized by Younkin (1973). Although these species produce small amounts of seed compared with *Betula nana, Carex bigelowii,* and *Eriophorum vaginatum*, species important in undisturbed tundra, their seed has germination percentages over 80% and germination rates that are relatively rapid over a range of temperature and moisture conditions.

Recovery of the native plant community total annual production at a burned cottongrass tussock tundra site in a subalpine location in interior Alaskan taiga had reached 51% of total annual production in the unburned area 2 years after the fire (Wein and Bliss, 1973). Improved nutrient contents were encountered in plant tissue collected from burned sites. With regard to ground stabilization after fire, Viereck (1973) concludes that depth of thaw of permafrost on burned black spruce sites in the taiga of Alaska should return to preburn levels after about 50 years. After five growing seasons 5% to 10% of the plant mass had recovered on a burned area of the northern taiga and tundra zone northeast of Inuvik, Northwest Territory (Wein, 1974). Higher recovery rates were found in tundra vegetation. Shrubs were recovering slowly, but tree seedlings could not be found in any of the burned areas. Grasses and fireweed increased dramatically in the burned areas. Revegetation of exposed river bars on the Tanana River floodplain in central Alaska may occur within a period of 2 to 5 years by alder (*Alnus incana*), poplar (*Populus balsamifera*), and various species of willow (Viereck*, pers. comm.; Van Cleve et al., 1971).

* Dr. L. A. Viereck, Institute of Northern Forestry, U.S. Forest Service, Fairbanks, Alaska.

Recommendations for Future Work

Future work with introduced species should continue to emphasize the longevity of cover production on disturbed surfaces. Observations on growth responses should be obtained from a wider variety of substrates than examined in the past in order to more fully evaluate winter and summer hardiness of the plants. For example, resistance to snow abrasion and desiccation during winter months, moisture, and nutrient stress during the growing season will undoubtedly be critical factors in revegetation of raised sand and gravel fill construction pads.

Extensive soil testing has been conducted along the pipeline corridor to estimate fertilizer requirements for revegetation of various substrates. Confirmation of these recommendations should be carried out using appropriately designed field trials with introduced species and recommended nutrient applications in conjunction with estimates of yield and plant tissue nutrient content. Interspecific competition studies should be initiated at selected arctic and taiga sites and in the laboratory in order to more fully understand growth response of members of species mixes proposed for surface restoration.

With regard to primary succession, continued work is needed to evaluate, on a variety of disturbed surfaces, the rates of growth of colonizing plant species. At the same time a clear definition of physical and chemical properties of the substrate should be obtained, and response of primary succession to fertilization should be examined.

Work should continue on seed increase programs and ecological studies of native species such as blue joint and arctic grass which show promise in surface restoration. More attention should be paid to use of woody plant species which show rapid growth and reproductive capability by sexual and vegetative means (willows) and which may provide long-term site amelioration through nitrogen fixation (alder). A clearer picture should be obtained of the impact of introduced species on rates of reinvasion by native species. Laboratory studies indicate a possible inhibitory effect of rapid-growing annual and perennial grasses on development of slower-growing native herbaceous and woody plants. Competition for moisture, light, and nutrients is undoubtedly intense among seedlings of introduced and colonizing native plants.

Summary

Increasing levels of resource development in the arctic and taiga regions of Alaska necessitate a better understanding of the impact of disturbance on the cold-dominated soil-plant-animal ecosystems.

Studies conducted during the past 5 to 7 years in a variety of locations on the arctic coastal plain and in the foothills region have qualitatively documented the history of disturbance and rates of disturbed surface recovery. Time for recovery of disturbed surfaces by primary succession will primarily depend on the extent of disturbance and associated impact on the soil as a physical, chemical, and biological system. Twenty to 30 years after severe disturbance, effects of surface disruption are still visible in altered microrelief, possible erosion, and altered plant communities. Lesser degrees of disturbance show more rapid recovery. The amount of organic matter remaining on the soil, from the standpoint of moisture-nutrient relations and viable rootstock, and the available seed sources appear to be important factors in rate of primary succession.

Success in growth of introduced species for surface restoration has depended on the use of fertilizers to improve soil nutrient status. Restoration of disturbed surfaces using introduced species has shown greatest promise for the most extended time periods at sites south of Prudhoe Bay. At the present time it appears that use of introduced plant species may provide a means of covering disturbed surfaces for periods up to six years, but with declining levels of plant production from approximately the third year following seeding.

Additional research is needed to more clearly establish the short- and long-term roles of introduced and native species in surface restoration in these cold-dominated environments.

Acknowledgments

The author wishes to acknowledge support of the Alyeska Pipeline Service Company, Inc., the McIntire Stennis Cooperative Forestry Research Program, the U.S.I.B.P. Tundra Biome Program, and the Institute of Agricultural Sciences, University of Alaska, in conduct of this research.

Literature Cited

Barney, R. J., and A. L. Comiskey. 1973. Wildfires and thunderstorms on Alaska's north slopes. U.S.D.A. Forest Service Research Note. PNW-212. 8 pp.

Bonde, E. K., M. F. Foreman, T. D. Babb, S. Kjeliuk, J. D. McKendrick, W. W. Mitchell, F. J. Wooding, L. L. Tieszen, and W. Younkin. 1973. Growth and development of three agronomic species in pots ("phytometers"). Pages 99-110 *in* E. C. Bliss and F. E. Wielgolaski, eds. Primary Production and Production Processes, Tundra Biome. Proceedings of the Conference, Dublin, Ireland, April 1973.

Brown, J., W. Rickard, and D. Vietor. 1969. The effect of disturbance on perma-frost terrain. U.S. Army Cold Regions Research and Engineering Laboratory, Hanover, N.H. Special Rep. 138. 13 pp.

Burt, G. R. 1970. Summer travel on the tundra with low ground pressure vehicles. Field Studies Institute of Arctic Environmental Engineering, Univ. of Alaska. Test Rep., June, 1970. 9 pp.

Campbell, W. B., R. W. Harris, and R. E. Benoit. 1973. Response of Alaskan tundra microflora to a crude oil spill. Pages 53-62 *in* Proceedings of Symposium on the Impact of Oil Resource Development in Northern Plant Communities. Institute of Arctic Biology, Univ. of Alaska. Occ. Publ. on Northern Life No. 1.

Challinor, J. L., and P. L. Gersper. 1975. Vehicle perturbation effects upon a tundra soil plant system. II. Effects on the chemical regime. Soil Science Society America Proceedings 39:689-694.

Chapin, F., III. 1974. Morphological and physiological mechanisms of temperature compensation in phosphate adsorption along a latitudinal gradient. Ecology 55:1180-1198.

Child, K. N. 1974. Reaction of caribou to various types of simulated pipelines at Prudhoe Bay, Alaska. Pages 805-812 *in* Proceedings of Symposium on the Behavior of Ungulates and Its Relation to Management. Univ. of Calgary, Alberta, Canada, Nov. 2-5, 1971. International Union for Conservation of Nature and Natural Resources, Morges, Switzerland.

Dabbs, L., W. Friesen, and S. Mitchell. 1974. Pipeline Revegetation. Northern Engineering Services Company Limited, Calgary, Alberta, Canada. Biology Rep. Series, vol. 2. 67 pp.

Gersper, P. L., and J. L. Challinor. 1975. Vehicle perturbation effects upon a tundra soil plant system. I. Effects of morphological, physical and environmental properties of the soils. Soil Science Society America Proceedings 39:737-743.

Heginbottom, J. A. 1973. Some effects of surface disturbance on the permafrost active layer at Inuvik, N.W.T. Environmental-Social Committee Northern Pipelines, Task Force on Northern Oil Development, Rep. No. 73-16. 29 pp.

Hernandez, H. 1973*a*. Revegetation Studies, Norman Wells, Inuvik and Tuktoyaktuk, N.W.T. and Prudhoe Bay, Alaska. Interdisciplinary Systems Ltd. for Canadian Arctic Gas Study Limited, Winnipeg, Manitoba. 127 pp.

——. 1973*b*. Natural plant recolonization of surficial disturbance, Tuktoyaktuk Peninsula region, Northwest Territories. Can. J. Bot. 51:2177-2196.

Hok, J. R. 1971. Some effects of vehicle operation on Alaskan arctic tundra. M.S. thesis, Univ. of Alaska. 85 pp.

Klebesadel, L. J., and W. M. Laughlin. 1964. Utilization of Native Bluepoint Grass in Alaska. Univ. of Alaska Agr. Exp. Station. Forage Research Rep. No. 2. 22 pp.

Klein, D. R. 1971. Reaction of reindeer to obstructions and disturbances. Science 173:393-398.

——. 1972. Problems in conservation of mammals in the north. Biol. Conserv. 4:97-101.

Lutz, H. J. 1956. Ecological Effects of Forest Fires in the Interior of Alaska. U.S. Dep. Agriculture Tech. Bull. No. 1133. 121 pp.

McCown, B. H. 1973. The influence of soil temperatures on plant growth and survival in Alaska. Pages 12-33 *in* Proceedings of Symposium on the Impact of Oil Resource Development in Northern Plant Communities. Inst. of Arctic Biol., Univ. of Alaska. Occ. Publ. on Northern Life No. 1.

McCown, B. H., F. J. Deneke, W. E. Rickard, and L. L. Tieszen. 1973. The response of Alaskan terrestrial plant communities to the presence of petroleum. Pages 34-43 *in* Proceedings of Symposium on the Impact of Oil Resource Development in Northern Plant Communities. Inst. of Arctic Biol., Univ. of Alaska. Occ. Publ. on Northern Life No. 1.

McCowan, D. D., and F. J. Deneke. 1973. Plant germination and seedling growth as affected by the presence of crude petroleum. Pages 44-51 *in* Proceedings of Symposium on the Impact of Oil Resource Development in Northern Plant Communities. Inst. of Arctic Biol., Univ. of Alaska. Occ. Publ. on Northern Life No. 1.

Mitchell, W. W., and J. D. McKendrick. 1974. Tundra rehabilitation research: Prudhoe Bay and Palmer Research Center. Report to Alyeska Pipeline Service Company, Atlantic Richfield Co., Canadian Arctic Gas Study Ltd., Exxon Co., Shell Oil Co., Union Oil Co., by the Institute of Agricultural Sciences, Univ. of Alaska, Palmer Research Center.

Mooney, H. P., and W. D. Billings. 1961. Comparative physiological ecology of arctic and alpine populations of *Oxyria digyna*. Ecol. Monogr. 31:1-29.

Radforth, J. C. 1973. Long Term Effects of Summer Traffic by Tracked Vehicles on Tundra. Environmental-Social Committee, Northern Pipelines, Task Force on Northern Oil Development. Rep. No. 73-22. 60 pp.

Rickard, W. E., and J. Brown. 1974. Effects of vehicles on arctic tundra. Environ. Conserv. 1:55-62.

Rickard, W. E., and C. W. Slaughter. 1973. Thaw and erosion on vehicular trails in permafrost landscapes. J. Soil Water Conserv. 28:263-266.

Saville, D. B. O. 1972. Arctic Adaptation in Plants. Research Branch Canada Dep. Agric. Monograph No. 6. 81 pp.

Scarborough, A., and P. W. Flanagan. 1973. Observations on the effects of mechanical disturbance and oil on soil microbial populations. Pages 63-71 *in* Proceedings of Symposium on the Impact of Oil Resource Development in Northern Plant Communities. Inst. of Arctic Biol., Univ. of Alaska, Occ. Publ. on Northern Life No. 1.

Scotter, G. W. 1971. Fire, vegetation, soil and barren-ground caribou relations in northern Canada. Pages 209-230 *in* C. W. Slaughter, R. J. Barney, and G. M. Hansen, eds. Fire in the Northern Environment–A Symposium. Pacific Northwest Forest and Range Exp. Station.

Tieszen, L. L. 1972. The seasonal course of aboveground production and chlorophyll distribution in wet arctic tundra at Barrow, Alaska. Arctic and Alpine Res. 4:307-324.

Van Cleve, K. 1973*a*. Revegetation of disturbed tundra and taiga surfaces by introduced and native plant species. Pages 7-11 *in* Proceedings of Symposium on the Impact of Oil Resource Development in Northern Plant Communities. Inst. of Arctic Biol., Univ. of Alaska, Occ. Publ. on Northern Life No. 1.

_____. 1973*b*. Short-term growth response to fertilization in young growth quaking aspen. J. For. 71:12.

Van Cleve, K., L. A. Viereck, and R. L. Schlentner. 1971. Accumulation of nitrogen in alder (*Alnus*) ecosystems near Fairbanks, Alaska. Arctic and Alpine Res. 3:101-114.

Viereck, L. A. 1973. Ecological effects of river flooding and forest fires on permafrost in the taiga of Alaska. Pages 60-67 *in* Permafrost: The North American Contribution to the Second International Conference.

Wein, R. W. 1974. Recovery of Vegetation in Arctic Regions after Burning. Environmental Social Committee Northern Pipelines, Tack Force on Northern Oil Development. Rep. No. 74-76. 63 pp.

Wein, R. W., and L. C. Bliss. 1973. Changes in arctic *Eriophorum* tussock communities following fire. Ecology 54:845-852.

_____. 1974. Primary production in arctic cottongrass tussock tundra communities. Arctic and Alpine Res. 6:261-274.

Younkin, W. 1973. Autecological studies of native species potentially useful for revegetation, Tuktoyaktuk region, N.W.T. Pages 45-74 *in* L. C. Bliss, ed. Botanical Studies of Natural and Man-modified Habitats in the McKenzie Valley, Eastern McKenzie Delta Region and the Arctic Islands. Environmental-Social Committee Northern Pipelines, Task Force on Northern Oil Developments. Rep. No. 73-43.

Zasada, J. C., and R. A. Gregory. 1969. Revegetation of white spruce with reference to interior Alaska: a literature review. U.S.D.A. Forest Service Res. Paper PNW-79. 37 pp.

Recovering the American City: Minneapolis as a Case Study

David R. Goldfield

Abstract

The American city was once an efficient ecosystem that functioned as the economic and demographic focus of the region. Through a combination of public policy and technological innovation, the city has become an obsolete and deteriorating environment. The Minneapolis case study demonstrates, however, that the process of urban decay is reversible. Close cooperation between the public and private sectors, constructive intergovernmental relations, and the scale and imagination of planning and implementation have begun to arrest the decline of downtown, the decay of a historic neighborhood, and the erosion of the city's economic base.

Rise and Decline of the City

Long ago, when the air was clear and the water was pure, the American city was born. Succored by water which gave life-giving commerce and people and nourished by natural resources which produced wealth and industry, the American city grew to manhood a strapping, confident, and triumphant replica of the environment from which it came. The city's downtown was the focus of a hopeful metropolis in a hopeful nation. It was the Greek agora, the Roman forum, and the medieval marketplace combined into one teeming mass of shops, theaters, schools, and humanity. It was the arena for success or failure; it was the place where the city met; it was variety; it was choice; it was the very essence of urban civilization.

Downtown in nineteenth-century America was controlled chaos. Since cities were primarily "walking cities" of roughly a mile and a half square, a multitude of activities crowded into the business district jostling for choice frontage. The conglomeration of land use would present a severe challenge to the orderly city planners of our present day. Factories, hotels, general stores, single-family homes, churches, and bars were thrown together as if some miscreant tornado had deposited these structures at random. Admittedly, not all land use was so chaotic. Some juxtaposition of activities was evidently well thought out, such as the brothel which adjoined the Globe Theater in downtown Boston in the 1820s. Nowadays one is merely treated to an orange drink during intermission. Such economic linkages notwithstanding, most downtowns presented a mélange that was impossible to classify.

Therein lay one of the major lures of the downtown. Since the downtown was synonymous with the city's economy, a variety of business activities existed there. Capitalists, immigrant workers, shoppers, and youngsters mingled in the street, each pursuing his or her objectives and dreams. The centripetal force which drew people to downtown to conduct business worked in other ways as well. Theaters sprouted beside the countinghouses. The theater was an essential perquisite for urbanity as well as for entertainment. Since downtown was already the business center, convenience and attraction assured that downtown would become a cultural district, too. The theater became a popular urban institution. In an era when entertainment generally meant an evening stroll, a glass of beer, or several hours by the fire, it was not surprising that the stage proved attractive to a growing urban populace. In fact, before organized sports, the theater was probably the best-attended democratic diversion in urban America. Instead of loyalty to teams, citizens professed loyalty to actors. Sometimes the avidity of the fans surpassed reason. In 1849 a New York City mob—mostly workers—that championed actor Edwin Forrest attacked the Astor Place Theater where Forrest's rival, aristocratic William Macready, was playing. Twenty-two people were killed before troops quelled the disorder (Glaab and Brown, 1967).

The excitement of downtown which sometimes erupted into violence came from another source besides business and culture: the people who lived there. Executives built and occupied brownstones and relished their proximity to their growing commercial and financial concerns; workers crowded into substandard dwellings near the wharves or factories which defined the boundaries of downtown. Their streets were downtown streets and their routes were through downtown. The residential component of downtown insured that the area would be a 24-hour activity center.

In addition to being a people center, downtown was also the hub of government. Local, county, state, and even for a few years in downtown New York federal buildings gravitated to downtown to impart official recognition to the area's importance. Since business and government leaders formed an interlocking directorate, proximity to the city's commercial and financial core was essential for competent administration. The booster character of nineteenth-century urban government doubtless arose from the fact that a bustling downtown street was directly outside the window of city council chambers. These city builders constructed elaborate, if somewhat fallacious, images of their metropolis to attract business and to promote growth. The role of the entrepreneur was an essential ingredient for urban success in nineteenth-century America, and downtown was often the business and the residence of the civic elite (Abbott, 1970).

If urban boosters prayed at the altar of progress, their gods were the river and steam. The wharf and/or railroad tracks were adjacent to downtown and, in the case of the latter, often ran through it. Riverfront and railroad depot supported downtown and the city in a number of ways. The water or iron highway was the major trade outlet. Commerce was the catalyst of growth–the "goddess of Christianity," as one merchant put it. In the intense urban rivalry that marked nineteenth-century America, access to the hinterland meant the difference between cityhood and oblivion. Entrepreneurs plotted the strategies of growth in downtown offices, and the wharves and depots reflected the success or failure of their policies (Belcher, 1947; Wade, 1959).

Downtown was the major beneficiary of growth, and growth altered the configuration of downtown and of the city. Besides bringing and taking produce and manufactured goods, the wharves and depots brought people–immigrants from Europe and from rural America. Since downtown was the area nearest to the point of entry and the city's employment center, most newcomers settled there or nearby. Downtown became even more crowded than before–just at a time when businesses were expanding. Fortunately, technological innovations in the transportation field allowed this potentially strangling situation to rectify itself. The horse-drawn railway and later the electric trolley enabled both downtown and the city to expand.

Expansion, not surprisingly, followed the trolley tracks. Since the trolley tracks converged on downtown, the electric era opened an unprecedented period of prosperity for downtown. In downtown Atlanta, for example, five trolley lines came together at or near Rich's Department Store. Expansion also brought segregation. Downtown became exclusively a place to work rather than a place to live. Downtown's strategic location increased ground rents to such a degree that residential land use, at the densities possible then, became financially infeasible. In losing its residential component, downtown was losing a vital part of its unique heterogeneity. Downtown was no longer a 24-hour show (Vance, 1971).

The neighborhood as a distinct geographic area outside of the downtown became the primary residential repository of the city. A half century ago, Robert Park, an urban ecologist from the University of Chicago, identified the neighborhood as the essential unit of city life (Park et al., 1925). In late nineteenth and early twentieth-century America, the neighborhood shared urban glory with downtown. The neighborhood, in many respects, was downtown in microcosm. Variety was also the keynote of neighborhood living. Social historian Richard Sennett averred that the multitude of "contact points" lent a distinctive character to the turn-of-the-century neighborhood. Contact points included bars, meeting halls, brothels, grocery stores, and stoops—often within the same block. These contact points, Sennett contended, provided a variety of experiences and reference points for residents that made city living unique and exciting (Sennett, 1971).

The political system reinforced the growing importance of the neighborhood in urban life. Professional politicians, often operating through political machines, replaced business leaders in local government. They took advantage of the ward system to forge strong neighborhood organizations that provided jobs, welfare services, and access to the political system. The machine was the ultimate in citizen participation. It was sensitive to neighborhood needs and formulated policy according to the recommendations of ward heelers. Although urban machines were rife with corruption, they extended the urban infrastructure to all citizens and directed the American city to its greatest era (Hays, 1974).

The discrete neighborhood units had direct access to downtown through the numerous private-franchise trolley companies the eager city council provided citizens. When residents arrived downtown, the scene scarcely resembled the unconnected hodgepodge of an earlier era. Just as there were discrete neighborhoods, discrete business districts had developed. There was a financial neighborhood—within a five-block radius of downtown Atlanta there were 13 banks; there was a commercial district, a warehouse district, and a theater district. Growth, expanding affluence, and transportation technology spawned the segregated metropolis.

Downtown still reflected the image of a vibrant, prosperous metropolis. Wharf activity was declining, but railroad business continued to burgeon. It was the skyscraper, however, that lent a particular dynamism to the turn-of-the-century American city. In Chicago especially, the skyscraper was the epitome of confidence and success. Dominating downtown, it announced the arrival of a new urban power to the world. Louis Sullivan, one of the masters of the Chicago School of Architecture, perceived these concrete creations "as a social act. In such act, we read that which cannot escape our analysis, for it is indelibly fixed in the building, namely the nature of the thoughts of the individual and the people whose image the building is or was" (English, 1963). The "social act" of the vertical monuments to success had its dark counterparts, however. While downtown basked in the glow of limestone and lights, its periphery was beginning to rot. Carl W. Condit, chronicler of the architectural rise of Chicago, observed that "in Chicago it was no longer possible to hide the fact that there were two cities, one consisting of the skyscraper core and the luxury apartments that stretched along the lake front, the other its exact opposite—the tangled rail trackage, the deteriorating factories and warehouses, the slums and spreading ghettos, the miles of gray areas bordering the frantic commercial strips" (Condit, 1973).

Though residents, politicians, and businessmen did not know it, the halcyon era of the American city was drawing to a close. The raiments of downtown, the variety of neighborhoods, and the volume of commercial, financial, and industrial activity masked for the moment the deteriorating quality of life, the decaying waterfront, and the noxious air. New technologies and new affluence were springing out of the city to sap its strength. The skyscrapers would remain, but business, population, and excitement drained from the metropolis as blood from a corpse.

The impact of the automobile on the American city is only beginning to be understood. Its appearance as a means of common transportation coincided with the decline of downtown, the weakening of neighborhoods, the decay in the natural environment, and the exodus of people and business. The auto alone certainly was not responsible for all of these misfortunes, but it was a primary catalyst. The auto increased the mobility of the urban resident significantly. Not limited by tracks or schedules, the auto went in all directions and out of the city as well. If all trolley tracks led to downtown, all roads did not. In terms of commuting time to work, the auto made suburban living a possibility to millions of middle-class Americans. Widespread affluence aided by trade unionism and a growing civil service made suburban homesteads a distinct probability. The visible decay of the city, the loss of its attractions, and the success of ethnic groups which brought both discrete neighborhoods and machine politics to the brink of extinction, contributed to the suburban stream which became a torrent following World War II.

The truck—the commercial counterpart of the auto—impacted the city in a similar manner. The new transportation technology killed whatever river trade was left and dealt a severe blow to the railroads. Downtown's strategic proximity to the wharf and the depot meant considerably less. They were no longer the founts of growth and prosperity. Like the auto, the truck went wherever the road led. With the introduction of controlled access freeways in the 1930s, proximity to the highway became the new key to economic growth. Suburban communities directly off the freeways provided ideal utilization for truck technology without the drawback of clogged city streets. Industry and commerce went to the suburbs with the truck and the auto. Cheaper land, room for expansion, affluent customers, and transportation access were attributes that downtown was unable to offer (Rae, 1965).

Though the auto functioned as the Pied Piper to the suburbs, its work in the city was not yet complete. Decay might have been tugging at the fringes of downtown and shoppers might have preferred spacious comfortable suburban malls, but downtown remained the government and business center. During the daylight hours, downtown's affliction was not immediately apparent. Even in this cauldron of activity, though, lay disturbing trends precipitated by the auto.

The auto degenerated space. Parking lots proliferated in downtown, yielding huge profits for owners and meager taxes for the city. The auto's spatial pollution resulted from the fact that it was tied to its owner. Since the auto occupies the same lifezone as the human being—the first six feet above the earth's surface—it is, as architectural historian John Marsten Fitch has observed, "in constant irreconcilable kinesthetic and aesthetic conflict with the pedestrian." In motion, the auto displaces 600 cu ft, compared with 12 cu ft for an upright individual; when parked it occupies 120 sq ft, compared with 20 sq ft for a reclining individual. The parking process requires three times as much space. The auto, by restricting the amount of living space, reduces the vibrancy of downtown and adds to its visual pollution as well (Fitch, 1967).

Air pollution is one of the most nefarious results of widespread auto use. The vast numbers of autos in the downtown area and the tall buildings combine to produce an unhealthy carbon monoxide trap. Cities without large-scale heavy industry such as Los Angeles and Washington,

D.C., still suffer severe pollution, primarily from the automobile. The attraction of the relatively clean suburbs becomes more evident. Flowers wilt and respiratory ailments multiply in the city. The city, once the most desirable environment, has become a milieu to be avoided.

To attribute the contemporary urban condition entirely to the auto is too mechanistic an explanation. Other alterations contributed, perhaps not so dramatically, to the decline of downtown and the city as a whole. In a wave of missionary zeal, political reformers, usually from the business community, retook the city from the machine. Experts and special-purpose agencies fluttered about city hall concocting plans and policies for the metropolis. The carefully constructed fabric of neighborhood life was ignored, and citizen input ceased. The age of the technocrat had arrived (Brownell and Stickle, 1973).

The neighborhood itself, bisected by freeways, inundated by autos, and riddled by out-migration, took on a listless aspect. Bars, churches, brothels, and union halls closed. The rich tapestry of neighborhood life melted into suburban similitude. Affluence provided a visa out of the neighborhood and allowed urban residents to pursue the Golden Fleece of the pastoral ideal. Contact points disappeared and neighborhoods lost their distinction. Halstead Street in Chicago struggles on as a variegated reminder of the past, and the Lower East Side of New York is still an interesting place to visit, but the somber tones of adjacent districts have tarnished the gaiety of this vibrant arena. Admittedly the dwellings were, in some cases, dilapidated, and disease lurked in the dark alleys, but the spirit of community life and the joy of social interaction infused the neighborhood and spilled over to the rest of the city.

The neighborhoods still attract newcomers. The city in its tottering condition still provides employment and diversion. The blacks and Latins who comprise an increasing percentage of the urban population have inherited a shell, however. Local government responds barely, poverty has become institutionalized, and the disappearance of mass transit in the wake of the auto, mismanagement, and misplaced priorities have increased the isolation of minorities. A declining tax base, inflation, and growing demands for services have placed unbearable burdens on already hard-pressed cities. Services such as health delivery, garbage collection, and police protection are, as a result, haphazard and at times inhumane. The questionable quality of drinking water, the dying rivers, the moribund railroad lines, the declining downtowns, the disappearing open space, the triumph of the auto over man and common sense, and the decay of neighborhood institutions have cast a pall over the American city. Writers have already drawn the shroud over urban civilization and buried it with varying degrees of lamentation befitting a fallen giant (Jacobs, 1961).

Reports of the city's demise, to borrow a phrase, have been greatly exaggerated. The American city, to be sure, has undergone severe trauma in the past few decades, but the same spirit which built urban empires is returning to save it. The recovery of the American city has taken numerous forms, as the case study which follows demonstrates. The common feature of all regenerative efforts, though, has been the restoration of the positive features of the historic metropolis. Projects have concentrated on restoring the economic and cultural importance of downtown, regaining the economic base lost to the suburbs, bolstering mass transit while limiting the auto, improving the natural environment, and reconstituting the vitality of the urban neighborhood. The movement to recover the historic city is still in its infancy. It is still too early to draw conclusions about the ultimate success or failure to regain urban preeminence. Nevertheless, private and public individuals and agencies are making concerted efforts toward this end. The very attempt itself is significant if only because it serves as a model for cities in similar situations.

The Recovery of Minneapolis

Like so many other cities across the country, Minneapolis owes its existence to its strategic location on a river. The city is at the head of Mississippi River navigation. Real estate developers, excited over the town's potential as a lumber center, laid out a traditional gridiron street pattern in 1857 to facilitate the sale of lots. By 1872 the city expanded to 12½ square miles and flour milling was beginning to challenge lumber as the city's prime industry. In 1881 the mills of John and Charles Pillsbury were producing over 15 million barrels of flour annually. Soon, the Great Northern Railroad complemented the fine river access of the city by erecting a depot along the riverfront. Hardworking Scandinavian and Eastern European immigrants joined a growing populace to provide the sinews for progress. Their ethnic neighborhoods added a cosmopolitan atmosphere to the old frontier settlement. With an economic base firmly established by 1900 and with effective transportation links, the growth and prosperity of the "Mill City" seemed assured. By the 1950s, however, the same malaise that was striking other American cities began to affect Minneapolis. The waterfront area had become a refuge for winos and rats; rust was the only presence on the railroad tracks; auto traffic clogged and polluted downtown; the recently opened Southdale Mall in suburban Edina challenged downtown business prosperity; the old ethnic neighborhoods were in decay and depopulation; and industrial flight threatened the city's economic base. It was a sad but familiar urban story. The city, however, still possessed some valuable resources, and these elements attacked the problems besetting their urban environment.

The automobile had been a primary culprit in the decline of downtown Minneapolis. During an average workday more than a quarter of a million autos entered and left downtown. Since the retail and office core of the city is relatively compact, the air, noise, and spatial pollution engendered by the auto's presence was considerable. The parking facilities downtown consisted of single-level lots that were unable to meet the demands created by the daily influx of cars. The result was twofold. First, the land occupied by the lots was among the most valuable in the city. Second, those who arrived after 9 A.M. on weekdays–primarily shoppers–faced long walks to the shopping district because of inadequate parking facilities immediately adjacent to the area. To compound the problem, the mass transit system was outmoded, uncomfortable, and unreliable. In short, getting downtown required effort, stamina, and a great deal of patience. The historic attribute of downtown–easy access–disappeared in a cloud of auto exhaust fumes (Lu, 1974).

Downtown Minneapolis faced a more unusual problem than auto contamination. Natives call the climate "rigorous," though "severe" would probably be more descriptive. Weather Bureau records indicate that Minneapolis has temperatures below 50°F for 150 days of the year and a measurable amount of rain and snow on 125 days. Visitors, shoppers, and workers moved about downtown unprotected. Considering the outdoor treks some shoppers had to make, it was a wonder that retail sales did not decline significantly during the winter months. The alternative–the enclosed, all-weather suburban shopping mall with ample parking–was understandably preferable (Lu, 1974).

The weather was not the only chill in downtown Minneapolis. Taxes were rising and retail sales were sagging with no new construction to relieve the declining tax base of the area. Moreover, a skid row and a deteriorating neighborhood framed downtown, portending an ominous future for the business district. The removal of General Mills to the suburbs in 1955 seemed to signal the ultimate doom of the preeminent business district in the Upper Midwest (Lu, 1974).

In the tradition of the nineteenth-century city, local entrepreneurs organized to attack the nemeses of downtown. In response to the exodus of General Mills, businessmen formed the Downtown Council—a well-funded, well-staffed booster group that triggered the Minneapolis Renaissance (Downtown Council, 1974). The following year, 1956, the Downtown Council formulated a plan to revive downtown and restore it to its historic function as the economic and cultural center of the region (Minneapolis Planning & Development Department, 1973*c*).

The planning process that the Downtown Council developed should serve as a model for similar endeavors. The council's methodology was thorough and wide-ranging. Working through a Nicollet Avenue Survey Committee, the council viewed its primary goal as the restoration of the diversity and accessibility of downtown. Further, the council hoped that eventually, regeneration of the entire city would accompany downtown recovery. The council retained Barton-Aschman Associates to develop a plan for downtown. The consulting firm presented five alternatives to the council in 1960 and the Mall and Transitway approach seemed most feasible to the council and to city authorities who became involved in the planning process through the urging of Nicollet Committee members. The plan received city approval, and this opened the way for public funding with the city acting as general contractor (Aschman, 1971).

One of the major emphases of the Mall and Transitway Plan, and its great attraction to local business leaders, was its proposed transportation system. Just as turn-of-the-century trolley tracks converged on downtown, so the plan called for bus routes from every Minneapolis neighborhood to culminate at the retail district. To facilitate the bus system, the Mall area—an eight-block stretch along Nicollet Avenue—would be closed to all vehicles except buses and taxicabs. With one stroke, the major downtown polluter and congestor—the auto—had been removed, while accessibility to the shopping district was maintained (Aschman, 1971).

The plan did not eliminate the auto, but merely put it in its proper place—on the periphery. Staging areas on the outskirts of the Mall would hold multistory parking ramps to conserve land and to provide shelter during inclement weather. The next problem was to transport the automobile owners from the parking facilities to the department stores, specialty shops, or restaurants lining the urban Mall. It was doubtful, not to say uneconomical, that these individuals would take a bus or a cab five or six blocks. On the other hand, the brutal Minnesota winters precluded strolling even short distances.

The plan looked to the past again for the solution to pedestrian transit and borrowed an idea from a new adversary as well. The obvious solution was to enclose the Mall to maximize pedestrian use. One suggestion recommended a geodesic dome over the entire area. The $6-billion price tag, however, was too expensive. The creation of a skyway system connecting the second level of buildings accomplished a similar purpose—returning the downtown to its nineteenth-century pedestrian character—at a considerably smaller cost, $20 million (Fig. 1; Minneapolis Planning & Development Department, 1973*e*).

The skyways are glass-enclosed bridges between buildings and across streets that keep the pedestrian cool during the summer and warm during the long winter months. They also separate pedestrians from the vehicle level. The skyways are part of the show that downtown should be. They house art displays, fashion shows, and exhibits. Eventually, more than 60 such structures will honeycomb the entire downtown area, providing continuous pedestrian access for 54 blocks of downtown activity. The "walking city" with all its intimate excitement and variety has returned.

With the problem of pedestrian and vehicular access solved to some degree, the restoration of the diversity and visual excitement of the nineteenth-century downtown presented even greater planning difficulties. The first reform was to eliminate large, overhanging signs that cheapened downtown visually, change uncoordinated street furniture, and prohibit construction of look-alike buildings on reclaimed parking lots and other available sites. Following these precepts would avoid what one urban designer termed "sensory underload"—a common affliction of most postindustrial downtowns that attempt to substitute steel and glass for vibrancy (Lu, 1974).

The gridiron street pattern was a major contributor to "sensory underload" with its monotonous vistas. The planners tranformed Nicollet Avenue into a serpentine roadway. Lawrence Halprin, a noted landscape architect retained by the city and the Downtown Council, took his inspiration from the medieval city with its irregular street system offering the perambulator sharply contrasting vistas and surprises at every wind in

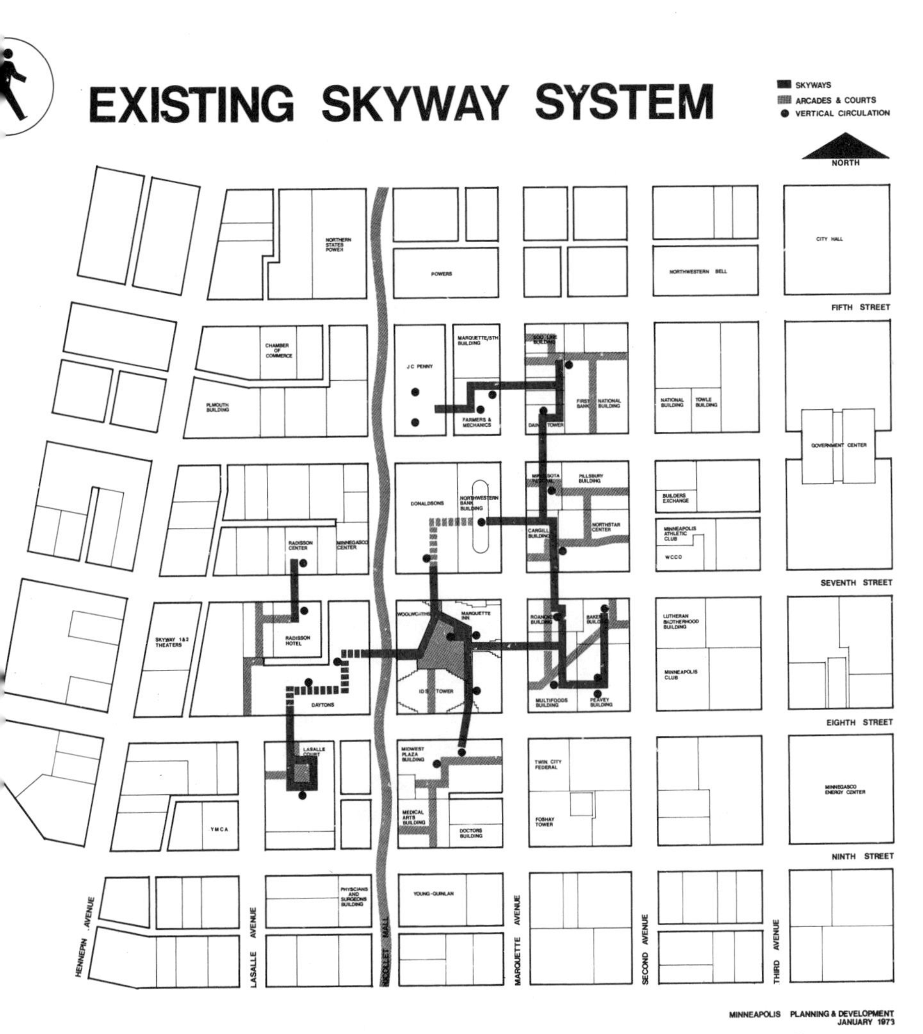

Fig. 1. Existing skyway system, Minneapolis, January 1973. (Minneapolis Planning and Development Department, 1973)

the roadway. To maximize the pedestrian aspect of the Mall, Halprin cut the four traffic lanes to two, utilizing 24 feet–just enough for two buses to pass–instead of the customary 120 feet. The sidewalks, correspondingly, were as much as 36 feet wide on one side at some bends in the curvilinear roadway. Thus, Halprin effectively reconciled the pedestrian emphasis of the Mall with the objective of diversity (Aschman, 1971; Minneapolis Planning & Development Department, 1973*c*).

In another attempt to rejuvenate downtown excitement, the design team planned for each block to have a distinctive character. Landscaping was the primary mode of differentiation. The sidewalks, resurfaced in earth tones of rustic terrazzo, included different patterns. Plants abound on each block in various arrangements. There are seasonal floral arrangements in pots and planters. All trees have removable base covers. Lighting on each block is unique, depending on the fronting establishments for type and intensity. Finally, street furniture adds a casual, almost European atmosphere to the Mall. The furniture is encased in placements and includes trash containers, benches, and planters. Even the traffic signals have a distinctive quality–bronze-anodized signal housings with integrated street name signs and traffic instructions. Police and fire call boxes and mailboxes were relocated on side streets (Aschman, 1971).

Some design features are common throughout the Mall and add to the visual and physical pleasure of the shopper. Heated bus shelters provide comfort for those who choose to use mass transit. There are six fountains, some with heating coils to extend their usefulness; a number of flagpoles to display colorful seasonal banners after the fashion of some European business districts; 17-ft high, four-face clock; and a weather station (Minneapolis Planning & Development Department, 1973*c*).

The success of these design elements, as well as of the transit plans, depended on adequate funding. The creation of the Mall, excluding the skyway system, ran close to $4 million–an investment that transcended the financial capabilities of the Downtown Council. Financing the project indicated not only the resourcefulness of the council but also the efficacy of close public-private interaction–a relationship that has characterized most recovery programs in Minneapolis. The construction cost of $3.8 million was financed as follows:

Urban Mass Transportation Grant (federal)	$512,000
Urban Beautification Grant (federal)	$483,000
City bond issue	$2,751,785
Total	$3,746,785

The line score, though, reveals only a portion of the financial story. First, only $1.3 million of the cost is visible above ground. The entire space under Nicollet Avenue's 80-ft right-of-way is occupied by utilities. The private companies, in a display of civil consciousness that is characteristic of the business community in Minneapolis, agreed to inspect their systems and to rework them to conform to the aboveground construction. The city undertook extensive replacement of the water system to accommodate fountains and plants that would adorn the new Mall. Fire hydrants were relocated as the sidewalk expanded and this required further adjustments underground. Finally, the city installed a new system of traffic signalization. Electrical circuitry was expanded to service the infrared radiation system that heated the 16 bus shelters along the eight blocks of the Mall. Separate controls for snow-melting grids were embedded in the sidewalks to provide secure footing for the pedestrians (Minneapolis Planning & Development Department, 1973*c*).

Perhaps the most unusual aspect of the utilities' reconstruction was the method the city employed to finance the alterations. Issuing city bonds was not, of course, a unique remedy. The manner of amortizing the bonds, however, was. The city employed a "benefit assessment" tax to insure that the burden of paying for the Mall would not fall on all of the citizens. The benefit assessment is a good political maneuver for local governments. Under the plan, the city created two benefit zones—on the Mall and off the Mall—covering an 18-block area. Each zone included sectors providing for 100%, 100%-75%, 75%-59%, and 50% allocations of cost so that properties in the heart of the Mall would bear the greatest construction and maintenance expenses. The benefit assessment, in keeping with the restoration theme of the Mall, was actually a common funding method in nineteenth-century cities when local government lacked sufficient resources to undertake massive street paving and repairing programs. The city levied the tax with the approbation of the Downtown Council, emphasizing the unanimity of purpose among the city's business community (Aschman, 1971).

Since Nicollet Mall opened in November 1967, the benefit assessment has proved a worthwhile investment for the Mall-area entrepreneurs. Businesses reported a 14% increase in volume through 1973. Considering the condition of the economy and the precipitous decline in business in downtowns elsewhere, the increase is impressive. The source of the Mall's clientele is further cause for optimism about the future vigor of downtown Minneapolis. Over 56% of the Mall's sales are to families residing more than five miles from downtown. This figure signifies not only that downtown has become, once again, an attraction to people outside the immediate area, but that Nicollet Mall is cutting into the suburban shopping center. Indeed, in recent assessment of the Mall's drawing power, the Downtown Council claimed that the proliferation of suburban malls has impacted those malls more than downtown, which offers shoppers a unique experience in addition to a wider variety of merchandise (Downtown Council, 1974). Despite professions of regional cooperation, individuals responsible for Minneapolis's regeneration clearly view the suburbs as adversaries and plan diligently to recoup the city's fortunes at the expense of other parts of the region.

The business renaissance in the Mall area has helped a struggling mass transit system. The refined routing system and the exclusive busway helped to increase ridership by 8% in the downtown area during the past year. In addition, the city added a fleet of mini buses to the Mall transportation system. The QT system, as it is called, runs on six-minute schedules and offers rides throughout downtown for ten cents. Last year the mini buses transported over one million passengers. This, in addition to the 8% increase in conventional mass transit, indicates the resurgence of public transportation in downtown (Downtown Council, 1974).

The skyway system remains the backbone of the Mall's unique transit system, and has played a significant role in enhancing the distinctiveness of downtown. Seventy-six skyways connecting 64 city blocks are planned for 1985. At present, there are a dozen skyways, all built with private capital. If the system is to be completed, a source of public funding must be tapped. The bridge portion of the skyway costs $100,000, while renovation of building interiors to accommodate the structure runs upwards from $150,000. The city, in conjunction with the council, is exploring the possibility of levying a benefit assessment tax to finance the completion of the skyway system. The best argument in favor of funding the network is use. On an average winter day 20,000 people travel the skyway routes. When mutilevel parking facilities on the periphery of the Mall are completed to connect with the skyways, there will be twice that number using the system. Already, second-floor rents have risen to first-floor levels with no diminution in first-floor business. The system, even though it is incomplete, has proven comfortable for pedestrians and profitable for businessmen (Minneapolis Planning & Development Department, 1973*e*).

When Nicollet Mall opened in 1967, the Downtown Council was rightfully proud of its accomplishment. The Mall, however, could be only a beginning in the long process of restoring downtown and in rejuvenating the entire city. Several problems remained. First, the residential component which imparted a 24-hour festival atmosphere to nineteenth-century downtown was missing. Second, although the Mall injected considerable diversity into downtown, the area still lacked entertainment activities. Third, broken people and run-down structures still encircled downtown, especially along the Mississippi River to the east and in Loring Park to the southwest. Finally, the city's economic base was still threatened by the continuing flight of industries to the suburbs. The cooperative spirit engendered by the Mall project retooled to attack problems beyond Nicollet Avenue.

A residential component was essential to restore complete vitality to downtown. In choosing the Loring Park neighborhood as downtown's residential district, city and private interests hoped to solve several problems plaguing the area, while bolstering the new success of Nicollet Mall. The plan, as outlined by the city Planning and Development Department in 1970, called for the extension of Nicollet Mall to Grant Street, four blocks from its present termination. At that point, the city would construct a finger park, or greenway, to link the Loring Park community with the Mall. The greenway would not be a mere ornamental swath of grass and flowers but would include 2,700 apartments, townhouses, and condominiums to inject a high-density residential factor characteristic of nineteenth-century downtown areas. By using the greenway in this manner, and by extending the Mall, the planners brought the residential component into close contact with the city's economic center (Fig. 2; Minneapolis Planning & Development Department, 1973*d*).

The project has the added benefit of linking the city's major cultural attractions to the heart of downtown. Entertainment and cultural activities that were so characteristic of historic downtown centers were absent in Minneapolis. The Mall extension and the connection with Loring Park filled the cultural void. Orchestra Hall, the new home of the Minnesota Symphony Orchestra, is located on the Mall extension, while the nationally famous Tyrone Guthrie Theater and the Walker Art Center are in the Loring Park neighborhood. In addition, the city is aiding the expansion of Metropolitan Community College in Loring Park to be of greater service to the area (Office of City Coordinator, 1974).

Besides increasing the diversity and attraction of downtown, the city hopes to reap considerable financial rewards from the project. In 1971 the Minnesota legislature passed the Development District Act. The act allows a city to purchase and improve land, thereby producing an increase in value, and ultimately a greater tax return. The city is developing Loring Park under this concept. The city is acquiring land, making public improvements such as eliminating roads to maximize pedestrian travel and to minimize the impact of the auto, and selling parcels to private contractors who build according to the city's plan. The city places the

Fig. 2. Loring Park development. (Minneapolis Planning and Development Department, 1973)

additional tax money generated by the increased value of the property into a special fund until all public costs including relocation and administration expenses have been recovered. In 1973 the assessed value of the area was $12 million. When the project is completed the estimated assessed value will be close to $60 million. Since restoring the city's tax base is a prerequisite to recovery, such a program benefits the entire metropolis (Minneapolis Planning and Development Department, 1973*d*).

The other components of the Loring Park plan follow some of the planning principles utilized in the original Mall project and indicate an appreciation for recovering the quality of life in the historic city. First, the 36-acre Loring Park will include new play equipment and walkways. Since it is Minneapolis's only center-city park, its open spaces are a welcome addition to the variety of downtown life. Second, planners have sought to restrict the automobile not only on the greenway but throughout the neighborhood. Pedestrian bridges, closed streets, and landscaped courts underneath which are the neighborhood parking facilities insure that the automobile will maintain a low profile in the Development District. In order to diversify the residential character of the neighborhood, commercial services including food stores, branch banks, and restaurants are planned. The idea is to increase community contact points in the fashion of turn-of-the-century neighborhoods (Office of City Coordinator, 1974).

By the end of 1974 the city had committed $13 million to the Development District through the sale of general obligation bonds. The project should take about 15 years to complete at a total cost of $60 million. City planners feel that it is a wise investment because it (1) increases the city's tax base, (2) introduces residential and cultural components into downtown, (3) rehabilitates a declining neighborhood, and (4) follows the successful planning precepts pioneered by the Mall, such as the maximization of pedestrian access and the diversity of activity (Minneapolis Planning & Development Department, 1973*d*).

A more serious problem existed east of the Mall adjacent to the river. In the mid-nineteenth century this area was the hub of the city's commercial activity. By the 1950s the area had deteriorated into a skid row consisting of abandoned structures and railroad tracks. The Mississippi River, once the lifeblood of the community, was as devoid of meaningful life and activity as its banks. The city became aware of the situation as early as 1956, when it designated the area for urban renewal. By the end of 1975 more than $150 million will have been allocated to the Gateway Center Urban Renewal project. Nicollet Mall begins where Gateway Center ends, although some Gateway projects such as the Northern States Power Company building are actually on the Mall.

As usual, there was a sizable input from the private sector. In addition to the power company, the Northwestern Life Insurance Building with its pillared facade and reflecting pool, and the Sheraton-Ritz Hotel with its seven-level parking ramp, shops, and restaurants have added to the visual vitality of the area. IBM has a regional office in the project area in the shape of a giant computer card. The Federal Reserve Bank with its huge plaza and hard rubber furniture is the federal contribution to the area's physical rehabilitation. The architectural styles of these structures have won acclaim from such noted critics as Wolf Von Eckhardt (Von Eckhardt, 1974).

Gateway Center is a traditional urban renewal area in that the emphasis is on physical planning. Although it is visually pleasing and provides for ample public space, it is somewhat antiseptic when compared with the diversified planning that has characterized other sections of the downtown area. In 1966 the city added a much-needed residential structure to the area by erecting two high-rise, upper-income apartment dwellings—one 17 stories and the other 28 stories—near the Northwestern Life building. A park along the riverfront adds a pastoral aspect to the proliferation of steel and glass in some of the urban renewal projects.

Gateway Center, as a planning entity, is not on a design or a planning par with other downtown projects, which admittedly set high standards in both categories. There is minimal attention to the features that made historic downtowns such attractive areas. The fascination with bulk and height is too reminiscent of the skyscraper era of the early twentieth century when facade replaced substance and diversity in urban downtowns. Nevertheless, the project has been a financial success for the city, and this is an important factor. Before renewal the area yielded the city $554,000 in tax revenue; by 1970 the return was $2,400,000. The project helped to reverse the general decline in tax revenues throughout the city (Minneapolis Planning & Development Department, 1973*b*).

With all the activity centered around downtown, it was not surprising that eventually the city would concoct a comprehensive plan for the area in order to establish criteria and objectives for future development. Weiming Lu, an urban designer for the city, headed the interdisciplinary team that produced the comprehensive framework, Metro Center '85, in 1971. The plan emphasizes the planning features that have been successful in the downtown to date: continued cooperation between business and local government, pedestrian access, and diversity of activity including expanded health, education, cultural, and industrial facilities. The plan also underscores the necessity of carrying out the specific objectives of ongoing projects like the elimination of single-level parking lots in favor of multilevel ramps in the Mall area and the transformation of the entire Mississippi riverfront in the downtown area (1.7 miles) into a park.

Finally, Metro Center '85 charts some new planning areas. First, the plan urges the resurrection of the Mississippi River as a recreation facility. Although its recovery is nowhere near the magnitude of the Thames, latest reports are that after many years' absence, fish are swimming in the river again. Second, the plan suggests a rapid rail transit system to serve the revitalized downtown area. Considering the difficulty San Francisco has had with BART and the dim financial prospects facing Washington's Metro, however, it would probably be wiser for Minneapolis to continue to upgrade its bus system. Finally, the plan calls for a vigorous citywide effort, with special concentration in the downtown area, to conserve old buildings and selectively blend in new structures with rigid design controls (Lu, 1974).

Metro Center '85 is fairly similar to comprehensive plans currently the vogue in cities across the country. The major and significant difference, however, is that components of the plan have already been set in motion and a strong legal framework has evolved to insure future implementation. The Development District Act, passed by the state legislature at the recommendation of the Minneapolis Planning and Development Department, is an effective financial mechanism, as the Loring Park case demonstrates. Second, enabling state legislation led to the passage of a Municipal Heritage Ordinance in 1971. Under the ordinance, Minneapolis established a heritage commission with the authority to designate and protect landmarks. The commission may use a variety of legal tools in the performance of its duties, including the use and sale of air rights, the granting of use variances, and the power of eminent domain. Just as it is imperative to recover the historical features of urban life, so it is vital to preserve the city's physical legacy. Finally, there is a design review committee which does not apply abstract criteria of similitude but, rather, encourages diversity and the use of history and geography to make design an active part of downtown restoration (Lu, 1974).

The exodus of industry and the dearth of available office space so necessary in the postindustrial city were two serious deficiencies cited by Metro Center '85. The decay of cities and industrial flight occurred simultaneously. Industries represent jobs, tax base, and economic base. Further, the multiplier effect of industry is far-reaching—from luncheonettes to dry-cleaning establishments. Although environmentalists attack industry as a primary polluter, many industries are not, and many, with the benefit of technology, can improve the quality of effluent pumped into air and water. Minneapolis's industries have not been severe air polluters—the automobile is the major culprit there. The sources of Mississippi River pollution are primarily upstream, and these are being controlled. The loss of industry represented a genuine setback to many sectors of the urban economy, with little or no improvement in the air or water quality.

The erosion of the city's industrial base was a steady flow rather than a sudden flood. Between 1962 and 1970, 186 industries left Minneapolis, with as many as 36 firms leaving during 1967, the year the Nicollet Mall opened (Table 1). The companies moved to the suburbs, especially Bloomington, Golden Valley, and Edina. For Minneapolis, the serious aspect of the flight was what the companies left behind. They occupied a total of 180 acres, supplied 11,000 jobs, or 4% of the city's work force, and paid $1,666,000 in property taxes annually (Minneapolis Industrial Development Commission, 1973*b*). It was evident that the era when industry and city were synonymous had passed rudely by.

The reasons offered by company spokesmen for leaving Minneapolis revealed a number of deficiencies in the urban environment. The

Table 1. Companies migrating from Minneapolis

	5 10 15 20 25 30 35
1962	XXXXXXXXXXXXXXXX
1963	XXXXXXXXXXXXXXXXXXX
1964	XXXXXXXXXXXXXXXXXXXXX
1965	XXXXXXXXXXXXXXXXXXXXXX
1966	XXXXXXXXXXXXXXXXXXXXXXXX
1967	XXXXXXXXXXXXXXXXXXXXXXXXXXXXXXXXXXZZ
1968	XXXXXXXXXXXXXXXXXXXXXXXXXXZ
1969	XXXXXXXXXXXXZZZ
1970	XXXZZZ
1971	ZZZZZZZZZ
1972	ZZZZZZZZZZZZZZZ
1973	ZZZ

Sources: *X*: Minnegasco data; *Z*: MIDC data.
Note: One letter = one company.

Minneapolis Industrial Development Commission (MIDC) conducted a survey of firms that had left Minneapolis between 1962 and 1970. The study indicated that the most common complaint of migrating companies was lack of space for expansion. Such widely speculated reasons for leaving as lower taxes in the suburbs and industrial transportation problems in the city generally were not named. In fact, some respondents complained about high suburban taxes, which in some instances were twice as high as in the city (Minneapolis Industrial Development Commission, 1973*b*).

Lack of local government cooperation was a second major problem encountered by migrating firms. One executive cited the "apparent disinterest of city government in its small industries, with the exception of the City Assessor, Fire Inspector, and ticket sellers for the Police/Firemen's Ball" (Minneapolis Industrial Development Commission, 1973*b*). Several industrialists maintained that they had contacted city officials repeatedly in efforts to locate new sites or secure variances on adjacent property to allow for expansion. Except for an occasional alderman, the response at City Hall was limited and generally ineffective. It was, ironically, the removal of a major industry to the suburbs in 1955 that resulted in the formation of the Downtown Council. Evidently, the immediacy of the loss of the city's industrial base did not impress city leaders until more than a decade later, when they formed the Minneapolis Industrial Development Commission.

When the MIDC began operations in 1968 it announced two interrelated goals: to halt the migration of industry from the city and to develop long-range policies to maintain and to increase the city's industrial base. To date, the commission's objectives have been only partially fulfilled. Between 1967 and 1970 the number of firms leaving Minneapolis declined, with only six migrating to the suburbs in the latter year. Between 1971 and 1973 the number increased, and preliminary reports indicate that 1974 will show a decrease. Apparently, MIDC policies have had only a slight impact, the key variable being the condition of the economy. In short, it is not that fewer expanding firms have decided to leave the city during the past year, but rather that fewer companies have desired to expand. Surveys conducted after 1970 demonstrate that the availability and low cost of land are still the major factors influencing migration. A positive note was that complaints about uncooperative city officials dropped significantly as a reason for moving. On the other hand, the congested traffic situation around industrial sites evidently caused increasing disruptions to deliveries and to employee access (Minneapolis Industrial Development Commission, 1973*b*).

Since 1972 the MIDC has concentrated on securing land for expansion and subsidizing industrialization through various financial devices. The commission received a local appropriation of $400,000 to institute a land reserve program. Land purchased from this fund generated immediate dividends to the city. The Gresen Manufacturing Company constructed a research facility on one of the MIDC sites. Over the next few years, the company expects to hire more than 1,200 employees. The second major development in 1972 was the issuance, for the first time, of Municipal Industrial Development Bonds to finance the expansion of a plastics firm specializing in erosion control. (Minneapolis Industrial Development Commission, 1972*a*).

Industrial Development Bonds are probably the most effective financial weapon cities can marshal in defending their industrial base. Mississippi was the first political entity to issue these bonds in 1936 under its Balance Agriculture with Industry program. It was twelve years before another state (Kentucky) followed Mississippi's lead, but 42 other states were utilizing the device by the end of the 1960s. The original concept of the bonds was to develop industries in struggling rural areas. In the mid-1960s, however, cities became alarmed by declining industry. Local administrators, seeking to restore the traditional relationship between manufacturing and the city, began to issue the bonds. A city may issue up to $5 million worth of the tax-free bonds–more if the particular industry is engaged in air and/or water pollution control. The development bonds differ from general obligation bonds in that the city does not pledge its credit. This means that local government cannot assess property owners to pay the principal and interest.

The bonds finance land acquisitions, an expanding or relocating company's construction and equipment costs, and legal and consulting fees. The company occupies its new location as a lessee. The rents received by the city amortize the bond issue. After a stated period of time (usually 20 years), the firm may exercise its option to buy the property from the city at a nominal fee. Since the interest on the bonds is anywhere from 0.5% to 2% lower than the commercial lending rate, the financial advantages to the company are obvious. The business may also depreciate the building as if they owned it. In event of default, which has never occurred, the city may take possession of the plant or re-lease it. The city, through the development bonds, is able to compete on a more equal footing with suburban communities that might be able to offer cheaper land (Minneapolis Department of Economic Development, 1974; Minneapolis Industrial Development Commission, 1972*b*).

The major problem for the MIDC remains the collection of suitable industrial land. The city's first industrial park, the 39-acre Seward South, assembled by MIDC in conjunction with the Minneapolis Housing and Redevelopment Authority in 1968, was virtually full four years later. Since then, the MHRA has added Industry Square, North Washington, North Loop, and West Broadway industrial parks. The MIDC has a staff-consulting contract with the MHRA to develop these areas. The city is acquiring land in these areas through the tax increment method of financing, which is more effective for assembling large parcels of land than the bond issue, which is limited to specific relocation or expansion projects. The city issues general obligation bonds to purchase the land and sells it to a private developer at a "written down" price. In essence, the city is subsidizing industrial development by providing land to firms on a competitive basis with suburbs. The differences between the city's purchase and selling prices, plus public improvements, i.e., roads and utilities, are paid off through the increased taxes (tax increment) on the redeveloped property. (Minneapolis Industrial Development Commission, 1973*a*).

The city expects to utilize these financial mechanisms extensively during the next five years. In late 1974 Mayor Albert J. Hofstede announced the results of a survey of 1,800 Minneapolis-based businesses conducted by the MIDC. The commission discovered that nearly one-sixth of these firms plan either to relocate or to expand during the next half decade. Since the federal government provides no funding for industrial renewal, the city must bear the temporary financial burden of what must become a concerted effort to expand the current 61-acre land reserve. The mayor indicated that he hoped to triple that figure by 1976. The ultimate benefit to the city would be considerable. The land should generate $4.9 million in tax revenues and 8,200 jobs, not including the 5,300 jobs resulting from the development of the properties. (Hofstede, 1974*b*).

The MIDC has recently expanded its activities to make Minneapolis more alluring to resident industries. One of the major attractions of the nineteenth-century metropolis was the existence of a large and diverse labor pool. Transportation and industrial technology increased the mobility of workers significantly and released factories from dependence upon urban power sources. The MIDC has embarked on a program of vocational training to meet demands created by industries suddenly requiring inputs of skilled labor. For example, the metal fabrication industry has been expanding rapidly despite economic recession. A survey of metal fabrication firms in Minneapolis indicated that the industry could employ 150% of its present labor force.

Reacting to the energy shortage, or at least to increased fuel costs, the

MIDC is hoping for rebirth of the Mississippi as a major commercial port. Since barge transportation uses much less energy than any other transportation means, the commission is directing the development of the Municipal Harbor Terminal. It has already installed a granular handling device to facilitate the efficient transfer of bulky commodities from rail to barge. As truck and auto transportation costs increase, companies located in low-density suburbs might have greater difficulty attracting labor and in keeping up with the increased costs of moving supplies and manufactured goods. The city, meanwhile, hopes to turn the energy problem into an energy advantage.

The resurrection of the Mississippi is only one example of the attempt to broaden the city's economic base. Whether the MIDC has been stage-struck, or there in fact exists a genuine potential, there is a concerted effort to encourage the growth of the filmmaking industry in Minneapolis. Portions of *The Heartbreak Kid, Airport,* and *Slaughterhouse Five*, were filmed in Minneapolis. Although it is doubtful that the city will become the Hollywood of the Upper Midwest, a pool of expertise exists to generate a sizable industrial and documentary filmmaking capability. In any case, it is a worthwhile attempt to diversify the city's industrial pursuits as a hedge against recession and migration (Minneapolis Industrial Development Commission, 1973*a*).

In the attempt to restore the attributes of the historic city, there is no effort fraught with more complexities than the task of creating a neighborhood. The Loring Park project was not so much one of neighborhood building as of interfacing a residential community with a commercial district. The major difficulty with planning neighborhoods is that historic neighborhoods were not planned, they just happened. The creation of neighborhoods in new towns is an artificial process that results in an area little resembling the socially interactive urban neighborhood. This is the essential planning dilemma confronting cities that hope to recover the vitality of neighborhood life in an urban setting: to develop a plan that will incorporate the components necessary for neighborhood vibrancy without becoming another static overplanned community. If the neighborhood is, as Robert Park and others believed, the nucleus of urban living, its recovery must become a high priority for urban planning objectives.

Minneapolis's major attempt at neighborhood reconstruction—Cedar-Riverside—demonstrates both the possibilities and the difficulties involved in the re-creation process. A century ago, Cedar-Riverside was a growing Scandinavian and Eastern European neighborhood. Its community institutions bore the stamp of its immigrant inhabitants from the vast

Dania Hall to the small storefront coffee shops. It was a strong, self-contained community much like urban neighborhoods across a thriving urban America. By 1910 Cedar-Riverside, bordering on the river and a half mile from downtown, included 20,000 inhabitants. The history of the community thereafter was a recapitulation of the decay of similar neighborhoods elsewhere. The area's economic base—the lumber industry—declined, sons and daughters of immigrants left, and the elderly remained interspersed with several broken lumberjacks, prostitutes, and winos. Cedar Avenue, once the heart of the ethnic community, became derisively known as Snoose Boulevard after the Swedish word for the tobacco juice that splattered its sidewalks (Fischer, 1973).

In what has become a characteristic theme in the recovery of Minneapolis, private initiative began to transform a deteriorating neighborhood into what it was hoped would become a model community. Gloria Segal sought a small real estate investment in the city in 1962. Land in Cedar-Riverside was relatively inexpensive, and her immigrant uncle once operated a store there. Sentiment and financial advice from a tax expert and professor at the University of Minnesota, Keith Heller, convinced Segal that here was a sound investment. She held some hope for at least the partial redemption of the neighborhood, especially since the University of Minnesota located its West Bank campus in the community in 1959. The Segal-Heller investment team looked forward to erecting a small apartment building for students, university employees, and staff members of nearby St. Mary's Hospital.

Ralph Rapson, a friend of both Segal and Heller, and the head of the university's School of Architecture, was the catalyst who transformed a routine real estate operation into a project rivaling the Seven Cities of Cibola. In 1963 he suggested that since land was cheap and available, why not plan a larger development. Accordingly, Segal and Heller continued to purchase property in the southwest sector of the neighborhood. In 1965 they formed the Cedar Village Associates and decided to begin planning operations. Imitating the success of the Downtown Council in forming a unique multidisciplinary planning team, the Associates hired Barton-Aschman and Lawrence Halprin, the two major figures in the Nicollet Mall project. In 1968 the city designated the entire area for urban renewal and developed a renewal plan in conjunction with the private consortium. From a small apartment building, the project had grown to include the entire community (Cedar-Riverside Associates, 1973).

Cedar-Riverside is a significant departure from traditional urban renewal projects in two ways. First, the private input is overwhelming.

Second, the planning team has proposed a unique plan that incorporates many aspects of the historic neighborhood. The plan called for high density—a major characteristic of the discrete turn-of-the-century neighborhoods. High-density residential development seemed the optimal land use plan for a number of reasons. The institutions in the neighborhood, including the university, Augsburg College, and St. Mary's Hospital, required numerous housing units. A major objective of the planning team was to combine the working and living experience within the neighborhood as historic communities had functioned. High density was necessary also from a purely logistical standpoint. Two freeways had circumscribed the original neighborhood—a typical, though unfortunate, occurrence in the modern city— and nearly 250 acres were given over to institutions, commercial sites, and Mississippi riverfront park land. This left barely 100 acres for residential development. Since the Associates projected a population of 30,000 by 1990, the density per acre surpasses some of the more congested sections of Manhattan. Finally, the developers believed that high density would facilitate social interaction and conserve land for recreational and cultural activities (Fig. 3; Cedar-Riverside Associates, 1972, 1973).

The Associates are committed to re-creating the diversity of the historic neighborhood. In another unique maneuver, they proposed to mix residents of widely varying incomes in their apartment dwellings. There would be nothing to distinguish which unit is low income, federally subsidized, and which unit is renting at the private market value. The Minneapolis Housing and Redevelopment Authority would choose which units to subsidize. Thus, there would be no segregation as to building or floor. In another attempt to encourage diversity as well as to take advantage of the proximity of the university, there is a strong emphasis on cultural activities. Leaving the formal entertainment services to the university, the Associates have built sculpture blocks for local talent to display their wares; the plazas connecting dwelling units will provide space and cover for shows and displays; and a portion of the residential design plan calls for the erection of artists' lofts and studios to facilitate the creative process (Fischer, 1973).

Unlike traditional renewal projects, Cedar-Riverside will emphasize structural conservation and is making the old community a functional part of the new. The focus of the neighborhood's social activity will be the Centrum and, more specifically, Dania Hall, just as it was the focal point of community life in the 1890s. The hall includes a theater and a ballroom. Nearby, a nineteenth-century firehouse is serving as the neighborhood social center. Since its rehabilitation in 1972, four

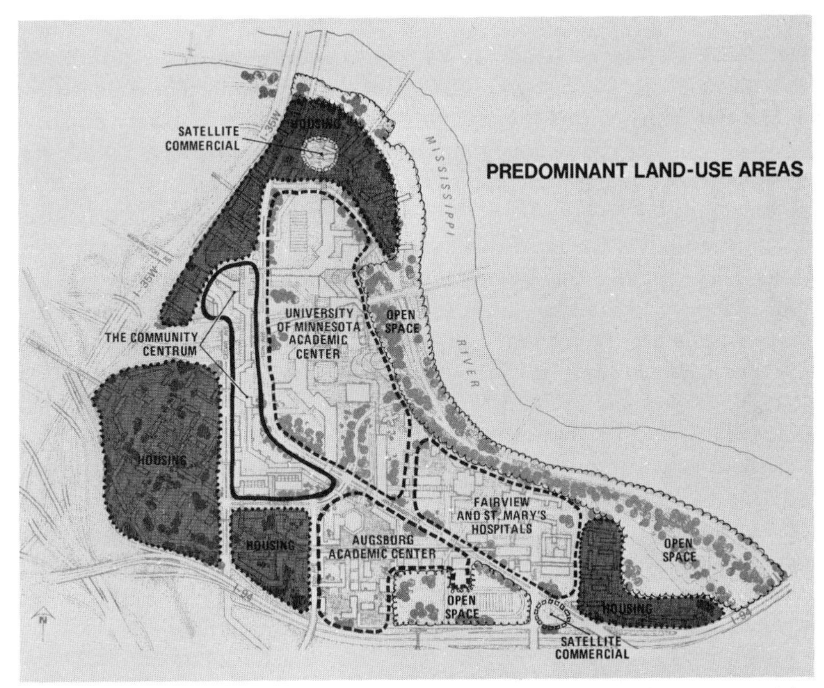

Fig. 3. Predominant land-use areas: Cedar-Riverside. (Cedar-Riverside Associates, 1973)

weddings have taken place there. The Centrum incorporates the major shopping facilities of the neighborhood, though there will be convenience stores throughout the community to avoid the segregated functions that destroyed the continuity of historic neighborhoods (Cedar-Riverside Associates, 1974; Minneapolis Planning & Development Department, 1973*a*).

Perhaps the most unusual planning feature of the neighborhood is the communications system. Technology, of course, has played a leading role in stripping cities of their unique identity. Perhaps it is fairer to say, the use to which people have put technology has been a destroyer of urban life. Cedar-Riverside's communications system employs technology to personalize relationships, facilitate health delivery, and liberate women from the tediums associated with house and child care responsibilities. The medium is cable television, which is not, of course, new, but it will carry a two-way network. It will allow a doctor to make "house call," a parent on the thirtieth floor of an apartment dwelling to supervise a child in a community tot-lot a few blocks away, a student to ask a question of a professor, and police or fire emergency units to respond speedily to a crisis.

The system has raised the specter of surveillance in a society already sensitive to electronic snooping. There are also skeptics who see transmitted shopping lists for eggs and milk coming out pig's feet and champagne. There is no doubt about the potential for abuse and the possibilities for breakdowns more serious than the garbled grocery order. Nevertheless, the system represents an attempt to break down bureaucratic barriers that large institutions like governments, hospitals, and universities have placed in front of potential users. It will, it is hoped, make urban living less frustrating (Cedar-Riverside Associates, 1974).

In addition to the trend-setting plans to recover this neighborhood, the Associates are implementing principles that have worked in other sections of the city. The automobile, whose accoutrements helped to emasculate the neighborhood, has been relegated to subterranean parking areas in what it is hoped will be a permanent exile. Grade-separated walkways over existing roadways, the closing off of numerous streets (courtesy of the city), and the partial or complete covering of arterial walkways signal the rebirth of pedestrian travel and the walking neighborhood.

There has also been an attempt to vary residential design within the limitations imposed by high density. Diversity, which has become the keynote of recovery projects throughout the city, has taken root at Cedar-Riverside. Five different neighborhoods (actually sub-neighborhhoods) have been planned, with each having a distinctive design character. It probably would have been better in terms of recreating the historic neighborhood to integrate these design distinctions within each section, rather than segregating them; but perhaps the mixture of function will mitigate this uncharacteristic desire for order (Cedar-Riverside Associates, 1974).

Cedar-Riverside has benefited from several funding sources in the process of implementing the grandiose schemes. The Associates have sunk $25 million of private capital into the project. The scale which the project assumed after 1968, though, obviously outran the financial capabilities of Segal and her team. Since 1968 the city has worked closely with the Associates in funding the project. The city is upgrading Cedar Avenue by adding plantings, furnishings, and ornamental sidewalks. Owners of business property along the avenue, encouraged by city activity and the future of the community, are renovating their own structures. The city also has plans to perform cosmetic rehabilitation on Riverside Avenue so that the community's two major thoroughfares will reflect the upgrading of the neighborhood as a whole. In addition to facelifting operations, the city Housing and Redevelopment Authority, through its urban renewal powers, has purchased property that the Associates would have probably been unable to obtain. The Authority has overseen the necessary public utilities and street improvement work, as well as landscaping plazas and pedestrian bridges that traverse major arteries. Finally, the Authority has directed private rehabilitation efforts along Cedar Avenue (Cedar-Riverside Associates, 1972).

The role of the federal government as a financial partner has increased dramatically since 1968, though there is some uncertainty at present as to the future of federal funding on the project. Under the various sections of urban renewal there are 117 units of low-income housing, 552 federally subsidized moderate-income housing (FHA 236 program), 408 market rent units (FHA 236b program), and 222 semiluxury units (FHA 220 deluxe program). The funds are channeled through the city's Housing Authority. The project received a significant financial boost in late 1971 when the federal government designated Cedar-Riverside as its first new-town-in-town project under the New Communities Program. This allowed the Associates to issue $24 million in federally guaranteed bonds to implement their plans. In addition, the city became eligible for supplementary loans and grants up to 20% of facility costs. In all, about one half of the housing units for Cedar-Riverside will receive some form of federal financial support (Cedar-Riverside Associates, 1972, 1973).

The Associates have sought other forms of funding. With a strong emphasis on the arts, the community's search led naturally to the National Endowment for the Arts. The Associates received a grant of $20,000 from the Endowment through the cooperation of the Minnesota State Arts Council. The University of Minnesota has provided some aid-in-kind by coordinating its parking lot and street system with the plans of the Associates. Privately, project members complain that university cooperation, especially in terms of its available entertainment facilities, might be more widespread. Finally, the Hennepin County Public Works Department is examining the feasibility of using solid waste as fuel to cool and heat the community (Cedar-Riverside Associates, 1972).

The funding and the intergovernmental cooperation have been instrumental in allowing the Associates to transform their plans from the drawing board to reality. The first stage of development (there are five altogether), Cedar Square West, is partially finished. The "neighborhood" includes more than twenty styles and sizes of apartment dwellings in keeping with the planners' emphasis on design diversity. The rents vary from $41 to $550 a month in the same building. The apartments are generally smaller than average, and the balconies are too narrow for even a kitchen chair to fit. For young childless couples and singles (including students and the elderly)–a rapidly increasing segment of the urban population–the apartments seem functional enough. The rents for the nonsubsidized units are slightly higher than the metropolitan average–$240 for a two-bedroom compared with $210 for the rest of the area. The buildings will range, when completed, from 4 to 40 stories. The plaza for Cedar Square West is in place. There is an 850-car garage underneath the plaza, and federal funds are supplying trees, grass, furniture, and banners. The Associates financed the plaza at $135,000 (Minneapolis Planning & Development Department, 1973*a*; *Minneapolis Tribune*, 1973).

Cedar-Riverside Associates have been handling the delicate task of relocation during Stage 1. Early urban renewal projects generally minimized the social costs of relocation. The Federal Uniform Relocation and Real Property Acquisition Act of 1970, however, required the funded agency to offer financial aid and standard housing facilities to relocatees. Relocation in Cedar-Riverside has been, in contrast to earlier renewal efforts, a private affair. In many ways, the Associates have been more sensitive to human needs than most city agencies have been. Demolition has been kept at a minimum, in keeping with the project's policies of preserving old structures and maintaining the historic character of the neighborhood. Further, although the Associates acquired the right to condemn property in 1968, they have preferred to work according to the willing-buyer/willing-seller principle. When difficulty arises, the Associates attempt a compromise. The relatively slow pace of development (20 years) insures that acceptable dwellings will be available to those who must relocate. An important aspect of the relocation process has been that almost all affected residents have been relocated within the neighborhood at no increase in rent. The creation of a new community no longer signifies the destruction of the old. There was a precious social fabric in the historic neighborhood, and although now frayed, it should be allowed to enrich the recovery process.

The case of the Sixteen Old Men demonstrates the tact of the Associates. The group had lived together in an old frame house slated for clearance. The Associates offered the men apartments in a nearby project for the elderly. They refused, disdaining to be housed with a bunch of "chattering old women," as one put it. The Associates hammered out a compromise and invited the men to a dinner party at a nearby hotel, including an overnight stay, all at Associates' expense. When the men returned to their house, everything was in place except it had been moved to new foundations several hundred feet down the block (Fischer, 1973).

The diplomatic handling of relocation problems has carried over to the new tenants. Observers attribute the relatively good tenant-landlord relationship to Segal, whose modest office is in a converted ice-cream factory in the center of the community. She has been accessible to everyone. As one visitor noted: "If a student commune wants to repaint its house with purple flowers and green peace symbols—as several of them have—she comes up with paint and brushes the next morning" (Fischer, 1973).

Segal's tolerance, even encouragement of variance, is emphasized by the number of new contact points she has helped into existence and the number of such locations she has saved from extinction. The Triangle Bar is a popular student rock discotheque near Cedar Square West; the refurbished firehouse holds dance workshops, jazz concerts, and weddings; the New Riverside Cafe is a collective with emphasis on organic food and bluegrass; Coffehouse Extempore, a church-related poetry and nonalcoholic meeting place, resides virtually rent free in a location owned by the Associates; and the West Bank School of Music offers a wide variety of instruction from jazz clarinet to country fiddle (*Minneapolis Tribune*, 1973).

The initial successes of Stage 1 and the supporting contact points have increased optimism about the developments in the immediate future. The city, as part of its continuing efforts to improve the river and riverfront environments throughout the metropolis, is planning a major recreational facility on the portion of Cedar-Riverside fronting the river. The facility will include bicycle paths and walkways tied to similar systems in the community. Minneapolis has a Legacy of Parks and Neighborhood Facilities fund that will finance at least part of this project. Residents may begin to enjoy the scenic serenity of the Mississippi as their ancestors did (Cedar-Riverside Associates, 1972).

If the story of the recovery of Cedar-Riverside were to end here, we could fade out on a cherubic Gloria Segal dispensing paint-brushes and plans for a better tomorrow. For the past year, however, the hammers and bulldozers have lain silent while legal wrangles threaten to alter the comprehensive framework of the plan. In a sense, the present precarious condition of the entire project seems inevitable, at least from hindsight. Established residents and institutions in the community have had mixed emotions about the plans from the inception, and some have voiced strong opposition to what they perceive to be an invasion rather than a resurrection.

An ineluctable quality of middle-class morality pervades the Associates' leadership that is incongruous to some of the features of the old community. Spokesmen proudly point to the number of bars that have closed since the Associates moved into the area. Older residents reminisce about the good times and camaraderie that permeated those establishments. While some of this is undoubtedly a nostalgia trip, the loss of community contact points should not be an occasion of rejoicing. The bars added a distinction to the area, and while there may have been too many of them, a glass of beer and good conversation may be just as cultural to some individuals as a pirouette is to others.

While the Associates have dealt with the relocation problem with sensitivity for the most part, they have moved swiftly and stongly against those they consider undesirable tenants. The Electric Fetus was a record shop that also dabbled in posters and head supplies. In 1969 the shop displayed nude posters (simulated) of then-President and Mrs. Nixon. The final outrage was a record promotion campaign that promised a free record to anyone who disrobed in the store. When the Associates canceled their lease, the Electric Fetus reported that forty individuals had accepted the shop's offer. The Associates paid for the store's moving expenses out of the neighborhood (*Minneapolis Tribune*, 1973).

If a community is to be truly diverse and provide residents with choice–the major attraction of urban living–such activities should be allowed to continue. If community residents decide, however, that such activities are, in fact, inimical to community life as they perceive it, then sheer economics will force these enterprises from the neighborhood. In short, such decisions should be reserved for the residents, not for the planners. Planning represents an imposition of sorts in any case. That imposition or control should be maintained in a low profile. As English geographer Peter Hall asserted: "The planner cannot and should not prescribe any ideal pattern of social life; he should merely see that the opportunity for maximum choice is given" (Hall, 1975). Otherwise, planners court certain opposition.

The certain opposition to Cedar-Riverside originated among older residents and young counterculturists headquartered in the New Riverside Café. The café and the Associates have clashed repeatedly during the past few years over community objectives. Early last year, the opposition coalesced into the Environmental Defense Fund, which filed a suit against the Associates, the city Housing and Redevelopment Authority, and HUD. The brief contended that the environmental impact statement filed with HUD was incomplete and inaccurate. Specifically, the plaintiffs were concerned about the impact of high-density development on their property and life-style (Donald A. Jacobson, pers. comm.).

Whatever the outcome of the suit—it is still pending—there will be several changes in the project. First, and most immediate, the financial viability of the project has been threatened. There has been no work on the plans for more than a year. In the meantime the Associates have been encumbered by holding costs of several million dollars which are unrecoverable. Accordingly, the Associates have asked the city to assume a greater role in development. The city is still considering this option, but observers in both the public and private sectors agree that the city will exercise a greater input in future planning decisions. These sources also indicate that lower densities and even greater emphasis on rehabilitation rather than renewal will result from the city's participation (Larry Anderson, pers. comm.; Hofstede, 1974*a*).

The loss of the high-density component will be a severe setback to a major objective of neighborhood restoration. There could be a compromise, though, that would actually improve upon the original plan. High-rise structures were the major environmental threats cited in the legal action. High-rise and high density, however, are not necessarily synonymous. High density in the historic neighborhood community was not achieved through vertical imperialism. Structures rarely went beyond five or six stories. Further, there have been indications that crime and social isolation have strong correlations with high-rise buildings.

Lower structures, of course, mean less open space, but this is not so unfavorable a trade-off as it might seem. Historic neighborhoods created their own open spaces, and this in fact might be preferable to preplanned open spaces, which sometimes do not become places for social interaction anyway. In short, while Cedar-Riverside must change, it could seize the opportunity to turn what must be a disappointment to the Associates into a positive contribution to neighborhood recovery in particular, and urban life in general. In any case, it is important that neither the

Associates, the city, nor HUD abandon the project. It has been a unique endeavor in public-private cooperation that could, in future years, serve as an important inspiration to recovery projects (Whyte, 1968).

A Model for Recovering the Historic City

Minneapolis is a case study, but by no means the only example of the recovery of the American city. Programs to increase the tax base, restore diversity, and attract people exist in most moderate- and large-sized cities today. Urban river projects in Milwaukee, San Antonio, and Washington, D.C., are similar in scope and objective to the Mississippi River projects planned for Minneapolis. Downtown pedestrian malls in Philadelphia and Washington emphasize the same preference for the walking city and a similar disdain for the auto. As a consequence, our urban environments are becoming better places in which to live from both an environmental and a social aspect.

Although the Minneapolis projects—the recovery of downtown, the restoration of an industrial base, and the regeneration of neighborhood life—are not unique, the success of most of these efforts in recovering the city result from a set of factors that might prove beneficial if incorporated in similar efforts elsewhere. The initiative of the private sector has been the most obvious asset of the Minneapolis recovery. Nicollet Mall, the project which began the renaissance, was a conception of downtown businessmen. Local entrepreneurs contributed not only ideas and organization to the Mall effort but money as well in the form of benefit assessments and the skyway system. Although the industrial development program is primarily a city operation, it was local business leaders who sounded the alarm over the city's declining industrial base. For six years, as Cedar-Riverside evolved, it was primarily a private enterprise. Perhaps in retrospect it was a larger task than the private sector could hope to accomplish—land acquisition, comprehensive planning, and relocation. Nevertheless, whatever Cedar-Riverside becomes, its debt to private enterprise is enormous.

The ability of farsighted business leadership to work with local government has been another crucial element in the regeneration of Minneapolis. Local officials have been responsive to city recovery and have been innovative in their own right. Financial mechanisms such as the benefit assessment and the tax increment have increased capital flow without placing greater burdens on taxpayers. City government itself has been streamlined in recent years to handle the recovery program with greater efficiency. The expansion of the city coordinator's office in 1970 to include the planning division has resulted in closer cooperation between the planning commission and the city council—the city's governing body. The city coordinator is now the council's administrative and planning arm, indicating the importance of the planning function as a city priority. Finally, the personnel on the council as well as the vigorous young mayor, Albert J. Hofstede, have uniformly sought to project the city's role in its own recovery. Mayor Hofstede especially desires the city to assert itself in the planning process to a greater extent, especially in the area of social planning. As the Cedar-Riverside development extricates itself from its legal entanglements, it will be interesting to watch the city's progress in its more active planning capacity (Hofstede, 1974*a*).

The state has proved a cooperative partner in the endeavor, if only because it can see tangible results from its wise policies just by looking across the river. From financing industrial redevelopment to landmark preservation, state enabling legislation has benefited city recovery. The rural-urban split which has characterized legislatures in some states is relatively muted in Minnesota. Perhaps the hard climate, the common immigrant experience, and the tradition of supporting social programs has made Minnesotans indifferent to residential distinctions.

The state legislature played the dominant role in creating the Metropolitan Council. The council is probably the most effective regional organization in the country. Its role in the urban recovery process has been minimal. Since the council is the conduit for federal funds, all projects involving federal monies must receive its approval. Aside from that, which in the case of Minneapolis's major projects has been a formality, there is little Metropolitan Council input into recovery projects. In fact, the recovery programs generally run counter to the concept of regional cooperation, and understandably so.

The suburbs ringing the city are the enemy. The resources of Minneapolis—its people and tax base—are there. Even Cedar-Riverside, which is not as explicit in its competition with the suburbs as the Nicollet Mall or the industrial development program, is hoping to lure suburbanites to its more diverse and exciting confines. Some refugees from suburbia have, in fact, settled in the redeveloping community. The energy crisis has been another boon to urban recovery. The MIDC reports scarcely conceal their glee concerning the city's energy advantages over the suburbs. Suburban cooperation with the city is limited now to nonpolitical projects like airports and mosquito control. Years hence, the suburbs may be seeking the same close relationship that cities throughout the country are trying in vain to establish with suburbs. The cities may reassert their regional preeminence in all areas (Metropolitan Council, 1973).

Part of the reason for the successful public-private relationship to date has been the proclivity of participants to focus on relatively small projects. True, there is a comprehensive Mississippi River plan and a comprehensive downtown plan, but these have come after several components have been completed. The federal government recognized the efficacy of piecemeal planning through its block grant approach to urban recovery projects begun in 1974. Now funds may be applied to specific areas in the city with only the vaguest connection to a comprehensive plan (U.S. Dep. of Housing and Urban Development, 1974). Nicollet Mall, Loring Greenway, the creation of industrial parks in various sections of the city, and the Mississippi River project in downtown all indicate the benefits of spot developments that focus all the intensity and expertise of comprehensive planning to a small area with only one or two objectives in mind. (In this context, *comprehensive planning* refers to planning a large area with multiple design and planning components and objectives over a period of more than five years.)

The troubles of Cedar-Riverside stemmed in part from its comprehensiveness. Unless all parts of a comprehensive plan are developed simultaneously, which is clearly impossible, the development of one part of the plan often changes the nature of other portions to a degree different from the original plan. Thus, Stage 1 of Cedar-Riverside with its high-density megastructures will produce significant changes in the rest of the components not foreseen by the comprehensive plan. After the development of several piecemeal projects it is possible, as with Metro Center '85, to formulate a comprehensive plan from the data base generated by the structures and activities in place.

The scaling-down of plans to discrete levels may coincide with future energy needs. As fossil fuels become scarcer, there could be a reversion to settlement units that are smaller in scale, designed to make use of small-scale wind, solar, and other "natural" energy sources. With increased energy costs, there will be a greater substitution of electronic communication for energy-intensive transportation. Production units of all sorts are likely to become smaller in scale as well in order to reduce transportation costs to market. In other words, the future Minneapolis may be a series of mile-and-a-half-square walking cities reminiscent of the early days of American urbanization (this is based on initial formulations by Robert F. Wells of the U.S. Air Force, Albuquerque, and Dr. Robert G. Dyck, chairman of the Urban and Regional Planning Program, Virginia Polytechnic Institute and State University).

Though the discrete nature of Minneapolis recovery projects corresponds to sound planning principles and even perhaps future energy constraints, it is arguable that the success of these smaller programs is due, simply, to the fact that urban problems are smaller in Minneapolis than elsewhere. The Minneapolis region has the third highest family income of the 17 largest metropolitan areas. In terms of air pollution, it is the cleanest city in the nation. Finally, poverty is not a major problem in the Twin Cities area, nor is there severe racial trauma—1.4% of the area's population is black. The American Indian is probably lowest on the region's socioeconomic scale, but the magnitude of their difficulty is minute compared with black Detroit or Spanish Harlem, for example. The complex problems related to minority citizen participation are not evident in Minneapolis, since none of the recovery projects impact a large ethnic or racial community (Downtown Council, 1974).

The scope of a city's problems, however, should not impede the recovery process. In fact, the broader the low-income base, the more recovery projects will aid in removing inequities. Recovery generates jobs and increases the city treasury, which, in turn, will free funds for human recovery programs. Since numerous recovery programs aim to regenerate deteriorating neighborhoods, the physical environment improves. With sensitive relocation procedures, with a minimum of demolition and a maximum of rehabilitation, and with a program for active citizen participation, renewal is no longer anathema to low-income residents. In Washington, D.C., for example, residents of blighted areas clamor for recovery programs where a decade ago there was widespread animosity against the renewal process. Finally, the commitment to recovering the city is a statement that somebody cares. Resentment in the lower-income neighborhoods against white-collar suburbanites who speed in and out of town over freeways, unconcerned about city people or problems, taking the city's tax base and vitality with them, is understandable. Now, movement back to the city and city living for younger couples and singles has become more fashionable, especially since the energy crisis. The recovery programs with their emphasis on diversity rather than homogeneity are bound to improve understanding. The city is no longer set adrift, abandoned by region and state to shrivel and die in a welter of manicured lawns and shopping centers. The city is regenerating, and as in urban centers of the past, its citizens will take pride in knowing that they are part of a unique civilization.

The quality of planning is a considerably more significant reason for the success of Minneapolis recovery projects than are the relatively small size and scope of the city's problems. In the early stages of Nicollet Mall planning, the Downtown Council expressed its desire that the project would be a "first-class" operation or not happen at all. The Mall and Transitway Plan presented by Barton-Aschman was by no means the least expensive or the easiest plan to implement. It would have been cheaper in terms of money and time to apply some cosmetic changes to the area and leave it at that. Fortunately, the council chose a broader approach, and the Mall project has served as an admirable precedent for future recovery projects. The Downtown Council also established the efficacy of a multidisciplinary approach to planning. The planning team has been effective in getting input from different perspectives. The era when the planner merely concerned himself or herself with the physical dimensions of the project is gone. Minneapolis has been fortunate in the selection of its planning teams as well. Segal even borrowed Heikki von Hertzen from his famous new town of Tapiola in Finland. Quality, in short, has traditionally outweighed penny-pinching, and the results have proved the worth of this philosophy.

The historical perspective is perhaps one of the most important approaches Minneapolitans have brought to the planning process. The recovery of the city is the restoration of the attributes of the historic metropolis. The oft-repeated dictum, "You can't go home again" is essentially accurate. No one is suggesting the exact replication of the late nineteenth-century city. In terms of air and water pollution, cities are striving to make their natural environment considerably cleaner than the be-sooted cities of yesteryear. Poverty was more abject, disease was a constant companion, industry crowded open spaces, and plumbing was invariably outdoors in the city of a century ago.

The historic metropolis, however, possessed certain positive aspects that are worthy of re-creation. The planning process should proceed with these factors uppermost in mind because they helped to transform the city into an exciting and prosperous environment. Mass transit, pedestrian travel, industry, neighborhood cohesion, high density, an enterprising commercial elite, culture, education, security, a sense of place, and, above all, choice flourished in the historic city. The Minneapolis recovery projects aim at the restoration of these historic attributes.

Historical methodology should enter the planning process in a more specific manner, as the Minneapolis neighborhood recovery projects demonstrate. Cedar-Riverside represented a fine example of the necessity of learning the historic roots of a community. The Associates have taken care to preserve the Scandinavian flavor of the neighborhood, restoring such old structures as Dania Hall and the firehouse and making them part of the new plan. Careful research into land use patterns, socioeconomic status, age composition, and social institutions of the historic immigrant neighborhood would have provided the Associates with an even better understanding of the mechanisms of neighborhood life. They might have discovered, for example, that taverns were not a community blight but important contact points for the neighborhood residents. They also might have found a high mixture of functions and the absence of physical barriers that would question the advisability of creating distinctive subneighborhoods. Finally, the Associates could have discerned a relatively large young, single, and transient population–characteristic of immigrant communities. This would have reinforced the Associates' emphasis on providing services for the same demographic base that the planners hope to attract to Cedar-Riverside today. While no historic analysis should bind present planning, a restructuring of some of the major patterns of historic neighborhoods will facilitate a re-creation of the social interaction and excitement of community life. The city should be an educational experience in the broadest sense. The neighborhood is the classroom and diversity its most effective teaching tool.

The recovery of Minneapolis has only begun. It is an encouraging beginning. The process will be expensive, lengthy, controversial, and, at times, heartbreaking, but ultimately rewarding. The measure of great civilizations is their cities. For this alone, the commitment is worthwhile.

Acknowledgments

I would like to thank the following individuals for the help they gave me in preparing and revising my manuscript: Dr. Robert G. Dyck, Chairman, Urban and Regional Planning, Virginia Polytechnic

Institute and State University; Bill Hamm, Director of Public Relations for the City of Minneapolis; and William Issel, Town Planner for the Town of Blacksburg, Virginia.

Literature Cited

Abbott, Carl. 1970. Civic pride in Chicago, 1844–1860. J. Illinois State Historical Soc. 63:399-421.

Aschman, Frederick T. 1971. Nicollet Mall: civic cooperation to preserve downtown's vitality. Planners' Notebook 1:1-8.

Belcher, Wyatt W. 1947. The Economic Rivalry between St. Louis and Chicago, 1850–1880. AMS Press, New York. 421 pp.

Brownell, Blaine, A., and Warren E. Stickle, eds. 1973. Bosses and Reformers. Houghton Mifflin, Boston. 252 pp.

Cedar-Riverside Associates. 1972. Proposed Development Programs, 1972–1975. 22 pp.

Cedar-Riverside Associates. 1973. Cedar-Riverside New Community: Narrative Description. 18 pp.

Cedar-Riverside Associates. 1974. Cedar-Riverside Planning Framework. 17 pp.

Condit, Carl, W. 1973. Chicago, 1910–1928: Building, Planning, and Urban Technology. Univ. of Chicago Press, Chicago. 238 pp.

Downtown Council of Minneapolis. 1974. Nicollet Mall. 78 pp.

English, Maurice, ed. The Testament of Stone: Themes of Idealism and Indignation. Northwestern Univ. Press, Evanston, Ill. Page 44 from Writings of Louis Sullivan.

Fischer, John. 1973. The possibly glorious dream of Mrs. Gloria M. Segal. Harper's Magazine 247:14-19.

Fitch, John Marsten. 1967. Environment aspects of the preservation of historic urban areas. Monumentum 13:35-56.

Glaab, Charles N., and A. Theodore Brown, eds. 1967. A History of Urban America. Oxford Univ. Press, New York. 267 pp.

Hall, Peter. 1975. The urban culture and the suburban culture. Page 171 *in* Michael E. Eliot Hurst, ed. I Came to the City. Houghton Mifflin, Boston.

Hays, Samuel P. 1974. The changing political structure of the city in industrial America. J. Urban History 1:6-38.

Hofstede, Albert J. 1974*a*. Statement on community development needs.

Hofstede, Albert J. 1974*b*. Statement on a program for coordinating industrial/commercial development in the city of Minneapolis.

Jacobs, Jane. 1961. The Death and Life of Great American Cities. Random House, New York. 316 pp.

Lu, Weiming. 1974. Minneapolis: Metro Center '85, a Design Solution. City of Minneapolis. 12 pp.

Metropolitan Council. 1973. Citizens Guide to the Metropolitan Council.
Minneapolis Industrial Development Commission. 1972*a*. 1972 Annual Report. City of Minneapolis. 23 pp.
Minneapolis Industrial Development Commission. 1972*b*. Guidelines for Rating Municipal Revenue Bond Issues. City of Minneapolis. 56 pp.
Minneapolis Industrial Development Commission. 1973*a*. 1973 Annual Report. City of Minneapolis. 26 pp.
Minneapolis Industrial Development Commission. 1973*b*. Industrial Migration Study, 1962–1973. City of Minneapolis. 37 pp.
Minneapolis Planning and Development Department. 1973*a*. Cedar-Riverside. Minneapolis Today, Spring, 1973.
Minneapolis Planning and Development Department. 1973*b*. Gateway Center. Minneapolis Today, Spring, 1973.
Minneapolis Planning and Development Department. 1973*c*. Nicollet Mall. Minneapolis Today, Spring, 1973.
Minneapolis Planning and Development Department. 1973*d*. Nicollet Mall extension–Loring Park development. Minneapolis Today, Spring, 1973.
Minneapolis Planning and Development Department. 1973*e*. Skyway system. Minneapolis Today, Spring, 1973.
Minneapolis Tribune. 1973. The old west bank–and the new. Pages 2-26 *in* Picture Magazine, Dec. 9.
Minnesota Department of Economic Development. 1974. Municipal Industrial Development Bonds. City of Minneapolis. 41 pp.
Office of the City Coordinator. 1974. Loring Park development. City of Minneapolis.
Park, Robert E., Ernest W. Burgess, and Roderick D. McKenzie. 1925. The City. Univ. of Chicago Press, Chicago. 239 pp.
Rae, John B. 1965. The American Automobile: A Brief History. Univ. of Chicago Press, Chicago. 216 pp.
Sennett, Richard. 1971. The Uses of Disorder: Personal Identity and City Life. Vantage Press, New York. 198 pp.
U.S. Department of Housing and Urban Development. 1974. Summary of the Housing and Community Development Act of 1974. Aug. 22.
Vance, James E., Jr. 1971. Focus on downtown. Pages 112-119 *in* Larry S. Bourne, ed. Internal Structure of the City. Oxford Univ. Press, New York.
Von Eckhardt, Wolf. 1974. Minneapolis renaissance. Washington Post, Nov. 22.
Wade, Richard C. 1959. The Urban Frontier: The Rise of Western Cities, 1790–1830. Harvard Univ. Press, Cambridge, Mass. 360 pp.
Whyte, William H. 1968. The Last Landscape. Doubleday, Garden City, New York. 323 pp.

Political Problems Inherent in Environmental Protection Legislation and Implementation

Martha Sager

Abstract

The slow development of the recognition of the need for environmental protection began, possibly, with the early English riparian water use regulations and developed gradually until 1970 when the passage of the National Environmental Policy Act of the Congress of the United States of America formalized recognition of environmental imbalance and the need for its correction and became one of the dominant forces in the political action taking place in the United States.

A multitude of federal and local ordinances and statutes tumbled into the political arena, each bearing its own particular idiosyncratic problems of the power of the economics versus reason in the environmental protection field. One example, water pollution control legislation, is discussed in this chapter and presented in chart form from the English Rule of Natural Flow of the early eighteenth century through the various statutes of the U.S. Congress, i.e., those of 1833, 1899, 1948, 1965, 1972, and the latest Safe Drinking Water Act of 1974.

Discussion

> In this squalid, dirty dooryard
> where the chickens scratch and run
> White, incredible, the pear tree
> stands apart and takes the sun.

Environmental legislation, like the pear tree in Edna St. Vincent Millay's poem, "stands apart" because of its newness, its comprehensiveness, its scientific foundations, and its emphasis on public participation (Millay, 1956).

The environment, as a generic term, has perhaps as many true meanings as it has persons who speak them. Over time, these definitions have changed. The Greek philosopher Hippocrates in an essay entitled "Air, Water and Places" included not only the physical-chemical world but also some societal aspects: "We must also consider the qualities of waters . . . whether they be marshy and soft, or hard, and running from elevated and rocky situations, and then if saltish and unfit for cooking; and the ground, whether it be naked and deficient in water, or wooded and well watered, and whether it lies in a hollow, confined situation, or is elevated and cold; and the mode in which the inhabitants live, and what are their pursuits, whether they are fond of drinking and eating to excess, and given to indolence, or are fond of exercise and labour, and not given to excess in eating and drinking" (Aller et al., 1949).

As precepts developed, attitudes about the environment were adopted; as concepts changed, so did the attitudes. "For example, wilderness in Biblical references was the antithesis of a garden, and gardens were created by men overcoming nature. To the Puritans, wilderness was a hostile environment, a last refuge of sinners. Wilderness in 18th and 19th century western literature had an evil connotation. Often the phrase 'howling wilderness' was used. Today, we have a Wilderness Act, Wilderness Areas and a Wilderness Society—all more or less in praise of wilderness. Other concepts relative to the environment have been subject to similar changes of attitude" (Reitze, 1972).

A basic scientific description of the environment of the planet was given by La Fleur in 1941:

1. An available set of chemicals that will allow variation and reproduction and will carry on the complex processes of metabolism.
2. A suitable temperature; the high temperatures on the average star exclude the possibility of the organization of molecules of sufficient complexity to serve as the basis of life. Cold slows down chemical processes. . . . Practically, life is limited to the temperatures at which water is a relatively warm solid or a cool to warm liquid. Molten lava aside, life in some form can exist at most earth temperatures.
3. The proper range of density and pressure . . . .
4. There must be a source or sources of energy and of new materials. There is also a need for controlled reaction rates. Limited but renewable amounts of all needed materials and energy must be locally available to permit living processes to continue.

The sun's radiation is a source of energy but reaches the earth in unlimited amounts.

Carbon dioxide, rainfall for water needs, oxygen, etc., are also distributed in available but discrete amounts.

5. The adsorption of lethal (deadly) ultra violet rays of the atmosphere is of great importance. Life, again as we know it, could not occur on the earth today if these shorter, a-biotic rays were not screened out.

In general terms, the earth is a dense crusted body of sufficient size to have strong enough gravitational attraction to hold an extensive gaseous atmosphere, but not strong enough to hold more than a trace of free hydrogen. The presence of water and carbon dioxide in the atmosphere seems to be a normal result of the physical and chemical properties of water and carbon dioxide that have much to do with regulating the general environment of living things on earth. There is good reason to believe that water is the substance whose movement in the organic and in the inorganic world constitutes the first, the most fundamentally important activity in the world that we live in.

Having clarified the concepts of the environment as used in this context, the next resolution lies in the definition of the word *political.* The word *political* is defined in Webster's *Third International Dictionary* (1966) as: "of, relating to, or concerned with the *making* as distinguished from the administration of government policy." Politics has also been defined as the highest art in a democracy. Perhaps it is this art which could be defined as that one activity which has separated the human from the animal societies which together cohabit planet earth.

Ecologists have long described the activities of higher, postreptilean animal societies with such terms as *speciation, competition, domination, territorialization, predation, adaptation,* etc., each of which could also be assigned to certain human societal operations. Man, with his ability for abstract thought, combined and welded these activities into a functioning unit called government, whose operations embrace the various tangents of each of these societal activities.

Laws formed the foundations of government and helped man control the drives for domination, competition, and predation, and those primitive drives were detrimental to his survival. Thus, law and government were developed by man for his protection from both the internal pressures of society and the external forces of the environment.

The development of ecology as an academic discipline paralleled the rapid advances in technology experienced by this nation in the early part of the twentieth century. Scientific and technological development provided the knowledge necessary for natural resource exploitation. The industrial revolution surged forward. Industry, automated and nourished by the planet's supplies of water and fossil fuel, exploited all the resources of planet Earth to provide artificial and unnatural living conditions for man. This exploitation produced an undreamed-of luxurious life-style and simultaneously dominated, abused, destroyed, or parasitized all biotic and abiotic systems of the planet.

When these abuses of the environment became obvious and threatened the very survival of human beings, members of the highly industrialized, powerful, and rich societies turned to their great talent, the art of politics, based on law and implemented through legislation, to help with solutions to the environmental dilemma.

Water pollution control, in terms of water quantity regulations, was first attempted in England with the English Rule of Natural Flow. This regulation permitted the rightful use of water "as it was wont to flow" past the property of every waterfront owner, or riparian.

This rule was detrimental to the landowners downstream because of the variety of abuses to which the water was subjected before it flowed past their properties. As a law, the English Rule of Natural Flow was virtually unworkable.

Later, when colonial America embarked on industrialization, recognition of the needs for water in great quantities by industry was paramount. So was the recognition of the need for water use legislation. The enterprising founders of this industrial democracy restyled the English Rule of Natural Flow and called it the American Rule of Reasonable Use. More liberal in design, it was written for the protection of the user and not for the survival of the water itself. The legality of water use by nonlandowners was also included, and was subject to evaluation by the criterion of reasonableness, defined as early as 1883 in *Red River Roller Mills* vs. *Wright* as follows: "In determining what is a reasonable use, regard must be had to the subject-matter of the use; the occasion and manner of its application; the object, extent, necessity, and duration to which it is subservient; the importance and necessity of the use claimed by one party, and the extent of the injury to the other party; the state of improvement of the country in regard to mills and machinery, and the use of water as a propelling power; the general and established usages of the country in similar cases; and all the other and ever-varying circumstances of each particular case, bearing upon the question of the fitness and propriety of the use of the water under consideration" (Myers and Tarlock, 1971).

This liberal rule of water use resulted in physical, chemical, and biological contamination of the major watersheds in the United States of America because industry and agriculture felt no pangs of guilt in their water uses and/or abuses. The results of these abuses became obvious to all in the mid-nineteenth century. The nation's major waterways, the lifeblood of the industrial transportation system, were completely befouled and cluttered with debris, rendering even barge movement difficult, dangerous, and time-consuming.

Pressure groups–not yet called lobbies–formed. Some represented shipping companies whose operations were hampered around Sandy Hook and the New York Bight, off New York City, by waste emptied into the waters by construction companies who were busy building the metropolis of New York City. The U.S. Congress, subject to shipping pressure groups, constructed, passed, and implemented a water pollution control bill known as the Refuse Act of 1899. The central thrust of this first legislation for water pollution control was limited to navigation protection; concern for the quality of the nation's waters was not yet perceived as critical. The Refuse Act of 1899 had been preceded by an amendment in the Rivers and Harbors Act of 1888 which protected fishways at rivers and harbors projects, and indeed, the Refuse Act of 1899 remained the main piece of water pollution control legislation until 1972, although it was amended many times. Table 1 outlines these various amendments, referred to by date and title.

In 1945, when the Federal Water Pollution Control Act became the workable amendment, the emphasis shifted from the protection of navigation to one centered on overall water quality. Scientific analysis in the basic disciplines of physics, chemistry, and biology had progressed sufficiently to permit some of the new analysis and bioassay techniques necessary to water quality analysis. In 1947, through the Department of the Army Act, the Army Corps of Engineers entered the water management picture in relation to conservation practices.

Public pressure and interest in the environment, not only in water but in wildlife and regulation conservation, had been smoldering in America for some time. Theodore Roosevelt, president of the United States from 1901 to 1909, stated: "I recognize the right and duty of this generation to develop and use our natural resources; but I do not recognize the right to waste them, or to rob by wasteful use, the generations that come after us."

Dozens of pieces of environmental protection legislation gradually have been legislated and implemented in the past seventy-five years of this nation's history. For example:

Rivers and Harbors Appropriations Act of 1888
  Fishways at rivers and harbor projects

Rivers and Harbors Appropriations Act of 1899
Protection of navigable waters
Better known as the Rivers and Harbors Act of 1899

Federal Power Act of 1920
Amended in 1935 to require in Section 10(a) that dam or related project that they license must be adapted to a comprehensive plan which considers commerce, water power, "and beneficial public uses including recreation"
1965 *Scenic Hudson Preservation Conference* vs. *FPC*. Local group challenged the licensing of the Storm King Power Plant on the Hudson River–impact scenic beauty of river. Court defined recreational purposes as including conservation efforts of natural resources; directed FPC to reinvestigate alternatives.
1967 *Udall* vs. *FPC*. Supreme Court overturned the license of the High Mountain Sheep Dam on the Snake River because the FPC had failed to consider the impact on fish and wildlife and the desirability of federal development. They gave the same reading to the statute.

1947 Department of the Army Act
As amended in 1956
Conservation facilities at Water Resource Projects of the Corps of Engineers

Fish and Wildlife Coordination Act as amended in 1958
Consultation with other agencies when impounding, diverting, or controlling H_2O

Multiple Use Sustained Yield Act of 1960
Agriculture must weigh relative value of resources and maintain wilderness

Wilderness Act, 1964
Established a National Wilderness System under which every department or agency controls the land at the time it is included

Reorganization Plan No. 2 of 1965
Established the Environmental Science Services Administration
Transferred Weather Bureau and Coast and Geodetic Survey to Department of Commerce

Freedom of Information Act of 1966
Laid down a general rule that all agency data must be available

Table 1. Development of U.S. governmental activities in environmental protection legislation as exemplified in water pollution control legislation

Date promulgated	1833	1899	1948	1965	1972	1974
Title of act	American Rule of Reasonable Use	Refuse Act Sec. 13; Rivers & Harbors	Federal Water Pollution Control Act	Water Quality Act amended FWPCA	Federal Water Pollution Control Act amended FWPCA PL 92-500	Safe Drinking Water PL 93-523
Purpose/direction	Equitable use of nation's water resource	Protection of waterways for navigation	Water quality improvement for human health	Restoration of water quality for human health & welfare	Regulation of point & nonpoint source discharges for water quality recovery	Applies to (*a*) public water systems, (*b*) undergrown source of dirty water. To protect public from any contaminant in drinking water which may have any adverse effect on human health
Enforcement mechanism	Individual court action (demonstrate extent of injury by continued use)	Court action: civil & criminal actions possible	State-federal enforcement	State-federal responsibility for enforcement	Federal gov't. responsibility through EPA permit program	Court action: civil suit by EPA administrator against state (or municipality) for violators

Substantive issues	1. Subject matter for use. 2. Occasion & manner of use. 3. Duration of use. 4. Nature & size of stream. 5. Improvement of countryside by use. 6. Establish similar usages	Maintenance, protection, preservation of navigable waters to prevent impediments to navigation intrastate	Maintenance of navigable waters & directed toward quality of receiving waters to abate pollutants which endanger health & welfare of human persons intrastate	Addition of coastal waters. Three categories: 1. Drinking water quality. 2. Recreational water quality. 3. Industrial waters	Controls on discharges Point-source. Municipal. Industrial. Non-point-source. Agricultural. Other toxic products. $ awarded for treatment plant. Construction deadlines for compliance	Lists contaminant. Levels primary regulations focused on human health. Provides for R&D by the National Academy of Sciences
Results	? →	? →	? →	50% nation's waters grossly polluted. →	Court actions are delaying deadlines ? →	?

to the public unless the data is internal data on policy and advice
Predates the movement of citizen concern for environmental data
NEPA public review of decision-making process

National Historic Preservation Act of 1966

Another of the Department of the Army Acts amendment 1967
Wildlife conservation at water resource projects of the army
Due regard for conservation

Administrative Procedures Act of 1967
Section 10(e)
Under this act a court will set aside (not reverse) the action of an agency if it can be proved that the action was arbitrary, capricious, or an abuse of discretion. Agency action must be supported by statutory jurisdiction and substantial evidence to the contrary. This is the backbone of NEPA.

Federal-aid Highway Act Amendments of 1968
Preserve wildlife refuges and maintenance of natural beauty on lands traversed by highway projects

Yet, with all of this legislation and the accompanying dollars and energies spent, the nation's environmental degradation became apparent to all. Congress in the 1960s began what was to become a ten-year effort to enact environmental protection legislation which could be properly implemented. A Resources and Conservation Act was proposed in 1959 which called on the executive branch to consolidate conservation efforts. An Ecological Resources and Survey Bill was proposed in 1966 and was designed to improve the use of environmental data by federal agencies.

Confusion was beginning to reign in the federal bureaucracy as to where the central authority lay with regard to specific environmental issues. No one agency had overview authority. Environmental problems cut across federal agency lines because they are comprehensive and multidisciplinary and encompass health, sociology, economics, conservation, and a host of other issues.

Early in 1969 Henry Jackson introduced a bill, S. 1075, which was the precursor to the National Environmental Policy Act of 1970, destined to changed the course of this nation's attitudes toward environmental protection.

Angry and discontent with the United States continued interest and support of the Vietnam War and anxious for a national cause which could attain some positive results, students, with support from the Department of Interior, declared Earth Day in April 1970, in response to the public participation clause in the National Environmental Policy Act of January 1970.

NEPA, the acronym for the National Environmental Policy Act, contained concepts fundamentally different from any previous legislation. It gave the federal government the primary responsibility for the nation's environment; it provided a legal forum in which citizens could bring suit against either the government at any level or any private person or group which, in the citizen's opinion, was guilty of abusing any segment of the environment, biotic or abiotic. NEPA also created the President's Council for Environmental Quality, which, in turn, was to accept responsibility for environmental analysis and evaluation through the mechanism called an Environmental Impact Statement, a concept new to the governmental operations of this nation. The following goals and objectives of NEPA are stated in Section 101. (a)(b)(c) of the 1970 act:

Title I

Declaration of National Environmental Policy

Sec. 101. (a) The Congress, recognizing the profound impact of man's activity on the interrelations of all components of the natural environment, particularly the profound influences of population growth, high-density urbanization, industrial expansion, resource exploitation, and new and expanding technological advances and recognizing further the critical importance of restoring and maintaining environmental quality to the overall welfare and development of man, declares that it is the continuing policy of the Federal Government, in cooperation with State and local governments, and other concerned public and private organizations, to use all practicable means and measures, including financial and technical assistance, in a manner calculated to foster and promote the general welfare, to create and maintain conditions under which man and nature can exist in productive harmony, and fulfill the social, economic, and other requirements of present and future generations of Americans.

(b) In order to carry out the policy set forth in this Act, it is the continuing responsibility of the Federal Government to use all practicable means, consistent with other essential considerations of national policy, to improve and coordinate Federal plans, functions, programs, and resources to the end that the Nation may–

(1) fulfill the responsibilities of each generation as trustee of the environment for succeeding generations;

(2) assure for all Americans safe, healthful, productive, and esthetically and culturally pleasing surroundings;

(3) attain the widest range of beneficial uses of the environment without degradation, risk to health or safety, or other undesirable and unintended consequences;

(4) preserve important historic, cultural, and natural aspects of our national heritage, and maintain, wherever possible, an environment which supports diversity and variety of individual choice;

(5) achieve a balance between population and resource use which will permit high standards of living and a wide sharing of life's amenities; and

(6) enhance the quality of renewable resources and approach the maximum attainable recycling of depletable resources.

(c) The Congress recognizes that each person should enjoy a healthful environment and that each person has a responsibility to contribute to the preservation and enhancement of the environment.

Apparently, the goals and objectives stated by Congress in the National Environmental Policy Act had started the pendulum swinging back toward power for the people, at least in this one area; government by and for the people again became a reality for the man-in-the-street, the mother-in-the-home, the child-in-the-classroom, the industrial powers, and finally, for the lawmakers in Congress and the lawyers in our court system. Each could identify with his part in the protection of his unique and idiosyncratic definition of the environment. Political activities within all societal groups ranging from elementary school children's letters to the president of the United States about endangered species to professional industrial lobbyists protecting the GNP of their interest groups became rampant throughout the country. The National Environmental Policy Act of 1969-70 unified this nation once again and provided America's citizens with the power with which to protect their national habitat.

The national news media, particularly television, quick to see the developmental possibilities latent in the environmental movement, began educational forums in the form of documentaries, usually funded by special interest groups anxious to justify their rather nefarious past activities in environmental degradation and to proclaim their future plans for environmental enhancement.

The Environmental Protection Agency was established as the main environmental legislation enforcement agency for the federal government. It received as "transfer functions" the operations listed in Table 2. The National Oceanic and Atmospheric Administration was created at that same period, under the reorganization plan No. 4 of 1970. NOAA received transfer functions from the agencies listed in Table 3.

As public and private interest groups became active in striving for their individual needs to be protected in the legislation, each exerted pressure on Congress in his own behalf. Lobbying became a lucrative profession. There are several hundred specific lobbies listed in Washington, D.C., alone. Table 4 lists a few of these active groups. These groups are comprised, usually, of professionals whose efforts are directed at influencing lawmakers and others to include items in legislative statutes profitable to their respective economies. The key spots at which legislation is subject to pressure groups and individuals are shown in Table 5.

The pool of persons involved in drafting legislation in the environmental protection area gradually changed to include not only legislators, lawyers and economists but also persons from the ever-broadening scientific community, selected for expertise in given areas. Because these inputs are needed at specific times, during both the composition and the implementation stages of the process, task forces and committees became an integral part of environmental legislation. Some are specifically assigned and charged with definitive responsibilities and are called statutory committees; others become attached to a certain division of an agency as advisory groups based on needs as perceived from time to time. It has become standard operating procedure for federal agencies to establish the advisory committee mechanism to obtain the necessary scientific and technical personnel. Table 6 illustrates the number of advisory committees operating for the benefit of one federal agency during one reorganization.

The increasing dependence on scientific and technical information from the academic, scientific, and industrial communities is illustrated in Figure 1, a chart which shows the complexities of the communication process as it operates with just one committee under PL 92-500 (The Federal Water Pollution Control Act, as amended in 1972).

Monitoring the efficiency of environmental and other complicated legislation is accomplished by legislative review, or "oversight," committees. Congress established such oversight committees to insure smooth implementation of legislation. One such subcommittee is the Sub-Committee on Investigations and Review of the Committee on Public Works, House of Representatives, 93d Congress. The oversight committee interviews and interrogates individuals directly involved in the legislative operation with regard to a particular program under review. These committee proceedings are open to the public. The entire verbal

Table 2. Functions transferred to the Environmental Protection Agency

Functions	Old Agency
Federal Water Quality Administration	Interior
National Air Pollution Control Administration	HEW
Solid Waste Management Program	HEW
Federal Radiation Council	AEC
Standards for radiation hazards in general	AEC
Pesticide registration	Agriculture
Pesticide research and standard setting	HEW

Table 3. Functions transferred to the National Oceanic and Atmospheric Administration

Functions	Old agency
Environmental Science Services Administration	Commerce
Bureau of Commercial Fisheries	Interior
Bureau of Sport Fisheries and Wildlife	Interior
Office of Sea Grant Program	NSF
Great Lakes Survey	Corps of Eng.
Weather Bureau	Commerce
Coast & Geodetic Survey	Commerce

accounts are published in due time and become part of the public record. A part of the Table of Contents of the recent hearings on PL 92-500 is included here as an example of the comprehensive coverage of such oversight committee hearings (Fig. 2).

Enforcement of regulation by necessity demands standards; so EPA and NOAA developed standards-setting divisions. Standards must be monitored; so EPA and NOAA developed divisions engaged in standard setting. For example, the Effluent Guidelines Division of EPA was established particularly and uniquely to provide uniform effluent limitations for industrial point source discharges as specified under Sections 304(b), 306, and 307 of PL 92-500.

Thus has the federal bureaucracy grown in numbers and in power in direct response to the environmental legislative mandate. The political problems, both internal and external to governmental agencies (bureaucracies), inherent in this process became rampant growths and,

Table. 4. Major lobbies

Lobby	Interest
Friends of the Earth	Natural environment
Man-Made Fiber Producers Association	Forests & fibers
Institute for Rapid Transit	Transportation
Zero Population Growth	Population
National Rural Electric Cooperative Association	Energy
National Right to Work Committee	Societal economics
The American Bankers Association	Financial operations
United Transportation Union	Transportation
American Farm Bureau Federation	Farmers/agriculture
American Hospital Association	Hospitals
American Petroleum Institute	Oil interests
National Association of Food Chains	Foods
Associated Dairymen, Inc.	Dairy products
National Agricultural Chemicals Association	Chemicals, fertilizers, pesticides, etc.
Citizens Committee for Postal Reform	Citizens/public policy of mails

much like cancer, actually threatened to consume and thus destroy the legislation which gave rise to their being. A simple example can be seen in the gradual rise and fall of the Clean Air Act. (The following situational analysis is an extremely oversimplified example used as a fictional illustration of the complexities which underlie environmental protection legislation and implementation.)

Congress passed the Clean Air Act Amendments of 1970 (PL 91-604) which provided under section 304(a):

> Except as provided in subsection (b), any person may commence a civil action on his own behalf–
>
> (1) against any person (including (i) the United States, and (ii) any other governmental instrumentality or agency . . . who is alleged to be in violation of (A) an emission standard or limitation under this act or (B) an order issued by the Administrator or a State with respect to such a standard or limitation) or
>
> (2) against the Administrator (of EPA) when there is alleged a failure of the Administrator to perform any act or duty under this act which is not discretionary with the Administrator . . .

Table 5. Stages at which legislation is subject to pressure groups

Stages	Specific activity
Legislation drafting	Pressure groups jockeying for the "in positions" for the initial drafting for the legislation in order to make certain all their interests are protected
Legislation passage	Pressure groups lobbying in both Senate and House for passage or veto of a legislation, depending on their particular interest, and pressure at the executive branch for the same reasons
Legislation implementation	Pressure groups impinging power upon federal and state agencies responsible for implementation of the bill through enforcement or compliance
Legislation administration	Court suits initiated by persons with opposing views with regard to the legislation or with respect to the environmental impact statement, in whole or in part, which has given rise to the action

Table 6. Advisory committees within EPA as of March 1975

1. Air Pollution Chemistry and Physics Advisory Committee
2. Ecology Advisory Committee
3. Effluent Standards and Water Quality Information Advisory Committee
4. Environmental Radiation Exposure Advisory Committee
5. Hazardous Materials Advisory Committee
6. Lake Michigan Cooling Water Studies Panel
7. National Air Pollution Control Techniques Advisory Committee
8. National Air Pollution Manpower Development Advisory Committee
9. National Air Quality Criteria Advisory Committee
10. National Drinking Water Council
11. President's Advisory Committee on the Environmental Merit Awards program
12. Science Advisory Board
13. Technical Advisory Group for Municipal Waste Water Systems

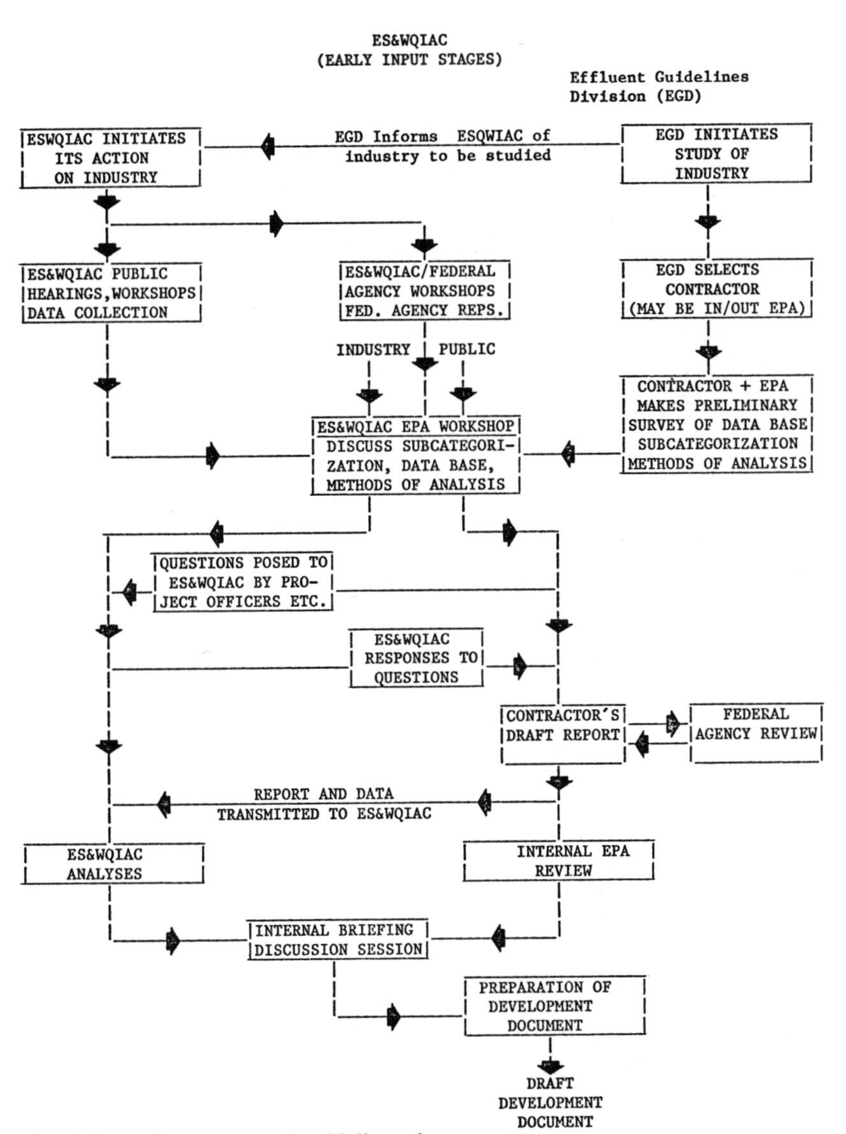

Fig. 1. Input into proposed guidelines document

CONTENTS

TESTIMONY

(V)

Fig. 2. Part of Table of Contents of recent hearings on PL 92-500 (93d Cong., 2d sess., 1974, House Committee on Public Works)

Air quality became the byword of the clean air protagonists. Air quality standards for air pollutants were required by the law. Citizens pressed for rapid and almost immediate announcement of those required standards. Human health and welfare were to be the criteria for Primary Ambient Air Standards for these United States. So, Primary Standards were established for the protection of human health. But the data documenting the negative effects of NO_X, SO_X, particulates, hydrocarbons, and carbon monoxide on human health were sparse and inconclusive. Arguments developed among scientists centered upon the validity of the data bases. Citizens pushed for air pollution controls on stationary and mobile sources of air pollutants. Lobbyists argued for both sides. Delays resulted. Citizens for clean air groups, armed with Section 304(a), sued the administrator of EPA. The regulations became law as a result of those actions.

Then, apartment landlords, no longer able to use their incinerators, were forced to hire solid waste companies to remove once burnable solid wastes, thereby increasing their operating costs, and were forced to use more expensive low sulfur fuels in their heating units, thereby increasing their operating costs. Rental landlords, in most major cities, were under strict rent control legislation; these increased operating costs necessitated by compliance with the Clean Air Act decreased their incomes sufficiently to make renting apartments a nonprofitable business endeavor. The result was a national movement to condominium concession of rental units to individually owned units in multiunit buildings, to the detriment and deprivation of lower-income persons who could not afford to purchase a condominium but were forced out of their rental units because of the concession. Chaotic conditions in the housing market are rampant throughout urban America today, not the least of which is this condominium movement.

Meanwhile, the Technical Advisory Board of the Department of Commerce established a panel to study sulfur oxide control technology. Again, part of that charge was taken from a statement submitted to Congress by the president of the United States on January 30, 1975, proposing 1975 amendments to the Clean Air Act of 1970: "Under no circumstances would extensions be granted in areas where the health related sulfur oxide standard would be violated" (which means primary ambient air standards).

Even as that CTAB panel on SO_2 was meeting in Washington, D.C., in March, 1975, the National Academy of Sciences report was released carrying the statement that "these and other currently available data on the health effects of sulfur oxides should be viewed with caution."

Five years after the passage of the Clean Air Act, no comprehensive planning for implementation has yet been developed. The question might well be at this point, have citizens properly deployed their power for individual action to press for Clean Air? Could not more progress have been made had the emphasis been placed on research and development programs in a search for valid data which would support the legislative mandates rather than on the application of pressure for legal implementation of PL 91-604? While the preceding simplified situational analysis illustrates the kind of real frustration inherent in the implementation of all legislation, it should be pointed out that, on the other hand, the Clean Air Act Amendments of 1970 have led (for better or for worse) to air pollution emission reductions throughout the United States which would have been highly unlikely to occur under a less mandatory and more voluntary strategy. Perhaps the political climate allowed Primary Standards for ambient air quality to be set at a level more stringent than the sparse and inconclusive data on negative effects, in hindsight, justify. If so, these overly stringent goals led to overly stringent emission standards and hence to the use of more strict controls than would otherwise have been forced on that large segment of society now in compliance with the act. Perhaps implementation deadlines were set so unrealistically short that the resulting implementation plans make it appear that no comprehensive planning has yet been developed. If so, at least these deadlines have provided legal grounds for forcing compliance with the above standards. Perhaps a federal program of research and development to provide a firm scientific basis for our ambient air quality standards would have provided only a semifirm basis. Or perhaps any scientific basis, no matter how firm, by the basically nonabsolute nature of our scientific knowledge, allows the same legal challenges by opponents of air pollution controls we are currently seeing. The point is not whether the Clean Air Act Amendments of 1970 have been "good" legislation; that is a qualitative judgment which must be based on individual as much as collective values. The real point is that legislation has a life cycle which is heavily affected by the political currents surrounding it. As the legislation ages, it and the political environment from which it sprang mature, and during the maturation the legislation goes through the growing pangs and pains of any child developing through adolescence into adulthood. The congressional process of writing and passing legislation represents only the labor pains. The hope for the future is that all our environmental legislative children will grow toward a more productive adult life.

Another statute, the Water Pollution Control Act, as amended in 1972 (PL 92-500), was passed by Congress over the president's veto. Because of this action, in one sense, this Clean Water Act "died aborning" or, at least, was severely handicapped, retarded, because of the imposition of party politics onto the already crowded environmental stage.

Factions from both political parties were anxious to maintain individual positions and responsibilities for which they had lobbied, persuaded, and managed to have included in that act as it was drafted. (The legislative history of PL 92-500 was initially bound in a draft copy which contained seven volumes.) Great political stands were won and lost by many members of Congress who were responding to the requests and demands of their respective constituents who, in turn, were responding to the public participation clause in NEPA and the newfound sense of contribution and involvement they were experiencing in the environmental movement in general.

Top administrative positions in federal agencies historically have been political appointments by the chief executive and remained so within the new environmentally oriented agencies of EPA and NOAA. This is not to imply that lack of skills or professional stature is evidenced by these politically appointed administrators—only to point out that political philosophies do, therefore, color the points of view and therefore influence the decisions made by these individuals. One final word here. Congressional decisions are made not only on sound scientific and technical inputs but also with an eye to the proper support of their constituents' economic advantages in relation to environmental benefits.

Advisory, statutory, and review committees have added still another dimension to the legislative operational structure. The Federal Water Pollution Control Act, as amended in 1972, was a complicated and lengthy piece of legislation which required the technical knowledge of individuals from many academic disciplines to insure its accurate scientific and legal implementation.

Congress included two statutory committees in that bill. The National Water Commission under Section 315 was to review, after two years, the entire operations mandated in the act and to report to Congress whether or not the nation was moving toward the achievement of the goals and objectives of the bill, i.e., "to restore and maintain the physical-chemical and biological integrity of the nation's waters." The Effluent Standards and Water Quality Information Advisory Committee, established under Section 515, was assigned, by statute, responsibility for industrial point source effluent limitations as regulated by Sections 304(b), 306, and 307 of PL 92-500. This committee was directed to provide, assess, and evaluate scientific and technical information pertinent to industrial point source effluent limitations and advise the administrator of EPA as to its findings and to report to Congress annually on the status of industrial point source limitations. Both committees are, in effect, on-the-spot congressional observers and act as communication links between EPA and Congress.

Sharing the responsibility for compliance surveillance between federal and state agencies represents still another complex target in the environmental protection matrix. A recent newspaper article, taken from the New York *Times*, March 16, 1975, is used here to illustrate this point:

EPA Proposes Rules for Water

Standards Are First for U.S. Drinking Supply

WASHINGTON, March 15 (AP)–The Environmental Protection Agency has proposed the first national quality standards for drinking water.

The proposed interim standards, published yesterday in the Federal Register, set maximum permissible levels for a variety of contaminants and specified requirements for monitoring most of the nation's 240,000 public water supplies.

The interim standards will take effect next June after a three-month period to receive and consider public comments; permanent standards will take effect in December, 1976.

"These proposals represent the first comprehensive effort to regulate uniformly and effectively the purity of the nation's drinking water," the E.P.A. Administrator, Russell E. Train, said in a statement.

Establishment of the regulations was required by the Safe Drinking Water Act, signed into law by President Ford last Dec. 17. The measure authorized $156.5-million for water quality improvement projects.

Primary responsibility for assuring safe drinking water still rests with the states, however, provided they demonstrate their ability to enforce standards at least as stringent as the Federal standards, the agency noted.

It said Federal officials would intervene only in cases where a state either fails to enforce the minimum standards or chooses not to accept responsibility.

The new law requires a public announcement if a local water system does not comply with the standards or does not perform proper monitoring.

In addition to accepting written comments on the regulations, the E.P.A. will conduct public hearings in Boston, April 15; Chicago, April 17; San Francisco, April 22, and Washington, April 25.

The implications of national environmental protection were recognized by Congress as being only a small part of the environmental mosaic of the entire planet. NEPA also contained instructions about United States cooperation in international environmental protection. Section 102(E) pledges to "recognize the worldwide and long-range character of environmental problems and, where consistent with the foreign policy of the United States, lend appropriate support to initiatives, resolutions, and programs designed to maximize international cooperation in anticipating and preventing a decline in the quality of mankind's world environment."

In the fifth report to Congress by the Council of Environmental Quality, released in January 1975, the following statements were made with respect to global environmental perspectives and interests.

In the perception of developing countries, the major environmental problems relate to the lack of economic development. . . . The developed countries, in contrast, are more concerned about the impact of man on natural systems. . . . The action Plan reflects the interests of both groups. [p. 432]

Over the last year, world attention has been focused on drought in the Sahel, a strip of land stretching across Africa south of the Sahara Desert . . . . The drought-stricken area is as large as the continental United States, with a population of around 25 million. . . . Only in the past year have the enormity and consequences of the drought begun to be fully realized [p. 437]

The environmental significance of [The Law of the Sea] Conference, held in Caracas this summer, cannot be overemphasized [p. 445]

Earthwatch is one of UNEP's major functional tasks. It is designed to provide a global environmental assessment so that decisions on the management of the environment are sound and rational. [p. 449]

Bilateral Cooperation (between the U.S. and its continental neighbors)

During the last year an environmental problem of great importance to Mexico and the United States moved toward resolution as the result of an agreement . . . on a "Permanent and Definitive Solution to the International Problems of the Salinity of the Colorado River [p. 453]

The United States and Canada are seeking mutually beneficial solutions to a number of environmental problems, ranging the length of the border from Puget Sound to the waters off Maine and New Brunswick [p. 454]

Multilateral Cooperation

[The OECD] has developed an "early warning system" to signal to other members actions taken in the environment that might significantly affect international trade. However, no clear cases of trade distortions attributable to differing environmental constraints or practices have been brought before the Committee. [p. 460]

Conclusion

This year's report has concentrated on the United Nations Environmental Program. The rapid development of this new organization is heartening. Its growth has encouraged nations in all stages of development to understand the need for environmental concern. UNEP is institutionalizing environmental concern on a global scale just as NEPA has done on a national scale in the United States. [p. 462]

From the onset of this chapter, environmental protection legislation has been described in terms of the complimentary functions of ecology, technolgy, economics, and politics. The success of future restoration of damaged ecosystems rests with the facility with which each of these facts meshes with the others to provide a smoothly functioning mechanism. Thus, the basis of environmental protection for the future is rooted in scientific/legal technology.

Werner Heisenburg, the atomic physicist who is also director of the Max Planck Institute in Germany, wrote in 1958:

> In an age, when the earth is becoming ever more densely populated, limitations of living possibilities, and thus the thrust, are primarily due to other men who also are claiming their rights to the goods of this world. . . . in our age we live in a world which man has changed so completely that in every sphere–whether we deal with the tools of daily life, whether we eat food which had been prepared by machines or whether we travel in a countryside radically changed by man, we are always meeting man-made creations. . . . we get the impression that it could not be too crude or over simplification to say that for the first time in the course of history modern man now confronts himself alone.

Acknowledgment

The lines from Edna St. Vincent Millay's poem "The Pear Tree" are reprinted from her *Collected Poems* (Harper & Row; copyright 1956 by Norma Millay Ellis).

Literature Cited

Allee, W. C., Orlando Park, Alfred E. Emerson, Thomas Park, and Karl P. Schmidt. 1949. Principles of Animal Ecology. W. B. Saunders Co., Philadelphia.

Heisenburg, Werner. 1958. The Physicist's Conception of Nature. Harcourt Brace & Company, New York.

LaFleur, L. J. 1949. Theoretical biochemistry. Page 73 *in* W. C. Allee, Orlando Park, Alfred E. Emerson, Thomas Park, and Karl P. Schmidt. Principles of Animal Ecology. W. B. Saunders Co., Philadelphia. Reprinted from Acte Bio-theoreticae 5:177-183.

Millay, Edna St. Vincent. 1956. The Pear Tree. *In* Collected Poems. Harper & Row, New York.

Myers, Charles J., and A. Dan Tarlock. 1971. Water Resources Management. Foundation Press, Mineola, N.Y.

Reitze, Arnold W. 1972. Foreword. *In* Environmental Law, Vol. 1. North American International, Washington, D.C.

Webster's Third International Dictionary. 1966. G. & C. Merriam Co., Springfield, Mass.

The Recovery and Restoration of Damaged Ecosystems, a Challenge for Action: Symposium Analysis

John Cairns, Jr., Kenneth L. Dickson, and Edwin E. Herricks

The characteristics of restored ecosystems are bound by two general constraints, the publicly perceived restoration and the scientifically documented restoration. For example, *recovery* may be defined as restoration to usefulness as perceived by the "users" of the resource. This is significantly different than restoration to either the original structure or the original function (or both) as rigorously determined by scientific methodology. Thus during the course of this symposium, *recovery* was defined three ways: (1) restoration to usefulness as perceived and defined by the general public, (2) restoration to original functional and structural conditions although the elements (species) that comprise this structure may be significantly different than those present originally, and (3) restoration to the original functional and structural condition with original elements (species) present. The semantic difficulties created by the difference between social perception and scientific documentation require restoration and recovery programs which integrate these diverse views.

The case histories of this symposium provided solid evidence that damaged ecosystems may recover naturally if given enough time. They also suggest that scientifically justifiable intervention may accelerate the recovery process. Scientists can provide the required information base and appropriate methodology, but policy makers must make restoration a reality by supporting programs aimed at understanding the recovery processes and the ways man can aid that process. Compelling evidence of the need for immediate action in developing the technology for and in implementing programs on restoration of damaged ecosystems has been included in nearly all of the presentations at this international symposium. Damaged ecosystems exist in every country of the world, and the potential for further damage from advancing technology and population growth is great. Global-scale ecological alterations, such as the oceans or the atmosphere, and multinational use of some resources, such as the Rhine, Danube, or Mekong rivers, all transcend traditional geopolitical boundaries. It is apparent, then, that an international organization is needed to inventory damaged ecosystems and prepare a global plan for their restoration. However, since international organizations dealing with environmental matters have not yet proved to be effective, short-term solutions must depend almost entirely on national organizations which deal with environmental matters within the boundaries of their political jurisdictions. Support for these national organizations (and the international organization when formed) must come from the group of "users"–industries which use streams and rivers to degrade wastes, populations which depend on ocean protein reserves or terrestrial natural resources, individuals who seek healthy and safe environmental conditions, etc. Support implies not only development of the proper political, scientific, and social climate but also provision of the financial resources to permit the scientific community to develop a detailed understanding of the consequences of damage and the methods of restoration.

A list of the most important needs of a program to facilitate the recovery and restoration of damaged ecosystems follows.

1. Groups either nationally or internationally charged with the responsibility for the direction of the restoration of damaged ecosystems must be organized.
2. Research activities designed to provide a scientific basis for restoration and recovery programs must be instigated.
3. More information on ecosystems where man-augmented or man-controlled restoration efforts were attempted must be generated.
4. Academic or federal assessment of existing information is mandatory.
5. A comprehensive assessment of the available information on the recovery and restoration of damaged ecosystems must be made, with a mandate that functional aspects of the recovery process be included in all studies of ecosystems.
6. The location and existing conditions of representative ecosystems which may act as epicenters for recolonization of damaged areas must be systematically assessed.
7. The role of physical environment factors in the recovery process must be elucidated.
8. There is a pressing need to develop an array of definitions of restoration and recovery.

Although this list can certainly be expanded, the implementation of a program to satisfy even a few of the needs will improve attempts to restore damaged ecosystems. The commendable efforts of state and federal regulation agencies (and other organizations as well) will not guarantee that the ecosystems will return to a desirable condition in an acceptable period of time. This requires an organization charged with the restoration of the ecosystems. Operation and maintenance of such a managing organization will require the combined efforts of political, scientific, and user groups. Some relevant information is already available (although form and interpretation must be changed). For example, the massive data collection efforts required in the United States by PL 92-500, the Federal Water Pollution Control Act, not only monitor present conditions but also supply a data base which will allow comparison of future changes. It may be possible to add to this data base and provide documentation of ecosystem function, damage, and recovery.

After the presentation of all papers at the symposium, many of the participants and speakers met to summarize the state-of-the-art of our knowledge about recovery of ecosystems. In order to organize the vast array of topics covered in a manner which addressed the purposes of the symposium, the group was requested to determine the rank order of various ecosystems in regard to the available information on recovery, the potential for man-influenced restoration, and the rate of natural recovery. Table 1 represents the results of this activity.

After a lengthy debate over the appropriateness of the breakdown of ecosystems into the categories found in Table 1, the group determined that more information was available on the recovery process for temperate terrestrial ecosystems. However, very little is known about the recovery of tropical terrestrial ecosystems. In general more is known about the recovery of marine ecosystems than of freshwater systems. The group also felt that the greatest potential for man-influenced restoration was found in the temperate terrestrial ecosystems; tropical terrestrial ecosystems offered the least potential. The group felt that it was not possible to determine the rate of natural recovery of water ecosystems. However, they felt that tropical ecosystems would be very slow to recover after stress, so that this would have to be measured in geological time since natural recovery was dependent on the development of new soils.

Based on the papers presented in the conference and subsequent discussion, the group made the following recommendations.

1. A real need exists for a comprehensive effort to synthesis the existing information on the restoration and recovery of ecosystems after damage.

2. A series of well planned and comprehensive baseline studies of representative ecosystems should be undertaken.

3. Future studies on the recovery process should include measurements of the functional characteristics as well as structural characteristics of the ecosystem.

Table 1. Evaluation of recovery characteristics of ecosystems

	Information available	Rated	Natural recovery	Potential for man-influenced restoration
Water ecosystems				
Fresh water flowing	4*			6
Fresh water impounded	5			7
Estuary	3			4
Marsh	6			8
Shoreline	7			2
Terrestrial Ecosystems				
Arctic	8	M†		5
Temperate	10	F‡		10
Tropic	2	S§		1

* 1-10 system where 10 represents most information or highest potential.
† *M*=Medium rate of natural recovery measured in geologic time.
‡ *F*=Fast rate of natural recovery measured in geologic time.
§ *S*=Slow rate of natural recovery measured in geologic time.

Conference Summary

Robert B. Platt

I know I express the keen feelings of all of us in recognizing the wisdom of John Cairns and his associates for organizing this symposium and then having it in this beautiful center with all of its comforts and away from the hustle and pressure which are generally associated with symposia at national meetings. The subject is of mutual interest to all of us and is of global significance. We have all been caught up in the spirit and substance of these sessions. They have been characterized by an openness of discussion and a willingness, on the part of participants, to "go out on a limb" with admitted skimpy data in making predictions and trying to probe for bases of common interests and of standards and procedures which will further our knowledge of this particular subject. This has been most refreshing. It is hard to come by in a symposium and has been one of the really bright spots of these three days.

In fact this symposium has had a great deal going for it. First, we were treated to a remarkable keynote address from Bob Curry. He emphasized our dependence on a mosaic of ecosystems with their genetic materials and their dependence upon a wide range of natural stresses. He pointed out that we must maintain this natural range of stresses in order to maintain our own welfare and to ultimately "rehabilitate the earth" and return to a more stable and a more primitive system of living.

The next morning John Cairns set the tone of the conference in another sense in that he gave us some very basic and very simple concepts regarding the stressing and recovery of ecosystems. These included (*a*) existence of near epicenters for reinvading organisms, (*b*) transportability or movement of dissemules, (*c*) existing conditions of habitat following stress, (*d*) presence of residual toxicants, (*e*) chemical-physical water quality following stress, (*f*) the number of indigenous organisms which have evolved resistance to the stress, (*g*) the extent of structural and functional similarity, (*h*) flushing capacity, (*i*) the distance from strong ecological threshold stress, and (*j*) management capabilities for control and recovery of damaged systems. Although these were used for aquatic systems, they have a wide applicability to all systems. Throughout the program these points were variously and frequently introduced. Martha Sager, then, at the end of the program gave us an insight into the political and legislative history and development of our environmental problems and opportunities.

I have put the key elements of this symposium into a little model.

```
              Understressed→Remove stress→Natural processes→Recovery
            ↗   (Reversible)
Ecosystems
            ↘ Overstressed→Remove stress→Man-assisted→Recovery
                (Irreversible)              natural
                                            processes
```

It seems to me that basically we have been talking about ecosystems in two ways. On the one hand are those systems which are understressed such that, when the stress is removed, recovery takes place by natural processes. We did not see many of these examples in the papers given, but I recall one which dealt with a stream in Pennsylvania polluted by acid mine drainage. By removing the stress and also by dilution and buffering from tributary streams, the downstream system returned to normal. We can deal directly with these kinds of understressed systems, but we are now more and more confronted with the other kind—the overstressed systems which are often related to irreversible changes. In these cases, when the stress is removed, natural processes alone cannot operate to restore the system to its original condition. Man must help; he must help sometimes in very substantive ways, and at least, he must exercise control. In this summation, it would be useful to take these components—the ecosystem, the stress, and the restoration or recovery—and look at them in terms of the various papers that have been given.

No one thought it necessary to define an ecosystem, and I would not do so except for the last paper which considered the city as an ecosystem. I often talk about urban ecosystems, and I think it is appropriate that we have this paper included for two reasons; one is because it provides an intrinsic challenge to man to restore something to a degree of quality on which his life has been dependent; and second, it highlights what an ecosystem really is and what it is not. Let me point out that a city is not an ecosystem in any true sense of the word. What is an ecosystem? We all know that it has energy flow, nutrient cycling, and the capacity for self-perpetuation, given a source of energy, the sun. Thus, it has the capacity to restore itself from damage due to stress, unless it is overstressed and the system tumbles. The city does not have these characteristics. It is not independent. On the one hand, the city constantly imports clean air, clean water, energy, and food. On the other hand, the city exports dirty air and dirty water; it exports solid waste; and it exports goods and services. What does it give you? Well, it gives a lot—entertainment, information, jobs, living quarters, and human interaction. Although we may go on calling the city an ecosystem, I do think it important that we recognize the distinction I have made, especially with reference to this symposium.

We have talked about all kinds of ecosystems—terrestrial, freshwater, and marine. Then we talked about these in various ways because of the tremendous range of environmental gradients on this planet. These extend from the poles and the tundra through the temperate regions into the tropics. Another thing of importance is the tremendous diversity of the ecosystems on which man feels he is dependent.

Now let's talk about the word *stress*. We get hung up on this word. I noticed that in one sense there was great unity throughout the papers as to what stress is, and then in another sense there was a lack of unity, particularly when we get to the words *reversible* and *irreversible*. We can handle the word *reversible* fairly well, but what do we mean by *irreversible*? Does it mean that you cannot ever bring it back, or that it is going to take a thousand years to bring it back, or five years, or a million years? We have dealt in a tremendous range of time frames and this greatly enriched the discussions because it gave us a perspective we rarely have the privilege of getting at one time. Then we talked about stresses in terms of their physical nature: oil, fire, chemicals, mining, urban wastes, etc. We did not talk about biological stresses very much, such as the introduction of new species or undue competition. We did consider the question several times of how much stress it takes to topple a system. I recall that in an early paper someone pointed out, with reference to a salt marsh in England, that it can be stressed with oil once, twice, or perhaps four times and it will come back, but if it is stressed eight times it will not come back. Thus, when the limit of the ability of the system to reestablish itself by natural processes is reached, the alternative is that man then has to come in and do it for nature. In other words, man has taken away the system's capacity for maintenance and repair.

From another point of view, ecosystems may be on the very edge of their climatic range. So, if they are severely stressed they flip over to something else, and the climate will not support reestablishment of the same system. It was emphasized in several papers, however, that in some cases of very great stress, the system can recover by simply removing the stress. An excellent example was the cleaning up of certain oil spills. In still another vein, it was shown very graphically that methods used to clean up and remove a stress can be worse than the stress itself. Also, many graphic examples were given showing that severe stresses, such as fire, are an essential part of ecosystem maintenance and stability.

With respect to the word *recovery*, it was, of course, in every paper, and it evoked a great range of responses. What does recovery mean? Recovery means something different in almost every situation which was developed. Thus, it can mean something different to the politician or engineer than to the ecologist or to a conservationist. For many people, if you can simply get grass to grow on some highly disturbed area, it is recovered. Maybe the fertilizer will carry it for a year or two and then this temporary system will collapse because an adequate soil is not there. But nevertheless, on this temporary short-term basis it has been "restored" in that person's mind. At the other extreme are some of the tropical forests. It was shown that if many of the tropical rain forests are destroyed, the only way they can be restored is by evolutionary processes. Here, then, we are talking about a million-year period for recovery. Again, with regard to tropical ecosystems, one paper showed recovery of vegetation in three years. However, this fast growth is from a different group of species than was in the natural forest. Thus, it may take a thousand years to bring back the same balance in diversity although only a few years to get an apparent restoration from secondary succession. From these examples we have seen that it is very dangerous to extrapolate from a particular condition and say, "Yes, since it happened here, it's going to happen over there." Each situation is different, and it is up to us to define for each particular situation what we mean by *recovery*.

It was pointed out that we may not want what was there before. Perhaps we want something different because it will serve our use better. This planet is now managed by man, so let's not get hung up on a "back-to-nature" kick because that is impossible. We have altered things such that if we cannot exert some kind of control, many systems cannot survive. This came out very strongly. We need to look at each situation in terms of what we mean by recovery, how far we want it to go back, in what form, and at what price in terms of management problems, etc. If we can get a system back to its true condition, nature will run the system for us, but if we put it back into a different form, then man must constantly spend money and energy to manage it. Agricultural systems are highly unnatural, and as Bob Curry said, "Modern agriculture is a suicidal system which is slowly deteriorating the soils of the earth." These points at least highlight the fascinating considerations and factors that came out of *recovery.*

Another kind of consideration that came out is the need for baseline data, as, for example, the Santa Barbara oil spill and the whooping crane. Although some data existed, they had not been organized. Good baseline data are one of our most urgent needs. Going back to Martha Sager's presentation, we have the NEPA and Environmental Impact Statements with which to work. Although these give us the opportunity to develop baseline data, the results are inadequate in a vast majority of cases. It is up to ecologists such as us to raise the standards and develop the means and procedures for getting the much better baseline data which we badly need for the future.

This leads to a sequel which several of us have discussed and to which we have not yet given much attention. This is that mechanisms do not exist, with few exceptions, for follow-up on impacts after the Impact Statement is accepted and the action or stress completed. The lack of follow-up on this kind of legislatively controlled stress is something we need to consider.

Another thing that we did not consider, with one exception, is the socio-economic aspects of ecosystem recovery. Ultimately this comes down to energy, and we are going to have to develop cost-benefit ratios in much greater depth than is customarily done and become actively involved in the business of assessing the cost of different kinds of recovery in terms of energy input and energy yield.

Now, how do we wind up? What can we do with this symposium? Several years ago I vowed that I was not going to attend any more conferences where people got together, discussed a fascinating subject for three days, and then went home with a pigeonholed report being the only germane record. But this symposium is to be published! And soon! This is of tremendous significance, because I do not believe there now exists in one place the kind and depth of information on this subject that we have been sharing with each other over the past three days. First, this must have tremendous significance to our colleagues in this country and abroad. It is an international program because it is of international significance. I think a person who goes through this volume when it is published will get an understanding of the structure and function of nature in terms of stress and recovery that I do not believe can be found elsewhere. Certainly, it has given me a much higher level of integration and understanding. Second, I think the proceedings of this symposium should be of great importance in various kinds of decision-making processes. Even though it is a beginning, a first step in attempting to bring these kinds of data together, there are enough kinds of basic information, guidelines, types of situations, case histories, etc., that we can go to these, as to a well, for information which will help in future studies in the interpretation of someone else's data, in the ordering of priorities, in the input for new legislation, and for many other things as they come along.

May I add one final comment. This symposium has also highlighted the fact that we need to be much more articulate with the public. What do we mean by an ecosystem, by its stress, and by its recovery? Since these have been often difficult for us to cope with, they must be very confusing terms to most people. When we go back home and share this symposium with others, it can come to life by providing a high level of clarity and understanding. It should help those who use it to translate to their colleagues, to conservation organizations, and to legislators what a particular issue means politically, economically, scientifically, and ultimately to the human quality of life.

Surely, with this challenge, the subject will be actively pursued. And what better way than by following this with a second conference, and then perhaps a third or more as the area develops.